Yeast Diversity in Human Welfare

Tulasi Satyanarayana · Gotthard Kunze
Editors

Yeast Diversity in Human Welfare

Springer

Editors
Tulasi Satyanarayana
Division of Biological Sciences and Engineering
Netaji Subhas Institute of Technology
Dwarka, Delhi
India

Gotthard Kunze
Yeast Genetics, Department of Physiology and Cell Biology
Leibniz Institute of Plant Genetics and Crop Plant Research
Gatersleben, Sachsen-Anhalt
Germany

ISBN 978-981-10-2620-1 ISBN 978-981-10-2621-8 (eBook)
DOI 10.1007/978-981-10-2621-8

Library of Congress Control Number: 2017930136

Printed on acid-free paper

This Springer imprint is published by Springer Nature
The registered company is Springer Nature Singapore Pte Ltd.
The registered company address is: 152 Beach Road, #21-01/04 Gateway East, Singapore 189721, Singapore

Preface

Yeasts are eukaryotic microbes placed in the kingdom Fungi, under the phyla Ascomycota and Basidiomycota with approximately 2000 species described till date. These are estimated to constitute 1–1.5% of the fungal species described, and the number of existing yeast species is expected to exceed that of the described ones. In case yeasts make up 1–1.5% of the estimated fungal species extant on Earth of three million species, the yeast species would be between 30,000 and 45,000. Extensive efforts are needed to understand the diversity of yet to be cultured yeast species. Yeasts are mostly unicellular, although some species develop multicellular characteristics by forming pseudohyphae. Most yeasts reproduce asexually by mitosis, and many do so by the asymmetric division process called budding and a few by fission.

By fermentation, the yeast species *Saccharomyces cerevisiae* and others have been converting carbohydrates to carbon dioxide and alcohols for thousands of years, and carbon dioxide has been used in baking and the alcohol in alcoholic beverages. It is also a centrally important model organism in modern cell biology research, and is one of the most thoroughly investigated eukaryotic microbes. Researchers have used it to gather information about the biology of the eukaryotic cell and ultimately human biology. Other species of yeasts like *Candida albicans* are opportunistic pathogens and known to cause infections to humans. Yeasts have recently been used to generate electricity in microbial fuel cells, and to produce ethanol for the biofuel industry.

Certain strains of some yeast species produce proteins called yeast killer toxins, which allow them to eliminate competing strains. This may cause problems for wine making, but could potentially be used to advantage by using killer toxin-producing strains to make wine. Yeast killer toxins may find medical applications in the treatment of yeast infections.

Yeasts occur in the environment, and particularly in sugar-rich materials. For instance, naturally occurring yeasts are found on the skins of fruits and berries and plant exudates. Some yeasts are also found in association with soil and insects. The ecological function and biodiversity of yeasts have not yet been adequately understood. Yeasts are also present in the gut flora of mammals and some insects.

Even deep-sea environments also host some yeasts. An Indian investigation on 7 bee species and 9 plant species found 45 yeast species belonging to 16 genera to colonize the nectaries of flowers and honey stomachs of bees. Most were members of the genus *Candida*; the most common species in honey was *Dekkera intermedia* and in flower nectaries, *Candida blankii*. Yeast-colonizing nectaries of the stinking hellebore have been found to raise the temperature of the flower, which may aid in attracting pollinators by increasing the evaporation of volatile organic compounds. Black yeast has been observed as a partner in a complex relationship between ants, their mutualistic fungus, a fungal parasite of the fungus and a bacterium that kills the parasite. The yeast has a negative effect on the bacteria that normally produce antibiotics to kill the parasite, so may affect the ants' health by allowing the parasite to spread.

Some species of yeasts are opportunistic pathogens, which can cause infection in people with compromised immune systems. *Cryptococcus neoformans* and *Cryptococcus gattii* are significant pathogens of immuno-compromised individuals. They are the species primarily responsible for cryptococcosis, a fungal disease that occurs in about one million HIV/AIDS patients, causing over 600,000 deaths annually. Yeasts of the genus *Candida* cause oral and vaginal infections in humans called candidiasis. The pathogenic yeasts of candidiasis in probable descending order of virulence for humans are: *C. albicans*, *C. tropicalis*, *C. stellatoidea*, *C. glabrata*, *C. krusei*, *C. parapsilosis*, *C. guilliermondii*, *C. viswanathii*, *C. lusitaniae*, and *Rhodotorula mucilaginosa*. *Candida glabrata* is the second most common pathogenic yeast after *C. albicans*, causing infections of the urogenital tract, and of the bloodstream (candidemia).

The useful physiological properties of yeast have led to their use in the field of biotechnology. Fermentation of sugars by yeast is the oldest and largest application of this technology. Many types of yeasts are used for making many foods: baker's yeast in bread production, brewer's yeast in beer fermentation, and yeast in wine fermentation and the production of xylitol.

Some yeasts can find potential application in the field of bioremediation. One such yeast, *Yarrowia lipolytica*, is known to degrade palm oil mill effluent, TNT (an explosive material), and other hydrocarbons such as alkanes, fatty acids, fats, and oils. It can also tolerate high concentrations of salt and heavy metals, and is being investigated for its potential as a heavy metal biosorbent. *Saccharomyces cerevisiae* has potential to bioremediate toxic pollutants like arsenic from the industrial effluents. Bronze statues are known to be degraded by certain species of yeast. Different yeasts from Brazilian gold mines accumulate free and complexed silver ions.

Yeast is used in nutritional supplements popular with health-conscious individuals and those following vegetarian diets. It is often referred to as "nutritional yeast" when sold as a dietary supplement. Nutritional yeast is deactivated yeast, usually *S. cerevisiae*. It is an excellent source of protein and vitamins, especially B-complex vitamins as well as other minerals and cofactors required for growth. It is also naturally low in fat and sodium. Some brands of nutritional yeast, though not all, are fortified with vitamin B_{12}, which is produced separately by bacteria.

In 1920, the Fleischmann Yeast Company began promoting yeast cakes in a successful "Yeast for Health" campaign. They initially emphasized the importance of yeast as a source of vitamins, good for skin and digestion. Their advertising later claimed a much broader range of health benefits. Nutritional yeast has a nutty, cheesy flavor that makes it popular as an ingredient in cheese substitutes. It is often used by vegetarians in the place of Parmesan cheese. Another popular use is as a topping for popcorn. It can also be used in mashed and fried potatoes, as well as in scrambled eggs. It comes in the form of flakes or as a yellow powder similar in texture to cornmeal. In Australia, it is sometimes sold as "savory yeast flakes." Though "nutritional yeast" usually refers to commercial products, inadequately fed prisoners have used "home-grown" yeast to prevent vitamin deficiency.

Some probiotic supplements use the yeast *Saccharomyces boulardii* for maintaining and restoring the natural flora in the gastrointestinal tract. *S. boulardii* has been shown to reduce the symptoms of acute diarrhea, reduce the chance of infection by *Clostridium difficile*, reduce bowel movements in diarrhea patients, and reduce the incidence of antibiotic-, traveler's-, and HIV/AIDS-associated diarrheas.

Yeasts are able to grow in foods with a low pH (5.0 or lower) and in the presence of sugars, organic acids, and other easily metabolized carbon sources. During their growth, yeasts metabolize some food components and produce metabolic end products. This causes the physical, chemical, and sensible properties of a food to change, and the food is spoiled. The growth of yeast within food products is often seen on their surfaces, as in cheeses or meats, or by the fermentation of sugars in beverages, such as juices, and semi-liquid products, such as syrups and jams. The yeast of the genus *Zygosaccharomyces* has had a long history as spoilage yeasts within the food industry. This is mainly because these species can grow in the presence of high sucrose, ethanol, acetic acid, sorbic acid, benzoic acid, and sulphur dioxide, representing some of the commonly used food preservation methods. The major spoilage yeast in enology is *Brettanomyces bruxellensis*.

Several yeasts, in particular *S. cerevisiae*, have been widely used in genetics and cell biology, largely because this is a simple eukaryotic cell, serving as a model for all eukaryotes including humans, for studying fundamental cellular processes such as the cell cycle, DNA replication, recombination, cell division, and metabolism. Yeasts are easily manipulated and cultured in the laboratory, which has allowed the development of powerful standard techniques, such as yeast two-hybrid, synthetic genetic array analysis, and tetrad analysis. Many proteins important in human biology were first discovered by studying their homologues in yeast, which include cell cycle proteins, signaling proteins and protein-processing enzymes.

Saccharomyces cerevisiae was announced to be the first eukaryote to have its genome on April 24, 1996, comprising 12 million base pairs, fully sequenced as part of the Genome Project. At that time, this was the most complex organism to have its full genome sequenced at that time, and took 7 years with the efforts of more than 100 laboratories. The second yeast species to have its genome sequenced was *Schizosaccharomyces pombe*, which was completed in 2002. It was the sixth eukaryotic genome sequenced that comprised 13.8 million base pairs. By 2012, over 30 yeast species have had their genomes sequenced and published. A total of approximately 24,200 novel genes were identified, the translation products of which were classified together with *S. cerevisiae* proteins into about 4700 families, forming the basis for interspecific comparisons. The analysis of chromosome maps and genome redundancies revealed that the different yeast lineages have evolved through a marked interplay between several distinct molecular mechanisms, including tandem gene repeat formation, segmental duplication, a massive genome duplication, and extensive gene loss.

Yeast species have been genetically engineered to efficiently produce various drugs by a technique called metabolic engineering. *S. cerevisiae* is easy to genetically engineer; its physiology, metabolism, and genetics are well known, and it is amenable for use in harsh industrial conditions. A wide variety of chemicals in different classes can be produced by engineered yeast, including phenolics, isoprenoids, alkaloids, and polyketides. About 20 biopharmaceuticals are produced in *S. cerevisiae*, including insulin, vaccines for hepatitis, and human serum albumin.

The advances in modeling and synthetic biology tools and how these tools can speed up the development of yeast cell factories have been recently made. Metabolic engineering strategies for developing yeast strains for the production of polymer monomers: lactic, succinic, and cis, cis-muconic acids have been attempted. *S. cerevisiae* has already firmly established itself as a cell factory in industrial biotechnology and the advances in yeast strain engineering will stimulate the development of novel yeast-based processes for production of chemicals in the near future. Strategies are being developed for metabolic engineering of ethanologenic yeasts for the production of bioethanol from complex lignocellulosic residues. Recent examples of yeast metabolic engineering have shown that evolutionary potential of cells should not be underestimated in strain improvement. Evolutionarily evolved strains can form suitable starting points for inverse metabolic engineering approaches too. For developing an understanding of the cell as a whole, sophisticated computational methods capable of integrating copious amounts of data/information are required.

This book is an attempt in bringing together the scattered information on various aspects of the utility of yeast diversity for human welfare into one volume. This includes recent developments made in the past few decades on these aspects. The chapters have been written by experts, who have done a commendable job of reviewing the developments made in recent years. We wish to thank all the contributors. The views expressed by authors are their own. We sincerely hope and wish that the book will be useful for teachers, scientists, researchers and students of biology, microbiology, mycology, and biotechnology.

We wish to appreciate and thank the efforts made by Springer in publishing the book for disseminating knowledge on the utility of yeast diversity for human welfare.

New Delhi, India — Prof. Tulasi Satyanarayana
Gatersleben, Germany — Prof. Gotthard Kunze

Contents

Editors and Contributors

About the Editors

Tulasi Satyanarayana after obtaining M.Sc. and Ph.D. at the University of Saugar (India), Tulasi Satyanarayana had postdoctoral stints at the Paul Sabatier University and National Institute of Applied Sciences, Toulouse, France. In 1988, he joined the Department of Microbiology, University of Delhi South Campus as Associate Professor and became Professor in 1998. His research efforts have been focused on understanding the diversity of yeasts, and thermophilic fungi and bacteria, their enzymes and potential applications, heterotrophic carbon sequestration and metagenomics. He has published over 250 scientific papers and reviews and edited six books. He has two Indian patents to his credit. He has been conferred with Dr. G.B. Manjrekar award of the Association of Microbiologists of India in 2004, Dr. V.S. Agnihotrudu award of Mycological Society of India in 2009 and Malaviya award of Biotech Research Society of India in 2012 for his distinguished contributions.

Gotthard Kunze studied biology at the Ernst Moritz Arndt University in Greifswald. He got a postdoctoral fellowship and a position as scientific assistant at the Department of Biology of the University. In 1986 he joined as a research associate at the Institute of Plant Genetics and Crop Plant Research (IPK), Gatersleben. Since 1998, he has been a Visiting Professor at the University of Greifswald and Professor at the Technical University Anhalt at Köthen since 1998. During this period, he focused his research activities on yeast genetics (construction of new yeast host vector systems, heterologous gene expression, thermo- and osmo-resistance in nonconventional yeasts and microbial yeast biosensors). Professor Gotthard Kunze is the author of about 182 publications, editor of two books, and teaches at the universities of Greifswald and Köthen.

Contributors

Adivitiya Department of Microbiology, University of Delhi South Campus, New Delhi, India

Bijender Kumar Bajaj School of Biotechnology, University of Jammu, Jammu, India

Tek Chand Bhalla Department of Biotechnology, Himachal Pradesh University, Shimla, Himachal Pradesh, India

Felix Bischoff Leibniz Institute of Plant Genetics and Crop Plant Research (IPK), Gatersleben, Germany

Bruno Blondin SPO, INRA, SupAgro, Université Montpellier, Montpellier, France

Erik Böer Leibniz Institute of Plant Genetics and Crop Plant Research (IPK), Gatersleben, Germany

Carole Camarasa SPO, INRA, SupAgro, Université Montpellier, Montpellier, France

Alexandre Chamas Leibniz Institute of Plant Genetics and Crop Plant Research (IPK), Gatersleben, Germany

V. Choudhary Department of Microbiology, Institute of Home Economics, University of Delhi, New Delhi, India

Preeti Dabas School of Biotechnology, G.G.S. Indraprastha University, Dwarka, Delhi, India

Vikas Kumar Dagar Department of Microbiology, University of Delhi South Campus, New Delhi, India

Sylvie Dequin SPO, INRA, SupAgro, Université Montpellier, Montpellier, France

Mukund V. Deshpande Biochemical Sciences Division, National Chemical Laboratory, Pune, India

Kostyantyn Dmytruk Institute of Cell Biology, NAS of Ukraine, Lviv, Ukraine

Sonali Kashinath Doke School of Pharmacy, SRTM University, Nanded, Maharashtra, India

Virginie Galeote SPO, INRA, SupAgro, Université Montpellier, Montpellier, France

Loughlin Gethins School of Microbiology, University College Cork, Cork, Ireland; Teagasc Research Centre, Fermoy, Co. Cork, Ireland

Vandana Ghormade Nanobioscience Group, Agharkar Research Institute, Pune, India

Shivkrupa Devrao Halbandge DST-FIST & UGC-SAP Sponsored School of Life Sciences, SRTM University NAAC Accredited with 'A' Grade, Nanded, Maharashtra, India

Olena Ishchuk Department of Biology, Lund University, Lund, Sweden

Kumaraswamy Jeyaram Microbial Resources Division, Institute of Bioresources and Sustainable Development (IBSD), Imphal, Manipur, India

Swati Joshi National Institute of Occupational Health (NIOH), Ahmedabad, Gujarat, India

Sankunny Mohan Karuppayil DST-FIST and UGC-SAP Sponsored School of Life Sciences, SRTM University (NAAC Acredited with a Grade), Nanded, Maharashtra, India

Yogender Pal Khasa Department of Microbiology, University of Delhi South Campus, New Delhi, India

Deepak Kumar School of Biotechnology, G.G.S. Indraprastha University, Dwarka, Delhi, India

Gotthard Kunze Leibniz Institute of Plant Genetics and Crop Plant Research (IPK), Gatersleben, Germany

Olena Kurylenko Institute of Cell Biology, NAS of Ukraine, Lviv, Ukraine

Jean-Luc Legras SPO, INRA, SupAgro, Université Montpellier, Montpellier, France

Ina Lemke Leibniz Institute of Plant Genetics and Crop Plant Research (IPK), Gatersleben, Germany

Katarzyna Litwińska Leibniz Institute of Plant Genetics and Crop Plant Research (IPK), Gatersleben, Germany

A. Madan Department of Microbiology, Institute of Home Economics, University of Delhi, New Delhi, India

Sarika S. Mane Biochemical Sciences Division, National Chemical Laboratory, Pune, India

Falko Matthes Leibniz Institute of Plant Genetics and Crop Plant Research (IPK), Gatersleben, Germany

John P. Morrissey School of Microbiology, University College Cork, Cork, Ireland

Volkmar Passoth Department of Molecular Sciences, Swedish University of Agricultural Sciences, Uppsala, Sweden

Reinhard Pätz Bioprocess Technology, University of Applied Sciences, Köthen, Germany

Amit Kumar Rai Institute of Bioresources and Sustainable Development, Sikkim Centre, Tadong, Sikkim, India

Jayant Shankar Raut UGC Center for Advanced Study, University Institute of Pharmaceutical Sciences (UIPS), Panjab University, Chandigarh, India

Jau-Henryk Richter-Listewnik Bioprocess Technology, University of Applied Sciences, Köthen, Germany

Paul Ross School of Microbiology, University College Cork, Cork, Ireland; Alimentary Pharmabiotic Centre, Microbiome Institute, University College Cork, Cork, Ireland

Justyna Ruchala Rzeszow University, Rzeszow, Poland

Tulasi Satyanarayana Division of Biological Sciences and Engineering, Netaji Subhas Institute of Technology, Dwarka, Delhi, India

Savitri Department of Biotechnology, Himachal Pradesh University, Shimla, Himachal Pradesh, India

Frieder Schauer Institute of Microbiology, Ernst-Moritz-Arndt-University of Greifswald, Greifswald, Germany

Rabea Schlüter Institute of Microbiology, Ernst-Moritz-Arndt-University of Greifswald, Greifswald, Germany

Nimisha Sharma School of Biotechnology, G.G.S. Indraprastha University, Dwarka, Delhi, India

Andriy Sibirny Institute of Cell Biology, NAS of Ukraine, Lviv, Ukraine; Rzeszow University, Rzeszow, Poland

Satbir Singh School of Biotechnology, University of Jammu, Jammu, India

Catherine Stanton Teagasc Research Centre, Fermoy, Co. Cork, Ireland; Alimentary Pharmabiotic Centre, Microbiome Institute, University College Cork, Cork, Ireland

Santosh G. Tupe Biochemical Sciences Division, National Chemical Laboratory, Pune, India

Javier A. Varela School of Microbiology, University College Cork, Cork, Ireland

Pandit B. Vidyasagar SRTM University, NAAC Accredited with 'A' Grade, Nanded, Maharashtra, India

A. Vohra Department of Microbiology, Institute of Home Economics, University of Delhi, New Delhi, India

Sebastian Worch Leibniz Institute of Plant Genetics and Crop Plant Research (IPK), Gatersleben, Germany

Diversity of Natural Yeast Flora of Grapes and Its Significance in Wine Making

Sarika S. Mane, Vandana Ghormade, Santosh G. Tupe and Mukund V. Deshpande

Abstract The biodiversity of yeasts associated with grapes has been studied in different regions of wine producing countries throughout the world. Most of the species associated with the wine environment are similar, while some species are specifically associated with specific regions. Though *Saccharomyces cerevisiae* is primarily used for fermentation of grape juice, its occurrence is low on grape berries. Non-*Saccharomyces* yeasts belonging to the genera *Torulaspora, Hanseniaspora, Pichia, Candida, Issatchenkia, Metschnikowia* etc. are in abundance in grape musts and may dominate the early stages of fermentation. Subsequently, *S. cerevisiae* proliferates, becomes dominant and completes the wine fermentation. Therefore, yeasts diversity associated with the grapes and must significantly contribute to the quality and varietal character of wine. In present review, the diversity of yeasts associated with vineyard, winery, succession of yeasts during fermentation and their role in wine quality is discussed. The knowledge will be useful to monitor and control the fermentation with respect to quality and spoilage.

Keywords Natural yeast flora of grapes · Non-*Saccharomyces* yeasts · *Saccharomyces cerevisiae* · Wine fermentation

1 Introduction

The earliest known wine was made in Mesopotamia around 3500 BC (Robinson 2006). However, chemical analyses of organic residues on ancient pot sherds indicated that grape juice was deliberately being fermented in China as early as

S.S. Mane · S.G. Tupe · M.V. Deshpande (✉)
Biochemical Sciences Division, National Chemical Laboratory, Pune 411008, India
e-mail: mvdeshpande1952@gmail.com

V. Ghormade
Nanobioscience Group, Agharkar Research Institute, Pune 411004, India

T. Satyanarayana and G. Kunze (eds.), *Yeast Diversity in Human Welfare*, DOI 10.1007/978-981-10-2621-8_1

7000 BC (McGovern et al. 2004). According to historical mural paintings and ancient pottery, the Egyptians, Phoenicians and Greeks were also quite willing winemakers and consumers. The Romans are assumed to have acquired the ability for cultivating grapes and winemaking from the Greeks and spread it into central and northern Europe. European pioneers in the 16th and 17th century introduced the grape vine into South, Middle and North America. Presently, France, Italy and Spain are the largest wine producing countries with total output of 84%, followed by Germany, Portugal, Greece, Romania and Austria. Italy tops the list with 4.49 billion liters of wine produced which is $\sim$17% of the world market share (Bettini 2014) and is followed closely by Spain (4.46 billion litres), and France (4.41 billion litres).

2 Grape Varieties Used for Winemaking

Worldwide different grape varieties are used for wine production. So far 1368 vine varieties have been documented across the globe. Wine is differentiated in to two types based on color as—red wine which is produced from grape varieties such as —Barbera, Cabernet Sauvignon, Carignan, Black Rieslin, Cabernet Franc, Cinsaut, Dornfelder, Gamay, Riesling, Sangiovese, Grenache, Malbec, Merlot, Shiraz, Syrah, Trollinger, Muscat, Montepulciano, Pinot Noir, Pinotage, Portugieser, Saperavi, and Zinfandel; and white wine made from grape varieties—Aligote, Sauvignon Blanc, Mueller-Thurgau, Chardonnay, Feteasca Alba, Chenin Blanc, Clairette, Feteasca Regala, Prosecco, Ugni Blanc, Pinot Blanc, Pinot Grigio, Semillon, Silvaner Garganega, Viognier and Vermantino. White wines are made without must (Skin and seeds) and are much lower in phenolics as compared to red wines. Other regionally important and aromatically distinctive varieties are Corvina, Dolcetto, Negro Amaro (red), Fiano, Garganega, and Torbato (white) from Italy; Malvasia, Parellada (white), and Graciano (red) from Spain; Arinto (white) and Ramisco (red) from Portugal and Rhoditis (white) from Greece; Furmint (white) from Hungary.

Grape variety used for wine making is an important factor determining wine quality as it imparts the "varietal character" to the wine, which is mainly because of the presence of different secondary metabolites responsible for the principal flavor compounds in grape must (Lambrechts and Pretorius 2000). For instance, the varietal differences impart characteristic flavor and aroma to the wine, like reminiscent of blackcurrants or cedar wood or firm tannins for Cabernet Sauvignon, herbal for Sauvignon Blanc, spicy with pepper and wild berry flavors for Zinfandel and soft and rich wine characterized by smoky and chocolaty aromas in case of Shiraz.

The red grape varieties predominantly used for wine making in India are Cabernet Sauvignon, Carignan, Grenache, Merlot, Pinot Noir, Saperavi, Shiraz, and Zinfandel; whereas, white varieties include Chardonnay, Chenin Blanc, Clairette, Garganega, Sauvignon Blanc, Ugni Blanc and Vermantino.

3 Red and White Wine Making Process

Alcoholic fermentation is an anaerobic process carried out mainly by *S. cerevisiae* in which sugars, glucose and fructose are converted into ethanol and carbon dioxide. Yeasts present on grapes reach there by wind and insect dispersal, increasing in number from the onset of fruit ripening (Lafon-Lafourcade 1983). After harvesting, the grapes are taken to winery, destemmed and crushed. In production of white wine, crushing is followed by limited maceration, pressing and extraction of juice for primary fermentation. Whereas, for red wine must obtained by crushing, which includes skin and seeds of red grapes along with the juice is directly fermented and macerated during fermentation to extract the phenolics, tannins, anthocyanins from skin and seeds into the must (Pretorius and Hoj 2005).

Primary fermentation is carried out by adding starter culture *S. cerevisiae* to the must containing other non-*Saccharomyces* yeasts coming from the berries and which takes ~15 days. After the primary fermentation of red grapes the wine is pumped off into tanks and the skin is pressed to extract the wine. White wines are generally fermented at 10–18 °C to improve the retention of aromas; whereas red wines are fermented at higher temperatures between 18–29 °C to achieve good extraction of phenolic compounds. An initial temperature of 20 °C is recommended for fermentation of both wines in order to stimulate initiation of yeast growth (Jackson 1994). For certain stylistic wines, secondary/malolactic fermentation is carried out in which lactic acid bacteria convert malic acid to lactic acid. The process decreases acidity of the wine and softens the taste. The wine is then clarified, allowed to mature (for certain wines), filtered and bottled.

Wines are also classified as dry wines (up to 4 g/L residual sugar), Semi sweet wines (up to 12 g/L residual sugar) and dessert wine (wines containing more than 45 g/L residual sugar). Based on manufacturing practices, wines are termed as sparkling wine (dissolved carbon dioxide in the wine held under pressure), fortified wine (wine blended with liquor) and spicy wine (Herb-flavored wine). Along with the vine variety and fermentation process followed, the yeast diversity of the grapes and must is an important factor contributing to the quality of wine (Barata et al. 2012a).

4 Microbial Diversity of Phylloplane

The microbial communities of phylloplane are diverse comprising of different genera of bacteria, filamentous fungi, yeasts, algae, and, less frequently, protozoa and nematodes too. The yeasts usually colonize rapidly on the leaves. Number of yeasts were reported by Chand-Goyal and Spotts (1996) from the apple and pear fruit surface. *Aureobasidium pullulans, Cryptococcus albidus* and *Rhodotorula glutinis* were found on fruits in most of the studied pear orchards. Other yeasts colonizing pear fruit surfaces were *Cryptococcus infirmo-miniatus, Cryptococcus laurentii, Debaryomyces hansenii, Rhodotorula aurantiaca, Rhodotorula fujisanensis, Rhodotorula minuta* and *Sporobolomyces roseus*. Slavikova et al. (2009) isolated 150 plus strains belonging to 11 genera from 5 fruit trees, namely apple, cherry, apricot, peach and plum leaves. Most common were *A. pullulans, C. laurentii* and *Metschnikowia pulcherima* while *Hanseniaspora uvarum, Pichia anomala, R. glutinis* and *Saccharomyces cervisiae* were less frequent.

Nakase et al. (2006) reported the presence of *H. uvarum, Kluyveromyces marxianus, Pichia amethioina, Pichia chambardii, Pichia farinose, Pichia kluyveri, Pichia membranaefaciens, S. cerevisiae, Lachancea kluyveri* (Synonyms: *Saccharomyces kluyveri, Torulaspora kluyveri*) and *Zygosaccharomyces rouxii* in 17 cultivars of bananas from Java, Indonesia. While Gana et al. (2014) observed different yeast species such as *Brandoniozyma complexa, Candida wangnamkhiaoensis, Debaryomyces nepalensis, Hypopichia burtonii, Kodamaea ohmerii, P. anomala, Pseudozyma hubeiensis, Psuedozyma prolific* and *Pseudozyma pruni* on the surface of banana from Philippines. The presence of different yeasts was attributed to the geographical differences.

Xue et al. (2006) isolated 8 *Metschnikowia* strains under 3 different species, *M. sinensis, M. zizyphicola* and *M. shanxiensis* from the surface of jujube fruits (*Zizyphus jujube*) collected in China. Phylogenetically by 26S rDNA D1/D2 domain sequence analysis, it was suggested that these three novel species could be clustered in a clade together with *M. fructicola, M. andauensis, M. pulcherrima* and *M. chrysoperlae*.

Janisiewicz et al. (2014) reported that there was a significant change in the natural yeast flora on plum surface during development/ripening. The presence of *Rhodotorula, Sporidiobolus* and *Aureobasidium* was significantly higher than *Cryptococcus* throughout the fruit development. However, on the mature fruit *Hanseniaspora, Pichia, Zygosaccharomyces* and *Wickerhamomyces* species were observed. The natural yeast flora of the fruit, especially *A. pullulans* and *R. phylloplana* exhibited antagonistic activity against *Monilinia fructicola*, a fungus that causes brown rot.

On grapes, bacteria and unicellular and filamentous fungi with different physiological characteristics have been reported. Some yeast species, lactic acid bacteria and acetic acid bacteria are unique to grapes which can survive and proliferate during fermentation, constituting the wine microbial consortium. The qualitative and quantitative differences of these microbes depend on the grape ripening stage and on the availability of nutrients. Furthermore, the microbial ecology is affected by grape health, abiotic and biotic factors which are involved in the primary damaging effect.

Different bacterial species found to be associated with grapes are *Bacillus* sp., *Enterobacter* sp., *Burkholderia* sp., *Serratia* sp., *Enterococcus* sp., and *Staphylococcus* sp. However, due to high acidity and ethanol concentration these bacterial species cannot grow in wine (Barata et al. 2012b), whereas lactic acid bacteria such as *Lactobacillus, Oenococcus, Leuconostoc,* and *Pediococcus* and acetic acid bacteria species of the genera *Acetobacter, Asaia, Acidomonas, Gluconobacter, Granulibacter, Neoasaia, Kozakia, Swaminathania, Saccharibacter* can grow and cause malolactic fermentation during wine making (Barata et al. 2012b; Gonzalez et al. 2005; Lonvaud-Funel 1999; Nisiotou et al. 2011; Osborne et al. 2005).

Valero et al. (2007) reported presence of filamentous fungi, like *Alternaria, Aspergillus, Cladosporium, Eurotium, Penicillium* and *Trichoderma* on grapes that are unable to grow in wine, similar to some bacterial genera. *Plasmopara viticola, Erysiphe necator* and *Botrytis cinerea* are the main pathogens on grapes which cause downy mildew, powdery mildew and grey rot, respectively (Barata et al. 2012a). Besides, *Erysiphe* and *Fusarium* were also observed on grapes (Diguta et al. 2011). Natural yeast flora of the grape plays an important role in imparting varietal character to the wine and is discussed in detail in following section.

5 Natural Yeast Flora of Grapes

5.1 Saccharomyces

Saccharomyces yeasts have a unicellular, globose, spheriodal shape. Multilateral (multipolar) budding is typical for vegetative reproduction (Vaughan-Matini and Martini 1998) and is one of the most studied organisms at biochemical and molecular level.

Saccharomyces and 15 plus genera of non-*Saccharomyces* yeasts are associated some time or other with wine fermentation. *S. cerevisiae* is not a common phyllosphere isolate; in fact it is prevalent on the surface of winery equipment (Fleet et al. 2002; Von Wallbrunn 2007). Earlier Mortimer and Polsinelli (1999) also reported the absence of *S. cerevisiae* on the grapes, in general. According to them, only one in one-thousand grape berries carried *S. cerevisiae.* Furthermore damaged berries were rich depositories of microorganisms including *S. cerevisiae.*

S. cerevisiae has enormous capacity to ferment sugars to ethanol and carbon dioxide. As a result this organism is one of the key players in baking, wine making, brewing, and bioethanol industry. Additionally, *Saccharomyces* has also been used as a transformation host for protein production (Nevoigt 2008). *S. cerevisiae* is relatively tolerant to low pH, high sugar and ethanol concentrations. Targets for wine yeast genetic improvements are: better fermentation performance, efficient wine processing, control of wine-spoiling microorganisms, and quality improvement.

Capallo et al. (2004) isolated *S. cerevisiae* strains from 12 grape varieties grown in the experimental vineyard of Apulia, South Italy. One of the important observations made was that these isolates were found to be well-adapted to the specific climatic conditions of the area and not the variety, per se. All these isolates were found to tolerate high ethanol concentration. Whereas, Capace et al. (2010) reported that different *Saccharomyces* isolates from Nero d'Avola grapes collected from different areas of the Sicily showed similar physiological characteristics such as high ethanol and SO_2 tolerance. Chavan et al. (2009) have isolated *Saccharomyces* strains from different grape varieties grown in two different geographical areas, Pune (18° 31′ N, 73° 55′ E) and Sangli (16° 52′ N, 74° 34′ E), India. Out of four varieties grown in Pune region, namely Banglore Blue, Zinfandel, Shiraz and Cabernet, *Saccharomyces* strains were found only on Zinfandel variety. Whereas, *Saccharomyces* strains were isolated from the berries of all four varieties grown in Sangli area namely Cabernet, Shiraz, Chenin Blanc and Sauvignon Blanc. These observations indeed suggest that no explicit role to either region (environmental factors) or variety could be assigned.

As the importance of role of *S. cerevisiae* in winemaking has long been established, the use of commercial strains of these yeast cultures in fermentation is a common practice in order to ensure a reproducible product and to reduce the risk of wine spoilage.

S. cerevisiae plays important role in wine fermentation mainly through metabolism of sugar to alcohol and CO_2 and it has an equally important role in the formation of secondary metabolites as well as in the conversion of grape aroma precursors to varietal aroma in wine. Molecular and biochemical studies have enabled researchers to develop sugar and alcohol tolerant, highly flocculent strains for wine production (Soares 2010). Flocculation contributes significantly in the brewing industry, in the production of renewal fuels (bio-ethanol), in modern biotechnology (production of heterologous proteins) and in environmental applications (bioremediation of heavy metals), etc. Barbosa et al. (2014) studied phenotypic and metabolic diversity of 20 commercial *Saccharomyces* strains used in different countries. According to their findings there was a relationship between nitrogen availability, yeast cell growth and sugar utilization during wine fermentation which can be additional criteria for strain selection. Brice et al. (2014) reported that the differences in nitrogen requirement between *S. cerevisiae* strains results from a complex allelic combination. They identified four genes namely *MDS3, GCN1, ARG81* and *BIO3* for which allelic variations were found to be associated with the differences in fermentation under nitrogen limiting conditions.

5.1.1 *Status of* Saccharomyces *During Wine Fermentation*

Various yeast species present on the berries and on winery equipments contribute significantly to wine fermentation. In the early stages of fermentation, genera like *Kloeckera, Hanseniaspora* and *Candida* were reported to be predominant followed by *Metschnikowia* and *Pichia*, when the ethanol concentration was 3–4%, while the later stages are dominated by alcohol tolerant strains of *Saccharomyces* species such as *S. cerevisiae, S. bayanus, S. paradoxus* and *S. pastorians* (Pretorius et al. 1999).

Two successive processes, namely, alcoholic fermentation of must by yeast and second, biological aging are involved in producing sherry wine. Species like *Candida stellata, Dekkera anomala, Hanseniaspora guilliermondii, Hanseniaspora uvarum, Issatchenkia terricola* and *S. cerevisiae* were observed at higher frequencies than other species like *Candida incommunis, Candida sorbosa* and *Zygosaccharomyces cidri* or *Z. fermentati* during alcoholic fermentation, while *S. cerevisiae, Pichia membranaefaciens, Pichia anomala* were found during biological aging. The *S. cerevisiae* strains involved in fermentation (*S. cerevisiae,*

S. bayanus, S. paradoxus and *S. pastorians*) are different from the strains responsible for biological aging (flor yeast, *S. cerevisiae* races *beticus, cheresiensis, montuliensis*, and *rouxii*) has been demonstrated by studying the *Saccharomyces* diversity using mtDNA restriction analysis and karyotyping of strains during sherry wine production (Esteve-Zarzoso et al. 2001). Diaz et al. (2013) using quantitative real-time PCR reported that *S. cerevisiae* remained active at the end of the fermentation along with *M. pulcherrima, R. mucilaginosa, P. kluyveri, P. membranifaciens*.

5.2 *Non*-Saccharomyces *Yeasts*

Grape berry surface provides physical environment suitable for the growth of microorganisms. *Rhodotorula, Cryptococcus* and *Candida* are the predominant candidates on unripe-grapes. With an increase in sugar concentration and decrease in acidity during maturation of berries, *Kloeckera/Hanseniaspora* become dominant, accounting for more than 50% of the total yeast flora. Other species of obligate aerobic or weakly fermentative yeasts with low alcohol tolerance are present in lesser proportions. These belong to the genera *Candida, Cryptococcus, Debaryomyces, Hansenula, Issatchenkia, Kluyveromyces, Metschnikowia, Pichia, Rhodotorula, Hanseniaspora, Saccharomyces, Torulaspora and Zygosaccharomyces* (Chavan et al. 2009; Ciani and Maccarelli 1998; Fleet 2003; Li et al. 2010; Loureiro and Malfeito-Ferreira 2003). Most of these yeasts belong to ascomycetes and may exist on the grapes as sexual (ascospore producing, teleomorphic) or asexual (non-spore forming, anamorphic) or both the forms depending on the environmental conditions. Hot regions, cooler regions and moderate climate regions favor growth as teleomorphic, anamorphic and both types, respectively.

5.2.1 *Non*-Saccharomyces *Yeasts Associated with Fermenting Must*

The solid portion of must is called pomace. The grape must i.e. grape juice with skin, seeds and stems of fruits, has low pH, high sugar content. Availability of the oxygen, and/or ethanol concentration affects the predominance of different species of yeasts in the fermenting must. During fermentation, due to low oxygen and increasing level of ethanol most of the non-*Saccharomyces* yeasts cannot survive (Combina et al. 2005; Fleet et al. 1984; Hansen et al. 2001; Henick-Kling 1998; Jackson 1994). The clarification of white must (centrifugation, enzyme treatments, cold settling) also reduces the initial population of yeasts (Fleet 1990; Lonvaud-Funel 1999; Pretorius 2000).

The non-*Saccharomyces* yeast population changes during cold maceration and alcoholic fermentation which can be attributed to the changes in micro-environment. For instance, Hierro et al. (2006) reported that *H. osmophila, C. tropicalis* and *Z. bisporus* species were predominantly found during cold maceration. Depending on the availability of the oxygen, and/or ethanol concentration different species of yeasts become predominant in the fermenting must. Combina et al. (2005) studied non-*Saccharomyces* flora of fermenting must of Malbec variety of grapes. They reported the ubiquitous presence of *Kloeckera apiculata, C. stellata* and *Metschnikowia pulcherrima* in the spontaneous fermentation.

Predominance of non-*Saccharomyces* yeasts in fermenting must at the later stages is influenced by barrels and post-fermentation spoilage (Loureiro and Malfeito-Ferreira 2003). *Brettanomyces* sp. and *Zygosaccharomyces* sp. are ethanol tolerant like *S. cerevisiae* and can be found in bottled wine. *Dekkera bruxellensis* was often found to be associated with wineries and less commonly on grape berries (Fugelsang 1997; Ibeas et al. 1996; Martorell et al. 2006). The highly diverse non-*Saccharomyces* microflora has been reported to be present at 10^4–10^5 CFU/mL during cold maceration and the population increases to a maximum of 10^6–10^7 CFU/mL at the beginning of alcoholic fermentation, which then declines to $\sim 10^3$–10^4 CFU/mL at the end of fermentation (Zott et al. 2008). Non-*Saccharomyces* yeasts have also been observed to grow to levels upto 10^4 cells/mL during malo-lactic fermentations.

Nemcova et al. (2015) reported that the grape variety, physical damage of the grapes, weather conditions and chemical composition of the must influenced *Sacharomyces* and non-*Saccharomyces* yeast diversity. The ascomycetes yeasts (*Aureobasidium, Candida, Hanseniaspora, Metschnikowia, Pichia, Saccharomyces* and *Saccharomycopsis*) and basidiomycetous yeasts (*Cryptococcus, Dioszegia, Filobasidium, Rhodotorula* and *Sporidiobolus)* were reported to be associated with fermenting must of three grape varieties namely Blue Frankish, Green Veltliner and Sauvignon Blanc, while *Hanseniaspora uvarum, Metschnikowia pulcherrima, Pichia kluyveri, Pichia kudriavzevii* and *Sporidiobolus pararoseus* were observed on the berries. However, damaged berries were found to support the growth of *P. kluyveri* and *P. kudriavzevii*. Assis et al. (2014) studied yeast flora of Chenin Blanc variety cultivated in the "Sao Francisco Valley" region of Brazil and observed that *Hanseniaspora opuntiae* and mixed cultures of *H. opuntiae* and *S. cerevisiae* influenced the wine quality.

Domizio et al. (2014) studied eight non-*Saccharomyces* strains, namely *Hansensiaspora osmophila, Lachancea thermotolerans, M. pulcherrima, Pichia fermentans, Saccharomycodes ludwigii, Starmerella bacillaris, Torulaspora delbrueckii* and *Zygosaccharomyces florentinus,* to check their potential to modulate the concentrations of various volatile compounds. Furthermore, these strains demonstrated a higher capacity to release polysaccharides such as mannoproteins compared to *S. cerevisiae*.

5.2.2 *Region Specific Non-*Saccharomyces *Yeasts*

The diversity of natural yeast flora of grapes changes significantly with geographical locations or regions and influenced by the grape varieties, and level of maritime (closeness of sea), temperature and rainfall. The vineyards from Italy, Spain and China show higher diversity of yeast flora followed by France, India,

Table 1 Diversity of yeasts associated with grapes from different countries

Country	Grape variety (red/white)	Associated yeast genera	References
Argentina	Malbec (red)	*Pichia, Kloeckera, Saccharomyces, Zygosaccharomyces, Rhodotorula, Metschnicowia, Issatchenkia, Kluyveromyces*	Combina et al. (2005)
Australia	Cabernet Sauvignon (red)	*Cryptococcus, Rhodotorula, Sporobolomyces, Hanseniaspora, Metschnikowia, Kluyveromyces, Torulaspora, Saccharomyces*	Prakitchaiwattana et al. (2004)
Brazil	Bordeaux (red) Isabel (red)	*Hanseniaspora, Saccharomyces, Issatchenkia, Sporidiobolus*	Baffi et al. (2011)
Canada	Icewine (red)	*Sporobolomyces, Cryptococcus, Rhodotorula, Hanseniaspora*	Subden et al. (2003)
China	Cabernet Sauvignon (red)	*Hanseniaspora, Cryptococcus, Pichia, Candida*	Li et al. (2010)
	Merlot (red)	*Hanseniaspora, Cryptococcus, Pichia, Candida, Zygosaccharomyces, Issatchenkia, Metschnikowia, Pichia*	
	Chardonnay (red)	*Hanseniaspora, Candida, Cryptococcus, Sporidiobolus*	
France	Merlot (red) Cabernet Sauvignon (red)	*Candida, Rhodotorula, Pichia, Sporidiobolus, Cryptococcus, Hanseniaspora, Rhodosporidium*	Renouf et al. (2005)
Greece	Mavroliatis, Sefka (red)	*Aureobasidium, Candida, Hanseniaspora, Issatchenkia, Metschnikowia, Zygosaccharomyces*	Nisiotou and Nychas (2007)
Italy	Sangiovese (red)	*Aureobasidium, Metschnikowia*	Guerzoni and Rosa (1987)
	Rossiola (red)	*Candida, Kloeckera, Issatchenkia, Pichia and others*	
	Catarratto (white)	*Candida, Hanseniaspora, Issatchenkia, Kluyveromyces, Metschnikowia, Zygoascus, Zygosaccharomyces*	Romancino et al. (2008)
	Muscat (white)	*Candida, Hanseniaspora, Kluyveromyces, Saccharomyces, Torulaspora*	
	Frappato (red)	*Hanseniaspora, Kluyveromyces, Metschnikowia, Zygosaccharomyces, Candida, Issatchenkia*	
	Nerod' Avola (red)	*Candida, Hanseniaspora, Issatchenkia, Metschnikowia, Zygoascus, Zygosaccharomyces*	
India	Banglore Blue (red)	*Candida, Hanseniaspora, Issatchenkia, Pichia*	Chavan et al. (2009)
	Cabernet Sauvignon (red)	*Candida, Hanseniaspora, Issatchenkia, Saccharomyces*	
	Zinfandel (red)	*Hanseniaspora, Issatchenkia, Saccharomyces, Zygoascus*	

(continued)

Table 1 (continued)

Country	Grape variety (red/white)	Associated yeast genera	References
	Shiraz (red)	*Debaryomyces, Hanseniaspora, Saccharomyces, Pichia*	
	Chenin Blanc (white)	*Hanseniaspora, Issatchenkia*	
	Sauvignon Blanc (white)	*Hanseniaspora, Pichia*	
Japan	Niagara (white)	*Kloeckera, Candida, Cryptococcus*	Yanagida et al. (1992)
	Chardonnay (white)	*Cryptococcus, Rhodotorula*	
	Zenkoji (white)	*Cryptococcus, Rhodotorula, Candida*	
	Koshu (white)	*Kloeckera, Cryptococcus*	
Portugal	Periquita (red)	*Metschnikowia, Kluyveromyces, Candida, Pichia, Hanseniaspora, Saccharomyces, Issatchenkia, Zygosaccharomyces, Zygoascus, Torulaspora*	Barata et al. (2008)
Slovenia	Žametovka, Modra Frankinja (red) and Kraljevina (white)	*Cryptococcus, Debaryomyces, Hanseniaspora, Metschnikowia, Pichia, Rhodotorula, Sporobolomyces*	Raspor et al. (2006)
Spain	Shiraz, Grenache, Barbera (red)	*Metschnikowia, Kluyveromyces, Candida, Pichia, Hanseniaspora, Torulaspora, Saccharomyces*	Cordero-Bueso et al. (2011)
	Abarino, Godello (white) and Mencia (red)	*Rhodotorula mucilaginosa*	Longo et al. (1991)
Spain (North)	Folle Blanche and Hondarrabi Zuri (white)	*Candida, Cryptococcus, Kloeckera, Rhodotorula, Saccharomyces*	Rementeria et al. (2003)
South Africa	Chardonnay (white)	*Kluyveromyces, Candida, Pichia, Kloeckera, Zygosaccharomyces, Rhodotorula*	Jolly et al. (2003)
Southern Slovakia	Frankovka (red) Veltlin (white)	*Pichia, Candida, Metschnikowia, Hanseniaspora, Issatchenkia*	Brezna et al. (2010)

Argentina and Portugal, while relatively low species diversity was observed in vineyards of Australia, Brazil, Canada, Greece and Japan (Table 1).

Longo et al. (1991) reported isolation of two hundred plus yeast strains from six wineries, all located in two wine regions of northwest Spain. The difference concerning yeast diversity between both regions was mainly due to their oxidative behavior. For instance, four species, *C. albidus, C. stellata, H. anomala*, and *H. silvicola* were predominant in the Atlantic region (near sea) where climate is moderate, while six species, *C. vini, H. canadensis, H. jadinii, P. carsoni, D. intermedia*, and *Sp. roseus*, were exclusive to the interior region (arid lifted plains with low lying river valleys).

Brilli et al. (2014) assessed the long-term relationship (1997–2012) between quantitative and qualitative yeast diversity and the meteorological variables such as air temperature, relative humidity, and rainfall at one location. *Kloeckera apiculata* and *Candida zemplinina* represented almost the totality of non-*Saccharomyces* yeasts in grape and fresh musts and quantitatively well correlated with temperature 10 days before grape harvest.

A significant change in the yeast diversity, species heterogeneity was observed in presence of *Botrytis cinerea* infection, with *Hanseniaspora opuntiae* being encountered as an inhabitant of the grape ecosystem (Longo et al. 1991). Nisiotou and Nychas (2007) also studied yeast species diversity using restriction fragment length polymorphism and sequence analyses of the 5.8S internal transcribed spacer and the D1/D2 ribosomal DNA (rDNA) regions of yeasts during the fermentation with and without *Botrytis*-affected grape juice from two regions in Greece, Attica and Arcadia. *Botrytis* infection significantly affected species heterogeneity. During initial phase of fermentation *Botrytis*-affected grape juice showed more biodiversity than grape juice without infection. The species such as *Zygosaccharomyces bailii* and *Issatchenkia* spp. or *Kluyveromyces dobzhanskii* and *Kazachstania* sp. were predominant.

Using PCR-RFLP and sequence analysis of ITS and rDNA regions, Li et al. (2010) evaluated the yeast diversity and its quantitative changes in three grape varieties cultivated in four different regions of China. Seventeen different yeast species belonging to eight genera were reported to be present on the grape berries. These include: *Hanseniaspora uvarum, Cryptococcus flavescens, Pichia fermentans, Candida zemplinina, Cryptococcus carnescens, Candida inconpicua, Zygosaccharomyces fermentati, Issatchenkia terricola, Candida quercitrusa, Hanseniaspora guilliermondii, Candida bombi, Zygosaccharomyces bailii, Sporidiobolus pararoseus, Cryptococcus magnus, Metschnikowia pulcherrima, Issatchenkia orientalis* and *Pichia guilliermondii*. Among these *H. uvarum* and *C. flavescens* were the dominant species with *Sporidiobolus pararoseus* being found for the first time.

To achieve unique regional qualities to the fermented wine, Sun et al. (2014) suggested the use of local strains. In this regard, the yeast flora of five grape varieties, namely Chardonnay, Cabernet Franc, Cabernet Sauvignon, Marselan, and Merlot were studied. The colony characteristics along with sequencing of the 26S rDNA D1/D2 domain were used to identify eight species of seven genera namely *A. pullulans, C. zemplinina, H. uvarum, H. occidentalis, I. terricola, M. pulcherrima, P. kluyveri*, and *S. cerevisiae*. The predominantly isolated species were *H. uvarum* and *S. cerevisiae*. They further reported the presence of six different genotypes of *S. cerevisiae* at different time points during the fermentation of Marselan variety. Earlier, Pallmann et al. (2001) used WL nutrient medium for qualitative and quantitative profiling of wine fermentation. Seventeen different colony morphologies were correlated with six different genera such as *Candida*, *Hanseniaspora*, *Issatchenkia*, *Pichia*, *Metschnikowia* and *Saccharomyces*. Interestingly, distinct colony sub-types were identified within a single species *M. pulcherrima* which produced antimicrobial pigment, the pulcherrimin.

5.3 Factors Affecting Yeast Diversity

Yeast diversity of grapes and must is quite important in wine production because of its influence on fermentation speed, wine flavour and wine quality. The density and diversity of the yeast population on grape berries is affected by numerous factors such as, grape variety (Cordero-Bueso et al. 2011), grape health (Barata et al. 2008; Loureiro and Malfeito-Ferreira 2003), grape ripeness (Martins et al. 2012), climatic condition and geographic location (Bezerra-Bussoli et al. 2013; Nicholas et al. 2013), application of different chemicals (Milanovic et al. 2013), use of different oenological practices (Andorra et al. 2008, 2011) as well as application of different farming systems (Cordero-Bueso et al. 2011; Martins et al. 2012). The numbers of yeast cells are greater close to the peduncle than it is at the centre and lower part of the bunch (Rosini 1984). The manner in which grapes are sampled (e.g. the berries or bunches) and processed (washing vs. crushing) also determines the yeast diversity in must (Martini et al. 1996). At harvest, grape temperature, method of harvest (manual vs. mechanical), method of transport to the cellar (picking crates/baskets, tipsters), time of transport to the cellar, time lapse before crushing, and sulphite and enzyme addition can all affect yeast populations (Lambrechts and Pretorius 2000; Pretorius et al. 1999). Despite all the variables in grape harvest and wine production, the yeast species generally found on grapes and in wines are similar throughout the world (Amerine and Kunkee 1968). However, the proportion (or population profile) of yeasts in different regions shows distinct differences (Longo et al. 1991). Cordero-Bueso et al. (2011) studied the biodiversity of yeasts in the conventional and organic viticulture in Spain. *K. thermotolerans, C. stellata, T. delbrueckii* and *P. anomala* were reported from the vineyard with both farming systems. However, the organic viticulture supported diversity of yeast species significantly more than conventional agriculture practices. For instance, in organic vineyard, in a must of a Shiraz variety, *K. thermotolerans* was the most abundant, while *S. cerevisiae, C. stellata, M. pulcherrima* and *H. guilliermondii* were also significant. While in Grenache must *H. guilliermondii* was more abundant than *K. thermotolerans, P. anomala, S. cerevisiae* and *C. stellata*. *S. cerevisiae* strains were reported to be in high number in Barbera must. Under conventional viticulture in the Barbera must *C. stellata* was in the highest proportion, followed by *T. delbrueckii* and *K. thermotolerans*. However, in Grenache must only two species, *K. thermotolerans* and *H. guilliermondii* were in significant number. *P. toletana, C. sorbosa* and *T. delbrueckii* were isolated from Shiraz variety from Spain (Cordero-Bueso et al. 2011).

6 Profiling of Yeast Flora, Enzyme Activities and Flavor Compounds During Fermentation

6.1 Profiling of Yeasts During Wine Fermentation

The qualitative and quantitative changes in *Saccharomyces* and non-*Sacharomyces* yeast strains during wine fermentation influence the wine quality. Traditionally the samples at different time intervals are analyzed using microbiological techniques of enrichment, isolation and identification. Combina et al. (2005) used the conventional microbiological techniques and showed the significant participation of non-*Saccharomyces* yeasts during spontaneous fermentation of Malbec musts, with the ubiquitous presence of three main species: *K. apiculata, C. stellata* and *M. pulcherrima*. In view of the advances in molecular techniques, denaturing gradient gel electrophoresis of PCR-amplified 26 rDNA genes was reported to be useful to analyze mixed yeast community during wine fermentation (Cocolin et al. 2000).

6.1.1 Succession of Yeast Flora

It was observed that the early stage of fermentation was always dominated by non-*Saccharomyces* yeast flora of grapes (Fleet 1990). For instance, *Candida* sp., *Hanseniaspora* sp., *Pichia* sp., *Rhodotorula* sp. and *Kluyveromyces* sp. were dominant in grape must during the early stages due to their low fermentative activity. Subsequently, as the ethanol level (5–7%) increased, most of the non-*Saccharomyces* yeasts did not survive and finally *S. cerevisiae* proliferated, became dominant and completed the wine fermentation (Fleet 2003; Fleet and Heard 1993; Gao et al. 2002; Heard and Fleet 1988). Hansen et al. (2001) reported that two wine related yeasts, *Kluyveromyces thermotolerans* and *Torulaspora delbrueckii* could not survive in the later stages due to the presence of ethanol, lack of oxygen, nutrient depletion or the presence of toxic compounds and cell-to-cell contact mechanism. Moreover, *S. cerevisiae* strains were reported to secrete peptides that inhibited the growth of some non-*Saccharomyces* yeast (Albergaria et al. 2010; Nissen and Arneborg 2003). However, some non-*Saccharomyces* yeast could survive till later stage of fermentation (up to 12 days) (Fleet 1990; Fleet et al. 1984). Heard and Fleet (1988) studied the effect of temperature and pH on the growth of the non-*Saccharomyces* yeasts during fermentation in mixed culture. It was observed that at low temperature (15–20 °C) the ethanol tolerance of *Candida* and *Hanseniaspora* was more and thus has more impact on the wine flavor at the end. On the other hand, species like *Schizosaccharomyces pombe, Zygosaccharomyces bailii* and *Zygosaccharomyces fermentati* were reported to survive in presence of high ethanol concentrations (>10%) (Fleet 2000; Romano and Suzzi 1993).

Furthermore, the ability of the yeasts to utilize malic acid was a positive attribute in many wine-making processes (Volschenkla et al. 2006). Usually commercially available *Saccharomyces* strains cannot degrade malic acid effectively during alcoholic fermentation. The expression of the malolactic pathway genes, i.e. the malate transporter *(mael)* of *S. pombe* and the malolactic enzyme *(mleA)* from *Oenococcus oeni* in *Saccharomyces*, can improve the malate utilization and thus improve the quality of wine. However, Volschenkla et al. (2006) suggested that the improper strain selection may give an off-flavor to the wine.

Jolly et al. (2013) have extensively reviewed the contributions and successions of non-*Saccharomyces* yeasts in wine fermentation. Ocon et al. (2010) analyzed the quantitative and qualitative changes of non-*Saccharomyces* yeasts present in spontaneous alcoholic fermentations of a tempranillo grape variety. Though qualitatively 17 different yeast species were reported, quantitatively *Candida stellata, Kloeckera apiculate* and *Saccharomyces cerevisiae*, appeared in large numbers.

Clemente-Jimenez et al. (2004) reported that in the initial phase of the natural fermentation in Macabeo grape varieties, *Kloeckera* and *Candida* genera appeared prominantly, followed by *Metschnikowia*, *Pichia* and sometimes, *Brettanomyces, Kluyveromyces, Schizosaccharomyces, Torulaspora, Rhodotorula* and *Zygosaccharomyces.* They further reported that the best profile of higher alcohols was given by *Saccharomyces cerevisiae* followed by *Hanseniaspora uvarum, Issatchenkia orientalis* and *Candida stellata.* While due to the presence of *Metschnikowia pulcherrima* and *Pichia fermentans* highest production of ethyl caprilate and 2-phenyl ethanol, compounds associated with pleasant aromas was seen.

The succession of non-*Saccharomyces* yeasts during natural fermentation of two varieties namely, Cabernet and Shiraz, was studied (Mane 2016). In natural fermentation of Cabernet variety *Pichia* and *Issatchenkia* were found in the initial phase (3 days) while in the fermentation of Shiraz variety both were present up to 6 days. In both the cases, *Hanseniapora* sp were observed up to 9th day of fermentation while *S. cerevisiae* up to 15th day (Mane 2016).

6.1.2 Factors Affecting Succession of Yeast Flora

The succession of yeast during fermentation is affected directly or indirectly, by a number of factors including grape variety, ripening stage, physical damage to berries, if any, climatic conditions, viticulture practices, etc. Renouf et al. (2005) observed qualitative and quantitative differences in yeast populations isolated from Merlot, Cabernet Sauvignon and Cabernet Franc varieties according to berry development stages, namely berry set, veraison and harvest. For instance, at berry set, *A. pullulans* was predominant which was never detected at harvest, while *Saccharomyces* was detected at harvest and not in the first stage of grape growth. The specific condition of the must with respect to the osmotic pressure, presence of SO_2 and temperature play a role in determining species which can survive and grow (Bisson and Kunkee 1991). The species of *Cryptococcus, Rhodotorula, Sporobolomyces, Candida* and *Hanseniaspora* which were low in number at the initial stage were seen in other two stages also, which can be attributed to their adaptive nature under environmental perturbations such as anaerobic condition, increased alcohol level etc. Excessive rainfall or even pesticide sprays especially during ripening stages affect the number of non-*Saccharomyces* yeasts in the initial stages and later in the fermentation (Guerra et al. 1999; Querol et al. 1990). *Botrytis cinerea* infection to grapes was found to increase the population of *C. krusei* and *K. apiculata* while decrease in *R. glutinis* (Le Roux et al. 1973). In fact, the methods of isolation and enumeration, type of growth media used are also important for the quantitative estimation. For example, on a medium containing lysine as a sole carbon source *S. cerevisiae* could not grow luxuriantly (Heard and Fleet 1986).

6.2 Profiling of Enzyme Activities During Fermentation

The quality of wine is mainly determined by aroma which is due to terpenes. The pivotal role of endogenous enzymes from grapes and also from natural flora in the wine making has been well emphasized (Van Rensburg and Pretorius 2000). The enzymes like pectinases, glucanases, xylanases and proteases are involved in the clarification and processing of wine and glucosidase plays a major role in release of aroma compounds (Pombo et al. 2011). The indigenous enzymes from grapes are not adequate in developing specific aroma by hydrolyzing non-volatile glycosidic precursors present in the grapes (Fia et al. 2005). The glycosidases from grapes have narrow substrate specificity, are inhibited by low pH (i.e. from 3 to 4) and glucose at concentrations >1%. Enzymes such as pectinases and glucanase increase juice extraction from grapes improve wine clarification and facilitate wine filtrations (Canal-Llauberes 1993, 1998; Villettaz and Duboudieu 1991), which however, are inactivated due to low pH and SO_2 conditions prevalent during wine fermentation. *S. cerevisiae* does not produce significant quantities of extracellular proteases, lipases or pectinolytic enzymes, while the non-*Saccharomyces* yeasts contribute significantly to a variety of enzyme reactions involved in aroma production during wine fermentation.

Van Rensburg and Pretorius (2000) emphasized the pivotal role of enzymes endogenous from grapes and also from natural flora of the berries in the wine making. The enzymes like pectinases, glucanases, xylanases and proteases are involved in the clarification and processing of wine. During the early stages of wine making there is substantial growth of non-*Saccharomyces* yeasts, which produce extracellular enzymes such as esterases, lipases, pectinases, proteases, β-1,3 glucanase and β-glucosidases (Strauss et al. 2001). These enzyme activities improve the process of winemaking and enhance wine quality. Pectinases and β-glucanases increased juice extraction from grapes, improved wine clarification and facilitated wine filtration (Canal-Llauberes 1993; Villettaz and Duboudieu 1991). Haze formation from proteins in the finished wine may be decreased by the use of proteolytic enzymes (Waters et al. 2005). The aroma and flavor properties of wine could be enhanced by glycosidases that hydrolyse non-volatile glycosidic precursors of the grape (Pombo et al. 2011). The reduction in ethyl carbamate as well as alcohol levels was catalysed by urease and glucose oxidase, respectively (Van Rensburg and Pretorius 2000). Esteve-Zarzoso et al. (1998) reported that non-*Saccharomyces* yeast species are important contributors to the final taste and flavor of wines due to their capacity to produce different enzymes such as protease, β-glucosidase, esterase, pectinase and lipase.

Enzymes of enological interest found in different non-*Saccharomyces* wine yeasts are presented in Table 2. The predominant genera which produce these enzymes are *Brettanomyces, Candida, Cryptococcus, Debaryomyces, Hanseniaspora, Hansenula, Kloeckera, Kluyveromyces, Metschnikowia, Pichia, Rhodotorula, Saccharomycodes, Schizosaccharomyces* and *Zygosaccharomyces.* Maturano et al. (2012) studied the enzymes from *Saccharomyces* and non-*Saccharomyces* species in pure and mixed culture during the fermentation. Pure cultures of *S. cerevisiae, H. vinae, T. delbrueckii* and mixed cultures of *Saccharomyces* with *H. vinae* or *T. delbrueckii* were used for fermentation of sterilized grape juice. In mixed cultures, *H. vinae* and *T. delbruckii* were detected in the initial half of the fermentation. Nevertheless β-glucosidase, protease and pectinase secreted by *H. vinae* and *T. delbruckii* in mixed culture could be detected up to the end of fermentation.

Enzyme profiling was carried out during Shiraz and Cabernet variety fermentations. Pectinase, β-1,3-glucanase and protease activities increased from 3–6 d while β-glucosidase activity decreased after 9 d. These enzymes correlated significantly with secondary metabolites, such as total phenolics, flavonoids and tannins that are important to wine quality (Mane 2016).

From the literature, it was seen that enzyme activities were influenced by pH and temperature, presence of sugars, SO_2 and ethanol. For instance, ethanol adversely affected β-glucosidase and pectinase activities during fermentation (Maturano et al. 2012).

The commercial wine yeast *S. cerevisiae* is not attributed with production of extracellular proteases, β-glucosidase or glucanases (Hernandez et al. 2003). The commercial β-glucanase preparations used in winemaking for clarification, filtration and maturation of wines were produced by *Trichoderma* species (Canal-Llauberes 1993). Mojsav et al. (2011) studied the effect of three commercial pectolytic enzyme preparations on the wine fermentation of white grape cultivar, Smederevka. These pectolytic preparations were found to be important in improving filtration rates, lees settling rates and clarity of wines. It was further suggested that such preparations can be used to increase sensory quality in a shorter time with cost effectiveness. However the activity of such exogenously added enzymes are compromised due the conditions prevailing during fermentation. Therefore, non-*Saccharomyces* yeasts as sources of these enzymes are important during wine fermentation. Alternately, expression of genes of polysaccharide degrading enzymes in *S. cerevisiae* was reported to be useful (Louw et al. 2006). Recombinant strains of *S. cerevisiae* were constructed using genes such as *T. reesei XYN2* xylanase, *Butyrivibrio fibrisolvens END1* glucanase, *A. niger XYN4* endo xylanase, *Erwinia chrysanthemi* pectate lyase *PEL5* and the polygalacturonase *PEH1* from *Erwinia carotovora*. The wine quality fermented with the recombinant strains was found to be comparable and acceptable (Louw et al. 2006).

Table 2 Enologically important enzymes found in non-*Saccharomyces* wine yeasts*

Yeast	β-glucosidase	Protease	β-1,3 glucanase	Pectinase	Esterase	Lipase
Brettanomyces	+	–	–	–	+	–
Candida famata	+	–	–	–	–	–
C. pulcherima	+	+	–	+	–	+
C. stellata	+	+	+	+	–	+
C. guilliermondii	+	+	–	–	–	–
C. valida	–	–	–	–	–	+
Debaryomyces hansenii	+	+	–	–	–	–
Hanseniaspora/Kloeckera	+	+	+	+	–	–
Hanseniaspora uvarum	+	+	+	+	–	+
Hansenula anomala	+	–	–	–	–	–
Issatchenkia orientalis	+	–	–	–	–	+
I. terricola	+	–	–	–	–	–
Metchnikowia pulcherima	+	+	+	+	–	+
Pichia anomala	+	+	–	+	–	+
Pichia fermentans/C. lambica	–	–	+	–	–	–
P. membranefaciens	+	+	+	+	–	–
P. kluyveri	+	–	+	+	–	–
Rhodotorula glutanis	–	–	–	+	–	+
Saccharomyces cerevisiae	+	+	–	+	+	+
Torulaspora delbrueckii	+	–	–	–	–	+
Schizosaccharomyces pombe	–	–	+	–	–	–
Zygoascus hellenicus/Candida hellenica	+	–	+	–	–	–
Zygosaccharomyces bailli	+	–	–	–	–	–

*Data compiled from—Barbagallo et al. (2004), Charoenchai et al. (1997), Esteve-Zarzoso et al. (1998), Fleet and Phaff (1974), Gonzalez et al. (2006), Jolly et al. (2006), Lagace and Bisson (1990), Otero et al. (2003), Rosi et al. (1987, 1994), Strauss et al. (2001)

6.3 *Profiling of Flavor and Aroma Compounds During Fermentation*

Different metabolites produced by *Saccharomyces* and non-*Saccharomyces* yeasts that contribute to wine quality are depicted in Fig. 1. *P. anomala* (*Hansenula anomala*), *K. apiculata* and *C. pulcherrima* were found to be high producer of esters (Bisson and Kunkee 1991; Clemente-Jimenez et al. 2004). *Hanseniaspora* isolates were reported to produce high levels of phenylethyl acetate and phenyl ethanol (Moreira et al. 2005; Rojas et al. 2001; Viana et al. 2008). Number of reports showed that glycerol was produced by *Candida, Hanseniaspora, Pichia* sp., *L. thermotolerans* and *C. zemplinina* (Soden et al. 2000; Comitini et al. 2011; Toro and Vazquez 2002). Acetic acid produced by *Hanseniaspora* spp., *Zygosaccharomyces* spp. and succinic acid producers were *Candida stellata, Saccharomyces ludwigii* and *T. delbrueckii* (Ciani and Maccarelli 1998).

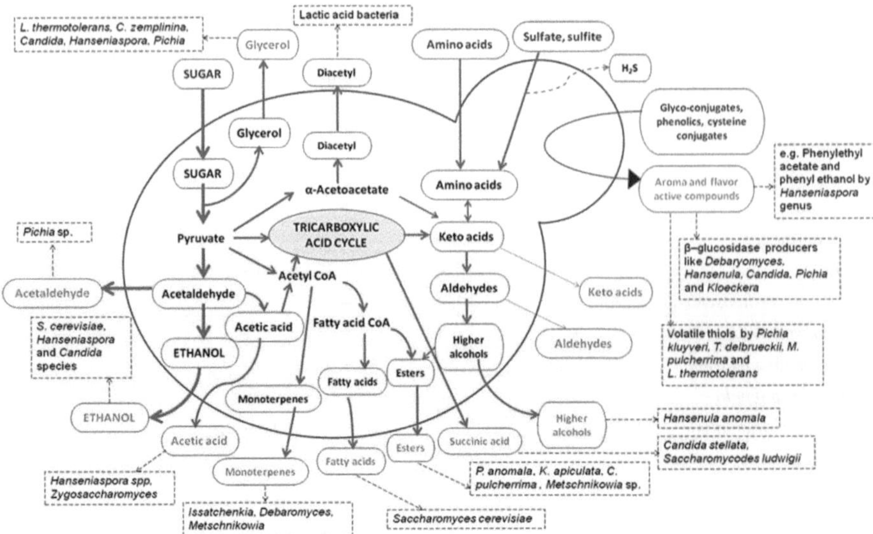

Fig. 1 The metabolites produced by *Saccharomyces* and non-*Saccharomyces* yeasts that contribute to wine quality (Modified from Swiegers et al. 2005)

Procopio et al. (2013) suggested that amino acid utilization by yeasts, *S. cerevisiae* and *S. pastorianus* var. *carlsbergenesis* was linked with flavour production. The most important amino acids involved in the synthesis of active flavour compounds were leucine, isoleucine, valine, glutamine, and proline. In addition, the cysteine was found to be important in case of *S. pastorianus* var. *carlsbergenesis* while histidine for *S. cerevisiae*. It was further suggested that a finger print of amino acid utilization and flavor compounds produced can be made.

It was earlier thought that the flavor of alcoholic beverages was due to a small number of volatile compounds. By 1985, the number of volatile compounds estimated was 1300 plus (Nykanen 1986). Many precursors of volatiles are present in grapes, which are processed more due to activities of non-*Saccharomyces* yeasts than *S. cerevisiae* (Cordente et al. 2012; Nykanen 1986). Non-*Saccharomyces* yeasts positively contribute to the analytical and sensorial composition of wine with production of hundreds of flavor active secondary metabolites such as acids, alcohols, esters, terpenoids, phenolic compounds, aldehydes, ketones, higher alcohols, glycerol, acetaldehyde, acetic acid and succinic acid volatile sulphur compounds (Jolly et al. 2013; Lambrechts and Pretorius 2000).

Lambrechts and Pretorius (2000) extensively reviewed the significance of both *Saccharomyces* and non-*Saccharomyces* yeasts in developing specific flavor to wine. The variety of grapes, conditioning of must, activities of microbial flora and aging contribute in to the flavor production. Further, the combination of yeast starter cultures can be used to have more or less predictable wine flavor. However, though it appears easy to have desired flavor, the understanding of physiological characteristics of each strain present and it's qualitative and quantitative interactions with other factors are important for avoiding undesirable flavor.

The common flavor compounds produced due to yeast enzymes are esters, fatty acids, fatty acid esters and higher alcohols (Maturano et al. 2015). Usually saturated, straight chain fatty acids along with one unsaturated fatty acid (palmitoleic acid) are produced during wine fermentation. The volatile fatty acid contents of wine are 10–15% of the total acid content which is mainly acetic acid. The predominant non-volatile organic acids are tartaric acid and malic acid in grape juice. Citric acid and lactic acid also add to acidity of the juice. The succinic and keto acids are present initially in traces but increase quantitatively during fermentation. These acids can influence wine flavor depending on concentration and type of wine (Swiegers et al. 2005).

The primary flavor of wine is derived from the grapes. However, secondary flavors are derived from ester formation by yeasts during fermentation (Lambrechts and Pretorius 2000; Nykanen 1986; Renault et al. 2015). *P. anomala* (*Hansenula anomala*), *K. apiculata* and *C. pulcherrima* are known to be high producer of esters (Bisson and Kunkee 1991; Clemente-Jimenez et al. 2004).

Phenolic compounds contribute to the color, flavor, bitterness and astringency of wine. The main types of phenolic compounds found in wine are phenolic acids (hydroxybenzoic and hydroxycinnamic acids), stilbenes, flavones, flavonols, flavanonols, flavanols, and anthocyanins (Monagas et al. 2007). Yeast strains in the *Hanseniaspora* genus produce high levels of phenylethyl acetate and phenyl ethanol that contribute to the complexity of wine aroma (Rojas et al. 2001; Viana et al. 2008; Moreira et al. 2005). Glycerol produced by *Candida, Hanseniaspora, Pichia* sp., *L. thermotolerans* and *C. zemplinina* contributes to smoothness (mouth-feel), sweetness and complexity in wines (Comitini et al. 2011; Toro and Vazquez 2002; Soden et al. 2000). Acetic acid produced by *Hanseniaspora* spp., *Zygosaccharomyces* spp. and succinic acid producers such as *Candida stellata, Saccharomyces ludwigii* and *T. delbrueckii* contribute to the total acidity of the wine (Ciani and Maccarelli 1998).

Interestingly, though not much studied, the role of non-*Saccharomyces* yeasts such as *Schizosaccharomyces pombe, P. guilliermondii* and *H. uvarum* in enriching wine color is also documented (Benito et al. 2013; Morata et al. 2012). For example, *P. guilliermondii* and *S. cerevisiae* were shown to increase the formation of vinylphenolic pyranoanthocyanins molecules which show greater color stability (Benito et al. 2013). *S. pombe* which can ferment grape must with high sugar contents also produced high levels of vitisin A type pigments-a natural phenol, while some strains with the help of hydroxycinnamate decarboxylase activity favored the formation of vinylphenolic pyranoanthocyanins (Morata et al. 2012).

The multi-starter fermentation with selected non-*Saccharomyces* yeasts and *S. cerevisiae* was found to be useful to avoid problems of natural fermentation, if any (Sadoudi et al. 2012). However, the interactions among the cultures used were complex and majority of the interactions were unpredictable.

To understand these interactions the volatile profiles of *Candida zemplinina, Torulaspora delbrueckii* and *Metschnikowia pulcherrima* either in mono-culture or in co-culture with *S. cerevisiae* were studied. It was reported that *C. zemplinina* produced high levels of terpenes and lactones which were decreased in co-culture with *S. cerevisiae*. On the other hand a synergistic effect on aromatic compound production by *M. pulcherrima* was reported in co-culture with *S. cerevisiae* (Sadoudi et al. 2012). No effect was seen in aromatic profiles of *T. delbrueckii* and *S. cerevisiae* in mono-culture and in co-culture. This study can be used to design a specific microbial profile for defined wine quality.

In addition to the role in the production of flavor compounds some of the non-*Saccharomyces* yeasts were also useful to reduce the alcohol levels indirectly enhancing the perception of wine aroma (Quiras et al. 2014). Contreras et al. (2013) earlier evaluated number of non-*Saccharomyces* strains for their capabilities to ferment sugar to ethanol. Although number of yeasts such as *Zygosaccharomyces bailii, Kluveromyces marxianus, Hanseniaspora uvarum, Pichia kluyveri, Issatchenkia terricola, Candida sake* and others, produce more ethanol (>0.45 g) per g of sugar under anaerobic condition, the % utilization of sugar for ethanol production in most of the strains was far lesser (<16%) as compared to *S. cerevisiae.* In case of *S. cerevisiae* 98.5% sugar was consumed for ethanol production (0.44 g/g sugar) from an initial concentration of 150 g/L sugar under anaerobic condition (Contreras et al. 2013). Therefore, the regulation of level of non-*Saccharomyces* inoculum along with *S. cerevisiae* can be a key feature to avoid masking of wine aroma due to ethanol.

7 Molecular Approaches to Improve Wine Fermentation

The wine quality and economics of production are the key features for the genetic improvement studies with wine yeast. In this regard, Pretorius and Bauer (2002) extensively reviewed the three areas for genetic improvement of the strains, namely to increase fermentation efficiency, wine processing and control of wine spoilage and most importantly improvement of wine quality, per se. The easiest way is selection of variant which can perform well under desired conditions. The induced mutagenesis, mating of strains, protoplast fusion, cytoduction have limited success while recombinant DNA technology is precise but commercially less explored because of GMO status to the developed strains. For example, to improve ethanol tolerance, membrane ATPase activity and sterol formation have been targeted to modify the expression of *PMA1* and *PMA2* (ATPase) genes. While for improved sugar utilization the genes involved in hexose transport and hexose kinases such as *HXT1-HXT18, SNF3, FSY1* were reported to be important (Pretorius and Bauer 2002). Late expression of *FLO1, FLO5, MUC1/FLO11* coding for cell wall hydrophobins and flocculins significantly contributed in the flocculation.

Joshi et al. (2010) developed a strategy to control flocculation using a model dimorphic fungus *Benjaminiella poitrasii.* As flocculation in yeasts is dependent on a lectin and cell surface mannan interaction, the regulation of mannan synthesis can be used to control flocculation. Fructose 6 phosphate (F6P) is a common precursor in the synthesis of mannan and chitin. Furthermore, the availability of F6P for mannan synthesis is regulated by NAD- and NADP-dependent glutamate dehydrogenases (GDHs) which significantly contributed in the synthesis of chitin (Khale et al. 1992). In other words, the modulation of GDHs could be significant in changing the surface properties. The rate of flocculation was positively affected in the presence of α-ketoglutarate (substrate for NADP-GDH) or isophthalic acid (NAD-GDH inhibitor) while decelerated by glutamate, a substrate for NAD-GDH. Thus, NAD-GDH modulating substances could have potential use as flocculation inducing agents. Furthermore, the regulation of expression of NAD-and NADP-GDH genes could be a potential approach to control flocculation.

To improve biocontrol potential the expression of genes for hen egg white lysozyme (*HEL1*), chitinase (*CTS1*), exoglucanase (*EXG1*) and other genes such as pediocin (*PED1*), leucocin (*LCA1*) have been studied in *S. cerevisiae*. The traits of non-*Saccharomyces* yeasts to improve wine flavor and other sensory characteristics can be achieved by the overexpression of *END1*, *EXG1*, *CEL1*, *BGL1*, *PEL5* and *PEH1*, *ABF2* genes for glycosidases, glucanases and arabinofuranosidases involved in the liberation of grape terpenoids, genes for esterases and phenolic acid metabolism (Pretorius and Bauer 2002). However, the main concern as stated before is the acceptability of the genetically modified organism for the fermentation.

8 Use of Yeast Flora for Value Addition to Winery

During wine fermentation yeast flora, natural as well as added as an inoculum, flourishes to an extent to be a component for value addition to winery.

8.1 Isolation of Cell Wall Polymers

Klis et al. (2006) extensively reviewed the cell wall architecture of *S. cerevisiae* and suggested that mannoproteins were 30–50%, 1-6-β-glucan 5–10%, 1,3-β-glucan 30–45% while chitin 1.5–6% of cell wall mass. The cell wall stress was important to increase the chitin content. Earlier it was reported that the dry weight and the % of cell wall components in *S. cerevisiae* could vary with the nature of the carbon and nitrogen sources, medium pH, temperature and aeration (Aguilar-Uscanga and Francois 2003). Interestingly, under controlled conditions, there was a drastic increase in mannan to β-glucan ratio when cells were grown at pH 5. Similarly the mixture of sugars also affected the proportion of cell wall polymers such as chitin, mannan and glucans. At the end of the fermentation, most of the times the yeast biomass is mainly consisted of *Saccharomyces* cells. After fermentation along with *Saccharomyces,* non-*Saccharomyces*, if any, yeast biomass can be used for the isolation of cell wall polymers. For instance, De Miranda Reis et al. (2013) used *P. pastoris* grown on glucose, mannose and/or galactose for the production of chitin and its derivatives.

8.2 Winery Yeasts for Biocontrol

Certain strains of *S. cerevisiae* known as killer strains secrete protein toxins that are lethal to sensitive strains of the same and/or other related species. These toxins are grouped into three types, K1, K2, and K28, genetically encoded by medium-size double-stranded RNA (dsRNA) viruses, based on their killing profiles. Rodríguez-Cousino et al. (2011) reported *S. cerevisiae* strains which produced a new killer toxin (Klus) that killed other killer strains of *S. cerevisiae* and also strains of *K. lactis* and *C. albicans*.

The killer toxin produced by *S. cerevisiae* was found to be useful to improve the process of wine making and wine quality with a narrow spectrum of biocontrol activity. However, genera such as *Pichia*, *Hansenula*, *Williopsis* and *Kluyveromyces* exhibited broad spectrum activities. *Kluyveromyces phaffii* killer toxin was reported to have a potential as a bio-preservative agent against apiculate wine yeast *H. uvarum* in wine making (Ciani and Fatichenti 2001). Sangorrín et al. (2007) studied the diversity and killer behaviour of the natural yeast isolates associated with surfaces of four Patagonian wineries. Mainly *P. anomala* and *S. cerevisiae* and to a certain extent *K. apiculata, K. thermotolerans, T. delbrueckii* and *C. albidus* showed activity against sensitive isolates of *S. cerevisiae* and *C. glabrata* by seeded agar plate technique on methylene blue containing yeast extract-peptone-glucose medium. Using same technique, recently the natural yeast isolates such as *C. quercitrusa, I. orientalis, C. azyma* and *P. membranifaciens* from Bangalore Blue variety, *S. cerevisiae* and *Z. steatolyticus* from Zinfindeal, *H. uvarum* from Cabernet and *D. hansenii* from Shiraz were reported to exhibit killing activity against *S. cerevisiae, I. terricola* and others. *P. fermentans* and *P. kluveri* from Sauvignon Blanc and Shiraz variety, respectively showed activity against *C. albicans* too (Mane 2016).

Other than wine industry, killer toxins were observed to be useful in preservation of food, biotyping of human pathogenic yeasts and also as antifungal agents. Natural yeasts isolated from different fruits have lot of biocontrol potential *in situ* as well as for applications in fields, in general. The non-*Sacharomyces* yeasts such as *P. guilliermondii, Candida musae, I. orientalis* and *C. quercitrusa* were reported to control post harvest chilli anthracnose caused by *Colletotrichum capsici* (Chanchaichaovivat et al. 2007).

9 Concluding Remarks

The success of wine making is dependent on the quality perception of the consumers. Therefore the fermentation process which gives desired quality, if monitored for the qualitative and quantitative changes of natural yeast flora, their enzyme profiles and changes in the secondary metabolites which impart flavor to wine will be useful. The early profiles will be useful to direct the process by controlling undesirable changes using preservatives, or by the addition of yeast flora or enzymes to get good quality wine.

Acknowledgements The authors are thankful to funding agencies, Department of Biotechnology, New Delhi (BT/PR8426/PID/06/355/2006) and Ministry of Environment and Forest, New Delhi (F.No.19-23/2010) for the financial support. SM and ST acknowledge the support as SRF and SRA, respectively, from Council of Scientific and Industrial Research, New Delhi.

References

Aguilar-Uscanga, B. and Francois, J.M. 2003. *Lett. Appl. Microbiol.* 37: 268–274.

Albergaria, H., Francisco, D., Gori, K., Arneborg, N. and Girio F. 2010. *Appl. Microbiol. Biotechnol.* 86: 965–972.

Amerine, M.A. and Kunkee R.E. 1968. *Annu. Rev. Microbiol.* 22: 323–358.

Andorra, I., Landi, S., Mas, A., Guillamon, J.M. and Esteve-Zarzoso B. 2008. *Food Microbiol.* 25: 849–856.

Andorra, I., Monteiro, M., Esteve-Zarzoso, B., Albergaria, H. and Mas A. 2011. *Food Microbiol.* 28: 1483–1491.

Assis, M.O., Santos, A.P.C., Rosa, C.A. and Mamede, M.E.O. 2014. *Food Nutr. Sci.* 5: 759–769.

Baffi, M.A., Bezerra, C.S., Arevalo-Villena, M., Briones-Perez, A., Gomes, E. and Silva R.D. 2011. *Ann. Microbiol.* 61: 75–78.

Barata, A., Malfeito-Ferreira, M. and Loureiro, V. 2012a. *Int. J. Food Microbiol.* 153: 243–259.

Barata, A., Malfeito-Ferreira, M. and Loureiro V. 2012b. *Int. J. Food Microbiol.* 154: 152–161.

Barata, A., Seborro, F., Belloch, C., Malfeito-Ferreira, M. and Loureiro, V. 2008. *J. Appl. Microbiol.* 104: 1182–1191.

Barbagallo, R.N., Spagna, G., Palmeri, R. and Torriani, S. 2004. *Enzyme Microb. Technol.* 34: 292–296.

Barbosa, C., Lage, P., Vilela, A., Mendes-Faia, A. and Mendes-Ferreira, A. 2014. *AMB Express* 4: 39–53.

Benito, S., Palomero, F., Morata, A., Calderan, F., Palmero, D., Suairez-Lepe, J.A. 2013. *Food Res. Technol.* 236: 29–36.

Bettini, O. 2014. GAIN Report Number: IT1414.

Bezerra-Bussoli, C.B.M., Gomes, E. and Da-Silva, R. 2013. *Curr. Microbiol.* 67: 356–361.

Bisson, L.F. and Kunkee, R.E. 1991. In: Mixed cultures in biotechnology (eds. Zeikus J.G. and Johnson E.A.), McGraw-Hill, New York, pp. 39–68.

Brezna, B., Zenisova, K., Chovanova, K., Chebenova, V., Krakova, L., Kuchta, T. and Pangallo, D. 2010. *Antonie van Leeuwenhoek* 98: 519–529.

Brice, C., Sanchez, I., Legras, J-L. and Blondin, B. 2014. *BMC Genomics* 15: 495–509.

Brilli, L., Buscioni, G., Moriondo, M., Bindi, M. and Vincenzini, M. 2014. *Am. J. Enol. Vitic.* 65: 375–380.

Canal-Llauberes, R.M. 1993. In: Wine Microbiology and Biotechnology (ed. G.H. Fleet), Harwood Academic Publishers, Chur, Switzerland, pp. 477–506.

Canal-Llauberes, R.M. 1998. In: Proceedings of 5th International Symposium - Innovations in wine technology, Stuttgart.

Capece, A., Romaniello, R., Siesto, G., Pietrafesa, R., Massari, C., Poeta, C. and Romano, P. 2010. *Int. J. Food Microbiol.* 144: 187–192.

Cappello, M.S., Bleve, G., Grieco, F., Dellaglio, F. and Zacheo, G. 2004. *J. Appl. Microbiol.* 97: 1274–1280.

Chanchaichaovivat, A., Ruenwongsa, P. and Panijpan, B. 2007. *Biol. Control* 42: 326–335.

Chand-Goyal, T. and Spotts, R.A. 1996. *Microbiol. Res.* 151: 427–432.

Charoenchai, C., Fleet, G.H., Henschke, P.A. and Todd, B.E.N. 1997. *Aust. J. Grape Wine Res.* 3: 2–8.
Chavan, P., Mane, S., Kulkarni, G., Shaikh, S., Ghormade, V., Nerkar, D.P., Shouche, Y. and Deshpande, M. V. 2009. *Food Microbiol.* 26: 801–808.
Ciani, M. and Fatichenti, F. 2001. *Appl. Environ. Microbiol.* 67: 3058–3063.
Ciani, M. and Maccarelli, F. 1998. *World J. Microbiol. Biotechnol.* 14: 199–203.
Clemente-Jimenez, J.M., Mingorance-Cazorla, L., Martinez-Rodriguez, S., Heras-Vazquez, F.J.L. and Rodrnguez-Vico, F. 2004. *Food Microbiol.* 21: 149–155.
Cocolin, L., Bisson, L.F. and Mills, D.A. 2000. *FEMS Microbiol. Lett.* 189: 81–87.
Combina, M., Elia, A., Mercado, L., Catania, C., Ganga, A. and Martinez, C. 2005. *Int. J. Food Microbiol.* 99: 237–243.
Comitini, F., Gobbi, M., Domizio, P., Romani, C., Lencioni, L., Mannazzu, I. and Ciani, M. 2011. *Food Microbiol.* 28: 873–882.
Contreras, A., Hidalgo, C., Henschke, P.A., Chambers, P.J., Curtin, C. and Valera, C. 2013. *Appl. Environ. Microbiol.* 80: 1670–1678.
Cordente, A.G., Curtin, C.D., Varela, C. and Pretorius, I.S. 2012. *Appl. Microbiol. Biotechnol.* 96: 601–618.
Cordero-Bueso, G., Arroyo, T., Serrano, A., Tello, J., Aporta, I., Vaolez, M.D. and Valero, E. 2011. *Int. J. Food Microbiol.* 145: 132–139.
De Miranda Reis, M.A.C.F., Oliveira, R.M.F., De Freitas, M.F.A., Chagas, B.F., Da Cruz, A.L. B., Pereira Da, A.E.P.B. and Clemente, J.J.V.M. 2013. *U.S. Patent No. 8,614,070.* Washington, DC: U.S. Patent and Trademark Office.
Díaz, C., Molina, A.M., Nähring, J. and Fischer R. 2013. *Biomed. Res. Int.* 2013: 2013:540465. doi: 10.1155/2013/540465.
Diguta, C.F., Vincent, B., Guilloux-Benatier, M., Alexandre, H. and Rousseaux, S. 2011. Food Microbiol. 28: 1145–1154.
Domizio, P., Liu, Y., Bisson, L.F. and Barile, D. 2014. *Food Microbiol.* 43: 5–15.
Esteve-Zarzoso, B., Manzanares, P., Ramón, D. and Querol, A. 1998. *Int. Microbiol.* 1: 143–148.
Esteve-Zarzoso, B., Peris-Toran, M., Garca-Maiquez, E., Uruburu, F. and Querol, A. 2001. *Appl. Environ. Microbiol.* 67: 2056–2061.
Fia, G., Giovani, G. and Rosi, I. 2005. *J. Appl. Microbiol.* 99: 509–517.
Fleet, G.H. 1990. *J. Wine Res.* 1: 211–223.
Fleet, G.H. 2000. In: Encyclopedia of Food Microbiology (eds. Robinson, R.K., Batt, C.A. and Patel, P.D.), Academic Press, London, Vol. 3, pp. 1984–1989.
Fleet, G.H. 2003. *Int. J. Food Microbiol.* 86: 11–22.
Fleet, G.H. and Phaff, H.J. 1974. *J. Biol. Chem.* 249: 1717–1728.
Fleet, G.H. and Heard, G.M. 1993. In: Wine Microbiology and Biotechnology (ed. G. H. Fleet), Hardwood Academic Publishers, Switzerland, pp. 27–57.
Fleet, G.H., Lafon-Lafourcade, S. and Ribereau-Gayon, P. 1984. *Appl. Environ. Microbiol.* 48: 1034–1038.
Fleet, G.H., Prakitchaiwattana, C., Beh, A.L. and Heard, G. 2002. In: Biodiversity and Biotechnology of Wine Yeasts. (ed. Ciani, M.), Research Signpost, Kerala, India, pp. 1–17.
Fugelsang, K.C. 1997. In: Wine Microbiology (ed. Fugelsang, K.C.), Chapman and Hall, New York, USA, pp. 68–116.
Gana, N.H.T., Mendoza, B.C. and Monsalud, R.G. 2014. *Philippine J. Sci.* 143: 81–87.
Gao, Y.C., Zhang, G., Krentz, S., Darius, S.U.E., Power, J. and Lagarde, G. 2002. *Aust. J. Grape Wine Res.* 8: 76–83.
Gonzalez, A., Hierro, N., Poblet, M., Mas, A. and Guillamon, J.M. 2005. *Int. J. Food Microbiol.* 102: 295–304.
Gonzalez, S.S., Barrio, E. and Querol, A. 2006. *J. Appl. Microbiol.* 102: 1018–1025.
Guerra, E., Sordi, G., Mannazzu, I., Clementi, F. and Fatichenti, F. 1999. *Int. J. Food Sci.* 11: 221–230.
Guerzoni, E. and Rosa, M. 1987. *Appl. Environ. Microbiol.* 53: 571–576.

Hansen, E.H., Nissen, P., Sommer, P., Nielsen, J.C. and Arneborg, N. 2001. *J. Appl. Microbiol.* 91: 541–547.

Heard, G.M. and Fleet, G.H. 1986. *J. Appl. Bacteriol.* 60: 477–481.

Heard, G.M. and Fleet, G.H. 1988. *J. Appl. Microbiol.* 65: 23–28.

Henick-Kling, T. 1998. *J. Appl. Microbiol.* 84: 865–876.

Hernandez, L.F., Espinosa, J.C., Fernandez-Gonzalez, M. and Briones, A. 2003. *Int. J. Food Microbiol.* 80: 171–176.

Hierro, N., Gonzalez, A., Mas, A. and Guillaman, J. 2006. *FEMS Yeast Research* 6: 102–111.

Ibeas, J.I., Lozano, I., Perdigones, F. and Jimcmez, J. 1996. *Appl. Environ. Microbiol.* 62: 998–1003.

Jackson, R.S. 1994. Wine Science – principles and applications, Academic Press, San Diego, USA.

Janisiewicz, W.J., Jurick, W.M., Peter, K.A., Kurtzman, C.P. and Buyer, J.S. 2014. *Yeast* 31: 207–218.

Jolly, N.P., Augustyn, O.P.H. and Pretorius, I.S. 2006. *S. Afr. J. Enol. Vitic.* 27: 15–39.

Jolly, N.P., Augustyn, O.P.H. and Pretorius, I.S. 2003. *S. Afr. J. Enol. Vitic.* 24: 35–42.

Jolly, N.P., Varela, C. and Pretorius, I.S. 2013. *FEMS Yeast Res.* 14: 215–237.

Joshi, C.V., Ghormade, V., Kunde, P., Kulkarni, P., Mamgain, H., Bhat, S., Paknikar, K. M. and Deshpande M. V. 2010. *Bioresource Technol.* 101: 1393–1395.

Khale, A., Srinivasan, M.C. and Deshpande M.V. 1992. *J. Bacteriol.* 174: 3723–3728.

Klis, F.M., Boorsma, A. and De Groot P.W.J. 2006. *Yeast* 23: 185–202.

Lafon-Lafourcade, S. 1983.In: Biotechnology (ed. Reed, G.), Verlag-Chemie, Heidelberg, Germany, pp. 81–163.

Lagace, L. and Bisson, L. 1990. *Am. J. Enol. Vitic.* 41(2): 147–155.

Lambrechts, M.G. and Pretorius, I.S. 2000. *S. Afr. J. Enol. Vitic.* 21: 97–129.

Le Roux, G., Eschenbruch, R. and de Bruin, S.I. 1973. *Phytophylactica* 5: 51–54.

Li, S.S., Cheng, C., Li, Z., Chen, J.Y., Yan, B., Han, B.Z. and Reeves, M. 2010. *Int. J. Food Microbiol.* 138: 85–90.

Longo, E., Cansado, J., Agrelo, D. and Villa T.G. 1991. *Am. J. Enol. Vitic.* 42: 141–144.

Lonvaud-Funel, A. 1999. *Antonie van Leeuwenhoek* 76: 317–331.

Loureiro, V. and Malfeito-Ferreira, M. 2003. *Int. J. Food Microbiol.* 86: 23–50.

Louw, C., La Grange, D., Pretorius, I.S., van Rensburg, P., 2006. *J. Biotechnol.* 125: 447–461.

Mane, S. 2016. A PhD thesis submitted to Bharati Vidyapeeth University, Pune.

Martini, A., Ciani, M. and Scorzetti, G. 1996. *Am. J. Enol. Vitic.* 47: 435–440.

Martins, G., Miot-Sertier, C., Lauga, B., Claisse, O., Lonvaud, A., Soulas, G. and Masneuf-Pomarade, I. 2012. *Int. J. Food Microbiol.* 158: 93–100.

Martorell, P., Barata, A., Malfeito-Ferreira, M., Fernandez-Espinar, T.M., Loureiro, V. and Querol A. 2006. *Int. J. Food Microbiol.* 106: 79–84.

Maturano, Y.P., Assaf, L.A.R., Toro, M.E., Nally, M.C., Vallejo, M., Figueroa, L.I.C., Combina, M. and Vazquez, F. 2012. *Int. J. Food Microbiol.* 155: 43–50.

Maturano, Y.P., Assof, M., Fabani, M.P., Nally, M.C., Jofré, V., Rodríguez Assaf, L.A., Toro, M. E., Castellanos de Figueroa, L.I. and Vazquez, F. 2015. *Antonie Van Leeuwenhoek.* 108:1239–1256.

McGovern, P.E., Zhang, J., Tang, J., Zhang, Z., Hall, G.R., Moreau, R.A., Nunez, A., Butrym, E. D., Richards, M.P., Wang, C.S, Cheng, G., Zhao, Z. and Wang, C. 2004. *Proc. Natl. Acad. Sci. USA* 101: 17593–17598.

Milanovic, V., Comitini, F. and Ciani M. 2013. *Int. J. Food Microbiol.* 161: 240–246.

Mojsav, K., Ziberoski, J., Bozinovic, Z., Ivanova, V. and Petreska, M. 2011. Proceedings of 46th Croatian and 6th International Symposium on Agriculture, Opatija, Croatia, pp. 963–967.

Monagas, M., Bartolome, B. and Gomez-Cordoves, C. 2007. *Crit. Rev. Food Sci. Nutr.* 45: 85–118.

Morata, A., Benito, S., Loira, I., Palomero, F., Gonzalez, C.M. and Suarez-Lepe, J.A. 2012. *Int. J. Food Microbiol.* 159: 47–53.

Moreira, N., Mendes, F., Hogg, T. and Vasconcelos, I. 2005. *Int. J. Food Microbiol.* 103: 285–294.

Mortimer, R. and Polsinelli, M. 1999. *Res. Microbiol.* 150: 199–204.

Nakase, T., Jindamorakot, S., Am-In, S., Potacharoen, W. and Tanticharoen, M. 2006. In: Biodiversity and Ecophysiology of Yeasts. Springer, Berlin Heidelberg, pp. 441–460.

Nemcova, K., Breierova, E., Vadkertiova, R. and Molnarova, J. 2015. *Folia Microbiol.* 60: 103–109.

Nevoigt, E. 2008. *Microbiol. Mol. Biol. Rev.* 72: 379–412.

Nicholas, A.B., Moe, O., Paul, M.R. and David A.M. 2013. *PLoS One* 8: e66437.

Nisiotou, A.A. and Nychas, G.J.E. 2007. *Appl. Environ. Microbiol.* 73: 2765–2768.

Nisiotou, A.A., Rantsiou, K., Iliopoulos, V., Cocolin, L. and Nychas, G.J.E. 2011. *Int. J. Food Microbiol.* 145: 432–436.

Nissen, P. and Arneborg, N. 2003. *Arch. Microbiol.* 180: 257–263.

Nykanen, L. 1986. *Am. J. Enol. Vitic.* 37: 84–96.

Ocon, E., Gutierrez, A., Garijo, P., Tenorio, C., Lopez, I., Lopez, R. and Santamaria, P. 2010. *Eur. Food Res. Technol.* 230: 885–891.

Osborne, J.P., Edwards, C.G. and Steve, L.T. 2005. In: Advances in Food and Nutrition Research, Academic Press, Vol. 50, pp. 139–177.

Otero, R.R.C., Iranzo, J.F.U., Briones-Perez, A.I., Potgieter, N., Villena, M.A., Pretorius, I.S. and Rensburg, P. 2003. *J. Food Sci.* 68: 2564–2569.

Pallmann, C.L., Brown, J.A., Olineka, T.L., Cocolin, L., Mills, D.A. and Bisson, L.F. 2001. *Am. J. Enol. Vitic.* 52: 198–203.

Pombo, P.G., Farina, L., Carrau, F., Viera, F.B. and Brena, B.M. 2011. *Process Biochem.* 46: 385–389.

Prakitchaiwattana, C.J., Fleet, G.H., Heard, G.M., 2004. *FEMS Yeast Res.* 4: 865 –877

Pretorius, I. 2000. *Yeast* 16: 675–729.

Pretorius, I.S., Bauer, F.F., 2002. *Trends Biotechnol.* 20: 426–432.

Pretorius, I.S., Van der Westhuizen, T.J. and Augustyn, O.P.H. 1999. *S. Afr. J. Enol. Vitic.* 20: 61–74.

Pretorius, I.S. and HoJ, P.B. 2005. *Aust.. J. Grape Wine Res.* 11: 83–108.

Procopio, S., Krause, D., Hofmann, T. and Becker, T. 2013. *Food Sci. Technol.* 51: 423–432.

Querol, A., Jimenez, M. and Huerta, T. 1990. *J. Food Sci.* 55: 1603–1606.

Quiras, M., Rojas, V., Gonzalez, R. and Morales, P. 2014. *Int. J. Food Microbiol.* 181: 85–91.

Raspor, P., Milek, D.M., Polanc, J., Smole Moaina, S. and Aoeadea, N. 2006. *Int. J. Food Microbiol.* 109: 97–102.

Rementeria, A., Rodriguez, J.A., Cadaval, A., Amenabar, R., Muguruza, J.R., Hernando, F. L. and Sevilla, M.J. 2003. *Int. J. Food Microbiol.* 86: 201–207.

Renault, P., Coulon, J., de Revel, G., Barbe, J.C. and Bely, M. 2015. *Int. J. Food Microbiol.* 207: 40–48.

Renouf, V., Olivier, C. and Aline, L.F. 2005. *Aust. J. Grape Wine Res.* 11: 316–327.

Robinson, J. 2006. The Oxford companion to wine. Oxford University Press, New York.

Rodríguez-Cousino, N., Maqueda, M., Ambrona, J., Zamora, E., Esteban, R. and Ramirez, M. 2011. *Appl. Environ. Microbiol.* 77: 1822–1832.

Rojas, V., Gil, J.V., Pinaga, F. and Manzanares, P. 2001. *Int. J. Food Microbiol.* 70: 283–289.

Romancino, D.P., Di, M.S., Muriella, R., Oliva, D. 2008. *J. Appl. Microbiol.* 105: 2248–2254.

Romano, P. and Suzzi, G. 1993. In: Food Microbiology and Biotechnology, Harwood Academic Publishers, Reading, UK, pp. 373–393.

Rosi, I., Costamagna, L. and Birtuccion, M. 1987. *J. Inst. Brew.* 93: 322–324.

Rosi, I., Vinella, M. and Domizio, P. 1994. *J. Appl. Bacteriol.* 77: 519–527.

Rosini, G. 1984. *J. Appl. Microbiol.* 30: 249–256.

Sadoudi, M., Tourdot-Marachal, R., Rousseaux, S., Steyer, D., Gallardo-Chacan, J.J., Ballester, J., Vichi, S., Guarin-Schneider, R., Caixach, J. and Alexandre, H. 2012. *Food Microbiol.* 32: 243–253.

Sangorrín, M.P., Lopes, C.A., Giraudo, M.R. and Caballero, A.C. 2007. *Int. J. Food Microbiol.* 119: 351–357.

Slavikova, E., Vadkertiova, R. and Vranova, D. 2009. *Ann. Microbiol.* 59: 419–424.

Soares, E.V. 2010. *J. Appl. Microbiol.* 110: 1–18.

Soden, A., Francis, I.L., Oakey, H. and Henschke, P.A. 2000. *Aust. J. Grape Wine Res.* 6: 21–30.

Strauss, M.L.A., Jolly, N.P., Lambrechts, M.G. and Rensburg, P.V. 2001. *J. Appl. Microbiol.* 91: 182–190.

Subden, R.E., Husnika, J.I., Twestb, R., Merwec, G. and Vuurenc, H.J. 2003. *Food Res. Int.* 36: 747–751.

Sun, Y., Guo, J., Liu, F. and Liu, Y. 2014. *Antonie van Leeuwenhoek* 105: 533–540.

Swiegers, J.H., Bartowsky, E.J., Henschke, P.A. and Pretorius, I.S. 2005. *Aust. J. Grape Wine Res.* 1: 139–173.

Toro, M.E. and Vazquez, F. 2002. *World J. Microbiol. Biotechnol.* 18: 351–358.

Valero, A., Sanchis, V., Ramos, A.J. and Marín, S. 2007. *Int. J. Food Microbiol.* 113: 271–276.

Van Rensburg, P. and Pretorius, I.S. 2000. *S. Afr. J. Enol. Vitic.* 21: 52–73.

Vaughan-Martini, A., and Martini, A. 1998. In: The yeast: a taxonomic study (eds. Kurtzman, C. P. and Fell J. W.), Elsevier, Amsterdam, pp. 358–371.

Viana, F., Gil, J.V., Genoves, S., Valle, S. and Manzanares, P. 2008. Food Microbiol. 25: 778–785.

Villettaz, J.C. and Duboudieu, D. 1991. In: Food Enzymology (ed. Fox P.F.), Elsevier Applied Science, London, Vol. 1, pp. 427–453.

Volschenkla, H., Vuuren, H.J. and Viljoen-Bloom, M. 2006. *S. Afr. J. Enol. Vitic.* 27: 123–136.

Von Wallbrunn, C. 2007. In: 8th International Symposium - Innovations in Enology, Stuttgart, pp. 125–132.

Waters, E., Alexander, G., Muhlack, R., Pocock, K.F., Colby, C., O'Neill, B.K., Hoj, P.B. and Jones, P. 2005. *Aust. J. Grape Wine Res.* 11: 215–225.

Xue, M.-L., Zhang, L.-Q., Wang, Q.-M., Zhang, J.-S. and Bai, F.-Y. 2006. *Int. J. Syst. Evol. Microbiol.* 56: 2245–2250.

Yanagida, F., Ichinose, F., Shinohara, T. and Goto, S. 1992. *J. Gen. Appl. Microbiol.* 38: 505–509.

Zott, K., Miot-Sertier, C., Claisse, O., Lonvaud-Funel, A., Masneuf-Pomarede, I. 2008. *Int. J. Food Microbiol.* 125: 197–203.

Saccharomyces cerevisiae as a Model for Space Biology

Shivkrupa Devrao Halbandge, Pandit B. Vidyasagar and Sankunny Mohan Karuppayil

Abstract Manned spaceflight continues to be in the agenda of most of the countries involved in space research. Development of human settlements in planets and sustainable space ecosystems where crops can be grown and waste recycled are the exciting aims of some of the future space missions. There is considerable concern on the health of the space travelers during long term travel and stay in these unexplored terrains. Astronauts may be exposed to ionizing radiations and weightlessness due to alterations in gravitational force. Studies on astronauts during and after space travel indicate effects on the immune system, cardiovascular system, bone density etc. It is not clearly known how the space missions may influence DNA replication, transcription, and translation and cell division cycle in humans. Information on these will be vital. Experiments on humans and animals could be cumbersome in space. As such use of eukaryotic models like *Saccharomyces cerevisiae* could be rewarding. The yeast *S. cerevisiae* is considered as an excellent model for studying eukaryotic biology and has contributed significantly to our understanding of cancer biology and fundamental metabolic processes in humans. In this review, the potential of *S. cerevisiae* as a model for space biology has been discussed.

Keywords *Saccharomyces cerevisiae* · Space biology · Model · Yeast · Gravitation · Space medicine

---

S.D. Halbandge · S.M. Karuppayil (✉)
DST-FIST & UGC-SAP Sponsored School of Life Sciences, SRTM University
NAAC Accredited with 'A' Grade, Nanded 431 606, Maharashtra, India
e-mail: prof.karuppayil@gmail.com

P.B. Vidyasagar
SRTM University, NAAC Accredited with 'A' Grade, Nanded 431606, Maharashtra, India

T. Satyanarayana and G. Kunze (eds.), *Yeast Diversity in Human Welfare*, DOI 10.1007/978-981-10-2621-8_2

1 Introduction

Life forms have evolved on Earth under the influence of gravitational force. Life forms possess the ability to perceive and respond to gravity (Volkmann and Baluska 2006). Gravity is a major player in the evolution of life from water to land and contributes to natural selection on the basis of suitable body sizes. To grow under normal gravity condition on Earth, living organisms need to grow defying gravity, uninterrupted fluid transport and structural support for rigidity and locomotion (Benoit and Klaus 2007). Gravity may contribute to changes in cellular behavior and cell structure (Bizzarri et al. 2015). During the 1960s, Russian scientists conducted experiments to reveal the changes under zero gravity conditions on human cells but failed to reveal any effect (Montgomery et al. 1978; Tairbekov et al. 1983). Later, several experiments carried out in space environments revealed that, biological functions are indeed affected when microgravity field is applied on it (Van Loon 2007; Hammond and Hammond 2001).

It is not easy to study the actual influence of gravitational force on physiology, metabolism and other life processes. Since it is hard to create gravitation free environment on Earth, such experiments need to be conducted on a space craft during space flight, in space stations or under simulated microgravity conditions. The advent of ground based simulation facilities have generated considerable interest in space biology. These facilities have contributed to studies on the role of gravity on growth, morphology, function of cells and developmental processes in animals and plants (Clément and Slenzka 2006).

Humans under space environment experience various unusual conditions which include absence of effective gravity and exposure to cosmic radiation. These conditions can be hazardous to the space travelers during spaceflight (Clément and Slenzka 2006). Space medicine is an area of space biology addressing studies on human health and combating diseases in the space environment. Space conditions provide unique environment to study the physiological responses in organisms ranging from unicellular organisms to the most organized species to microgravity. It is known that, humans can adapt to changes during spaceflight. Prolonged exposure to space environment can lead to nausea, disorientation, shift in body fluids, disruption of sleep pattern and reduced immune power. However, most of the changes may disappear on return to Earth. The changes may occur immediately, gradually, or may take several years to repair it or even some changes may remain permanent. It is difficult to produce real microgravity conditions and natural spectrum of space radiation effectively on the ground. Space environment is considered to be an ideal environment to study the effect of these conditions on living organisms (Clément and Slenzka 2006).

2 Successful Journey of Humans to Space

Humans are not the first living creatures who were sent to space. On 20 September 1951, the former Soviet Union launched a sounding rocket. A monkey and eleven mice were transported with this rocket. This became the first successful space travel for living beings. Yuri Gagarin, Soviet cosmonaut was the first human who travelled from Earth to orbit. He travelled in Vostok 1 to an orbit about 24,800 miles from Earth. On July 20, 1969, Sir Neil Armstrong became the first person to walk on moon through the Apollo 11 mission (Klaus et al. 2004).

In an orbiting space vehicle, microgravity conditions are generated and are typically in the range of 10^{-6}–10^{-4} g. This condition leads to abnormal physiological changes in humans (Clément 2011). For example, calcium level in the bone is reduced that can lead to 1–2% loss in bone density per month and the loss in muscle fiber results in up to 40% reduction in muscle function. In addition, other conditions such as space radiation, sensory deprivation and absence of circadian rhythms and the artificial environment can adversely affect human beings. These conditions may disappear on coming back to Earth but conditions such as bone calcium may take a long time to recover or may not be even recovered (Dayanandan 2011).

A long-term spaceflight may cause high health risks such as bone demineralization, skeletal muscle atrophy, and immune system suppression. Spaceflight condition which serves as closed environment causes another risk associated with pathogenesis because there are chances of the development of pathogens in a closed environment, where air, food, waste, and water are recycled. Pathogens can be transferred among astronauts which may lead to pathogenesis (Pierson et al. 1995, 1996).

A significant number of episodes of microbial infections, including conjunctivitis and acute respiratory and dental infections are reported among astronauts (Ball and Evans 2001). In-flight cross-contamination with opportunistic pathogens such as *Staphylococcus aureus* has been reported (Decelle and Taylor 1976; Pierson 2001). The threat of developing infectious diseases during space flight may become a serious concern in the future as the duration and frequency of space missions increase (Gueguinou et al. 2009; Horneck et al. 2010).

3 Gravity and Microgravity

During 1665–1666, Sir Isaac Newton proposed the law of gravity and motion which gave insights into understanding planetary revolution and space. Gravity is a well known but not a well understood physical phenomenon. It has been constant throughout the evolutionary history of Earth (Morey-Holton 2003). Any object having a mass on the surface of the Earth accelerates towards the Earth's centre approximately at 9.8 m/s^2 (Dayanandan 2011). This gravity acceleration on Earth is considered as 1 g. Even though gravitational force is the weakest force among the four fundamental forces including nuclear strong force, electromagnetic force and nuclear weak force, it has important roles in the evolution of life on Earth (Morey-Holton 2003).

The term microgravity is also referred to as ‘weightlessness’ and ‘zero gravity’. However, in microgravity condition, gravitational force is not exactly zero but less compared to gravity on the surface of Earth and is usually in the range of 10^{-4}–10^{-6} g. Real microgravity conditions can be achieved for a very short time by generating free fall conditions close to Earth’s surface with sounding rockets and airplanes in parabolic flight and drop facilities (Clément and Slenzka 2006). Real microgravity conditions can be achieved in drop towers or drop shaft (for 2–10 s), balloons (3–60 s), parabolic flights of aircrafts (20–25 s) or sounding rockets (up to 15 min). Since 1998, the International space station (ISS) has been established in space which offers laboratory conditions for systemic studies in microgravity (Herranz et al. 2013).

4 Tools Used in Space Biology

It is difficult to create spaceflight conditions on Earth and such studies are technically difficult and expensive. To achieve such conditions, different ground based facilities has been designed to simulate ‘weightlessness’ in laboratories on Earth. The environment created on Earth within ground based simulated microgravity condition is often considered as ‘simulated microgravity’ (Klaus 2001). Simulated microgravity conditions can be created on Earth by using different kind of simulators and are collectively referred as ground based microgravity simulators. Ground based facilities for simulation of microgravity includes simulators such as 2-D clinostat, 3-D clinostat, magnetic levitation and random positioning machines (Herranz et al. 2013).

4.1 Clinostat

Clinostat is used to create weightlessness on Earth even though 1 g force is exhibited. It cannot completely nullify the force of gravity but its influence can be changed by slowly by constantly rotating objects in a horizontal axis i.e. perpendicular to Earth’s gravitational field (Dayanandan 2011). Different models of clinostat have been established on the basis of speed, direction of rotation and number of rotation axes. 1D or 2D clinostat usually refers to clinostat with one rotation axis which rotates in horizontal axis (Klaus 2001).

Clinostat is used since 1965 to investigate effects of simulated microgravity. In 1958, Muller presented a human clinostat in which human test subject was rotated in a horizontal axis within a cylinder. On the basis of this human clinostat and many theories, Briegleb developed a rotating clinostat to investigate the influence of simulated microgravity on the behavior of microorganisms such as *Chlorella pyrenidosa* and blue green algae, *Phormiium autmnale*. However, there was no change observed in the metabolism of Chlorella or *Phormiium* compared to 1 g control. Briegleb extended his research to germ differentiation of the beetle *Tribolium castaneum* and revealed that the embroyonic development and the survival time of hatched larvae were not changed compared to 1 g control (Cogli 2007). Clinostats have been used to study the influence of gravity for a long period of time in the field of plant physiology (Yamada et al. 1993).

4.2 Random Positioning Machine

Two dimensional or three dimensional clinostat offer two rotation axes which can be operated at different speed with different directions. It is also referred as 'Random Positioning Machine' (RPM). 3 D clinostat was first developed in Japan and Netherlands (Van Loon 2007).

4.3 Rotating Wall Vessel (RWV)

Rotating wall vessel (RWV) is a bioreactor specially designed for the culture to create simulated microgravity condition. It was developed by the National Aeronautics and Space Administration (NASA) USA. High aspect ratio vessel (HARV) is one of the versions of rotating wall vessel (RWV). HARV have been used in gravitational biology research and typically used for the suspension culture and mammalian tissue growth (Hammond and Hammond 2001, Hammond et al. 2000; Klaus 2001; Sikavitsas et al. 2002). HARV is a rotating device which revolves around horizontal axis. Cells suspended in this vessel do not settle or are not constantly agitated but revolve around horizontal axis allowing exchange of gases, nutrients and wastes within culture medium in the vessel. This creates 'low gravity environment' by randomizing the unidirectional pull of gravity and reducing shear forces (Hammond and Hammond 2001; Klaus 2001; Unsworth and Lelkes 1998).

Low Earth orbits (LEO) is the most preferred and suitable area to carry out biology experiments which is at altitudes less than 1% of the distance from the Earth to the Moon. Space vehicles such as Sputnik I (215 km), ISS (350 km), Mir (390 km) and space shuttles (300–400 km) belongs to LEO era. At these distances, the force of gravity still remains high as 90% as on Earth but space vehicles moving in circular LEO experience actual microgravity environments (Dayanandan 2011; Hammond et al. 2000).

5 Studies Done on Models

Human blood, kidney, liver, tonsil cells and colon cancer cells are cultivated under microgravity conditions in bioreactors on ISS. These cell types and tissue were analyzed and compared to those cells and tissues in ground. It was found that cells function normally when cultivated in microgravity condition. Movement of motile flagellates such as *Chlamydomonas* and *Euglena* and ciliates such as *Paramecium* and *Loxodes* showed gravitactic response. Experiments done in the International Microgravity Laboratory on Columbia and parabolic rocket flight have revealed that gravitaxis in these organisms require a threshold of 0.16 g (Hader et al. 2003).

Various experiments were done during the past twenty years on the cultivation of plants in microgravity successfully. Plants were grown for more than one generation in microgravity condition. *Arabidopsis thaliana* was successfully grown in the 'Astroculture greenhouse' to obtain seeds of the third space generation (Souza et al. 2009; Wolverton and Kiss 2009). *Brassica repa* seeds germinated and produced normal plants and viable seeds in microgravity. Successful growth of cereal plants in space is reported (Levinskikh et al. 2000).

In plants, roots respond to gravitropic responses to get water and nutrients from soil. Land living organisms experience mechanical load on them due to gravity that is thousand times greater than the load that is experienced by living organisms in water. The production of anti gravitational substances such as lignin, cellulose and pectin support the increased load on plants while animals strengthen their bones with hydroxyapatite a mineral form of calcium associated with collagen in response to mechanical load exerted on them (Volkmann and Baluska 2006).

In the animal kingdom, gravitropic responses are studied in all major groups. Various organs in animal the kingdom have evolved sensory motors to recognize the gravity factor and to orient and move the organism, musculoskeletal system to support body mass and provide structural and postural stability to land animals and vestibular system for efficient swimming in fish (Highstein et al. 2004).

The nematode, *Caenorhabditis elegans* was found to reproduce and develop through several generations in microgravity without resulting in major structural differences. Studies have been carried out on *C. elegans* on ISS for eleven days. RNA interference treatment was used to regulate gene expression. Treated tissues functioned normally in microgravity, thus providing scope for treatment and control of muscle degradation (Etheridge et al. 2011).

Fish, birds, amphibians and small mammals are used model organisms for developmental studies in microgravity. Normally, upon sperm penetration, fertilized in frog egg rotates which may be essential for normal development. Upon fertilization, eggs divide and develop embryo that emerges from egg as a tadpole. When artificially inseminated female frogs were sent into space, it was found that eggs fail to rotate but normal tadpoles emerged. On coming back to Earth within 2–3 days, the tadpoles metamorphosed and matured into normal frogs. Although development appeared normal during Spaceflight, some morphological changes were observed in the embryo and the tadpoles such as thicker blastocoel roof of embryo. Flight tadpoles did not inflate their lungs until they returned to Earth (Souza et al. 1995; Morey-Holton 2003).

Microbes respond less to gravity than larger animals so should have less difficulties in adapting to different gravity than humans (Morey-Holton 2003). Studies are done on the bacterium *Escherichia coli* in culture. During space flight, *E. coli* shortens the lag phase and increases the time period of exponential phase and approximately doubles the final population density compared to ground control. Microorganisms such as *Salmonella enterica*, and *Bacillus subtilis* also changes their cell growth characteristics (Klaus et al. 1997; Horneck et al. 2010). It is reported that, microgravity has a key role in microbial physiology, regulation of gene expression and pathogenesis (Hammond et al. 2000; Collister et al. 2002). Bacteria are able to proliferate more readily in space, supporting that space environment is more suitable to initiate growth that could lead to contamination, colonization and infection. Under these conditions, microbes may become opportunistic pathogens and cause infectious diseases (Gueguinou et al. 2009).

6 The Yeast, *Saccharomyces cerevisiae* as a Model

The yeast, *S. cerevisiae* is a single celled eukaryotic organism. It is commonly known as baker's yeast (Sicard and Legras 2011). In the 1860s, Louis Pasteur discovered the involvement of *S. cerevisiae* in fermentation (Pasteur et al. 1860). Since then *S. cerevisiae* has been widely used as a model organism in biological sciences. *S. cerevisiae* has extensively contributed to our understanding of fundamental biological processes (Hartwell et al. 1970). Yeast has a ~12 Mb sized genome which comprises 6,607 open reading frames on 16 chromosomes (Forsburg 2007). The term yeast refers to any unicellular fungus. There are hundreds of different kinds of yeasts which may differ in their taxonomic status. For convenience in this review we will be using the term yeast to refer to the *S. cerevisiae.*

Yeast exhibits high level of conservation between its cellular processes and those of mammalian cells. Additionally, yeast is advantageous because of its simple growth requirements, rapid cell division, and ease of genetic manipulation and availability of experimental tools for genome-wide analysis (Botstein et al. 1997; Simon and Bedalov 2004). Yeast mutants have helped in the study of the biochemical function of gene products and the reasons for the failure of gene to function. Yeast strains were mutated in many studies by researchers as needed. Information on mutations in yeast strains concerning to diverse biological assays provides understanding of the biological functioning of genes (Ghaemmaghami et al. 2003; Huh et al. 2003). The *Saccharomyces* Genome Database (SGD) provides extensive information about systemic study of every *Saccharomyces* gene. Nearly, 5800 genes of *S. cerevisiae* are known for their biological function. About ~17% of yeast genes are orthologous with the genes associated with human disease (Heinicke et al. 2007). Among these orthologous genes, majority of genes are functional in yeast. These homologous genes can complement respective yeast deletion mutants (Dolinski and Botstein 2005, 2007). Yeast studies cannot give complete insights to all the biological processes in humans because of the complexities of human tissues, but it shows high level of homology with a number of human genes. About 70% of all essential yeast genes show homology with human genes (Botstein and Fink 2011). *S. cerevisiae* is the first eukaryotic organism whose complete genome database was studied and is available at SGD. Additionally, yeast whole genome wide knockdown collection is available. It carries a collection of strains of single gene deletion. These libraries have been widely used to study survival of *S. cerevisiae* at different environmental stress conditions and to reveal the genes required for survival (Giaever and Nislow 2014). Yeasts and their heterologous expression contribute to analyse the function of human proteins related with a specific disease state. Use of yeasts as models for study of human genes may be beneficial because of two reasons. First, a study on human defective proteins with the aid of yeast experiments reveal the diverged enzyme function that may not be obvious from assay in humans or from assessment of protein sequences. Second, studies of human gene function in yeast may be useful for the treatment of disease caused due to deficiencies (Botstein and Fink 2011). Physical and biochemical parameters of yeast can be controlled by using defined medium. Databases, plasmid and genomic libraries give information available on metabolic pathways of yeast (Giaever et al. 2002).

Studies on yeast have contributed significantly to cell cycle control, damage responses, chromosome segregation to protein secretion in humans and other eukaryotes (Forsburg 2007).

S. cerevisiae can be ideal model for Eukaryotes for Space Biology because:

(1) Although spaceflight can interfere with its physiological response in space, it grows well in space. Spaceflight studies done during last decade on yeast revealed changes related to morphology, survival rate, genomic and metabolic pathways. Additionally, ground based facilities helped researchers to reveal the changes specific to microgravity and spaceflight. However, comparative studies are essential to support these findings.
(2) It is a well studied eukaryotic model used for human diseases and human genes. Studies related to yeast spaceflight may be beneficial for newly emerged fields like space medicine that can contribute to astronaut's health in space.
(3) Genes and the regulatory mechanism in yeast are well conserved throughout evolution.

6.1 Spaceflight Induced Effects on S. cerevisiae

6.1.1 Effect of Spaceflight on Chronological Lifespan of Yeast

Yi et al. (2011) studied the viability of wild type *S. cerevisiae* during spaceflight. It was grown on onboard '*Practice* 8' a recoverable satellite in orbit for 15 days. The satellite was launched by China at Jiu-Quan Satellite Launch Center in September 2006 and recovered after 15 days from the central Szechwan province. Viability of yeast cells grown during spaceflight was enhanced to three folds compared to ground control. Viability of yeast cells were determined by counting the number of colony-forming units per ml (CFU/ml) on YPD agar plates. Spaceflight grown cells were cultured in water and YPD liquid media and viability was assessed. Spaceflight grown culture showed higher viability compared to control cells grown on ground. Spaceflight grown culture showed 17.6% of viability in water and 78.8% of viability in YPD whereas ground control culture exhibited only 5.7% in water and 25.7% in YPD.

Spaceflight environment affects the cell cycle of wild type *S. cerevisiae* grown during spaceflight. Post-mitotic *S. cerevisiae* cells grown during spaceflight exhibited increased G2/M cell population and decreased sub-G1 cell population. The postmitotic cells from saturated culture were 11% in sub-G1 cell population at day 0. After fifteen days, control yeast culture were 62% (in water) and 57.9% (in YPD) sub-G1 cell population indicating that apoptosis occurred in control cells. Yeast cells grown during spaceflight showed 32.5% sub-G1 cell population in water and 1.6% sub-G1 cell population in YPD. Spaceflight environment decreased apoptosis and increased viability in yeast cells (Yi et al. 2011).

Lipid peroxidation level decreased in the postmitotic *S. cerevisiae* grown during spaceflight supporting that decreased ROS level favors survival of yeast in space. Glycogen content was reduced in *S. cerevisiae* grown during spaceflight. The activity of enzymes involved in carbohydrate metabolism such as hexokinase (HK), malate dehydrogenase (MDH) and succinate dehydrogenase (SDH) in *S. cerevisiae* grown during spaceflight is reported (Yi et al. 2011). Yeast cells grown in spaceflight were having significantly reduced level of hexokinase and succinate dehydrogenase activities than control cells (Yi et al. 2011).

The activity of malate dehydrogenase (MDH) was enhanced by 1.24-folds in water and 1.34-folds in YPD as compared to control cells, respectively. Hexokinase act as a catalyst in the reaction of the ATP dependent phosphorylation which converts glucose to glucose-6-phosphate. This reaction is known to be the first major reaction of glucose utilization such as glycolysis and glucogenesis. MDH is an enzyme involved in the tricarboxylic Acid (Krebs) cycle that yields oxaloacetate from malate and vice versa. It is involved in gluconeogenesis for the synthesis of glucose from smaller molecules. SDH is involved in the conversion of succinate into fumarate in the Krebs cycle (Lemire and Oyedotun 2002). Metabolism of carbohydrates was restored during spaceflight that led to reduced glycogen accumulation (Yi et al. 2011).

Depending upon these findings, it is hypothesized that, spaceflight or microgravity environment may be responsible for enhanced chronological life span of yeast by maintaining carbohydrate metabolism, ROS (reactive oxygen species) level and cell cycle progression. Nevertheless, this hypothesis is limited to only single spaceflight studies so further investigation is called for confirming this hypothesis such as yeast chronological studies in ground based simulated microgravity devices (Yi et al. 2011).

6.1.2 Budding Pattern of *S. cerevisiae* on Solid Medium

Generally, budding pattern are widely typed into three: random, unipolar and axial. The influence of microgravity on the budding pattern of the *S. cerevisiae* was investigated by Van Mulders et al. (2011). In this study, $\sum$1278b strain and the industrial brewer's CMBSESA1 strain has been used and experiment was conducted in real microgravity condition on ISS. For S1278b, a haploid laboratory strain when grown on 2% YPD agar, most of cells appeared in an axial budding pattern in normal gravity whereas in microgravity condition the percentage of random budding pattern was enhanced by 5.2%. In case of an industrial brewer's CMBSESA1 strain, the percentage of random budding pattern increased by 6.0% (Van Mulders et al. 2011). However, the rate of random budding pattern is higher than that observed in liquid medium and which is about 12% in real microgravity condition (Walther et al. 1996) and in LSMMG conditions in diploid strain (Purevdorj-Gage et al. 2006). This difference may be because of different growth systems and ploidy conditions of strains i.e. growth on semisolid condition rather than of liquid condition and haploid strain (Van Mulders et al. 2011). Walther et al. (1996) has opined that, microgravity influences the cytoskeleton which may lead to change in bud scar position.

6.1.3 Reduced Relative Colony Growth Rate and Invasive Growth

Growth of *S. cerevisiae* $\sum$1278b and CMBSESA1 strain in real microgravity and normal gravity was studied on a semi-solid medium. These strains were grown on a semi-solid medium for twelve days in microgravity condition at the International Space Station (ISS) and normal microgravity condition on ground laboratory at Baikonur, Kazakhstan. *S. cerevisiae* $\sum$1278b is a haploid laboratory yeast strain. It has the ability to grow invasively on semisolid medium. *S. cerevisiae* CMBSESA1 is an industrial brewer's yeast. As this strain originates from brewery environment, it has adapted to survive in stress conditions such as high osmotic and ethanol concentration but does not show invasive growth. *S. cerevisiae* $\sum$1278b which is grown in microgravity condition at the ISS showed reduced colony surface and relative colony growth. Oppositely, *S. cerevisiae* $\sum$1278b grown under normal condition exhibited larger colony surface and increased relative colony growth. However, these observed effect was only limited for $\sum$1278b and not for non-invasive industrial brewer's CMBSESA1 strain.

In response to specific environmental condition such as nitrogen starvation, *S. cerevisiae* switches from a single cell morphology to elongated cell which forms short filaments. Haploid cells invade into the rich semi-solid medium after glucose depletion. Dipliod S. *cerevisiae* cells responds to nitrogen starvation and undergo dimorphic switching to psuedohyphal growth. The psuedohyphal growth allows cells to forage for nutrients. Pseudohyphae grow away from the colony and invade agar medium. This process is referred as invasive growth (Cullen and Sprague 2000; Prusty et al. 2004; Gimeno et al. 1992). Van Mulders et al. (2011) found that,

∑1278b, a Flo11p-dependent strain, showed reduced invasive growth in the middle of the colony on 2% agar under microgravity condition.

This strain dependant variation revealed that microgravity induced strain specific change. Strain specific change in colony spreading under microgravity may be due to change in colony growth morphology (Van Mulders et al. 2011). Similarly, *E. coli* exhibited thicker biofilm formation under low shear microgravity condition than normal microgravity conditions (Lynch et al. 2006). The Flo11 adhesin is involved in maintaining colony morphology, mat patterns and colony size in *S. cerevisiae* (Reynolds et al. 2008). FLO gene expression in S1278b *S. cerevisiae* was down regulated under microgravity condition. This down regulation may cause the smaller colony size. In *S. cerevisiae* CMBSESA1 strain, expression of FLO11 were down regulated but polyploidy condition in *S. cerevisiae* CMBSESA1 led to lower expression of FLO11. This remains unaffected to under microgravity conditions. A second hypothesis states that, yeast cells experience microgravity as a stress condition and alter its cellular metabolism and growth resulting in smaller colony size on semisolid medium (Van Mulders et al. 2011). *S. cerevisiae* ∑1278b is known for hyper activation of cAMP/PKA pathway. This leads *S. cerevisiae* ∑1278b for strong agar invasion but defective in stress-responsive gene induction (Stanhill et al. 1999). Microgravity is a stress condition but *S. cerevisiae* ∑1278b is not adapted for stress condition. So, it may be the reason for reduced growth in ∑1278b strain (Van Mulders et al. 2011).

6.1.4 Effect of Spaceflight Environment on the Cell Wall of *S. cerevisiae*

Fungal cell wall is composed of interconnected β-glucan, mannoprotein, and chitin. It creates protection layer against osmotic pressure and helps to maintain their different shape as per cell cycle progress (Inoue et al. 1995). The yeast cell wall comprises about 20% of the cell's weight. Mannan and β-glucan comprises 40–45 and 35–45% of the yeast cell wall, respectively (Aguilar and Francois 2003). β-Glucan is an abundantly found polysaccharide in the cell wall. It is a homopolymer of glucose bound by 1, 3-β- or 1, 6-β-D-glucosidic linkage. Two types of glucose linkages are found in *S. cerevisiae.* β-glucan is a long chain of approximately 1,500 β-1,3-D-glucose units which comprises about 85% of the total cell wall β-D-glucan and short chain of approximately 150 β-1,6-D-glucose units comprise about 15% of the cell wall β-D-glucan (Klis et al. 2002). β-glucans produced from yeast are having wide application in pharmaceutical, food, and feed industries (Thompson et al. 1987; Hofer and Pospisil 1997; Liu et al. 2007a) because of its immunostimulating, anticarcinogenic, hypocholesterolemic, and hypolipidemic properties (Gordon and Siamon 2003; Jamas et al. 1996; Peter et al. 2004).

Effect of spaceflight environment on polysaccharides of *S. cerevisiae* was studied by Liu et al. (2008). This study was done by using four *S. cerevisiae* strains, namely, FL01, FL03, 2.0016, and 2.1424. These were subjected to spaceflight by loading into a recoverable satellite. The satellite was launched at Jiu-Quan Satellite

Launch Center on 9 September 2006. After fifteen days in orbit, the satellite was successfully recovered from the central Szechwan province. After spaceflight, *S. cerevisiae* 2.0016 showed drastic increase in the biomass and cell wall mass from 591 ± 12.7 to 867 ± 7.1 mg/100 mL and 116 ± 6.69 to 204 ± 4.90 mg/100 mL, respectively. However, there was no significant change observed in biomass and cell wall mass between the spaceflight and ground samples of *S. cerevisiae* FL01, FL03, and 2.1424. All spaceflight S. *cerevisiae* strains did not exhibit significant difference in cell wall thickness. However, cell wall thickness of strain *S. cerevisiae* FL01, 2.0016, and 2.1424 increased whereas *S. cerevisiae* FL03 showed decreased cell wall thickness after spaceflight. Highest increment in cell wall thickness was observed *S. cerevisiae* 2.0016 i.e. 63% increase than control.

β-Glucan content was enhanced in cell wall of *S. cerevisiae* strain 2.0016 and this increase was three times more in spaceflight sample. Such increment was not seen in *S. cerevisiae* FL01, FL03, and 2.1424 strains. Mannan content was elevated in the cell wall of spaceflight *S. cerevisiae* FL01 and 2.0016 whereas spaceflight *S. cerevisiae* FL03 and 2.1424 strains showed decrease in glucan content. However, these changes were not statically significant (Liu et al. 2008).

β-Glucan content was evaluated on the basis of activity of enzymes involved in synthesis and degradation of β-glucan (Kim and Yun 2006). β-Glucanases, are enzymes involved in the cleavage of the β-O-glycosidic linkages of β-glucan chains which gives glucose and oligosaccharides. It is also involved in other cellular processes such as cell budding, wall growth, conjugation, ascus formation, and other morphogenetic events (Fleet 1991; Martín-Cuadrado et al. 2008). Production of glucanases was reduced in yeasts during spaceflight due to the low-shear and microgravity conditions. Reduced glucanases could lead to change in cell wall thickness and β-glucan content (Liu et al. 2008).

6.1.5 Increased Phosphate Uptake in Yeast During Spaceflight

Wild type *S. cerevisiae* was launched into space during the Trans Earth extra vehicular Activity of Appollo 16. Appollo Microbiology Ecology Evaluation Device (MEED) was used to maintain the culture during spaceflight. Additionally, ground control was maintained at room temperature and flight control with no light exposure. Viability and survival rate of spaceflight sample did not change compared to control. Phosphate uptake rate was enhanced for spaceflight sample than ground control sample. Berry and Volz (1979) speculated that, phospholipid content of cell membrane may be altered which leads to change in ion transport including uptake of phosphate by cell membrane.

6.1.6 Stress Response of *S. cerevisiae* to Microgravity Condition in Spaceflight

Proteomic analysis of *S. cerevisiae* ∑1278b grown at ground and space (in ISS) was carried out. The proteome map of the microgravity- grown yeast samples showed less visible proteins in the high-mass range in comparison to the proteome map of normal gravity-grown colonies. This could be because of increased protein degradation in microgravity (Van Mulders et al. 2011).

Key glycolyic enzymes, glyceraldehyde-3-phosphate dehydrogenase, pyruvate kinase, and a subunit from pyruvate decarboxylase are upregulated and tricarboxylic acid cycle and some oxidative phosphorylation enzymes are less abundantly found in microgravity conditions. Reduced production of gluconeogenic enzymes such as fructose 1, 6 biphosphate aldolase, and of acetyl-CoA hydrolase in yeast samples was also found in microgravity condition. These enzymes are known to be involved in oxidative metabolism during propagation (Kobi et al. 2004). There is also reduced abundance in several heat shock proteins involved in protein folding or protein translocation into mitochondria (Van Mulders et al. 2011). Some of the components of ribosome biogenesis were found to be down regulated in microgravity condition. Ribosome biogenesis is controlled by Sfp1, which was known for unique microgravity response (Coleman et al. 2008). Additionally, ribosomal genes show reduced expression in modeled microgravity condition (Sheehan et al. 2007). There is also down regulation observed in several components of the proteasome. Outcomes of this study found that protein abundance changes such as enzyme involved in glycosylation of cell wall proteins and microgravity is responsible to change the integrity of the cell wall. Proteins that deal with glycerol stress were over expressed which supports the induction of the HOG pathway. The induction of High Osmolarity Glycerol (HOG pathway) and CWI (Cell Wall Integrity) signaling pathway in response to microgravity underlies that microgravity is one of the stress conditions experienced by yeast cells which resembles with the osmotic stress. This stressful condition is investing cellular energy more into the protective measure such as cell wall biosynthesis (CWI pathway activation), and the production of compounds (such as glycerol and trehalose) to increase the osmotolerancy of the yeast (HOG pathway activation) (Nickerson et al. 2004; Sheehan et al. 2007; Johanson et al. 2002, 2007; Willaert 2013).

6.1.7 Oxidative Stress Response in Spaceflight

High energy free radicals are originated from various radiations in space environment. These high energy free radicals may be responsible for increased oxidative damage. In general, cells respond to oxidative damage by exhibiting different oxidative stress responses. Oxidative stress response includes detoxifying enzymes such as catalase, superoxidase dismutase and peroxidase and thiol systems i.e. glutathione (GSH) and thioredoxin (Jones 2008). Glutathoine is involved in many of the critical cellular processes such as differentiation, proliferation and apoptosis.

GSH homeostasis is associated with a number of human diseases such as cancer, cystic fibrosis, neurodegenerative disorders and aging related diseases (Ballatori et al. 2009).

S. cerevisiae releases endogenous GSH in culture medium upon exposure to oxidative stress in simulated microgravity condition (Bradamante et al. 2010a, b; Versari et al. 2005). The oxidative stress response of *S. cerevisiae* LS267 under real microgravity is investigated. The SCORE (*Saccharomyces cerevisiae* Oxidative Stress Response Evaluation) experiment was conducted during a 24 day FOTON-M3 space mission of Italy. Gravitational forces on space station or on a space craft downs to 10^{-4}–10^{-6} g which creates microgravity condition. SCORE experiment was aimed to investigate oxidative stress response specifically any changes in GSH homeostasis under real microgravity. Findings recorded in SCORE experiment were compared with ground experiment carried out in simulated microgravity condition (Bradamante et al. 2010a, b). To create simulated microgravity condition on ground, Rotating Wall Vessel bioreactor (RWV) was used. In this bioreactor, a rotating high aspect ratio vessel (HARV) revolves at 28 rpm around a horizontal vessel but permits gas diffusion across a semi permeable membrane. In RWV, an average gravitational force is found reduced to 10^{-2} g (Rucci et al. 2007; Unsworth and Lelkes 1998; Meaney et al. 1998).

SCORE experiment showed that, 24 h of hyperoxia enhances extracellular release of GSH (40%) in spaceflight samples. This result supports the hypothesis that microgravity is having a role in induction of glutathione production. Further, yeast cells showed stress response by activating high-osmolarity glycerol (HOG) MAP kinase and cell integrity/protein kinase C (PKC) pathways (Bradamante et al. 2010a, b). High-osmolarity glycerol regulates cell morphology by swelling or shrinking and responds to oxidative stress. Protein kinase C senses variation in cell morphology and regulates genes involved in cell wall biogenesis and the maintenance of the actin cytoskeleton in response to this variation (Hohmann 2002). In SCORE experiment, Hogl and slt2 varied in their activity on space station which may be a combined effect of microgravity and oxidative stress. Hogl was highly activated under hyperoxic conditions whereas Slt2 was activated in both hyperoxic and normoxic conditions on a space station (Bradamante et al. 2010a, b).

Bradamante et al. in 2010 carried out ground experiments to investigate the oxidative stress effect on *S. cerevisiae* under hyperoxic and normoxic conditions simulated microgravity condition. Ground experiment showed that the rate of extracellular release of GSH was increased (10%) with protein carbonylation in both hyperoxic and normoxic simulated microgravity condition. Increased extracellular release may be due to alteration of cytoskeleton induced by microgravity. To check this possibility, researchers treated cells with an inhibitor of actin polymerization i.e. Dihydrocytochalasin B (DHCB) and (R)-(+)-trans-4-(1-aminoethyl)-N-(4-pyridyl) cyclohexan-ecarboxamide dihydrochloride monohydrate (Y-27632), a potent and selective rho associated kinase inhibitor. DHCB and Y-27632 treated cells also exhibited high rate of extracellular release of glutathione (12%) with protein carbonylation under hyperoxic conditions and Hogl and Slt2 activation (Bradamante et al. 2010a, b).

Gene expression of *S. cerevisiae* under oxidative and simulated microgravity condition was studied. A gamma glutamylcysteine synthetase (GSH1) and glutathione synthetase (GSH2) catalyses main step in glutathione synthesis. GTT1, encodes GSH transferase 1; BPT1 and YCF1, encodes two vacuolar GSH S-conjugate transporters of the ATP binding cassette family. Heat shock protein 26 (HSP26) and CTT1 (cytosolic catalase T) are two genes involved in stress response pathway. HSP26 and CTT1 transcripts are highly upregulated between 4 and 9 h in samples grown under simulated microgravity condition. This remains noticeable after 11–12 h. Same effect was observed in DHCB treated cells. These findings support that cells may have activated MAP kinases pathway. Additionally, GSH1 and GSH2 transcripts were activated in simulated microgravity and DHCB cells after 4 h and were same until extracellular release stops. However, genes involved in vacuolar transport were not affected. GTT1, BPT1 and YCF1 mRNAs were present from initial point supporting that these genes remain same with high upregulated Bpt1 in the early stationary phase. It reveals that, simulated microgravity condition under hyperoxic conditions influence expression of genes involved in GSH biosynthesis and stress response but does not affect genes involved in vacuolar transport (Bradamante et al. 2010a, b).

6.1.8 Metabolic and Genomic Pathways Affected During Spaceflight

Nislow et al. (2015) identified genomic and metabolic pathways required during spaceflight by yeast. Strain fitness test was performed for two yeast deletion collection i.e. ~4800 homozygous strain and ~5900 heterozygous strains grown during spaceflight. These strains contain unique DNA barcode as strain identifiers. For spaceflight experiment, yeast samples were delivered on space shuttle mission via opticell processing module (OPM). OPM is a specially designed instrument used for spaceflight experiment in space. It can maintain yeast deletion pool for about 20 generation in microgravity. Simultaneously, same experiment was carried out as ground control at normal gravity condition in the orbital environmental simulator at Kennedy space centre (USA). Additionally, osmotic stress effect was also tested on survival of yeast in space by addition of 0.5 M NaCl (Nislow et al. 2015).

Growth rate of yeast grown in the opticell during spaceflight was quantified. Population doubling time of sample grown for ~21 generations in microgravity was ~100 min whereas population doubling time for ground based control was recorded as ~90 min. Morphological characters such as budding pattern, overall shape and size was not found different than ground control when observed by light microscopy. Differences were observed in budding pattern and polarity on scanning electron microscopy. However, these changes were not consistent (Nislow et al. 2015).

Gene ontology enrichment analysis is carried out for homozygous and heterozygous deletion strains. Some strains showed reduced abundance from pool in spaceflight. These strains showed significant requirement of genes involved in RNA metabolism and DNA integrity pathway. Different biological processes linked to RNA metabolism and DNA integrity pathway were required for survival of yeast in spaceflight. These processes include ribosome biogenesis, regulation of ribosomal protein, transcription, cytoplasmic RNA translation, rRNA processing, tRNA modification and mRNA decay from RNA metabolism, DNA repair, recombination, replication, chromatin modeling, mitochondrial maintenance and proper protein localization to the mitochondria (Nislow et al. 2015). Such kind of effects was also found when DNA damage was induced. Additionally, such enrichment was found with therapeutics that acts as nucleotide analogs such as 5-flurouridine and flurocytosine. Enrichment of genes needed for processing of DNA and RNA has been increased on combining spaceflight condition with hyperosmotic stress exerted by addition of 0.5 M NaCl in medium. Nislow et al. (2015) speculated that, higher osmotic stress enhances the DNA damage effect of space flight. Gene ontology profile showed strong concordance between pathway responded by yeast in spaceflight and effect of 5-flurouridine, a FDA approved anticancer drug on yeast. Similarly, marked concordance was found with carmoflur, 5-flurocytosine and 8 methoxysoralen. Strong concordance was also observed with diallyl disulfide which is involved in detoxification of cells with increase in glutathione—S-transferase (GST).

Survival of haploid deletion strains during spaceflight was studied. Yeast strains stored in distilled water were delivered on space through space shuttle Atlan mission STS 112 in October 2001. Yeast deletion strain without PEX 19 showed 133 fold survival advantages in space. PEX 19 is a chaperon and import receptor for class I peroxisomal membrane protein. Consequently, yeast deletion strain lacking components of aerobic respiration, isocitrate metabolism mitochondrial electron transport exhibited 77–40 fold advantage for survival in space. Strains deleted with hydrolases, oxidoreductase and transferase exhibited significant advantage for survival during spaceflight in rich medium (Johanson et al. 2007) (Table 1).

6.2 Low Shear Modeled Microgravity (LSMMG) Effect on S. cerevisiae*: Studies Done by Using Simulated Microgravity Device (Rotating Wall Vessel)*

6.2.1 Growth Rate, Viability and Growth Kinetics

Purevdorj-Gage et al. (2006) studied low shear microgravity induced effect on *S. cerevisiae* BY4743 in HARV (high-aspect-ratio vessel). Growth rate, overall metabolic activity and rate of viability did not differ significantly between *S. cerevisiae* grown under LSMMG (low shear modeled microgravity) condition in HARV (high-aspect-ratio vessel) and control. Samples from both conditions showed similar doubling time. However, growth kinetics was noticed different for LSMMG sample and control. In LSMMG sample, lag phase was shorter with slightly higher yield than control (Purevdorj-Gage et al. 2006). McPherson (1997)

Table 1 Spaceflight induced effects on *S. cerevisiae*

Sr. no.	Spaceflight induced effect on *S. cerevisiae*	Strain used	Period	References
1.	Higher viability of yeast cells	Wild type	15 days	Yi et al. (2011)
2.	Increased G_2/M population, Decreased sub G_1 population	Wild type	15 days	Yi et al. (2011)
3.	Decreased glycogen content	Wild type	15 days	Yi et al. (2011)
4.	Increased malate dehydrogenase level	Wild type	15 days	Yi et al. (2011)
5.	Decreased succinate dehydrogenase and hexokinase level	Wild type	15 days	Yi et al. (2011)
6.	Random budding pattern	$\sum$1278b	12 days	Van Mulders et al. (2011)
7.	Reduced colony growth rate	$\sum$1278b	12 days	Van Mulders et al. (2011)
8.	Reduced invasive growth	$\sum$1278b	12 days	Van Mulders et al. (2011)
9.	Increased in biomass and cell wall mass	2.0016	15 days	Liu et al. (2008)
10.	Increased β-glucan and mannan content in cell wall	2.0016	15 days	Liu et al. (2008)
11.	Increased phosphate uptake	Wild type	7 Min., 7 s	Berry and Volz (1979)
12.	Induction of CWI and HOG pathway	$\sum$1278b	12 days	Van Mulders et al. (2011)
13.	Increased gluthathoine content	LS267	24 days	Bradamante et al. (2010)

hypothesized that, low fluid shear environment of weightlessness may lead to unusual mass transfer of cellular byproducts in microenvironment surrounding cell and trigger rapid growth. It resulted in shorter lag phase (McPherson 1997; Mennigmann 1986). However, metabolic activity remained same in LSMMG condition and control during the experiment.

6.2.2 Random Budding Pattern in Liquid Medium

Under LSMMG condition *S. cerevisiae* BY4743 showed random budding pattern during early stationary phase and late stationary phase. Purevdorj-Gage et al. (2006) states that the disturbances in budding pattern in LSMMG condition may not be due to change in the cytoskeleton organization but rather due to changed expression of genes having role in budding processes and polarity establishment. They found that there is a significant difference in the expression level of genes involved in polarity establishment (*BUD5*) and bipolar budding phenotype (*RAX1*, *RAX2*, and *BUD25*). *BUD25* is involved in budding process, and its deletion resulted in random budding phenotype whereas BUD5 gene deletion or over expression lead to random budding phenotype (Ni 2001). *BUD25* is over expressed *in* LSMMG compared to the control. Reduced expression in both *RAX1* and *RAX2 in LSMMG* sample resulted in aberrant budding phenotype (Purevdorj-Gage et al. 2006). *BUD5 is* involved in maintaining cell polarity during budding process (Kang et al. 2001). Overexpression of *BUD5* in LSMMG sample lead into random budding pattern (Purevdorj-Gage et al. 2006). There was no significant difference observed on cell size of *S. cerevisiae* BY4743 between control and LSMMG sample in HARV when examined at logarithmic, late-logarithmic, early-stationary and late-stationary growth phases (Purevdorj-Gage et al. 2006).

6.2.3 Cell Clumping

Five percentage of the LSMMG cell population was clumped (of five or more than five yeast cells) in the late-logarithmic growth phase and ten times more clumped during early-stationary growth than control (Purevdorj-Gage et al. 2006). Clumped phenotypes in yeast are found due to the improper functioning of chitinases and glucanases that result into the defective mother-daughter cell separation process (Baladron et al. 2002; Doolin et al. 2001; Ufano et al. 2004). Chitinase activity was normal in LSMMG sample. *EGT2*, *DSE1*, and *DSE2* involved in mother-daughter cell separation were down regulated in LSMMG grown sample (Colman-Lerner et al. 2001). This process is dependent on the cell polarization process (Amon 1996; Chant 1999). Bud scarring and formation of aggregate in LSMMG may be because of cell polarity defects (Purevdorj-Gage et al. 2006).

6.2.4 Change in Gene Expression in Response to Microgravity

Johanson et al. (2002) provides first data on change in gene expression of *S. cerevisiae* (strain FE18984) in response LSMMG condition in RWV culture. Microarray analysis was carried out to reveal the changes in gene expression in response to LSMMG condition compared to ground control incubated in Rotating Wall Vessel (RWV) at 1 g condition at different time points (at 20, 60 and 180 min). Genes expressed differentially in response to LSMMG at different time periods were grouped on the basis of cluster analysis. Promoter analysis was performed for DNA sequences of identified genes. This analysis identified motifs that have similar core sequence with stress responsive element (STRE) and *Rap1* transcription factor (Johanson et al. 2002).

Genes identified with STRE sites were grouped into two categories. Stress responsive group of genes contains *HLJ1* and *SSA4* which are involved in stress response. Another group of genes are related to metabolic cycle and is comprised of *PGM2*, *GPM2*, and *COX5B*. Gene product of *PGM2*, phosphoglucomutase is involved in one of the steps in conversion of glucogen to glucose i.e. glucose-1-phosphate to glucose-6-phosphate. *GPM2* encodes for phosphoglycerate mutase which catalyses last step of glycolysis. *COX5B* gives product cytochrome-*c* oxidase chain Vb, and cytochrome oxidase and are involved in ATP synthesis. In addition to this, *Rap1* binding motifs were found in 13 upregulated genes after 20 min. *Rap1* regulates expression of genes which are differentially expressed in response to change in growth rate. Consequently, increased glucose utilization was observed in LSMMG yeast sample compared to control yeast sample. Authors speculated that, enhanced glucose utilization in RWP serve as a model for increased *Rap1*-mediated transcription and indicating that *Rap1* may be involved in RWP induced pathway of yeast (Johanson et al. 2002).

Sheehan et al. (2007) carried out microarray analysis of *S. cerevisiae* diploid strain BY4743 grown in High Aspect Ratio Vessel (HARV), which provides in Low shear modeled microgravity condition (LSMMG) in comparison with HARVs under normal gravity conditions. Cells were exposed to various stress factors in the environment. These conditions may lead cells to evolve stress specific pathways to survive in these conditions. There are various stress inducible genes studied in *S. cerevisiae.* Gasch et al. (2000) studied stress responsive genes from *S. cerevisiae* and these set of ~900 genes are referred as the 'environmental stress response' (ESR) genes. Microgravity may be as of the stress conditions experienced by an organism. Authors compared genes having altered expression in HARVs with Gasch et al. (2000) datasets and categorized microgravity responsive genes of which 26% of the genes were defined as ESR genes whereas remaining 74% genes may be specific in response to microgravity. Additionally, such microgravity related stress response was also observed in *E. coli* (Sheehan et al. 2007). It is revealed that, simulated microgravity conditions lead cells to be more resistant to other stress conditions such as hyperosmosis and low pH (Collister et al. 2002). Stress responsive genes were also identified in *Salmonella enteric* in a study conducted in HARVs (Wilson et al. 2002).

Sheehan and group identified the set of genes whose expression is altered significantly at 5 and 25 generations of growth. At 5 generations, 278 genes were significantly changed. Of these, 161 genes were up regulated and 117 were down-regulated whereas at 25 generation 197 genes were significantly changed in expression. Of these, 106 were up-regulated and 91 were down regulated in their expression. 897 genes did not alter their expression between 5 and 25 generations (Sheehan et al. 2007). Some of the genes involved in budding and cell wall integrity pathway are discussed below.

Normal yeast cells do not have the tendency for random budding pattern. NSR1 gene which is having role in budding pattern was found to be diminished in its expression in LSSMG grown cells in HARVs. On deletion of BUD21in yeast cell, cells exhibit random budding phenotype (Hahn and Thiele 2002). BUD21 was found to be down regulated in its expression in LSSMG grown yeast cells in HARVs. Additionally, 46% of bud localized transcripts were significantly altered in the same study. This include down regulation of genes EGT2, ASH1 which are involved in cell separation (Nickerson et al. 2004), daughter cell-specific transcription WSC2 (Martin et al. 2005). WSC2 encodes for heat shock sensor in MAP1 pathway (Saito and Tatebayashi 2004; Sheehan et al. 2007).

Table 2 Differential regulation of genes in *S. cerevisiae* under simulated microgravity condition

Function	Differentially regulated genes			
	Down regulated genes	Fold change	Up regulated genes	Fold change
Bud site selection	BUD 2 BUD 3 BUD 4 BUD 6 BUD 8 BUD 9 BUD 16 BUD 17 BUD 22 BUD 23 BUD 31 BUD 32 STE 20 RSR 1 RAX 1 RAX 2	1.1 1.4 2.0 1.4 1.3 1.9 1.0 1.0 2.0 2.4 1.3 1.3 1.0 2.0 5.1 3.2	BUD 5 BUD 7 BUD 13 BUD 20 BUD 25 BUD 27 AXL 1 THP 1	1.6 1.1 1.4 1.1 1.7 1.2 1.1 1.2
Axial bud site selection	–		ERV15	1.8
Cell bud growth	TOS2	1.7	LAS1	1.2
Cell wall integrity, stress response	–		SLG1	1.6
Regulation of cell size	PRS3	2.0	MUB1	1.2
Bud site selection, cell polarity	RSR1	2.0	–	–
Cell polarity	RRP14	1.4	–	–
Ribosomal protein	RSP18B	3.1	–	–
	RPL22A	4.4		
Lipid metabolism	FEN1 SUR4	3.2 2.5	–	
Nuclear protein	NSR1	7.5	–	
Cell wall	GAS1	2.0	–	
Hypothetical ORF	Unspecified	2.4	–	
Bipolar bud selection	–		TWF1	1.6

Reference Sheehan et al. (2007)

For the survival in stress condition such as exposure to heat, osmotic stress, and/or oxidative stress, yeast cells should retain its cell wall integrity (Saito and Tatebayashi 2004). The alterations in cell wall due to stress on cell wall is basically regulated by cell wall integrity (CWI) MAPK signaling Pathway. Sheehan et al. (2007) found that, SDP1 was up regulated at fold change of 6.71 at 5 generations and 8.12 at 25 generations. Sdp1 is a stress inducible negative regulator of MAP kinase cascade and have dual specificity as MAP kinase phosphatase. Another down regulator of MAP kinase pathway, PTP2 i.e. protein tyrosine phosphatase is highly expressed in LSMMG grown cells than control in HARVs (Sheehan et al. 2007). Sheehan et al. (2007) suggested that, simulated microgravity affects multiple signaling pathways and cells sense low shear microgravity environment through MAP kinase pathway (Sheehan et al. 2007) (Table 2).

7 Conclusions and Future Perspectives

Health of the astronauts during and after space flight, long term stay aboard in spaceships and future colonies is a major concern. Data available from experiments on humans and model organisms support this concern. Experiments on human and animals aboard a space ship are difficult to perform. Data available so far indicate major changes in human physiology in the space environment with impact on the astronaut's health. The way we treat Earth bound humans may not be valid for the astronauts considering the major changes they experience in space. A thorough understanding of the human biology in the space may be required for effectively treating astronauts or for maintaining their health. Information available on the impact of microgravity and space flight on basic eukaryotic biology is far from complete. Research using the yeast model is expected to give significant insights to space biology especially space medicine.

References

Aguilar U. B., Francois J. M. 2003. *Lett. Appl. Microbiol.* 37: 268–274.

Amon, A. 1996. *Cell* 84: 651–654.

Baladron, V., S. Ufano, E. Duenas, A. B. Martin-Cuadrado, F. del Rey, and C. R. Vazquez de Aldana. 2002. *Eukaryot. Cell* 1: 774–786.

Ball, J. R., Evans Jr., C. H. 2001. In: *Committee on Creating a Vision for Space Medicine During Travel Beyond Earth Orbit, Board on Health Sciences Policy*, Washington, DC, USA, National Academy.

Ballatori N., Krance S. M., Notenboom S., Shi S., Tieu K., Hammond C.L. 2009. *Biol. Chem.* 390: 191–214.

Benoit M. R,. Klaus D. M. 2007. *Adv. Space Res.* 39: 1225–1232.

Berry D., Volz P. A. 1979. *Appl. Environ. Microbiol.* 38: 751–753.

Bizzarri, M., Monici, M., & van Loon, J. J. 2015. *BioMed. res. int.*

Botstein, D., & Fink, G. R. 2011. *Genetics. 189*, 695–704.

Botstein, D., Chervitz, S. A., & Cherry, J. M. 1997. *Science (New York, NY). 277*, 1259.

Bradamante S., Barenghi L., Villa A. 2010a. Patent US 7695957 B2.

Bradamante, S., Villa, A., Versari, S., Barenghi, L., Orlandi, I., & Vai, M. 2010b. *Biochimica et Biophysica Acta (BBA)-Molecular Cell Research. 1803*, 1376–1385.

Chant, J. 1999. *Annu. Rev. Cell Dev. Biol.* 15: 365–391.

Clément, G. 2011. *Fundamentals of space medicine* (Vol. 23). Springer Science & Business Media. Germany.

Clément, G., & Slenzka, K. (Eds.). 2006. In: *Fundamentals of Space Biology: Research on Cells, Animals, and Plants in Space* (Vol. 18). Springer Science & Business Media. USA.

Cogli, M. 2007. *Gravitational and Space Research. 5* (2).
Coleman, C. B., Allen, P. L., Rupert, M., Goulart, C., Hoehn, A., Stodieck, L. S., and Hammond, T. G. 2008. *Astrobiology* 8: 1071–1078.
Collister M., Didmon M. P., MacIssac F., Stark M. J., MacDonald N. Q., Keyse S. M. 2002. *FEBS Lett.* 527: 186–192.
Colman-Lerner, A., T. E. Chin, and R. Brent. 2001. *Cell* 107: 739–750.
Cullen P. J. and Sprague G. F. Jr. 2000. *Proc. Natl. Acad. Sci. USA*. 97: 13619–13624.
Dayanandan, P. 2011. *J. Biosci. 36*: 911–919.
Decelle, J. G. and Taylor, G. R. 1976. *Appl. Environ. Microbiol.* 32: 659–665.
Dolinski, K., and D. Botstein. 2005. *Genome Res.* 15: 1611–1619.
Dolinski, K., and D. Botstein. 2007. *Annu. Rev. Genet.* 41: 465–507.
Doolin, M. T., A. L. Johnson, L. H. Johnston, and G. Butler. 2001. *Mol. Microbiol.* 40: 422–432.
Etheridge, T., Nemoto, K., Hashizume, T., Mori, C., Sugimoto, T., Suzuki, H., ... & Higashitani, A. 2011. *PloS One. 6*(6).
Fleet G. H. 1991. Cell walls. In: Rose A. H., Harrison J. D. (eds) *The yeasts.* Vol. 4. Academic, London, pp. 199–277.
Forsburg, S. L. 2007. *Gravitational and Space Research 18*(2).
Gasch A. P., Spellman P. T., Kao C. M., Carmel-Harel O., Eisen M. B., Storz G., Botstein D., Brown P. O. 2000. *Mol. Biol. Cell.* 11: 4241–4257.
Ghaemmaghami, S., Huh, W. K., Bower, K., Howson, R. W., Belle, A., Dephoure, N., ... & Weissman, J. S. 2003. *Nature 425*: 737–741.
Giaever G. and Nislow C. 2014. *Genetics* 197: 451–465.
Giaever, G., A. M. Chu, L. Ni, C. Connelly, L. Riles *et al.*, 2002. *Nature* 418: 387–391.
Gimeno, C. J., Ljungdahl, P. O., Styles, C. A., & Fink, G. R. 1992. *Cell. 68*: 1077–1090.
Gordon DB, Siamon G. 2003. *Immunity* 19: 311–315.
Gueguinou, N., Huin-Schohn, C., Bascove, M., Bueb, J. L., Tschirhart, E., Legrand-Frossi, C., & Frippiat, J. P. 2009. *J. Leukocyte Biology 86*: 1027–1103.
Hader D. P., Lebert M., Richter P. and Ntefidou M. 2003. *Adv. Space Res.* 31: 2181–2186.
Hahn J. S., Thiele D. J. 2002. *J. Biol. Chem.* 227: 21278–21284.
Hammond, T. G., & Hammond, J. M. 2001. *American J. Physiol-Renal Physiol. 281*: F12–F25.
Hammond, T. G., Benes, E., O'Reilly, K. C., Wolf, D. A., Linnehan, R. M., Taher, A., Kaysen, J. H., Allen, P. L., Goodwin, T. J. 2000. *Physiol. Genomics* 3: 163–173.
Hartwell L. H., Culotti J., Reid B. 1970. *Proc. Natl. Acad. Sci. USA*. 66: 352–9.
Heinicke, S., Livstone, M. S., Lu, C., Oughtred, R., Kang, F., Angiuoli, S. V., ... & Dolinski, K. 2007. *PLoS One. 2*: e766.
Herranz, R., Anken, R., Boonstra, J., Braun, M., Christianen, P. C., de Geest, M., ... & Hemmersbach, R. 2013. *Astrobiology 13*: 1–17.
Highstein S. M., Fay R. R. and Popper A. N. 2004. The vestibular system (New York: Springer-Verlag). USA.
Hofer M., Pospisil M. 1997. *International Immunopharmacology* 19: 607–609.
Hohmann S. 2002. *Microbiol. Mol. Biol. Rev.* 66: 300–372.
Horneck, G., Klaus, D. M., & Mancinelli, R. L. 2010. *Microbiol. Mol. Biol. Rev. 74*: 121–156.
Huh, W. K., Falvo, J. V., Gerke, L. C., Carroll, A. S., Howson, R. W., Weissman, J. S., & O'Shea, E. K. 2003. *Nature 425*: 686–691.
Inoue S. B., Takewaki N., Takasuka T., Mio T., Adachi M., Fujii Y., Miyamoto C., Arisawa M., Furuichi Y., Watanabe T. 1995. *Eur J Biochem.* 231: 845–854.
Jamas S., Easson J., Davidson D., Ostro G. R. 1996. US Patent. 5, 532, 223.
Johanson K., Allen P. L., Lewis F., Cubano L. A., Hyman L. E., Hammond T. G. 2002. *J. Appl. Physiol.* 93: 2171–80.
Johanson, K., Allen, P. L., Gonzalez-Villalobos, R. A., Nesbit, J., Nickerson, C. A., Höner Zu Bentrup, K., Wilson, J. W., Ramamurthy, R., D'Elia, R., Muse, K. E., Hammond, J., Freeman, J., Stodieck, L. S., and Hammond, T. G. 2007. *Acta Astronaut.* 60: 460–471.
Jones, D. P. 2008. *American J. Physiol. Cell Physiol. 295*: C849–C868.
Kang, P. J., A. Sanson, B. Lee, and H. O. Park. 2001. *Science.* 292: 1376–1378.

Kim K. S. and Yun H. S. 2006. *Enzyme Microb. Technol.* 39: 496–500.
Klaus D., Simske S., Todd P., Stodieck L. 1997. *Microbiology* 143: 449–55.
Klaus, D. M. 2001. *Gravit. Space Biol. Bull.* 14: 55–64.
Klaus, D. M., Benoit, M. R., Nelson, E. S., & Hammond, T. G. 2004. *Journal of Gravitational Physiology: a Journal of the International Society for Gravitational Physiology.* 11: 17–27.
Klis F. M., Mol P., Hellingwerf K., Brul S. 2002. *FEMS. Microbiol. Rev.* 26: 239–256.
Kobi, D., Zugmeyer, S., Potier, S., and Jaquet-Gutfreund, L. 2004. *FEMS Yeast Res.* 5: 210–230.
Lemire, B. D., Oyedotun, K. S. 2002. *Biochim. Biophys. Acta* 1553: 102–116.
Levinskikh M. A., Sychev V. N., Derendyaeva T. A., Signalova O. B., Salisbury F. B., Campbell W. F., Bingham G. E., Bubenheim D. L and Jahns G. 2000. *J. Plant Physiol.* 156: 522–529.
Liu H. Z., Wang Q., Liu X. Y., Tan S. S. 2008. *Appl. Microbiol. Biotechnol.* 81: 543–550.
Liu X. Y , Wang Q., Cui S. W., Liu H. Z. 2007a. *Food Hydrocoll.* 22: 239–247.
Lynch, S. V., Mukundakrishnan, K., Benoit, M. R., Ayyaswamy, P. S., and Matin, A. 2006. *Appl. Environ. Microbiol.* 72: 7701–7710.
Martin H., Flandez M., Nombela C., Molina M. 2005. *Mol. Microbiol.* 58: 6–16.
Martín-Cuadrado A. B., Fontaine T., Esteban P. F., Dedo J. E., Medina-Redondo M., Rey F., Latgé J. P., Aldana C. R. V. 2008. *Fungal Genet. Biol.* 45: 542–553.
McPherson, A. 1997. *Trends Biotechnol.* 15: 197–200.
Meaney D. F., Johnston E. D., Litt M., Pollack S. R. 1998. *Adv. Heat Mass Transf. Biotechnol.* HTD 362: 103–107.
Mennigmann, H. D., and M. Lange. 1986. *Naturwissenschaften* 73: 415–417.
Montgomery Jr, P. O., Cook, J. E., Reynolds, R. C., Paul, J. S., Hayflick, L., Stock, D., ... & Campbell, D. 1978. *In vitro.* 14: 165–173.
Morey-Holton E. R. 2003. In The impact of Gravity on Life; in Evolution on planet Earth: the impact of the physical environment (eds) L Rothschild and A Lister (New York: Academic Press) pp. 143–159.
Ni, L., and M. Snyder. 2001. *Mol. Biol. Cell.* 12: 2147–2170.
Nickerson C. A., Ott C. M., Wilson J. W., Ramamurthy R., Pierson D. L. 2004. *Microbiol. Mol. Biol. Rev.* 68: 345–361.
Nislow, C., Lee, A. Y., Allen, P. L., Giaever, G., Smith, A., Gebbia, M. & Hammond, T. G. 2015. *BioMed. Res. Int.*
Pasteur L. Mémoire.1860. *Annales de Chimie* 58: 323–426.
Peter J. R., Brent E. L., Luke A. B., Elizabeth L. A., Harry E. E., David L. W. 2004. *International Immunopharmacology* 4: 1209–1215.
Pierson, D. L. 2001. *Gravit. Space Biol. Bull.* 14: 1–6.
Pierson, D. L., Chidambaram, M., Heath, J. D., Mallary, L., Mishra, S. K., Sharma, B., Weinstock, G. M.1996. *FEMS Immunol. Med. Microbiol.* 16: 273–281.
Pierson, D. L., Mehta, S. K., Magee, B. B., Mishra, S. K. 1995. *J. Med. Vet. Mycol.* 33: 145–150.
Prusty, R., Grisafi, P., and Fink, G. R. 2004. *Proc. Natl. Acad. Sci. USA.* 101: 4153–4157.
Purevdorj-Gage, B., Sheehan, K. B., and Hyman, L. E. 2006. *Appl. Environ. Microbiol.* 72: 4569–4575.
Reynolds, T. B., Jansen, A., Peng, X., and Fink, G. R. 2008. *Eukaryot Cell* 7: 122–130.
Rucci N., Rufo A., Alamanou M., Teti A. 2007. *J. Cell. Biochem.* 100: 464–473.
Saito H., Tatebayashi K. 2004. *J. Biochem.* 136: 267–272.
Sheehan, K. B., McInnerney, K., Purevdorj-Gage, B., Altenburg, S. D., & Hyman, L. E. 2007. *BMC Genomics* 8: 3.
Sicard D., Legras J. L. 2011. *C R Biol.* 334: 229–36.
Sikavitsas, V. I., Bancroft, G. N., & Mikos, A. G. 2002. *J. Biomed. Materials Res.* 62: 136–148.
Simon, J. A., & Bedalov, A. 2004. *Nature Reviews Cancer.* 4: 481–487.
Souza K. A., Ilyin E. A., Sychev V. N. and Jahns G. C. 2009. In Biological Research in Space; in Space Biology and Medicine: Vol 5 U.S. and Russian Cooperation in Space Biology and Medicine (eds) A. E. Nicogossian, S. Mohler, O. Gazenko and A. Grigoriev (Reston: AIAA) pp. 1–44.

Souza, K. A., Black, S. D. and Wassersug, R. J. 1995. *Proc. Natl. Acad. Sci.* 92: 1975–1978.
Stanhill, A., Schick, N., and Engelberg, D. 1999. *Mol Cell Biol.* 19: 7529–7538.
Tairbekov, M. G., Parfyonov, G. P., Shepelev, E. Y., & Sushkov, F. V. 1983. *Adv. in Space Res.* 3: 153–158.
Thompson I. M., Spence C. R., Lamm D. L. 1987. *Am. J. Med. Sci.* 294: 294–300.
Ufano, S., M. E. Pablo, A. Calzada, F. del Rey, and C. R. Vazquez de Aldana. 2004. *J. Cell Sci.* 117: 545–557.
Unsworth, B. R., & Lelkes, P. I. 1998. *Nature Medicine* 4: 901–907.
Van Loon, J. J. W. A. 2007. *Biology in Space and Life on Earth: Effects of Spaceflight on Biological Systems* 17–32.
Van Mulders S. E., Stassen C., Daenen L., Devreese B., Siewers V., van Eijsden R. G., Nielsen J., Delvaux F. R., Willaert R. 2011. *Astrobiology* 11: 45–55.
Versari S., Villa A., Barenghi L., Bradamante S. 2005. *European Space Agency SP-585* versari/1–versari/2 [Special Publication].
Volkmann D. and Baluska F. 2006. *Protoplasma* 229: 143–148.
Walther I., Bechler B., Müller O., Hunzinger E., Cogoli A. 1996. *J. Biotechnol.* 47: 113–27.
Willaert G. R. 2013. *Current Biotechnology* 2: 226–234.
Wilson J. W., Ramamurthy R., Porwollik S., McClelland M., Hammond T. G., Allen P. L., Ott C. M., Pierson D. L., Nickerson C. A. 2002. *Proc. Natl. Acad. Sci.* 99: 13807–13812.
Wolverton B. C. and Kiss J. Z. 2009. *Gravitational Space Biol.* 22: 12–23.
Yamada, M., Takeuchi, Y., Kasahara, H., Murakami, S., Yamashita. 1993. *M. Biol. Sci. Space.* 7: 116–119.
Yi Z. C., Li X. F., Wang Y., Wang J., Sun Y., Zhuang F. Y. 2011. *Adv. Space Res.* 47: 2049–2057.

Yeasts and Traditional Fermented Foods and Beverages

Tek Chand Bhalla and Savitri

Abstract Fermented foods and beverages are the outcomes of metabolic activity of some microorganisms among the thousands of bacteria, yeasts and moulds. Yeast is one of the first domesticated microorganisms that ferment sugars to alcohol and carbon dioxide. *Saccharomyces cerevisiae*, *Candida* spp., *Debaryomyces* spp. and *Hansenula anomala* are the most common yeasts associated with the traditional fermentations and occur in a large number of fermented foods and beverages, prepared from raw materials of plant as well as animal origin. *Saccharomyces cerevisiae* is the most frequently involved yeast species in alcoholic fermentation which leads to the production of ethanol. This yeast predominates in most of the traditional fermented foods and beverages such as *burukutu, merissa, bhatooru, seera, chhang, fufu, tape, ogi, puto, dosa, idli, papdam, kecap, laochao, warri*, etc. In these traditional foods, yeast either alone or in association with bacteria and mould has substantial influence on taste, aroma, texture and nutritional value of the fermented products. Thus, yeasts play very important role in improving the nutrition and socio-cultural life of people living in rural and tribal areas of various countries across the world.

Keywords Fermented foods · Beverages · Yeasts · Traditional · *Saccharomyces cerevisiae*

1 Introduction

Fermentation is considered to be the oldest method of food preservation man devised several thousands years ago during the course of evolution of human civilizations. It plays a very significant role in safeguarding the food security of millions of people mainly the marginalized and vulnerable groups around the world

---

T.C. Bhalla (✉) · Savitri
Department of Biotechnology, Himachal Pradesh University,
Summer Hill, Shimla 171005, Himachal Pradesh, India
e-mail: bhallatc@rediffmail.com

T. Satyanarayana and G. Kunze (eds.), *Yeast Diversity in Human Welfare*, DOI 10.1007/978-981-10-2621-8_3

(Battcock and Azam-Ali 1998). It is an efficient, low energy consuming process which needs no sophisticated methods of preservation such as refrigeration or other energy dependent food preservation technologies and prolongs the shelf life of foods (Borgstrom 1968).

The food fermentation involves the concomitant action of one or more microorganisms, and in some, it may act sequentially with a change in dominant microflora as the fermentation progresses (Haard 1999). Among the various microbes involved in food fermentation, yeasts play a pivitol role in the transformation of the various substrates as yeasts grow and survive under stress conditions. Most of the indigenous or traditional fermented foods are prepared by solid-substrate fermentation in which substrates are allowed to ferment either naturally or by the addition of starter cultures. These foods and beverages have formed an integral part of the diet in most of the civilizations around the world since antiquity as these can be prepared at household level using locally available raw materials and simple processing techniques and equipments.

In food fermentation, yeasts transform the original raw materials physically, nutritionally and organoleptically into more desirable products. Amongst different yeasts, the genus *Saccharomyces* has a long history of traditional applications particularly in bread, beer and wine fermentations. Apart from the genus *Saccharomyces,* a variety of other yeast species also play important role in the production of traditional fermented foods and beverages across the world. In addition, many important food ingredients, additives and processing aids are also obtained from yeasts and these are also being explored as novel biocontrol agents in curtailing food spoilage as these possess very strong antifungal activities. The probiotic activity of yeast is another interesting aspect that is attracting the attention of both academia and industry. In this chapter yeast diversity in traditional fermented foods and beverages and their role in driving traditional fermentations will be discussed.

2 History of Yeast in Fermentation

Yeast was used to carry out fermentation from the very starting of brewing and thus considered as one of the earliest domesticated microorganisms. The role of yeasts or even microorganisms in food fermentation was not known in Middle Ages, but it was well experienced and perceived by the people practicing fermentation that the best beers were produced near the bakeries (Kruif 1935). Although Antonie van Leeuwenhoek in 1680 first observed both yeast and bacteria microscopically, it was not until the year 1837 when Cagniard-Latour suggested that the process of fermentation was closely associated with the budding (i.e. growth/division) of yeast cells. Schwann further extended the work of Cagniard-Latour and proposed that the growth of yeast cells depend on the presence of sugar which serves as a food material. Yeast as a living agent in fermentation or transformation of sugars remained a subject of debate among the chemists or biologists until Louis Pasteur who convincingly proved the association of yeast with fermentation (Stewart and Russell 1986).

The long history of the association of yeast with human society could be traced from the evolution of bread, beer and wine as well accepted foods and beverages across the world that started about 5000 years ago (Fleet 2006).

3 Yeast: Morphology, Physiology and Metabolism

Yeasts are unicellular, heterotrophic eukaryotes classified in the kingdom fungi. The term 'yeast' is generally used as a synonym for *Saccharomyces cerevisiae*, but the phylogenetic diversity of yeasts is so vast that these are placed in two separate phyla: the Ascomycota and the Basidiomycota (Kurtzman and Fell 1998).

Yeasts are fairly simple in their nutritional requirements as they are able to metabolize a range of carbohydrates as carbon source. They require a reduced carbon source, various minerals, a supply of nitrogen, vitamins including biotin, pantothenic acid, thiamine and relatively small amount of oxygen. Metabolic activities (aerobic respiration or fermentation) in yeasts are dependent on the concentration of sugar present in the medium e.g. when sugar concentration is high, the fermentative mode of sugar metabolism is switched on in yeast cell, and the pyruvate is broken down to ethanol yielding two molecules of ATP per sugar molecule. At low sugar concentrations, the pyruvate shunts into acetyl-CoA and the respiratory chain with net generation of 32 molecules of ATP per glucose molecule. This well-known phenomenon in yeasts is known as Crabtree effect and this has a great significance in commercial fermentation processes (Bamforth 2005). Yeasts mostly reproduce by budding; however, they exhibit both asexual and sexual stages in their life cycle (Kurtzman and Fell 2005).

4 Traditional Fermented Foods

The traditional fermented foods and beverages have social, religious, cultural and nutritional importance and their preparation and consumption is largely confined to specific region. The diversity of such fermented products is as an outcome of the heterogeneity of traditions and cultures prevailing in the world, geographical distribution and the raw materials used in the fermentation. These traditional fermented foods and beverages are prepared from a variety of substrates viz. cereals, pulses, vegetable, fruits, meat and milk (Table 1).

Most of the traditional fermented foods of the world are made by fermentation of legumes either alone or in combination with cereals as in the case of *idli* and *dosa* prepared in Southern parts of India (Soni and Sandhu 1999). *Warries*, *sepubari* and *borhe* are some traditional fermented legume based products prepared in Northern regions of India (Savitri and Bhalla 2007). Fermented soybean pastes, *miso,*

Table 1 Traditional fermented foods and beverages prepared in various parts of the world

Fermented product	Substrate	Nature of product/Use	Country/Area
Afitin	African locust bean	Condiment	Africa
Aisa	*Albizia saman*	Condiment	Africa
Amazake	Rice	Sweet, low alcoholic drink	Japan
Ambali	Millet flour	Semi solid/all time food	India
Ambulthiyal	Tuna fish	Fermented fish curry	Sri Lanka
Am-som	Sorghum, millet and maize	Non-alcoholic beverage	Northern Nigeria
Ang-kak	Rice	Dry red powder used as a colorant	China
Apong	Rice	Alcoholic drink	India
Appa/appam	Rice or white wheat flour	Semi solid breakfast food	Sri Lanka, India
Atole	Maize	Sour porridge	Southern Mexico
Bagni	Millet	Liquid drink	Caucasus
Bagoong	Fish	Thick paste used as seasoning agent	Philippines
Banku	Maize, cassava	Dough used as staple food	Ghana
Belacan	Shrimps	Fermented shrimp paste or sauce used as condiment	Southeast Asia
Bhatura/ kachauri	Wheat flour	Deep fried *roties* used as breakfast food	India
Bogobe	Sorghum	Porridge	Botswana
Bongkrek	Coconut press cake	Roasted in oil, used as a meat substitute	Indonesia
Bouza	Wheat	Thick acidic beverage	Egypt
Braga	Millet	Liquid drink	Romania
Brem	Glutinous rice	Cake	Indonesia
Brembali	Glutinous rice	Dark brown sour alcoholic beverage	Indonesia
Burukutu	Sorghum and cassava	Cream colored drink with suspended solid	Savannah regions of Nigeria
Busa	Rice/millet, sugar	Liquid drink	Tartars of Krim, Turkistan, Egypt
Chakka	Milk	Concentrated milk solids used as sweet dish	India
Chhang	Rice, barley	Alcoholic drink	India
Chee-fan	Soybean whey curd	Cheese like, eaten fresh	China
Chhash	Milk	Acidic refreshing drink	India

(continued)

Table 1 (continued)

Fermented product	Substrate	Nature of product/Use	Country/Area
Chicha	Maize	Alcoholic beverage	Peru
Chickwangue	Cassava roots	Paste used as staple food	Congo
Churpi	Buttermilk	Hard dried solids, eaten fresh as well as used in soups	India, Nepal
Darassum	Millet	Liquid drink	Mongolia
Dawadawa (iru, kpalugu, kinda)	African locust bean	Sun dried solids used as supplement to soups	West Africa, Nigeria
Dhokla	Bengal gram and wheat	Spongy cake used as staple food	India
Dosa (dosai)	Black gram and rice	Spongy cake used as breakfast food	India
Fufu	Cassava roots	Paste, eaten with soup	Africa
Gari	Cassava roots	Flour eaten boiled with stews, vegetables	West Africa
Gundruk	*Brassica campestris* leaves	Cured and sun-dried green leafy vegetable	Nepal
Hamanatto (Douchi)	Whole soybeans, wheat flour	Flavoring agent for meat and fish, also eaten as snack	Japan, China
Handwa	Rice, red gram and Bengal gram flour	Baked food	Western India
Hom-dong	Red onions	Pickled red onions	Thailand
Hua-chai-po-tai-tan-soi	Carrots and turnips	Pickled product as food adjunct	Thailand and China
Idli	Rice and black gram	Spongy steamed bread	Southern India
Imrati	Wheat flour	Sweet snack food	India, Pakistan, Nepal
Injera	*Teff*/maize, wheat, barley, sorghum, etc.	Moist bread like pancake	Ethiopia
Inyu	Black soybeans	Liquid flavour enhancer	Taiwan, China, Hong Kong
I-sushi	Fish	Fermented fish eaten with rice	Japan
Jaadi	Fish	Cured fish used as pickle	Sri Lanka
Jaanr	Finger millet	Sweet and sour alcoholic paste	India
Jalebi	Wheat flour	Deep fried pretzels, confection food	India, Pakistan and Nepal
Jamin-bang	Maize flour	Bread or cake like product	Brazil

(continued)

Table 1 (continued)

Fermented product	Substrate	Nature of product/Use	Country/Area
Jand	Wheat, barley	Alcoholic beverage	Nepal
Jeruk	Ginger and papaya	Pickled vegetable	Malaysia
Kaanga-kopuwai	Maize	Soft, slimy, eaten as vegetable	New Zealand
Kanji	Rice and carrots	Sour liquid added to vegetables	India
Katsuobushi	Whole fish	Dry solids used as seasoning agent	Japan
Kecap	Soybeans and wheat	Liquid used as condiment or seasoning agent	Indonesia and vicinity
Kefir	Cow/goat/sheep milk	Acidic alcoholic beverage	Russia
Kenkey	Maize	Steamed eaten with vegetables	Ghana
Ketjap	Black soybeans	Syrup used as seasoning agent	Indonesia
Khalpi	Cucumber, spices and salt	Pickle	Nepal
Khaman	Bengal gram	Solid cake used as breakfast food	India
Khanomjeen	Rice	Noodles	Thailand
Khaomak	Glutinous rice	Semi solid sweet alcoholic paste	Thailand
Kichudok	Rice, *takju*	Steamed cake	Korea
Kimchi	Cabbage, vegetable, nuts	Solid used as condiment	Korea
Kinema	Soybeans	Solids used as snack	Nepal, Sikkim and Darjeeling
Kishk	Wheat, milk	Solid dried balls eaten with soups	Egypt, Syria and Arab world
Kisra	Sorghum flour	Thin pancake used as staple	Sudan
Koko	Maize, sorghum	Porridge	Ghana
Koozhu	Millet	Breakfast food	India
Koumiss	Mare's milk	Alcoholic milk used in pulmonary tuberculosis	Russia
Kulcha	White wheat flour	Flat bread used as staple food	Northern India
Kunuzaki	Millet	Non alcoholic fermented beverage	Northern parts of Nigeria
Lafun	Cassava root	Paste used as staple food	West Africa, Nigeria
Lao-chao	Rice	Soft, juicy, glutinous food eaten as dessert	China, Indonesia

(continued)

Table 1 (continued)

Fermented product	Substrate	Nature of product/Use	Country/Area
Lambic beer	Barley	Alcoholic beverage	Belgium
Mahewu	Maize	Sour, nonalcoholic drink	South Africa
Mantou	Wheat flour, meat and vegetables	Steamed bread filled with meat, vegetables, etc. and eaten as staple food	China
Marchu	Wheat flour	Deep fried *roties* used as breakfast food	India
Masauyra	Black gram	Ball-like hollow product consumed as spicy condiment	Nepal
Me	Rice	Sour food ingredient	Vietnam
Meitauza	Soybean cake	Solids fried in oil or cooked with vegetables	China, Taiwan
Meju	Black soybean	Paste used as seasoning agent	Korea
Merissa	Sorghum	Liquid drink	Sudan
Mesu	Tender bamboo shoot	Pickle	Nepal
Minchin	Wheat gluten	Solids used as condiment	China
Mirin	Rice and *koji*	Clear liquid used as seasoning agent	Japan
Miso	Rice and soybeans/barley	Thick paste used as soup base	Japan, China
Munkoyo	Millet, maize or *kaffir* corn and roots of munkoyo (*Rhynchosia heterophylla*)	Liquid drink	Africa
Nam	Meat	Fermented meat sausage	Thailand
Natto	Soybeans	Moist, mucilaginous product used as meat substitute	Northern Japan
Nukamiso-zuke	Rice bran, vegetables	Vegetables fermented in rice bran and salt water	Japan
Ogi	Maize	Paste, porridge eaten for breakfast, weaning babies	Nigeria, West Africa
Ontjom (oncom)	Peanut press cake	Solids roasted in oil used as meat substitute	Indonesia
Opo	Rice	Cream coloured alcoholic beverage	Northeastern India
Pak-gard-dong	Mustard leaves	Pickled leafy vegetable	Thailand
Peujeum	Banana, cassava roots	Acidic product with alcoholic flavour, eaten as such or after baking	Java
Pitha	Legumes and cereals	Fermented cakes	India
Pito	Guinea corn or maize	Liquid drink	Nigeria
Poi	Taro corms	Side dish with fish and meat	Hawaii
Pozol	Maize	Beverage or porridge taken as basic food	Southeastern Mexico

(continued)

Table 1 (continued)

Fermented product	Substrate	Nature of product/Use	Country/Area
Prahoc	Fish	Paste used as flavour enhancer	Cambodia
Pulque	Cacti (*Agave atrovirens*)	Milky viscous beverage	Mexico
Puto	Rice	Solid as snack food	Philippines
Rabadi	Maize and buttermilk	Semi solid product eaten with vegetables	India
Rakshi	Rice, barley	Alcoholic beverage	Nepal
Sake	Rice	Alcoholic beverage	Japan
Salami	Meat	Fermented and air dried sausage	All over world
Seera	Wheat grains	Starchy food consumed during fast	India
Selroti	Rice flour	Deep fried, spongy, pretzel-like product commonly consumed as confectionery	Nepal
Shaosingjiu	Rice	Alcoholic beverage	China
Shamsy bread	Wheat flour	Spongy bread used as staple food	Egypt
Sher/shergun	Buttermilk	Soft cheese eaten with rice and bread	Nepal
Shrikhand	Curd	Sweet dish	Western and Southern India
Sierra rice	Unhusked rice	Brownish yellow solid used as seasoning agent	Ecuador
Sinki	Leaves of *Raphnus sativus*	Fermented sun dried product	Nepal
Sinnamani	Radish stem	Pickle	Nepal
Sufu	Soybean whey curd	Paste used as soybean cheese, condiment	China, Taiwan
Sujen	Rice	Alcoholic beer	Assam (India)
Sura	Millet flour	Alcoholic drink	India
Taette	Milk	Fermented milk used as beverage	Scandinavian Peninsula
Takju	Rice	Turbid liquid consumed as beverage	Korea
Tao-si	Soybeans and wheat flour	Semi solid, seasoning agent	Philippines
Taotjo	Soybeans, roasted wheat or glutinous rice	Semi solid used as condiment	East Indies
Tama	Succulent bamboo shoot	Bamboo shoot product with sour-acidic taste eaten as pickle	Nepal, Bhutan
Tape ketan	Cassava or rice	Soft solid eaten fresh as staple	Indonesia and vicinity

(continued)

Table 1 (continued)

Fermented product	Substrate	Nature of product/Use	Country/Area
Tapuy	Rice, glutinous rice	Sour sweet liquid or paste	Philippines
Tarhana	Parboiled wheat meal and yogurt	Dried cakes or powder used as seasoning for soups	Turkey
Tauco	Soybeans and cereals	Thick paste used as condiment	West Java
Teekwass	Tea	Fermented tea used as beverage	China
Tempeh	Soybeans	Solids fried in oil used as meat substitute	Indonesia, Surinam
Tesguino	Germinated maize or maize stalk juice	Thick paste, alcoholic beverage	Mexico
Thumba	Millet	Mildly alcoholic beverage	West Bengal, Nepal
Tibi	Fruits	Culture of bacteria and yeast held in a polysaccharide matrix formed by fruits	Mexico
Toddy	Sap from coconut, *palmyrah* and *kithu*l palm	Alcoholic beverage	Sri Lanka, Maldives
Togwa	Cassava, maize, sorghum and millet	Fermented gruel or beverage	Tanzania
Torani	Rice	Liquid used for seasoning for vegetables	India
Torshi felfel	Sweet peppers	Fermented sweet peppers	West Asia and Africa
Ugba	African oil bean	Solids used as flavoring agent	Nigeria, West and Central Africa
Uji	Maize, sorghum and mil	Semi solid used in breakfast	Kenya, Uganda and Tanzania
Vadai	Black gram	Deep fried patties used as snack food	India

Compiled from: Beuchat (1983), Padmaja and George (1999), Beuchat (2000), Adams and Moss (1996), Dahal et al. (2005) Mugula et al. (2003), Savitri and Bhalla (2007) *doenjang* and *tauco* are very popular traditional fermented food products of the oriental countries and used in soups and stews as a base to improve flavour (Shieh and Beuchat 1982). Peanuts and locust beans are also fermented in some parts of the world to prepare fermented products like *ontjom* and *dawadawa. Ontjom, tempeh, bongkrek,* etc. are popular fermented foods of Indonesia which are considered as meat analogues (Steinkraus 1978). *Natto/hamanatto* is a popular soybean based Japanese fermented food. *Sufu* is a mould fermented soybean curd mostly prepared in China.

In addition to legumes, various cereals based traditional fermented foods and beverages are prepared and consumed all over the world. *Bhatooru, chilra, seera, siddu, gulgule, marchu,* and beverages like *arak, sra, chhang, sura, behmi*, etc. are some of the traditional cereal based fermented products of Himachal Pradesh in India (Savitri and Bhalla 2007). *Lao chao* (slightly alcoholic fermented product), *ang-kak* (fermented red rice) used in the production of red rice wine, *sufu*, fish sauce and red soybean curd are prepared and consumed in China (Wong and Koehler 1981).

Puto is a fermented rice cake made in Philippines from rice by natural fermentation and is somewhat similar to *idli* in India. *Banku* is a fermented starchy food of Ghana prepared from maize or mixture of maize and cassava (Owusu-Ansah et al. 1980). Similar product called *seera* is prepared in rural areas of Himachal Pradesh in India (Savitri et al. 2012). In Nigeria, maize is fermented under natural conditions and used in porridge called as *ogi* or *kenkey. Tape ketan,* a fermented, partially liquefied, mildly alcoholic rice paste having sweet and sour taste is a very popular dish in Indonesia. *Kinema, gundruk, sinki, tama, dahi, mohi, sher, shergum, chhurpi, selroti, rakshi, tumba*, etc. are some popular fermented foods and beverages of Nepal, Bhutan and several parts of India (Thapa et al. 2003; Dahal et al. 2005).

5 Role of Fermentation

The food fermentation exploits the ability of growth of microorganisms on various substrates for the production of fermented foods and beverages and serves several functions. The benefits of fermentation in production of traditional foods and beverages (Joshi and Pandey 1999; Battcock and Azam-Ali 1998) are summarized below:

i. It adds variety to diets with the development of various flavors, aroma and textures in food substrates.
ii. Apart from improving digestibility of foods, the fermentation augments the content of vitamins, proteins and amino acids which enhance the nutritional value of a food.
iii. Some metabolites produced during fermentation have inhibitory action on spoilage/pathogenic organisms and thus helps in preservation of foods. It

prolongs the shelf life of food without refrigeration or use of other food preservation techniques and thus it is a cost effective and energy saving method of preservation of perishable food materials.

iv. It converts waste or cheap raw materials into valuable products.
v. It detoxifies or removes toxins and anti-nutritional compounds present in some fruits and vegetables.
vi. It improves economic, cultural and social well-being of millions of people around the world.

6 Yeasts in Traditional Fermentations

A wide range of yeast species are involved in the preparation of traditional fermented foods and beverages. The foods in which yeasts prevail ranges from leavened bread-like products such as *nan* and *idli*, to alcoholic beverages such as rice and palm wines, and condiments such as *papads* and soy sauce. Most of the indigenous/traditional fermented foods and beverages are invariably prepared at home level by rural, tribal or ethnic people and thus their preparation largely remained as household art. However, the preparation of some traditional fermented products such as soy sauce or sake is carried out on a large commercial scale (Bol and de Vos 1997). Yeasts either alone or often in association with lactic acid bacteria come across during the production of fermented foods and beverages (Gobbetti 1998). The establishment of yeast population in these foods mostly depends on the processing conditions and the presence of preexisting microorganisms in raw materials at the beginning of fermentation. The most common yeast species involved in preparation of traditional fermented cereal based foods and beverages is *Saccharomyces cerevisiae* (Jespersen 2003). A list of yeast species isolated from traditional fermented foods and beverages is given in Table 2.

As evident from Table 2, yeasts dominate among the microbes involved in the traditional fermented foods and beverages of the world. These play a significant role in providing safety and imparting organoleptic properties to food (Foligné et al. 2010). The genera of yeasts reported from fermented foods and beverages comprise *Brettanomyces* (*Dekkera*), *Candida*, *Cryptococcus*, *Debaryomyces*, *Galactomyces*, *Geotrichum*, *Hansenula*, *Hanseniaspora* (*Kloeckera*), *Hyphopichia*, *Kluyveromyces*, *Metschnikowia*, *Naumovozyma*, *Pichia*, *Rhodotorula*, *Saccharomyces*, *Saccharomycodes*, *Saccharomycopsis*, *Schizosaccharomyces*, *Torulospora*, *Torulopsis*, *Trichosporon*, *Yarrowia*, and *Zygosaccharomyces* (Kurtzman and Fell 1998; Pretorius 2000; Romano et al. 2006, Tamang and Fleet 2009). *S. cerevisiae* is the predominant yeast in Ghanaian *pito*, whereas *Candida* sp., *Candida tropicalis*, *Kluyveromyces* sp. and *Torulaspora delbrueckii* are observed in this traditional food product (Demuyakor and Ohta 1991; Sefa-Deheh et al. 1999).

Similarly, *S. cerevisiae*, *Saccharomycopsis fibuligera*, *Pichia kudriavzevii*, *Candida tropicalis*, *Cryptococcus laurentii* and *Torulospora delbrueckii* have been

Table 2 Some important yeast species isolated from various traditional fermented foods and beverages

Yeast	Fermented food	Reference
Candida boidini	*Dosa*	Sandhu and Waraich (1984)
Candida cacaoi	*Idli*	Sandhu and Waraich (1984)
Candida curvata	*Warri*	Sandhu and Waraich (1984)
Candida etchellsii	*Miso,* soysauce	Sandhu and Waraich (1984)
Candida famata	*Warri*	Sandhu and Waraich (1984)
Candida fragicola	*Idli*	Sandhu and Waraich (1984)
Candida glabrata	*Idli, dosa, kodo ko jaanr*	Venkatasubbaiah et al. (1985) Thapa and Tamang (2006)
Candida guillermondii	*Injera, torani*	Padmaja and George (1999)
Candida kefyr	*Idli*	Sandhu and Waraich (1984)
Candida krusei	Punjabi *warri,* cocoa beans	Sandhu and Soni (1989)
Candida lactose	*Tape ketan*	Cronk et al. (1977)
Candida melinii	*Tape ketan*	Cronk et al. (1977)
Candida menbranafaciens	*Bhallae*	Sandhu and Waraich (1984)
Candida milleri	Sourdough	Stolz (2003)
Candida parapsilosis	*Tape ketan*, Punjabi *warri*	Cronk et al. (1977) Sandhu and Waraich (1984)
Candida pseudotropicalis	*Idli, dosa*	Sandhu and Waraich (1984)
Candida sake	*Idli*	Venkatasubbaiah et al. (1985)
Candida tropicalis	*Torani, idli, pito, jau chhang*	Venkatasubbaiah et al. (1985), Sefa-Deheh et al. (1999), Thakur et al. (2015)
Candida vartiovaarai	*Warri*	Sandhu et al. (1986)
Candida vini	*Poi*	Padmaja and George (1999)
Cryptococcus sp.	*Seera*	Savitri et al. (2012)
Debaryomyces hansenii	*Warri*	Sandhu et al. (1986)
Debaryomyces tamari	*Idli, warri*	Sandhu and Waraich (1984)
Endomycopsis fibuliges	*Thumba, tape*	Padmaja and George (1999)
Hansenula anomala	*Idli*, Punjabi *warri, kanji, kecap, torani*	Soni and Sandhu (1999)
Hansenula malanga or *H. subpelliculosa*	*Tape ketan*	Cronk et al. (1977)
Hansenula polymorpha	*Dosa*	Sandhu and Waraich (1984)
Issatchenkia torricola (Pichia terricola, Saccharomyces terricolus)	*Dosa, idli*	Sandhu and Waraich (1984)
Kluyveromyces marxianus	*Warri*	Soni and Sandhu (1990)

(continued)

Table 2 (continued)

Yeast	Fermented food	Reference
Pichia anomala	*Kodo ko jaanr*	Thapa and Tamang (2006)
Pichia kudriavzevii	*Sura*	Thakur et al. (2015)
Pichia membranifaciens	*Warri*	Sandhu et al. (1986)
Rhodotorula flava	*Doenjang*	Chang et al. (1977)
Saccharomyces bayanus	*Jalebies*	Padmaja and George (1999)
Saccharomyces bisporus	*Teekwass*	Hesseltine (1983)
Saccharomyces carbajali	*Pulque*	Steinkraus (1998)
Saccharomyces cerevisiae	Beer, *burukutu*, cider, *fufu, ogi, puto, dosa, idli, papdam, lao chao*, scotch whiskey, *pito, bhatooru*, rice *chhang*	Padmaja and George (1999), Batra and Milner (1974), Soni and Sandhu (1990), Demuyakor and Ohta (1991), Savitri and Bhalla (2012), Thakur et al. (2015)
Saccharomyces intermedium	*Tibi*	Jay (1991)
Saccharomyces rouxii	*Miso*, Punjabi *warri*, Soysauce	Batra (1981), Wang and Hesseltine (1982), Yokotsuka (1960)
Saccharomyces sake	*Sake* (rice beer)	Jay (1991)
Saccharomyces soyae	*Miso*	Winarno et al. (1977)
Saccharomycopsis fibuligera	*Tape ketan, lao chao, kodo ko jaanr, dheli*, rice *chhang*	Hesseltine (1983), Thapa and Tamang (2006), Thakur et al. (2015)
Saccharomycopsis malanga	*Tape ketan, lao chao*	Hesseltine (1983)
Schizosaccharomyces pombe	*Teekwass, traditional African beverages*	Hesseltine (1983)
Torula sp.	*Kefir, kumiss*	Jay (1991)
Torulopsis candida	*Idli*	Batra and Milner (1974)
Torulopsis dattila	*Doenjang*	Chang et al. (1977)
Torulopsis etchellsii	*Miso*	Padmaja and George (1999)
Torulopsis holmii	*Idli*	Venkatasubbaiah et al. (1985)
Torulospora delbrueckii	*Seera*	Savitri et al. (2012)
Zygosaccharomyces soyae	*Tauco*	Winarno et al. (1977)

isolated from traditional fermented foods and beverages (*seera, jau chhang,* rice *chhang, dheli, sura*) of Himachal Pradesh in India (Savitri et al. 2012; Thakur et al. 2015). *S. cerevisiae* var. *sake* along with other yeast strains is used in sake (traditional alcoholic drink of Japan) fermentation along with *Aspergillus oryzae*. The fungus saccharifies rice starch while yeast converts the resultant sugars into ethanol

(Jay 1991; Azumi et al. 2001). This yeast possesses many characteristics such as good aroma and the production of a high concentration of ethanol at low temperature which makes it appropriate for sake brewing (Hosaka et al. 1998).

6.1 Yeasts in Traditional Milk Based Fermented Products

Most of the milk based fermented products are prepared using lactic acid bacteria (LAB). The growth and interaction of yeasts and LAB in traditional fermented milk products result in the production of different levels of metabolites which improve the quality and flavor of these products. However, the role of yeasts in development of flavour and texture in cheese and in the production of traditional fermented products such as *kefir* and *koumiss* is very well explored (Fleet 1990; Frohlich-Wyder 2003; Marshall et al. 1984). The low pH of fermented milk favours the growth of yeasts which is unsuitable for most bacteria (Fleet 1990; Rohm et al. 1992; Deak and Beuchat 1996). Among the yeasts, *S. cerevisiae* was reported to be the most common isolate in indigenous fermented milk prepared in Zimbabwe and Uganda (Narvhus and Gadaga 2003). *S. cerevisiae* together with *Geotrichum candidum* and *Kluyveromyces marxianus* have been reported as the predominant yeasts in the Ugandan traditional fermented milk called *makamo*, however other yeasts such as *Candida holmii, Naumovozyma dairenensis, Candida stellata* and *Zygosaccharomyces* spp. were also reported (Sserunjogi 1999).

Debaryomyces hansenii, Torulaspora delbrueckii and *K. marxianus* var. *marxianus* and its anamorph, *Candida kefyr* were the main yeast species found in traditional fermented milk (Loretan et al. 1998, Fleet 1990). *Saccharomyces unisporus* also called as *Kazachstania unispora* is found to be the dominant yeast in traditional dairy products (Lu et al. 2004; Rahman et al. 2009; Yildiz 2010). This yeast being a slow producer of ethanol carries a clean fermentation in milk and alongwith *K. marxianus* has been observed in a diversity of fermented milk products, such as *dahi, gariss, kefir, koumiss, shubat* and *suusac* (Montanari et al. 1996; Narvhus and Gadaga 2003; Abdelgadir et al. 2008; Rizk et al. 2008). *K. marxianus* (lactose fermenting yeast) was isolated as predominant yeast from *Airag* (fermented milk of mare prepared and consumed in Mongolia). Non lactose fermenting yeasts such as *Saccharomyces cerevisiae, Issatchenkia orientalis* and *Kazachstania unispora* were also reported (Watanabe et al. 2008) from *Tarag* (traditional fermented milk of cows, yaks, goats, or camels from Mongolia).

6.2 Yeast in Traditional Fermented Cereals

Most of the fermented foods and beverages all over the world are made using cereals as raw material. Yeasts are well associated with cereal based fermented products such as bread, beer, wine and other traditional alcoholic beverages.

A number of yeast species along with bacteria, are involved in the production of cereal based traditional fermented foods and beverages in Asia, Africa, and Latin America (Nout 2003). *S. cerevisiae* has been reported from many traditional fermented cereal based foods such as *nan, bhatooru, seera, kulcha*, *chilra,* etc. (Savitri and Bhalla 2012; Savitri et al. 2012; Soni and Sandhu 1990). Many species other than those of *Saccharomyces* e.g. *Candida, Debaryomyces, Hansenula, Kluyveromyces, Pichia, Rhodotorula, Saccharomycopsis, Torulopsis* and *Trichosporon* have been reported in cereal based fermented foods (Sandhu et al. 1986). The yeasts, *Geotrichum candidum*, *Torulopsis holmii*, *Torulopsis candida* and *Trichosporon pullulans* have also been implicated with *idli* fermentation (Chavan and Kadam 1989). In preparation of *jalebies* (a sweet dish of India), the yeast *S. bayanus* was found to be involved in fermentation of wheat/rice flour slurry, where it hydrolyses starch into maltose and glucose by producing extracellular amylolytic enzymes vis-a-vis metabolises sugars to produce carbon dioxide. Growth and metabolic activities of yeasts in the fermented products increase the levels of B vitamins and free amino acids (Venkatasubbaiah et al. 1985).

Yeasts other than *S. cerevisiae* are also observed to drive sourdough fermentation alongwith lactic acid bacteria, where they impact the flavour and rheology of the product. Sourdough is a dough containing lactic acid bacteria in symbiotic association with yeasts. The prevalence of yeasts or bacteria in sourdough depends upon the properties of dough e.g. dense dough promotes the growth of lactic acid bacteria, whereas, soft dough promotes the development of yeast (Gobbetti et al. 1995). Table 3 shows the different yeast species isolated from sourdough. Prominent yeasts in sourdough fermentations include *Candida boidinii, C. guillermondii, C. humicola, C. milleri, S. chevalieri, S. exiguus*, *T. delbrueckii* and various *Pichia* species (Spicher and Stephan 1993; Jenson 1998; Meroth et al. 2003; Maloney and Foy 2003 and Hammes et al. 2005). In sourdough, yeasts produce carbon dioxide and play a significant role in leavening, while the lactic acid bacteria are responsible for its souring. In addition to this, yeast also produces a variety of metabolites such as alcohols, acids, aldehydes and esters which impart specific and unique flavours to bread (Makoto et al. 1990).

In traditional Greek wheat sourdoughs, *S. cerevisiae*, *Pichia membranifaciens* or *Yarrowia lipolytica* were found to be predominant yeast species carrying out the fermentation (Paramithiotis et al. 2000). *Bhaturas* and *kulchas* of northern India and Pakistan are prepared from similar leavened doughs fermented by yeasts and lactic acid bacteria (Sandhu et al. 1986).

The yeasts play key role in the fermentation of traditional cereal based beverages. *Dekkera* (*Brettanomyces*) spp., *S. pastorianus* and *S. cerevisiae* have been reported to be involved in the production of some types of beer (Dufour et al. 2003), while *Schizosaccharomyces pombe* was found to be an important yeast involved in the fermentation of sugarcane/molasses leading to the production of rum (Fahrasmane and Ganou-Parfait 1998).

Table 3 Yeast species identified in traditional sourdoughs (Maloney and Foy 2003)

Yeast species	Sources
Candida boidinii	Rye and wheat sourdough
Candida guillermondii	Wheat sourdough
Candida humicola	Wheat sourdough
Candida krusei	Wheat sourdough
Candida milleri	Wheat sourdough
Candida norvegensis	Wheat sourdough
Hansenula anomala	Wheat sourdough
Hansenula subpelliculosa	Wheat sourdough
Pichia saitoi	Wheat sourdough
Pichia salvi	Rye sourdough
Pichia membranifaciens	Greek wheat sourdough
Saccharomyces cerevisiae	Rye, corn and wheat sourdough
Saccharomyces chevalieri	Rye and wheat sourdough
Saccharomyces curvatus	Rye and wheat sourdough
Saccharomyces exiguous	Rye and wheat sourdough
Saccharomyces fructuum	Wheat sourdough
Saccharomyces inusitalus	San Francisco wheat sourdough
Saccharomyces panis fermentati	Rye and wheat sourdough
Torulospora delbrueckii	Corn and rye sourdough
Yarrowia lipolytica	Greek wheat sourdough

6.3 Yeasts in Alcoholic Fermentation

The involvement of yeasts in the production of fermented alcoholic beverages is found in the earliest human records. The low pH and chemical composition of the substrate itself favours the selection of yeasts and species of *Saccharomyces* predominates during the course of alcoholic fermentation. Yeasts transform the sugars present in the substrate to ethanol and other metabolites. It also produces a number of volatile molecules, such as aldehydes, esters, organic acids, higher alcohols and carbonyl compounds responsible for sensory characteristics of alcoholic beverages (Hazelwood et al. 2008).

Korean alcoholic beverages (*yakju*) are made using *nuruk*, a traditional Korean fermentation starter, which enables the hydrolysis and fermentation of the raw materials to produce alcohol and other flavoring compounds imparting the final product a pleasant aroma and taste (Jin et al. 2007). The metabolites of yeasts responsible for particular flavor or aroma in the traditional Korean alcoholic beverages include ethyl caproate (exhibiting an apple like flavor), isoamyl alcohol (producing a banana like flavor) and 2-phenyl alcohol (which is a rose like essence) (Asano et al. 1999). *Pichia, Candida, Hanseniaspora, Hansenula* etc. have also been isolated from these traditional alcoholic beverages which produce metabolites, such as acetic acid, esters and acetoin (Ho et al. 2013).

Sm. fibuligera, S. cerevisiae, Pichia sp. and *Candida* sp. have also been reported to be associated with rice based alcoholic beverage, *bhaati jaanr* of North East Himalayas by Tamang and Thapa (2006). The prevalence of yeasts in this beverage is implicated with their presence in the starter i.e. *murcha* used for the preparation of beverage (Tsuyoshi et al. 2005). The yeasts isolated from fermented beverages viz. *chhang, jau chhang* and *sura* and starter (*phab*) from Himachal Pradesh belonged to genera *S. cerevisiae, Sm. fibuligera, P. kudriavzevii* and *C. tropicalis. S. cerevisiae* and *Sm. fibuligera* were the predominant yeast species commonly associated with most of the beverages prepared using *phab* as a starter (Thakur et al. 2015).

Rice wines ranging from simple Thai rice wine to highly sophisticated Japanese *sake* are prepared by the hydrolysis of starches and other polysaccharides present in cereals to fermentable sugars with simultaneous fermentation of the latter to alcohol. The main yeasts which ferment saccharified rice starch to alcohol in *sake* are *Pichia burtonii, Saccharomycopsis fibuligera, S. cerevisiae, Candida glabrata* and *Candida lactose* (Tsuyoshi et al. 2005). Other yeast species viz. *Hansenula* spp., *Pichia* spp. and *Torulopsis* spp. have also been isolated from rice wines. The role of *Sm. fibuligera* in these alcoholic beverages is two-fold i.e. being amylolytic yeast, it breaks glycosidic bonds of starch (Dansakul et al. 2004) and also produce ethanol though at low levels (Limtong et al. 2002). In addition, some yeast species such as *Zygosaccharomyces* and *Brettanomyces* cause spoilage of wine (Loureiro and Malfeito-Ferreira 2003).

Boza is a Turkish traditional alcoholic beverage prepared by fermenting cooked maize, wheat and rice flours with yeast and lactic acid bacteria. *Saccharomyces uvarum* and *S. cerevisiae* have been isolated and identified in *boza* (Hancioğlu and Mehmet 1997). Major yeast species identified during cider fermentation are *Metschnikowia pulcherrima, P. anomola, B. anomalus, B. bruxellensis, Debaryomyces polymorphus, H. uvarum, P. fermentas, P. guilliermondii, Saccharomycodes ludwigii* and *S. cerevisiae* (Morrissey et al. 2004).

Sanni and Lönner (1993) observed various species of yeast with a dominance of two or more species of *Candida* spp., *Geotrichum candidum, S. cerevisiae, Kloeckera apiculata* or T. *delbrueckii* in Nigerian sorghum beer, however, the prevalence of yeast was dependent on the type of beer.

6.4 Yeasts in Fermentation of Traditional Fermented Meats

Fermentation is a crucial stage in the curing and preservation of meat based products because it is responsible for major physical, biochemical and microbiological transformations in the meats. The most frequently isolated yeasts from traditional sausages of southern Italy comprise species of *Debaryomyces, Candida, Rhodotorula* and *Yarrowia. Candida* (*C. intermedia* or *C. curvata, C. parapsilosis, C. zeylanoides*), *Citeromyces matritensis, Cryptococcus, Debaryomyces hansenii, Debaryomyces kloeckeri*. Several species of *Pichia, Rhodotorula, Trichosporon ovoides* and *Yarrowia lipolytica* ferment meat sausages and immensely contribute to the development of flavor and colour in the fermented meat products (Encinas et al. 2000; Bolumar et al. 2003 and Samelis and Sofos 2003). The enzymes viz. lipase and protease secreted by these yeasts contribute to the development of flavour in the products by neutralizing the drop in pH due to the degradation of lipids to free fatty acids and glycerol and also catalyses the breakdown of nitrogenous compounds to corresponding amino acids. Studies on the isolation and characterization of enzymes from *D. hansenii* which are involved in imparting flavor and aroma to fermented meats have been undertaken by several researchers (Bolumar et al. 2003; Durá et al. 2002). There are several reports on the yeast populations in various meat products, but information on yeast biodiversity in traditional fermented meat products is scanty.

6.5 Yeast in Traditional Soft Beverages

Kombucha is a traditional fermented beverage made by the fermentation of sweetened black tea by acetic acid bacteria and yeast for 10–12 days (Anken and Kappel 1992). The association of yeasts and acetic acid bacteria is given the term 'tea fungus' where the cellulosic pellicle formed by the bacteria is called as the 'fungus' (Kappel and Anken 1993; Timmons 1994; Steinkraus 1996). The yeasts ferment the sugar in tea to ethanol, which is consequently oxidised to acetic acid by bacteria. The low pH resulted by the production of acids and antimicrobial metabolites produced during fermentation inhibit the growth of other bacteria, yeasts and filamentous fungi. Species of *Brettanomyces/Dekkera, Candida, Kloeckera, Pichia, Saccharomyces, Saccharomycoides, Schizosaccharomyces, Torulospora* and *Zygosaccharomyces* have been reported to be associated with the traditional *kombucha* fermentation (Jankovic and Stojanovic 1994; Frank 1995; Mayser et al. 1995; Liu et al. 1996).

6.6 Yeasts in Traditional Soy Sauce Fermentation

In the early stages of soy sauce fermentation, halophilic *Pediococcus halophilus* play a pivitol role in converting sugars to lactic acid and thus lower the pH of the product and subsequently osmotolerant yeast species e.g. *C. versatilis*, *C. etchellsii Zygosaccharomyces rouxii*, *Torulopsis* drive the fermentation (Hanya and Nakadai 2003). Yeasts generate some metabolites e.g. ethanol, ethyl acetate, 4-ethylguiaicol, 4-ethylphenol, furanones and pyrazones which are responsible for development of typical flavor in soy sauce (Bamforth 2005).

7 Traditional Yeast Starters

In fermentation, microorganisms or enzymes secreted by them convert the raw ingredients to products with improved characteristics. Most of the traditional fermentation processes are spontaneous fermentations that are initiated without the use of a starter or inoculum and have been applied to food preservation since antiquity. However, the natural microfloras of the raw materials are inefficient, uncontrollable, and unpredictable, or is inactivated by the heat treatments given to the raw materials prior to fermentation. There are reports of use of starter cultures (bacteria, yeast and moulds) in food fermentations to bring about preferred changes such as addition of new functions, increased shelf life, improved nutritional or health promoting benefits and better sensory qualities making the fermented product to fetch much higher price than the raw material (Kolawole et al. 2007). Since the beginning of the 1980s, the practice of using *S. cerevisiae* yeast starters has largely become common in the industrial and homemade beverage production processes. Srivastava et al. (1997) have documented that the wine made with selected yeast starter has a better quality than wines made by spontaneous fermentation. Studies on the microflora present in the starters such as Chinese *chiu-chu* and Indonesian *ragi* have been undertaken as early as towards the end of 19th century. The principal yeasts in many of the commercially available starters are *Candida* spp., *Hansenula* spp., *Sm. fibuligera* and *S. cerevisiae* (Saono et al. 1996). Starter cultures are selected either as single organism or multiple organisms based on the suitability or adaptation of microorganisms to raw material. A wine starter culture of yeast is usually able to dominate intrinsic yeasts in the grape must during fermentation (Pretorius 2000) as it complement the quality of raw material and some special characteristics of the wine, making the wine a more desirable, acceptable and valued product (Swiegers et al. 2005).

The starter cultures of yeasts used in wine making invariably produce mycocin which prevent the growth of spoilage yeasts and secondary fermentation in wines (Boone et al. 1990; Comitini et al. 2004). Calmette (1892) was the first to report the presence of several wild yeast species alongwith moulds such as *Amylomyces*, *Mucor*, *Aspergillus* and some bacteria in starters used in India and China to produce alcohol.

Traditional fermentation starters are referred to as *chu* in Chinese, *nuruk* in Korean, *koji* in Japanese, *ragi* in Southeast Asian countries, *bakhar ranu* or *marchaar/murcha* in North East India (Batra and Millner 1974) and *phab* (Fig. 1) in North Western Himalayan region of India (Thakur et al. 2004). *C. glabrata*, *P. anomala, P. burtonii, S. bayanus*, *Sm. fibuligera* and *Sm. capsularis* have been isolated from *murcha* which hydrolyse starch and subsequently produce alcohol (Tsuyoshi et al. 2005). *Hamei* is used in preparation of alcoholic beverages e.g. *bhaati Jaanr*, *aitanga* and *kodo ko jaanr* of Eastern Himalayas (Tamang et al. 2007). *Phabs* is an indigenous inoculum of the North Western part of India and used in fermentation of two traditional barley based alcoholic beverages i.e. *chhang* and *arrak* (Angmo and Bhalla 2014). A study on almost 100 amylolytic yeast

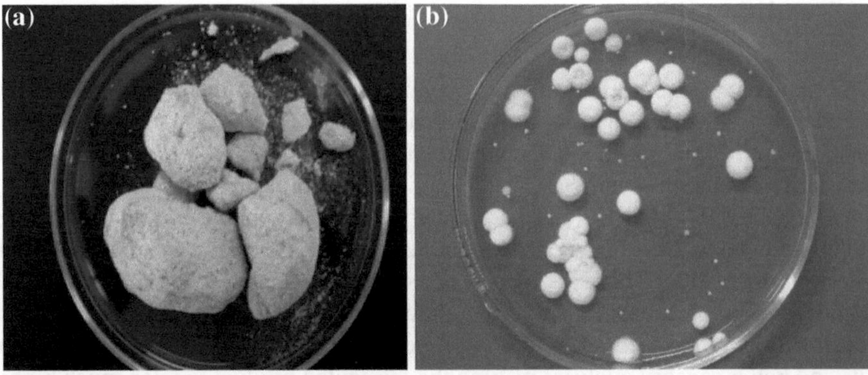

Fig. 1 **a** *Phab* (traditional starter used in preparation of *chhang*) from Himachal Pradesh **b** yeast isolated from *phab*

strains isolated from *ragi* and other starters have shown that *Sm. fibuligera* is the chief amylolytic yeast while *Sm. malanga* is the minor yeast component of *ragi* (Hesseltine and Kurtzman 1990).

In a study of yeast diversity in Thai traditional alcohol starters, *Sm. fibuligera, P. anomala, Issatchenkia orientalis, P. burtonii, P. fabianii, C. rhagii, C. glabrata, Torulaspora globosa, P. Mexicaua, P. heimii, Rhodotorula philyla, S. cerevisiae, T. delbrueckii* and *Trichosporon faecale* have been documented. These have low amylolytic activity but possess high or moderately high ethanol production potential (Limtong et al. 2002; Aidoo et al. 2006).

8 Role of Yeast in Fermentation

In general, yeasts are involved in the fermentation of sugars and in the production of aroma compounds. However, depending on the raw materials and fermentation conditions, yeasts also stimulate the growth of lactic acid bacteria, produce different enzymes viz. lipase, protease, pectinase, glycosidase and urease activities during fermentation which improve the nutritional value of products, inhibit mycotoxin-producing moulds, and degrade cyanogenic glucosides while some strains of yeasts also have probiotic properties (Table 4).

In the fermentation of any substrate, *Saccharomyces* ferments sugar, produces a range of secondary metabolites (Tamang and Fleet 2009), *Debaryomyces* contributes to sugar fermentation, increase the pH of the substrates and produces metabolites promoting growth of bacteria. *Hanseniaspora* and *Candida* ferment sugar, produce secondary metabolites and have some enzymes that have their influence the organoleptic properties of the final products of the fermentation. *Yarrowia lipolytica* plays a role in fermentation of sugar to alcohol, has lipolytic,

Table 4 Role of yeast in production of fermented food and beverages (modified from Romano et al. 2006)

Fermented Food/Beverage	Yeast	Function
Beer, wine, sourdough, bread, cheese, indigenous fermented foods and beverages	*Saccharomyces* sp.	Sugar fermentation and formation of alcohol Improvement of flavor and texture Production of secondary metabolites Inhibitory effect on undesirable microorganisms
Sourdough, cocoa beans, soysauce, indigenous fermented foods and beverages	*Candida* sp.	Production of different enzymes (protease, galactosidase and pectinase) Inhibition of undesirable organisms Secondary metabolite production
Cheese	*Clavispora lusitaniae*	Production of flavor compounds in cheese
Fresh fruits and fermented meat products	*Cryptococcus* sp.	Biocontrol agent against fungal pathogens (some species) Spoilage activity
Wine, sourdough, indigenous fermented foods and beverages	*Hansenula/Pichia* sp.	Production of volatile/aroma compounds in wine Inhibition of various moulds
Cheese	*Torulopsis* sp.	Flavour enhancement
Cheese, salami, Rye sourdough	*Debaryomyces* sp.	Production of flavor compounds in cheese (nutty or malty flavor) Increase in pH Help in ripening of cheese
Cheese, salami	*Yarrowia lipolytica*	Lipolytic and proteolytic activities Reduction in fat rancidity
Fermented meat products, wine	*Rhodotorula* sp.	Lipolytic activities Production of carotenoids Spoilage of dairy foods
Cheese, cocoa beans	*Kluyveromyces* sp.	Fermentation of lactose Production of pectinase, lipase and rennet
Soy sauce	*Zygosaccharomyces* sp.	Osmotolerance

proteolytic and urease activities, decreases fat rancidity, and influences the texture and flavor of the final product (Tamang and Fleet 2009).

8.1 Role of Bioactive Compounds of Yeast in Food and Fermented Beverages

The use of biologicals having antifungal and antibacterial activities is becoming popular to inhibit the growth of undesired fungi, yeasts or bacteria during and after the fermentation. The yeasts belonging to the genus *Candida, Cryptococcus, Wickerhamomyces, Tetrapisispora, Kluyveromyces, Pichia, Debaryomyces* and *Metschnikowia* have been explored as biological control agents (Janisiewicz and Korsten 2002). These yeasts along with bacteria inhibit or reduce the growth of filamentous fungi such as *Botrytis* spp., *Aspergillus* spp., *Penicillium* spp. *Monilia* spp. and are employed as natural antifungal in agriculture as an alternative to chemical treatments. Even *S. cerevisiae* which has been popularly associated with bread, beer and wine production is being explored as a biocontrol agent to inhibit the growth of *Aspergillus carbonarius* and *A. ochraceus* which produces ochratoxin A (Cubaiu et al. 2012).

Due to the ease of manipulation and cultivation, yeasts are increasingly used as models in biological control of undesired microorganisms in food and fermented beverages (Pimenta et al. 2009).

The killer toxins are generally proteins or glycoproteins which interfere with the DNA replication (Schmitt et al. 1996), membrane permeability (Santos and Marquina 2004) and the cell cycle in G1 and G2 phases ultimately leading to the death of sensitive cells (Butler et al. 1991). Moreover, in some cases the toxin inhibits the synthesis of β-1, 3 glucans (Takasuka et al. 1995) or hydrolyzes the major components of the cell wall β-1, 3 or 1,6 glucans (Izgü et al. 2011). Apart from fermented foods and beverages, yeasts are also used for the production of enzymes, fine chemicals, single-cell protein, and flavoring compounds (Gatto and Torriani 2004; Wang 2008).

8.2 Yeasts as Source of Food Additives

Increase in consumer demand for more natural foods has triggered academia and industry in the use of microorganisms including yeasts, as sources of food ingredients and additives, such as flavors, colours, antioxidants and vitamins. In food processing industries, yeasts are considered as a safe source of food additives due to their wider acceptability among consumers (Demain et al. 1998). The yeast cell wall is composed mainly of β-(1 → 3) and β-(1 → 6)-glucans that have thickening, gelling and fat lowering characteristics and therefore these biopolymer of yeast origin have potential applications in food processing (Seeley 1977). Since yeasts possess high contents of B vitamins, proteins, peptides, amino acids and trace minerals, baker's and brewer's yeasts have been used as dietary and nutrient supplements and are also considered as alternative source of protein for human consumption (Harrison 1993).

Yeasts are also source of antioxidants, aromas, colours, flavors, glutamic acid, nucleotides and vitamins. The most extensively used and commercially significant products extracted from yeasts include flavor ingredients that impart savoury, roasted, nutty, cheesy, meaty and chicken flavours. In addition, some extracts of yeasts are rich in glutamic acid and nucleotides that act as strong flavour enhancers (Stam et al. 1998). Although, the baker's and brewer's yeasts have been the traditional sources of these products, some other yeasts such as *C. utilis* and *K. marxianus* are also used for this purpose (Lukondeh et al. 2003).

In addition to the above mentioned compounds, yeasts are also reported as potential sources of high value aromatic compounds (Vandamme and Soetaert 2002) such as vanillin (*S. cerevisiae*, *Rhodotorula glutinis*), citronellol, geraniol and linalool (*K. marxianus*), and γ- and δ-decalactones (*Sporidiobolus sulmonicolor*, *Y. lipolytica*). Some yeasts such as *Rhodotorula* sp., *Yarrowia lipolytica*, *Cryptococcus* sp. and *Phaffia rhodozyma* have emerged as good source of pigments (Joshi et al. 2003). Food colorants such as astaxanthin and other carotenoid pigments (Lyons et al. 1993; Johnson and Schroeder 1995; Joshi et al. 2003) and several vitamins (Sauer et al. 2004) can also be obtained from yeasts. Yeast *Xanthophyllomyces dendrorhous* produces astaxanthin which is widely used as food colorants (Mata-Gómez et al. 2014). Species of the genus *Rhodotorula* viz. *R. glutinis*, *R. minuta*, *R. mucilaginosa*, *R. acheniorum* and *R. graminis* have been recognized as potential carotenoid producers. Carotenoids impart yellow to red colors to the foods and thus improve the acceptability of many foods. *Sporobolomyces roseus*, *S. salmonicolor* and *S. patagonicus* are some other yeasts which also produce carotenoid pigments. Most of these yeasts predominantly produce β–carotene however, some other carotenoids e.g. torulene, torularodine, and γ–carotene are also produced by the yeasts (Tinoi et al. 2005; Moliné et al. 2010). Some of these carotenoids serve as precursors of vitamin A and are known to mitigate the risks for degenerative diseases like cancer, cardiovascular disorders, cataract and macular degeneration.

9 Probiotic Yeasts

The lactic acid bacteria have been extensively explored and are being used as the main probiotic organisms, however, interest in probiotic yeasts has immensely increased in these years (Foligné et al. 2010; Hatoum et al. 2012; Živković et al. 2015) especially for application in animal feed. Live preparations of *S. cerevisiae* are in use as feed supplements to improve the growth and health of the animals in general (Lyons et al. 1993) and milk production by cows in particular (Maamouri et al. 2014). The use of yeasts as probiotics in the aquaculture industry has also been reported (Gatesoupe 1995). In humans, *S. cerevisiae* var. *boulardii*, has been successfully used over the last 20 years as an oral, biotherapeutic agent to treat patients suffering with diarrhea and other gastrointestinal diseases (Czerucka et al. 2007). A novel culture medium formulation for large scale biomass production of probiotic yeast *S. boulardii* has been designed by Chin et al. (2015).

The probiotic activities of yeast are mainly based on secretion of proteases and other inhibitory proteins, stimulation of immunoglobulins and elimination of toxins secreted by other microorganisms (Fooks and Gibsen 2002). Some yeast strains isolated from traditional fermented foods of Ethiopia have been reported to exhibit antimicrobial activities against *Listeria monocytogenes*, *Salmonella* spp. and *Staphylococcus aureus* (Mariam et al. 2014). Certain dairy yeast isolates have also been reported to exhibit strain-specific probiotic potential, since they are able to survive in the simulated conditions of the intestinal tract (Živković et al. 2015).

10 Health Significance of Yeasts in Food and Beverage

Humans invariably consume large populations of yeasts as part of fermented foods and beverages without adverse effect on their health. The occurrence of yeasts in foods and beverages is a matter of public health concern as some yeasts could serve as probiotic microorganisms while others may be infectious and have adverse effects on the consumer. Unlike many bacteria and viruses, yeasts are not aggressive infectious agents (Schulze and Sonnenborn 2009). However, several species of yeasts especially *C. albicans* and *Cryptococcus neoformans* are opportunistic pathogens and cause infections in humans (Hazen and Howell 2003). Generally, individuals with weak health and immune function especially persons with AIDS, cancer and those undergoing treatments with immunosuppressive drugs, broad-spectrum bacterial antibiotics, radio and chemotherapies are at higher risks.

11 Use of Genetically Modified Yeasts in Foods and Beverages

Generally yeasts for various purposes are obtained from breweries, distilleries, culture collections and commercial vendors but with the increase in their industrial use and scientific and technical awareness among the yeast users, there is an increase in the demand of yeasts with higher yield of the desired product and better technical characteristics. Classical genetic improvement methods and recombinant DNA technology have been widely used to construct novel yeast strains with following desirable characteristics (Verstrepen et al. 2006, Zhang et al. 2014):

- Improved performance and product quality
- Ability to ferment a wide range of carbohydrates
- Produce products with modified flavors
- Antimicrobial properties
- Altered flocculation properties
- Better oligosaccharide utilization
- Fermentation of branched oligosaccharides and polysaccharides
- Improved stress tolerance
- Improved sensory qualities of fermentation products
- Reduction in diacetyl levels in alcoholic beverages
- Improved flavour profiles of alcoholic beverages

The genetically engineered yeasts have improved production of ethanol and glycerol, control in acid levels in wine; shelf life of alcoholic beverages, antioxidant level and have decreased levels of hydrogen sulfide and ethyl carbamate. Attempts have also been made to alter fermentation rates using genetically modified yeasts which have either higher titre of glycolytic enzymes or their hyperactive forms (Verstrepen et al. 2006).

12 Safety of Yeasts

The impact of yeasts on the quality and safety of traditional fermented foods and beverages is intimately associated with their biological activities. These activities mostly depend on the physical and chemical properties of the ecosystem from where the yeasts have been isolated and the physiology, biochemistry and genetics of yeasts in use. The use of *S. cerevisiae* is associated with food and beverage production over millennia supports the fact that it is safe to work with/consume this yeast and thus declared as Generally Recognized as Safe (GRAS) by the United States' Food and Drug Administration (Verstrepen et al. 2006).

13 Future Outlook

Although considerable progress has been made in isolation, identification and characterization of yeast from traditional fermented foods and beverages, these require more precise characterization by using advanced techniques of genomics, proteomics and metabolomics. Continued research in these fields will further our understanding and eventually lead to the development of yeast strains that can produce predictable levels of products with specific metabolic profiles and thus allow the producers to 'shape' their fermented products to suit consumer preferences and add value to existing fermented foods. This will lead to develop novel fermented products based on the exploitation of new and genetically modified strains of yeasts of fermented food origin.

References

Abdelgadir, W.S., Nennis, D.S., Hamad, S.H.P. and Jakobsen, M. 2008. *Int. J. Food Microbiol.* 127: 215–219.

Adams, M.R. and Moss, M.O. 1996. In: *Food Microbiology* (eds. M.R. Adams and M.O. Moss) 1st edition, New Age International Publishers, New Delhi, pp. 252–302.

Aidoo, K.E., Nout, M.J.R. and Sarkar, P.K. 2006. *FEMS Yeast Res.* 6: 30–39.

Angmo, K. and Bhalla, T.C. 2014. *Indian J. Tradit. Knowl.* 13: 347–351.

Anken, R.H. and Kappel, T. 1992. *Eur. Arch. Biol.* 103: 219–222.

Asano, T., Inoue, T., Kurose, N., Hiraoka, N. and Kawakita, S. 1999. *J. Biosci. Bioeng.* 87: 697–699.

Azumi, M. and Goto-Yamamoto, N. 2001. *Yeast* 18: 1145–1154.

Bamforth, C.W. 2005. In: *Food, Fermentation and Micro-organisms.* Blackwell Publishing Ltd., UK.

Batra, L.R. 1981. In: *Advances in Biotechnology* (eds. M. Moo-Young and C.W. Robinson), Pergamon Press, Toronto, p. 547.

Batra, L.R. and Millner, P.D. 1974. *Mycologia* 66: 942–950.

Battcock, M. and Azam-Ali, S. 1998. *Bulletin of FAO Agricultural Services for United Nations,* Rome, 1–4.

Beuchat, L.R. 1983. In: *Biotechnology: Food and feed production with microorganisms* (ed. G. Reed) Vol. 5 Verlag Chemie, Weinheim, pp. 477–528.

Beuchat, L.R. 2000. In: *Food Microbiology: Fundamentals and Frontiers* (eds. M.P. Doyle, L.R. Beuchat and T.J. Montville) ASM Press, Washington D C, pp. 701–719.

Bol, J. and de Vos, W.M. 1997. In *Biotechnological Innovations in Food Processing* (ed. J. Green), Butterworth-Heinemann, Oxford, pp. 45–76.

Bolumar, T., Sanz, Y., Aristoy, M.C. and Toldrá, F. 2003. *Int. J. Food Microbiol.* 86: 141–151.

Boone, C., Sdicu, A., Degri, R., Sanchez, C. and Bussey, H. 1990. *Yeast* 41: 37–42.

Borgstrom, G. 1968. In: *Food Microbiology and Biochemistry*, Vol 2, Macmillan, New York, pp. 9–15.

Butler, A.R., White, J.H., Stark, M.J.R. 1991. *J. Gen. Microbiol.* 137: 1749–1757.

Calmette, L.C.A. 1892. *Ann. Inst. Pasteur* 6: 604–620.

Chang, C.H., Lee, S.R., Lee, K.H., Mheen, T.I., Kwon, T.W. and Park, K.J. 1977. *Symposium on Indigenous Fermented Foods*, Bangkok, Thailand.

Chavan, J.K. and Kadam, S.S. 1989. *CRC Crit. Rev. Food Sci. Nutr.* 28: 349–400.

Chin, T.S., Othman, N.Z., Malek, R.A., Elmarzugi, N., Leng, O.M., Ramli, S., Musa, N.F., Aziz, R. and Enshasy, H.E. 2015. *J. Chem. Pharm. Res.* 7: 122–132.

Comitini, F., De Ingeniis, J., Pepe, L., Mannazzu, I. and Ciani, M. 2004. *FEMS Microbiol. Lett.* 238: 235–240.

Cronk, T.C., Steinkraus, K.H., Hackler, L.R. and Mattick, L.R. 1977. *Appl. Environ. Microbiol.* 33: 1067–1073.

Cubaiu, L., Abbas, H., Dobson, A.D.W., Budroni, M. and Quirico, M. 2012. *Toxins* 4: 1468–1481

Czerucka, D., Piche, T. and Rampal, P. 2007. *Aliment. Pharmacol. Ther.* 26: 767–778.

Dahal, N.R., Karki, T.B., Swamylingappa, B., Li, Q. and Gu, G. 2005. *Food Rev. Int.* 21: 1–25.

Dansakul, S., Charoenchai, C., Urairong, H. and Leelawatcharamas, V. 2004. *FoodMicro 2004*: 19th International ICFMH Symposium. New tools for improving microbial food safety and quality. Portoroz, Slovenia, Poster 11–12, p. 263.

Deak, T. and Beuchat, L.R. 1996. *Handbook of food spoilage yeasts.* Boca Raton, FL: CRC Press.

Demain, A.L., Phaff, H.J. and Kurtzman, C.P. 1998. In: *The yeasts—a taxonomic study* (eds. C. P. Kurtzman and J.W. Fell), 4th edn. Elsevier, Amsterdam, pp. 13–19.
Demuyakor, B. and Ohta, Y. 1991. *Food Microbiol.* 8: 183–193.
Dufour, J.P., Verstrepen, K. and Derdelinckx, G. 2003. In: *Yeasts in food: Beneficial and detrimental aspects* (eds. T. Boekhout and V. Robert), Behr, Hamburg, pp. 347–388.
Durá, M.A., Flores, M. and Toldrá, F. 2002. *Int. J. Food Microbiol.* 76: 117–126.
Encinas, J.P., López-Díaz, T.M., García-López, M.L., Otero, A. and Moreno, B. 2000. *Meat Sci.* 54: 203–208.
Fahrasmane, L. and Ganou-Parfait, B. 1998. *J. Appl. Microbiol.* 84: 921–928.
Fleet, G.H. 1990. *J. Appl. Bacteriol.* 68: 199–211.
Fleet, G.H. 2006. In: *The Yeast Handbook: Yeasts in Food and Beverages* (eds. A. Querol and G. H. Fleet), Springer-Verlag, Berlin Heidelberg, Germany, pp. 1–12.
Foligné, B., Dewulf, J., Vandekerckove, P., Pignède, G. and Pot, B. 2010. *World J. Gastroenterol.* 16(17): 2134–2145.
Fooks, L.J. and Gibsen, G.R. 2002. *Br. J. Nutr.* 88: 39–49.
Frank, G.W. 1995. *Kombucha*: Healthy Beverage and Natural Remedy from the Far East. Wilhelm Ennsthaler, Austria.
Frohlich-Wyder, M.T. 2003. In: *Yeasts in food: Beneficial and detrimental aspects* (eds. T. Boekhout and V. Robert), Behr, Hamburg, pp. 209–237.
Gatesoupe, F.J. 1995. *Aquaculture* 180: 147–165.
Gatto, V. and Torriani, S. 2004. *Ann. Microbiol.* 54: 31–42.
Gobbetti, M. 1998. *Trends Food Sci. Technol.* 9: 267–274.
Gobbetti, M., Simonetti, M.S., Corsetti, A., Santinelli, F., Rossi, J. and Damiani, P. 1995. *Food Microbiol.* 26: 353–363.
Haard, N.F. 1999. *FAO Agricultural Services Bulletin no. 138. Agricultural Organization of United Nations,* Rome, pp. 15–21.
Hammes, W.P., Brandt, M.J., Francis, K.L., Rosenheim, J., Seitter, M.F.H. and Vogelmann, S.A. 2005. *Trends Food Sci. Technol.* 16: 4–11.
Hancioğlu, Ö. and Mehmet, K. 1997. *Int. J. Food Microbiol.* 35: 271–274.
Hanya, Y. and Nakadai, T. 2003. In: *Yeasts in food: Beneficial and detrimental aspects* (eds. T. Boekhout and V. Robert), Behr, Hamburg, pp. 413–428.
Harrison, J.S. 1993. In: *The yeasts, Yeast Technology* (eds. A.H. Rose and J.S. Harrison), vol 3. Academic, London, pp. 399–434.
Hatoum, R., Labrie, S. and Fliss, I. 2012. *Front. Microbiol.* 3: 1–12.
Hazelwood, L.A., Daran, J.M., Maris, A.J.A.van., Pronk, J. and Dickinson, J.R. 2008. *Appl. Environ. Microbiol.* 74: 2259–2266.
Hazen, K.C. and Howell, S.A. 2003. In: *Manual of clinical microbiology* (ed. P.R. Murray), 8th edn. American Society for Microbiology, Washington, DC, pp. 1693–1711.
Hesseltine, C.W. and Kurtzman, C.P. 1990. *An Inst. Biol. Univ. Nac. Auton. Mexico, Ser. Bot.* 60: 1–7.
Hesseltine, C.W. 1983. *Ann. Rev. Microbiol.* 37: 575–601.
Ho, C.J. Soo, H.Y., Ji-Hye, P. Choi, H.S., Ji-Eun, G., Soo, I.K., Jeong, S.T. and Kim, S.R. 2013. *Agr. Sci.* 4(5B): 136–140.
Hosaka, M., Komuro, Y. and Nakayama, M, *et al.* 1998. *J. Brew. Soc. Japan* 93: 833–840.
Izgü, D.A., Kepekci, R.A. and Izgu, F. 2011. *Antonie van Leeuwenhoek* 99: 85–91.
Janisiewicz, W.J. and Korsten, L. 2002. *Annu. Rev. Phytopathol.* 40: 411–441.
Jankovic, I. and Stojanovic, M. 1994. *Mikrobiol.* 33: 25–34.
Jay, J.M. 1991. In: *Modern Food Microbiology* (ed. J.M. Jay), Van Nostrand, Remheld, New York, pp. 362–407.
Jenson, I. 1998. In: *Microbiology of fermented foods* (ed. B.J. Wood), vol 1, 2nd edn. Blackie, London, pp. 172–198.
Jespersen, L. 2003. *FEMS Yeast Res.* 3:191–200.
Jin, T.Y., Kim, E.S., Eun, J.B., Wang, S.J. and Wang. M.H. 2007. *Korean J. Food Sci. Technol.* 39: 309–314.

Johnson, E.A. and Schroeder, W. 1995. *Studies Mycology* 38: 81–90.
Joshi, V.K. and Pandey, A. 1999. In: *Biotechnology: Food Fermentation, Microbiology, Biochemistry and Technology* (eds. V.K. Joshi and A. Pandey), Vol 1. Educational Publishers and Distributors, New Delhi, pp. 1–24.
Joshi, V.K., Attri, D., Bala, A. and Bhushan, S. 2003. *Indian J. Biotechnol.* 2: 362–369.
Kappel, T. and Anken, R.H. 1993. *The Mycologist* 7: 12–13.
Kolawole, O.M., Kayode, R.M.O. and Akinduyo, B. 2007. *Afr. J. Biotechnol.* 6: 587–90.
Kruif, P.D. 1935. *Mikrobenj ä ger,* Orell Füssli, Zurich.
Kurtzman, C.P. and Fell, J.W. 2005. In: *The Yeast Handbook* (eds. P. Gábor and C.L. de la Rosa), Springer, Berlin, pp. 11–30.
Kurtzman, C.P. and Fell, J.W. 1998. In: *The Yeasts, A Taxonomic Study* (eds. C.P. Kurtzman and J.W. Fell), 4th edn. Elsevier Science, Amsterdam, pp. 3–5.
Limtong, S., Sintara, S., Suwannarit, P. and Lotong, N. 2002. *Kasetsart J.: Nat. Sci.* 36: 149–158.
Liu, C.H., Hsu, W.H., Lee, F.L. and Liao, C.C. 1996. *Food Microbiol.* 13: 407–415.
Loretan, T., Viljoen, B.C., Mostert, J.F., Vogel, A.M. and Jordaan, H.F. du.P. 1998. In: Yeasts in the dairy industry: Positive and negative aspects, (eds. M. Jakobsen, J.A. Narvhus and B.C. Viljoen) *Proceedings of the symposium organized by Group F47*, Brussels, Belgium, pp. 178–182.
Loureiro, V. and Malfeito-Ferreira, M. 2003. *Int. J. Food Microbiol.* 86: 23–50.
Lu, D., Mausel, P., Brondizio, E. and Moran, E. 2004. *Int. J. Remote Sens.* 25: 2365–2407.
Lukondeh, T., Ashbolt, N.J. and Rogers, P.L. 2003. *J. Ind. Microbiol. Biotechnol.* 30: 52–56.
Lyons, T.P., Jacques, K.A. and Dawson, K.A. 1993. In: *The yeasts* (eds. A.H. Rose and J.S. Harrison), vol 5, 2nd edn. Academic, London, pp. 293–325.
Maamouri, O., Selmi, H. and M'hamdi, N. 2014. *Scientia. Agriculturae Bohemica.* 45: 170–174.
Makoto, W., Kazuro, F., Kozo, A. and Shigenori, O. 1990. *Appl. Microbiol. Biotechnol.* 34: 154.
Maloney, D.H. and Foy, J.J. 2003. In: *Handbook of Dough Fermentations* (eds. K. Kulp and K. Lorenz), Marcel Dekker, New York, pp. 45.
Mariam, S.H., Zegeye, N., Tariku, T., Andargie, E., Endalafer, N. and Aseffa, A. 2014. *BMC Res. Notes* 7:606.
Marshall, V.M., Cole, W.M. and Brooker, B.E. 1984. *J. Appl. Bacteriol.* 57: 491–497.
Mata-Gómez L.C., Montañez J.C., Méndez-Zavala A and Aguilar C.N. 2014. *Microb. Cell Fact.* 13:12.
Mayser, P., Gromme, S., Leitzmann, C. and Grunder, K. 1995. *Mycoses* 38: 289–295.
Meroth, C., Hammes, W. and Hertel, C. 2003. *Appl. Environ. Microbiol.* 69: 7453–7461.
Moliné, M., Flores, M.R., Libkind, D., del Carmen, D.M., Farías, M.E. and van Broock, M. 2010. *Photochem. Photobiol. Sci.* 9: 1145–1151.
Montanari, G., Zambonelli, C., Grazia, L., Kamesheva, G.K. and Shigaeva, M.K. 1996. *J. Dairy Res.* 63: 327–331.
Morrissey, W.F., Davenport, B., Querol, A. and Dobson, A.D.W. 2004. *J. Appl. Microbiol.* 97: 647–655
Mugula, J.K., Narvhus, J.A. and Sorhaug, T. 2003. *Int. J. Food Microbiol.* 83: 307–318.
Narvhus, J.A. and Gadaga, T.H. 2003. *Int. J. Food Microbiol.* 86: 51–60.
Nout, M.J.R. 2003. In: *Yeasts in food: Beneficial and detrimental aspects* (eds. T. Boekhout and V. Robert), Behr, Hamburg, pp. 451–473.
Owusu-Ansah, J., Van De Voort, F.R. and Bediako-Amoa, B. 1980. *Can. Inst. Food Sci. Technol.* 13: 131.
Padmaja, G. and George, M. 1999. In: *Food Processing: Biotechnological Applications* (eds. S.S. Marwaha and J.K. Arora), Asiatech Publishers Inc., New Delhi, pp. 143–189.
Paramithiotis, S., Müller, M.R.A., Ehrmann, M.A., Tsakalidou, E., Seiler, H., Vogel, R. and Kalantzopoulos, G. 2000. *Syst. Appl. Microbiol.* 23: 156–164.
Pimenta, R.S., Morais, P.B., Rosa, C.A. and Corrêa, A. 2009. In: *Yeast Biotechnology: Diversity and Applications*, (eds. T. Satyanarayana and G. Kunze), Springer, Dordrecht, Netherlands, pp. 199–214.
Pretorius, I. 2000. *Yeast* 16: 675–729.

Rahman, N., Xiaohong, C., Meiqin, F. and Mingsheng, D. 2009. *World J. Microbiol. Biotechnol.* 25: 1941–1946.

Rizk, S., Maalouf, K. and Baydoun, E. 2008. *Clin. Lymphoma Myeloma* 9: 198–203.

Rohm, H., Eliskases-Lechner, F. and Brauer, M. 1992. *J. Appl. Bacteriol.* 72: 370–376.

Romano, P., Capace, A. and Jespersen, L. 2006. In: *The yeast Handbook—Yeasts in Food and Beverages* (eds. A. Querol and G.H. Fleet), Springer-Verlag, Germany, pp. 13–53.

Samelis, J. and Sofos, J.N. 2003. In: *Yeasts in food: Beneficial and detrimental aspects* (eds. T. Boekhout and V. Robert), Behr, Hamburg, pp. 239–266.

Sandhu, D.K. and Soni, S.K. 1989. *J. Food Sci. Technol.* 26: 21–25.

Sandhu, D.K. and Waraich, M.K. 1984. *Kavaka* 12: 73–85.

Sandhu, D.K., Soni, S.K. and Vilkhu, K.S. 1986. In: *Yeast Biotechnology* (eds. R.K. Vashisht and P. Tauro), H A U Press, Hissar, p. 142.

Sanni, A.I. and Lo¨nner, C. 1993. *Food Microbiol.* 10: 517–523.

Santos A. and Marquina D. (2004) *Yeast* 21:151–162

Saono, S., Gandjar, I. and Basuki, T. 1996. In: *Handbook of Indigenous Fermented Foods* (ed. K. H. Steinkraus) Marcel Dekker, New York, pp. 363–508.

Sauer, M., Branduardi, P., Valli, M. and Porro, D. 2004. *Appl. Environ. Microbiol.* 70: 6086–6091.

Savitri and Bhalla, T.C. 2007. *Indian J. Tradit. Knowl.* 6: 17–24.

Savitri and Bhalla, T.C. 2012. *3 Biotech* 3: 247–254.

Savitri, Thakur, N. and Bhalla, T.C. 2012.. *Intl. J. Food Ferment. Technol.* 2: 49–56.

Schmitt, M.J., Klavehn, P., Wang, J., Schonig, I. and Tipper, D. 1996. *Microbiol.* 142: 2655–2662

Schulze, J. and Sonnenborn, U. 2009. *Dtsch. Arztebl. Int.* 106: 837–42.

Seeley, R.D. 1977. *Master Brewers Association of Americas* 14: 35–39.

Sefa-Deheh, S., Sanni, A.I., Tetteh, G. and Sakyi-Dawson, E. 1999. *World J. Microbiol. Biotechnol.* 15: 593–597.

Shieh, Y.S.C. and Beuchat, L.R. 1982. *J. Food Sci.* 47: 518–522.

Soni, S.K. and Sandhu, D.K. 1990. *Indian J. Microbiol.* 30: 135–157.

Soni, S.K. and Sandhu, D.K. 1999. In: *Biotechnology: Food Fermentation, Microbiology, Biochemistry and Technology* (eds. V.K. Joshi and A. Pandey), vol II. Educational Publishers and Distributors, New Delhi, pp. 895–949.

Spicher, G. and Stephan, H. 1993. *Handbuch Sauerteig. Biologie, Biochemie, Technologie.* Hamburg: Behr's Verlag.

Srivastava, S., Modi, D.R. and Garg, S.K. 1997. *Bioresour. Technol.* 60: 263–265.

Sserunjogi, M. L. (1999). Ugandan indigenous fermented dairy products, with particular focus on ghee. Ph.D. thesis, Agricultural University of Norway, Ass, Norway.

Stam, H., Hoogland, M. and Laane, C. 1998. In: *Microbiology offermented foods* (ed. B.J. Wood), vol 2, 2nd edn. Blackie, London, pp. 505–542.

Steinkraus, K.H. 1978. *Food Technol.* 32: 79.

Steinkraus, K.H. 1998. In: *Microbiology of Fermented Foods* (ed. B.J.B. Wood), 2nd ed. Blackie Academic and Professional, London, pp. 602–621.

Steinkraus, K.H., 1996. Handbook of Indigenous Fermented Foods, 2nd ed. Marcel Dekker, New York.

Stewart, G.G. and Russell, I. 1986. *J. Inst. Brew.* 91: 537–558.

Stolz, P. 2003. In: *Handbook of Dough Fermentations* (eds. K. Kulp and K. Lorenz) Marcel Dekker Inc., New York, pp. 23–43.

Swiegers, J.H., Bartowsky, E.J., Henschke, P.A. and Pretorius, I.S. 2005. *Aust. J. Grape Wine Res.* 11: 139–173.

Takasuka, T., Komiyama, T., Furuichi, Y. and Watanabe, T. 1995. *Cell Mol. Biol. Res.* 41: 575–581.

Tamang, J.P. and Fleet, G.H. 2009. In: *Yeasts Biotechnology: Diversity and Applications* (eds. T. Satyanarayana and G. Kunze), Springer, New York, pp. 169–198.

Tamang, J.P. and Thapa, S. 2006. *Food Biotechnol.* 20: 251–261.

Tamang, J.P., Dewan, S., Tamang, B., Rai, A., Schillinger, U. and Holzapfel, W.H. 2007. *Indian J. Microbiol.* 47: 119–125.

Thakur, N., Savitri and Bhalla, T.C. 2004. *Indian J. Tradit. Knowl.* 3: 325–335.

Thakur, N., Savitri, Saris, P.E.J., Bhalla, T.C. 2015. *Food Bioscience* 11: 92–96.

Thapa, S. and Tamang, J.P. 2006. *Indian J. Microbiol.* 46: 333–341.

Thapa, T. B., Jagat Bahadur, K.C., Gyamtsho, P., Kark, D.B., Rai, B.K. and Limbu, D.K. 2003. In: *Proceedings of International seminar and workshop on social well being*, Anand, India, 13–14, November 2003, pp. 9–10.

Timmons, S., 1994. *New Age* 78–96.

Tinoi, J., Rakariyatham, N. and Deming, R.L. 2005. *Process Biochem.* 40: 2551–2557.

Tsuyoshi, N., Fudou, R., Yamanaka, S, Kozaki, M., Tamang, N., Thapa, S. and Tamang, J. P. 2005. *Int. J. Food Microbiol.* 99: 135–146.

Vandamme, E.J. and Soetaert, W. 2002. *J. Chem. Technol. Biotechnol.* 77: 1323–1332.

Venkatasubbaiah, P., Dwarkanath, C.T. and Sreenivasamurthy, V. 1985. *J. Food Sci. Technol.* 22: 88–90.

Verstrepen, K.J., Paul, J., Chamber, S. and Pretorius, I.S. 2006. In: *The Yeast Handbook: Yeasts in Food and Beverages* (eds. A. Querol and G.H. Fleet), Springer-Verlag, Berlin Heidelberg, pp. 339–444.

Wang, H.L. and Hesseltine, C.W. 1982. In: *Prescott and Dunn's Industrial Microbiology* (ed. G. Reed), AVI Publishing Co. Inc., Westport, p. 492.

Wang, Y. 2008. Mutations that uncouple the yeast *Saccharomyces cerevisiae* ATP synthase. [PhD thesis]. North Chicago, Ill.: Rosalind Franklin Univ. of Medicine and Sciences. 164 p.

Watanabe, K., Fujimoto, J., Sasamoto, M., Dugersuren, J., Tumursuh, T. and Demberel, S. 2008. *World J. Microbiol. Biotechnol.* 24: 1313–1325.

Winarno, F.G., Muchatadi, D., Lakshmi, B.S., Rahman, A., Swastomo, W., Zainuddin, D. and Santaso, S.N. 1977. In: *Symposium on Indigenous Fermented Foods*, Bangkok, Thailand.

Wong, H.C. and Koehler, P.E. 1981. *J. Food Sci.* 46: 589–592.

Yildiz, F. 2010. *Development and manufacture of yogurt and other functional dairy products.* 1st ed. CRC Press.

Yokotsuka, T. 1960. *Adv. Food Res.* 10: 75–134.

Zhang, G.C., Kong, I.L., Kim, H., Liu, J.J., Cate, J.H.D. and Jin, Y.S. 2014. *Appl. Environ. Microbiol.* 80: 7694 –7701.

Živković, M., Čadež, N. Uroić, K. Miljković, M, Tolinački, M., Doušova, P. Kos, B., Šušković, J., Raspor, P., Topisirović, L. and Golić, N. 2015. *J. Intercult. Ethnopharmacol.* 4: 12–18.

Role of Yeasts in Food Fermentation

Amit Kumar Rai and Kumaraswamy Jeyaram

Abstract Yeasts are predominant in several fermented foods prepared from ingredients of plant as well as animal origin. The diversity of foods in which, yeasts predominate ranges from alcoholic beverages such as wines (e.g., fruit, palm and rice wines), cereal based leavened products (e.g., sourdough and idli), milk products (e.g., cheese and dahi) and condiments such as soy sauce and papads. In natural food fermentation, yeasts are either dominant alone or mixed with lactic acid bacteria or mycelial fungi. Many yeast strains have been selected from the natural fermentation and successfully utilised as starter culture for industrial food production. They have a significant impact on food quality by improving the taste, flavour, texture, nutritive values, reduction of anti-nutritional factors and improving the functionality (health promoting properties). This chapter focuses on the beneficial role of yeast in fermented foods with special reference in improving the functionality in fermented food products.

Keywords Yeast · Fermentation · Foods · Interaction · Metabolites · Health benefits

1 Introduction

Traditionally food fermentation is practised worldwide in the human society since many generations with the primary objectives to increase the shelf life, taste, aroma and nutritional properties of the fermented products. The art of fermentation, particularly with a particular starter is increasingly recognised recently, which has also

A.K. Rai
Institute of Bioresources and Sustainable Development, Sikkim Centre, Tadong, Sikkim, India

K. Jeyaram (✉)
Microbial Resources Division, Institute of Bioresources and Sustainable Development (IBSD), Imphal 795001, Manipur, India
e-mail: jeyaram.ibsd@nic.in

T. Satyanarayana and G. Kunze (eds.), *Yeast Diversity in Human Welfare*, DOI 10.1007/978-981-10-2621-8_4

led in application of specific culture and increased the importance of controlled fermentation (Poutanen et al. 2009). Among different microorganisms, yeast and lactic acid bacteria (LAB) are key players in acidic traditionally fermented food products, which are responsible for the reduction of pH during fermentation (Gobbetti et al. 2002). Yeast are abundant in many indigenous fermented foods, either alone or in a stable association of mixed population and play a significant role in defining the quality parameters of the fermented food products such as texture, taste, nutritive values, odour and functional properties (Aidoo et al. 2006; Tamang and Fleet 2009). Yeasts are present in diversity of fermented foods and beverages, which ranges from alcoholic beverages (e.g., fruit, palm and rice wines), non alcoholic beverages (fermented tea), cereal based leavened products, fermented legume products, milk products, fermented meat and fermented fish products (Tamang and Fleet 2009).

The recent developments in advanced techniques has led in better understanding the taxonomy, biochemistry, physiology and ecology of yeasts, which has created interest among researchers to study their effect on several fermented foods and beverages (Fleet 2007; Aidoo et al. 2006; Poutanen et al. 2009). In some of the fermented products yeast may not play a primary role of alcohol production, flavour or texture development of the fermented products, but its presence may improve the functional properties of the product (Chaves-López et al. 2012; Rai et al. 2015). The primary and secondary role of yeast depends not only at genus or species level but may differ within strains of a particular species (Chaves-López et al. 2012; Herjavec et al. 2003; Swiegers et al. 2009). Apart from beneficial properties some of the yeast species associated with fermented products can also affect the product in an adverse way (Jakobsen and Narvhus 1996). Thus it is necessary to understand the role of individual yeast in the diverse population of a naturally fermented food and application of selected species to obtain a fermented product of desired quality and functional properties.

In recent years researchers are focusing more on the effect of individual yeast on the reduction of anti-nutritional factors (Turk et al. 1996; Rai et al. 2010; Rai and Anu-Appaiah 2014) and production of novel metabolites responsible for specific health related benefits (Chaves-López et al. 2012; Li et al. 2015; Rai et al. 2015). The metabolites responsible for health benefits in yeast fermented foods are either produced by the yeast itself or produced by the action of enzymes produced by yeast on the food substrate (Jakobsen and Narvhus 1996; Chaves-López et al. 2012; Rai et al. 2015). The current chapter focuses on the presence of yeast in wide range of fermented products and their role in improving the quality and functional properties of the fermented food products.

2 Yeast in Fermented Food Products

Archaeological evidences have demonstrated the domestication of yeast in food fermentation since 7000 BC for rice wine production in China, 7000–4000 BC for grape wine production in Iran, 1500–1300 BC for beer and bread making in Egypt (Sicard and Legras 2011) and 1980–1450 BC for cheese production in China (Yang et al. 2014). The role of yeast in food fermentation includes alcohol production, improving texture by leavening, preservation by acidification and killer toxin production, improve nutritive values and removing anti-nutritional factors, value addition by developing bioactive peptides and vitamins production (Fig. 1). The major food products developed by yeast fermentation can be broadly categorised into fermented beverages, leavened products, milk products and soy sauce products.

2.1 *Fermented Beverages*

Alcoholic beverages by yeast fermentation are mostly produced from fruit juices, cereal grains and sap of palm tree. Yeast converts the sugar present in the raw material into ethanol. In cereal based fermentation, conversion of polysaccharides (starch) into simple sugar is the prerequisite. This has been achieved by malting (endogenous amylase produced during seed germination) in beer fermentation and addition of amylolytic starter of mold and yeast mixtures (*koji* technology) in case of rice wine fermentation (Hammes et al. 2005; Dung et al. 2007; Jeyaram et al. 2008; Zhu and Tramper 2013). The key player of this alcoholic fermentation is

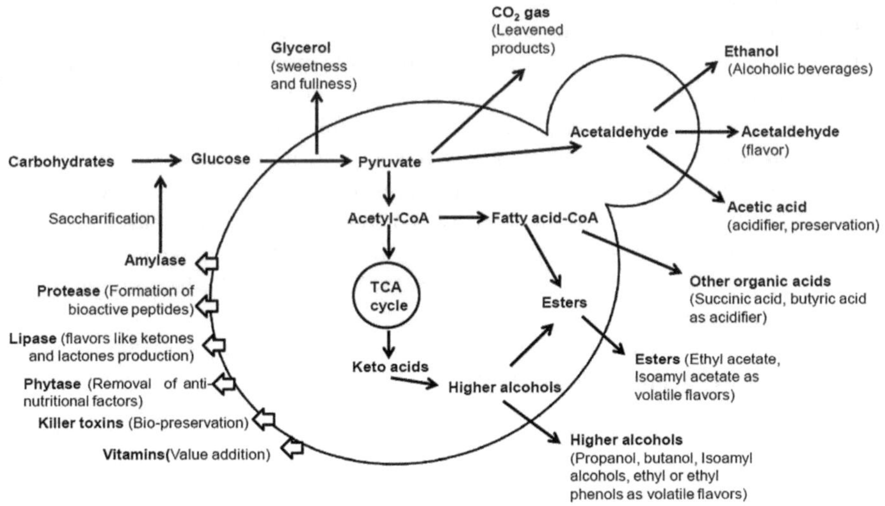

Fig. 1 Role of yeast in food fermentation

Saccharomyces sensu stricto complex especially *Saccharomyces cereus* and *Saccharomyces bayanus* (Louis 2011; Jeyaram et al. 2011) (Table 1). The native strains of *S. cerevisiae* are being selected with superior oenological and technological properties for industrial production of alcoholic beverages (Suárez-Lepe and Morata 2012). Out of 25 parameters studied, three key indicative parameters namely fermentation rate, low H_2S production and ethanol tolerance (8% v/v) were responsible for the superior quality of the wine yeast (Jeyaram 2002). The non-*Sacchromyces* yeast play an important role in improving the wine flavour by producing flavour components such as isoamyl acetate (fruity flavour) by *Pichia anomala*, glycerol by *Candida stellata*, methyl or ethyl phenol by *Pichia guuilliermondii* (Romi et al. 2014) and *Dekkera bruxellensi* (Schifferdeker et al. 2014). *Saccharomycoides fibuligera* and *Pichia anomala* are important amylolytic yeast of rice wine fermentation. These amylolytic traditional starters (*ragi* of Indonesia, *koji* of Japan, *mem* of Vietnam, *hamei* and *marcha* of India) are sustainably maintained through generations by indigenous process of sub-culturing and preservation as rice flat cakes by different ethnic communities in Asian countries (Fig. 2). This treasure house of yeast resources is yet to be explored for its biotechnological potential. In addition, production of many indigenous non-alcoholic or low alcoholic beverages (*kombucha, pozol, boza*) also involves yeast fermentation (Marsh et al. 2014).

2.2 Leavened Products

Fermentation in leavened products can be broadly categorised into sour dough fermentation for baked goods production and batter fermentation for pancakes production. Bread is the typical biologically leavened product of dough fermentation and Indian *dosa* is a typical product of batter fermentation. Hydrolytic activity of LAB and yeast are involved in this fermentation (Hammes et al. 2005). The gas CO_2 formed during fermentation is responsible for this leavening process (Verheyen et al. 2015). The major yeast species isolated from this leavened products are *S. cerevisiae* and *Pichia kudraiavzevii* (Table 2). Recently, the microbiota involved in sourdough fermentation was linked with its source plant (wheat) life cycle (Minervini et al. 2015). The gluten present in these cereal based foods is associated with celiac disease incidences. To overcome this problem, researchers are selecting proteolytic starter strains to get gluten free diet (Poutanen et al. 2009). However, gluten is responsible for the viscoelastic property of wheat dough, which is critical for expansion of gas bubbles and retention of CO_2 in the matrix (Verheyen et al. 2015). Gluten free *Amaranthus* breads are also available nowadays in the market (Vogelmann et al. 2009).

Table 1 Yeast species diversity in fermented beverages

Food	Substrate	Country	Yeast species									References
			Cg	Mp	Mg	Pa	Sc	Sk	Sf	Sp	O[b]	
Alcoholic beverages												
Mezcal	Agave juice	Mexico					X				Km, Td	Nuñez-Guerrero et al. (2016)
Caxiri	Cassava/ Sweet potato	Brazil			X		X				Pm, Rm	do-Amaral-Santos et al. (2012)
Kodo ko jaanr	Finger millet	India	X[a]			X	X		X			Thapa and Tamang (2004)
Wine	Grape	Spain		X							Bc	Fernández et al. (2000)
Wine	Grape	Europe					X				Db	Schifferdecker et al. (2014)
Wine	Grape	Europe						X			Sb	Louis (2011)
Nero di Troia	Grape	Italy					X				Cze, Hg	Garofalo et al. (2015)
Wine	Grape	Italy		X			X			X	Cs, Ka	Capece et al. (2003)
Wine	Grape	Italy									Dh	Capece (2009)
Grappa	Grape marc	Italy					X				Ho	Bovo et al. (2009)
Pito	Maize, Sorghum	Nigeria, Ghana										Blandino et al. (2003)
Hong Qu	Glutinous rice	China					X		X			Lv et al. (2013)
Ruou	Glutinous rice	Vietnam	X			X	X					Dung et al. (2007)
Ruou repthan	Purple glutinous rice	Vietnam					X					Dung et al. (2006)
Sake	Rice	China, Japan				X					Ss	Rhee et al. (2003)
Tapuy	Rice	Philippines							X		Cp, Dh, Rg, Tf	Kozaki and Uchimura (1990)

(continued)

Table 1 (continued)

Food	Substrate	Country	Yeast species									References
			Cg	Mp	Mg	Pa	Sc	Sk	Sf	Sp	O[b]	
Athingba	Rice	India				X	X					Jeyaram et al. (2008)
Bhaati Jaanr	Rice	India							X		Gc	Tamang and Thapa (2006)
Yakju and takjis	Rice, wheat, barley, maize, millet	Korea				X	X				Csa, Pp, Ti	Rhee et al. (2003)
Palm wines	Sap of coconut, date or palmyra palm	India and Southeast Asian countries					X			X		Joshi et al. (1999)
Non-alcoholic beverages												
Pozol	Maize	Mexico			X							Wacher et al. (2000)
Kombucha	Tea	China, Japan, Russia, Indonesia						X			Ck, Sl	Mayser et al. (1995)
Boza	Cereals and millets	Turkey, Bulgaria					X				Pf, Su	Marsh et al. (2014)

[a]*X*—indicates the presence of the specified yeast species in the respective fermented beverage; [b]*O*—indicates other species

Bc: Brettanomyces clausenii, *Cg*: Candida glabrata, *Ck*: Candida kefyr, *Cs*: Candida sake, *Csp*: Candida sphaerica, *Cze*: Candida zemplinia, *Dh*: Debaryomyces hansenii, *Db*: Dekkera bruxellensis, *Gc*: Geotrichum candidum, *Hg*: Hanseniospora gulilliermondii, *Ho*: Hanseniospora opuntiae, *Ka*: Kleoeckera apiculata, *Km*: Kluyveromyces marxianus, *Mp*: Metschnikowia pulcherima, *Pa*: Pichia anomala, *Pf*: Pichia fermentans, *Pm*: Pichia membranifaciens, *Pp*: Pichia polymorpha, *Rg*: Rhodotorula graminis, *Rm*: Rhodotorula mucilaginosa, *Sb*: Saccharomyces bayanus, *Sc*: Saccharomyces cerevisiae, *Sk*: Saccharomyces kudriavzevii, *Ss*: Saccharomyces sake, *Su*: Saccharomyces uvum, *Sl*: Saccharomycoides ludwigii, *Sf*: Saccharomycopsis fibuligera, *Sp*: Schizosaccharomyces pombe, *Td*: Torulaspora delbrueckii, *Ti*: Torulaspora incomp, *Tf*: Trichosporon fennicum

Fig. 2 Traditional amylolytic starters as source of yeast used for rice wine production in Northeast India (**a** Starter used for apongthobai production in Arunachal Pradesh; **b** Starter—hamei used for Atingba production in Manipur, **c**–**d**: Fermented beverages produced by using the traditional starters)

2.3 *Fermented Milk Products*

Current worth of fermented milk products market is Euro 46 billion (Marsh et al. 2014). The milk products can be classified into cheese type and yogurt type based on its production process. Fungi, LAB and yeast are involved in cheese solid state fermentation and ripening, while LAB and yeast predominate in yogurt fermentation. A batch to batch consistency in the microbiota of cheese rind between geographical locations made it as a model for studying the microbial ecology and substrate specific adaptive evolution (Wolfe et al. 2014). *Debaryomyces hansenii*, *Kluyveromyces lactis*, *Kluyveromyces marxianus* and *Yarrowia lipolytica* are the most frequently isolated yeast species from cheese fermentation (Table 3). Even though diverse yeast species has been reported in yogurt type milk fermentation, one of the frequently isolated is *Saccharomyces cerevisiae*. It is surprising to notice that there is no record of yeast characterised from Indian fermented milk 'Dahi' or 'Doi', the major fermented food of the region. The lipolytic yeast cultures play

Table 2 Yeast species diversity in leavened food products

Food	Substrate	Country	Yeast species										References
			Ch	Cp	Ct	Ke	Ku	Pa	Pk	Sc	Td	O[b]	
Dough fermentation for bread													
Amaranths bread	Amaranths	Germany							X[a]	X		Cg	Vogelmann et al. (2009)
Sourdough	Maize	Ghana			X				X	X		Km	Vogelmann et al. (2009)
Pozol	Maize	Mexico										Gg	Nuraida et al. (1995)
Sourdough	Quinoa	Germany							X	X			Vogelmann et al. (2009)
Sourdough	Rice	Germany							X	X			Meroth et al. (2004)
Sour dough	Rye	Finaland										Cm	Häggman and Salovaara (2008)
Sour dough	Rye	Denmark								X			Rosenquist and Hansen (2000)
Sourdough	Rye	Finland	X			X	X	X				Cs	Salovaara and Savolainen (1984)
Sourdough	Rye	Germany				X			X	X			Spicher and Schröder (1978)
Sourdough	Rye (bran)	Germany	X						X	X		Cg	Meroth et al. (2004)
Kisra	Sorghum	Sudan							X				Hamad et al. (1997)
Sourdough	Teff	Ireland								X			Moroni et al. (2010)
Sour dough	Wheat	China	X	X		X		X	X	X	X	Mg	Zhang et al. (2011)
Sourdough	Wheat	France				X		X		X	X		Infantes and Schmidt (1992)
Sourdough	Wheat	Greece								X	X	Pm	Paramithiotis et al. (2000)
Sourdough	Wheat	Italy	X			X			X	X	X	Kb	Valmorri et al. (2010)

(continued)

Table 2 (continued)

Food	Substrate	Country	Yeast species										References
			Ch	Cp	Ct	Ke	Ku	Pa	Pk	Sc	Td	O[b]	
Sourdough	Wheat	Morocco	X							X			Boraam et al. (1993)
Steamed bun	Wheat	Thailand		X	X				X	X			Luangsakul et al. (2009)
Steamed buns	Wheat	China		X	X					X		Io, Pst	Luangsakul et al. (2009)
Pizza	Wheat	Italy								X		Cc	Pepe et al. (2003)
Sour dough	Wheat	Finalnd										Cm	Mäntynen et al. (1999)
Sweet dough	Wheat	Spain									X		Hernández-Lopez et al. (2007)
Nan, Kulcha, Bhatura	Wheat	India, Pakistan, Afghanistan, Iran								X			Sandhu et al. (1986)
Organic bread	Wheat, Rye	France					X						Lhomme et al. (2016)
Sour dough	Wheat, Rye	Belgium					X	X		X	X		Vrancken et al. (2010)
Batter fermentation for pancakes													
Dhokla	Rice, Bengal gram	India										Ps	Kanekar and Joshi (1993)
Idli/Dosa	Rice, Black gram	South India, Sri Lanka								X		Dh,Tb	Soni et al. (1986)
Idli/Dosa	Rice, Black gram	South India, Sri Lanka										Gc, Rg, Th, Tc,Tp	Blandino et al. (2003)
Enjera	Teff	Ethiopia	X		X	X				X		Pn	Desiye and Abegaz (2013)

[a]*X*—indicates the presence of the specified yeast species in the respective fermented beverage; [b]*O*—indicates other species

Cc: Candida catenulata, *Cg*: Candida glabrata, *Ch*: Candida humilis, *Cm*: Candida milleri, *Cp*: Candida parapsilosis, *Cs*: Candida sake, *Ct*: Candida tropicalis, *Dh*: Debaryomyces hansenii, *Gg*: Galactomyces geotrichum, *Gc*: Geotrichum candidum, *Io*: Issatchenkia orientalis, *Kb*: Kazachstania barnettii, *Ke*: Kazachstania exigua, *Ku*: Kazachstania unispora, *Km*: Kluyveromyces marxianus, *Mg*: Meyerozyma guilliermondii, *Pa*: Pichia anomala, *Pk*: Pichia kudriavzevii, *Pm*: Pichia membranifaciens, *Pn*: Pichia norvegensis, *Ps*: Pichia silvicola, *Pst:* Pichia stipitis, *Rg*: Rhodotorula graminis, *Sc*: Saccharomyces cerevisiae, *Td*: Torulaspora delbrueckii, *Th*: Torulopsis holmii, *Tb*: Trichosporon beigelii, *Tc*: Trichosporon cutaneum, *Tp*: Trichosporon pullulans

Table 3 Yeast species diversity in fermented milk products

Food	Substrate	Country	Yeast species												References
			Ck	Ckr	Clu	Dh	Gc	Ku	Kl	Km	Pk	Sc	Yl	O[b]	
Cheese type															
Mozzaerella	Buffalo milk	Italy	X[a]			X			X	X					Romano et al. (2001)
Garis	Camel milk	Sudan, Somalia								X				Io	Shori (2012)
Suusac	Camel milk	Kenya, Somalia		X											Shori (2012)
Shubat	Camel milk	Turkey, Kazakhstan						X		X					Shori (2012)
Cheddar cheese	Cow milk	South Africa				X							X		De Wit et al. (2005)
Camembert	Cow milk	Australia				X							X		Addis et al. (2001)
Blue veined cheese	Cow milk	Denmark				X									Hansen and Jakobsen (2001)
Rokpol cheese	Cow milk	Poland												Csp, Cf	Wojtatowicz et al. (2001)
Cheddar cheese	Cow milk	South Africa				X							X		Ferreira and Viljoen (2003)
Stilton cheese	Cow milk	UK				X							X		Gkatzionis et al. (2014)
Soft red smeared	Cow milk	France											X		Monnet et al. (2010)
Saint-Nectaire	Cow milk	France				X							X		Callon et al. (2014)
Blue cheese	Cow milk	UK							X				X		Price et al. (2014)
Salers cheese	Cow milk	France				X			X	X		X		Cz	Callon et al. (2014)
Cheese	Cow milk	Spain				X			X				X		Atanassova et al. (2015)
Kefir	Cow milk	Caucasian region						X		X				N	Marsh et al. (2014)
Water Kefir	Cow milk	Belgium										X		Db	Laureys and De Vuyst (2014)
Tomme d'orchies	Goat milk	France				X			X	X					Ceugniez et al. (2015)

(continued)

Table 3 (continued)

Food	Substrate	Country	Yeast species												References
			Ck	Ckr	Clu	Dh	Gc	Ku	Kl	Km	Pk	Sc	Yl	O[b]	
Armada cheese	Goat milk	Sapin					X							Cl	Tornadijo et al. (1998)
Pecorino di Filiano	Sheep milk	Southern Italy				X									Capece and Romano (2009)
Bryndza cheese	Sheep milk	Solovakia					X								Pangallo et al. (2014)
Pecorino de Farindola	Sheep milk	Italy								X					Tofalo et al. (2014)
Semi-hard cheese	Sheep milk	Portugal				X								Ci	Pereira-Dias et al. (2000)
Fiore Sardo cheese	Sheep milk	Italy				X	X		X					Cz	Fadda et al. (2004)
Pecorino Crotonnese	Sheep milk	Southern Italy				X			X			X	X		Gardini et al. (2006)
Danish feta	Sheep milk	Denmark				X				X				Cs, Td	Westall and Filtenborg (1998)
Divlecave	Sheep milk	Turkey				X									Budak et al. (2015)
Apulian chesse	Cow/sheep/buffalo milk	Italy											X	Cc, Tc	Corbo et al. (2001)
Cheese	Sheep/goat milk	Spain				X							X		Padilla et al. (2014)
Gamembert	Goat/sheep milk	France					X								Boutrou et al. (2006)
Yogurt type															
Suusac	Camel milk	Kenya		X											Lore et al. (2005)
Shubat	Camel milk	China						X		X				Ce	Marsh et al. (2014)
Suusac	Camel milk	Kenya		X										Rm, Gp	Marsh et al. (2014)
Amasi/hodzeko	Cow milk	Zimbabwe	X												Gadaga et al. (2001)

(continued)

Table 3 (continued)

Food	Substrate	Country	Yeast species												References
			Ck	Ckr	Clu	Dh	Gc	Ku	Kl	Km	Pk	Sc	Yl	O[b]	
Amasi	Cow milk	Africa				X				X		X		Td	Narvhus and Gadaga (2003)
Yogurt	Cow milk	Switzer land			X	X	X								Lopandic et al. (2006)
nunu	Cow milk	Ghana									X	X			Akabanda et al. (2013)
Rob	Cow milk	Sudan	X									X			Marsh et al. (2014)
Nyarmie	Cow milk	Ghana										X			Marsh et al. (2014)
Amabere amaruram	Cow milk	Kenya										X		Tm, Cf	Nyambane et al. (2014)
Fermented milk	Cow milk	Tibet								X		X		Pf	Bai et al. (2010)
Koumiss	Horse milk	China						X		X		X			Mu et al. (2012)
Kumis	Horse milk	Italy			X						X			Gg, Ct	Chaves-López et al. (2012)
Kumis	Horse milk	Columbia			X						X			Gg, Ct	Marsh et al. (2014)

[a]*X*—indicates the presence of the specified yeast species in the respective fermented beverage; [b]*O*—indicates other species

Cc: Candida catenulata, *Ce*: Candida ethanolica, *Cf*: Candida famata, *Ci*: Candida intermedia, *Ck*: Candida kefyr, *Ckr*: Candida krusei, *Cl*: Candida lambica, *Cs*: Candida sake, *Csp*: Candida sphaerica, *Ct*: Candida tropicalis, *Cz*: Candida zeylanoides, *Clu*: Clavispora lusitaniae, *Dh*: Debaryomyces hansenii, *Db*: Dekkera bruxellensis, *Gg*: Galactomyces geotrichum, *Gc*: Geotrichum candidum, *Gp*: Geotrichum penicillatum, *Io*: Issatchenkia orientalis, *Ku*: Kazachstania unispora, *Kl*: Kluyveromyces lactis, *Km*: Kluyveromyces marxianus, *N*: Naumovozyma spp., *Pf*: Pichia fermentans, *Pk*: Pichia kudriavzevii, *Rm*: Rhodotorula mucilaginosa, *Sc*: Saccharomyces cerevisiae, *Td*: Torulaspora delbrueckii, *Tc*: Trichosporon cutaneum, *Tm*: Trichosporon mucoides, *Yl*: Yarrowia lipolytica important role in flavour (methyl ketones, alcohols and lactones) production during cheese ripening. In addition, proteolysis activity generates bioactive peptides with health promotion (De Wit et al. 2005; Rai et al. 2015). The selected yeast cultures along with LAB are used as starter culture for industrial production of fermented milk products.

2.4 Soy Sauces and Other Products

Preparation of soy sauces, table olive, fermented bamboo shoots, papad, and wadi also involve yeast fermentation (Table 4). The salt tolerant yeast species *Zygosaccharomyces rouxii* is predominately involved in soy sauce fermentation (Wah et al. 2013).

3 Role of Yeast in Food Fermentation

The important roles of yeast in fermented foods are production of alcohol, production and utilisation of organic acids, leavening agents in bakery, improvement of nutritional properties, flavour, aroma and texture, reduction of anti-nutritional factors, and production of health promoting metabolites (Fig. 1). In this section the different roles of yeast in food fermentation are briefly discussed.

3.1 Alcohol Production

Role of yeast in production of alcoholic beverages by bioconversion of sugar into alcohol and carbon dioxide is well known fact since many centuries. The principal yeast for the production of alcoholic beverages belongs to strains of *Saccharomyces cerevisiae.* However not all *Saccharomyces cerevisiae* strains are good producers of ethyl alcohol and high quality fermented beverage (Carrau et al. 2008). Alcoholic fermentation is characterized by production of alcohol and other secondary metabolites during the growth of various yeast species and strains, where the ecology and yeast–yeast interaction determine the metabolites and quality of the final product (Rai et al. 2012). However along with ethanol a wide range of metabolites are also produced which includes, glycerol, acetaldehyde, acetic acid, lactic acid and pyruvate. The proportion of alcohol and these metabolites is determined the yeast involved during fermentation, which determines the final quality of the alcoholic beverage. Fermentation carried out by different *Saccharomyces* strains has resulted in wines with different types and proportion of secondary metabolites (Herjavec et al. 2003; Perez-Coello et al. 1999).

Table 4 Yeast species diversity in soy sauce and other fermented foods

Food	Substrate	Country	Yeast species												References
			Cb	Ck	Ch	Co	Dh	K	Mg	Pa	Pm	Rg	Sc	Zr	
Soy sauce															
Miso	Soybean, rice	Japan												X[a]	Ebine (1989)
Shoyu (Soy sauce)	Soybean	Japan							X					X	Nunez-Guerrero et al. (2016)
Soysauce	Soybean	Thailand							X						Aryuman et al. (2015)
Thai soy sauce	Soybean	Thailand												X	Wah et al. (2013)
Soysauce	Soybean, wheat	Japan, China												X	Aidoo et al. (1994)
Others															
Table Olive	Olive fruit	Spain	X				X			X	X	X	X		Arroyo-López et al. (2008)
Papd	Black gram	India		X									X		Shurpalekar (1986)
Wadi	Black gram	India, Pakistan		X											Sandhu and Soni (1989)
Soibum	Bamboo shoot	India							X						Romi et al. (2014)
Soidon	Bamboo shoot	India			X	X		X							Romi et al (2015)

[a]*X*—indicates the presence of the specified yeast species in the respective fermented beverage

Cb: Candida boidinii, *Ck*: Candida kefyr, *Ch*: Candida humilis, *Dh*: Debaryomyces hansenii, *K*: Kazachstania spp., *Mg*: Meyerozyma guilliermondii, *Pm*: Pichia membranifaciens, *Rg*: Rhodotorula graminis, *Sc*: Saccharomyces cerevisiae

Fruit wines are fermented alcoholic beverages made up of variety of fruits including apples, strawberry, papaya, peach, mango, cherry, orange, red currant, bilberry, gooseberry, cranberry, raspberry, pomegranate, garcinia, marula, sand pear (Fernandez-Pachon et al. 2014; Hidalgo et al. 2013; Herrero et al. 1999; Li et al. 2013; Berenguer et al. 2016; Rai and Anu-Appaiah 2014). Alcohol content in these fruit wines were reported to be in the range of 5–13% (v/v), which dependent on the strain of a particular starter used for fermentation (Berenguer et al. 2016). Rice wines are very popular and traditional alcoholic beverage consumed in many Asian countries and are fermented using amylolytic starters in the form of flat cakes, which contain mixed cultures of moulds having starch degrading ability and fermenting yeasts (Dung et al. 2006; Jeyaram et al. 2008). The moulds in the starters degrade starch into simple sugars, whereas yeasts convert the simple sugars (glucose) into alcohol (Dung et al. 2006; Nout and Aidoo 2002). Rice wine prepared from glutinous rice is one of the popular alcoholic beverages consumed in different states of North-Eastern India (Jeyaram et al. 2008). They are prepared with amylolytic starters, which are in the form of flat cakes, e.g.: *Marcha* in Sikkim and *Hamei* in Manipur (Jeyaram et al. 2008). Alcohol content and quality of rice wine depends on the rice variety and starter used for fermentation (Palaniveloo and Vairappan 2013). In their study, Palaniveloo and Vairappan (2013) have shown that rice wines prepared from common rice resulted in higher content of alcohol (9.9–13.9%) in comparison to wines prepared using glutinous rice (7.0–8.4%) as a substrate. Vietnamese rice wines alcohol content has been reported to be maximum up to 15 ml/100 ml (approximately 12 g/100 ml) (Dung et al. 2007). Palm wine is another *S. cerevisiae* fermented alcoholic beverage, which after fermentation has ethanol content of 9% (v/v) (Aidoo et al. 2006; Joshi et al. 1999). In a recent study, it has been shown that alcohol production by yeast also has a strong impact on dough properties (Jayaram et al. 2014a). Low levels of ethanol formed during dough fermentation decrease dough extensibility and make it more stiff and tenacious by a different mechanism than organic acid metabolites (Jayaram et al. 2014b).

3.2 Organic Acids Production and Utilisation

Organic acids are metabolic products of yeast fermentation, which not only reduce the pH of the products but also affect the aroma and flavour of the food and beverages (Alvarez-Martin et al. 2008). The flavour and aroma depends on the type of organic acid produced by the yeast during fermentation (Rezaei et al. 2015). Yeasts are involved both in production and utilisation of selected organic acids, which affect the quality of final fermented product (Alvarez-Martin et al. 2008; Rai and Anu-Appaiah 2014; Yoshida and Yokoyama 2012). The organic acids such as acetate, malate, citrate, pyruvate, and succinate are derived during fermentation from carbohydrate metabolism (Coote and Kirsop 1974). The majority of pyruvate, acetate, and succinate in beer are produced through yeast metabolism, whereas malate and citrate are derived from malt (Yoshida and Yokoyama 2012).

The organic acid content and composition of fruits is also important because the changes by the yeast have an influence on the sensory properties of a fermented beverage. Utilisations of organic acids have also been reported in several alcoholic fermented beverages (Rai and Anu-Appaiah 2014). Orange juice fermentation for production of wine led in reduction of citric acid, ascorbic acid and malic acid. In a another study, fermentation of *Garcinia* must with *Saccharomyces cerevisiae* and *Hansioniospora* sp. resulted in reduction of citric acid and oxalic acid (Rai et al. 2010; Rai and Anu-Appaiah 2014) and simultaneous synthesis of aspartic and glutamic acid. Complete and partial utilisation of uric acid was also found in milk fermented with *Candida pararugosa* and *Geotrichum candidum*, respectively. Succinic acid is also one of the major organic acids produced by yeast species during production of alcoholic beverages (Jayaram et al. 2014c). Succninic acid production by yeast has also been shown to be responsible factor for affecting the dough rheology and flavour (Jayaram et al. 2014a). Apart from dough rheology, it is also predicted that the concentration and ratio of acetic and succinic acid formed during dough fermentation may influence the organoleptic properties of the final product (Rezaei et al. 2015). In contrast, complete utilisation of succninic acid was found in milk fermented with *Pichia membranifaciens* and *Geotrichum candidum* (Alvarez-Martin et al. 2008). Some of the yeasts starters were also found to be associated with production of lactic acid, pyruvic acid, propionic acid, acetic acid and butyric acid during milk fermentation (Alvarez-Martin et al. 2008). Thus selection of a proper yeast co-starter is necessary as to have controlled organic acid production in a desired product.

3.3 Leavening Agent

Yeast is also one of the commonly used leavening agents to raise bakery and pastry products. During dough fermentation process yeast converts sugar into alcohol and carbon dioxide, which determine the textural property of the baked product. The most widely used yeast in bakery product as a leavening agent is *Saccharomyces cerevisiae* or baker's yeast (Liao et al. 1998; Newberry et al. 2002). Production of carbon dioxide (CO_2) is responsible for the aerated structure of the product (Randez-Gil et al. 1999). Type of yeast and concentration of individual yeast strain are the important factors that affect the rate of gas production (Akdogan and Ozilgen 1992; Gobbetti 1998). The morphological structure of wheat dough is also affected by the volume of CO_2 and its rate of production by yeast during dough fermentation. The production of CO_2 during dough fermentation usually follows a sigmoidal function, which is due to the growth retarding factors occurring during fermentation such as inhibitory substances and substrate availability (Chiotellis and Campbell 2003a, b; Verheyen et al. 2015). In a recent study, Verheyen et al. (2015) studied the characteristics of gas release by using rheofermentometer with varied concentrations of two types of yeast starter (compressed yeast and instant dry yeast). Interestingly, even though colony forming units of yeast after 180 min fermentation was higher in dough fermented with compressed yeast starter, there was no significant different in CO_2 production. However, CO_2 production increased with the increasing concentration of both the starter. Their results also indicated that the structural integrity of dough matrix is affected by an increase in the maximum gas formation rate during fermentation. Apart from gas production yeasts are also responsible for bread flavor and dough rheology (Jayaram et al. 2014b; Liao et al. 1998; Randez-Gil et al. 1999).

3.4 Hydroytic Enzyme Production

Enzymes play an important role in production or quality improvement of fermented foods or beverages. Enzymes acting during food fermentation originate mainly from the microorganism associated and also from the food substrate used for fermentation. Yeasts are capable of producing both intracellular and extracellular hydrolytic enzymes including proteases, amylase, invertase, xylanase, cellulases, lipases, phytases, β-glucosidases, pectinases (Blanco et al. 1999; Charoenchai et al. 1997; Maturano et al. 2012; Strauss et al. 2001). In particular, yeast carboxypeptidases and aminopeptidases play important role in hydrolysis of milk protein (Ferreira and Viljoen 2003), which play important role in bioactive peptide production. Apart from bioactive peptides, proteolytic activity of yeasts is also studied in reduction of protein haze in wine and beer (Nelson and Young 1986; Bilinski et al. 1988; Lagace and Bisson 1990; Conterno and Delfini 1994). Proteolytic and lipolytic enzymes are associated in releasing products, which affects the flavour in cheese (Fox and Law 1991; Farag et al. 1992). In wine industry, β-glucosidases are the hydrolases, which catalyse the breakdown the linkage in aryl-β-glucosides and release of aromatic compounds which is improving the flavour of the fermented beverage (Blasco et al. 2006; Charoenchai et al. 1997; Fernández et al. 2000; Maturano et al. 2012). Higher β-glucosidase activity has been observed in the strains of *Pichia anomala*/*Candida pelliculosa* (formerly *Hansenula anomala*) and *Hanseniaspora uvarum*/*Kloeckera apiculata* in comparison to other yeast species (Charoenchai et al. 1997). β-Glucosidases is also associated with conversion of bound polyphenols to free polyphenols, which increases the antioxidant activity (Sanjukta et al. 2015). Enzymes like phytases are related in reduction of antinutritional phytate (Moslehi-Jenabian et al. 2010).

The principal wine yeast, *Saccharomyces cerevisiae* is not well known for producing highly active extracellular enzymes, although McKay (1990) has reported *S. cerevisiae* strains degrading polygalacturonate. However, several reports suggests that non-*Saccharomyces* wine yeasts have the ability to produce extracellular enzymes (Dizy and Bisson 2000; Masoud and Jespersen 2006; Strauss et al. 2001). Some of the strains of *Kloeckera apiculata* during wine fermentation have shown to produce extracellular protease (Dizy and Bisson 2000; Lagace and Bisson 1990). Strains of *Torulaspora* and *Hanseniaspora* genera were also reported for the production of enzymes like proteases, β-glucosidases, pectinases and xylan degrading enzymes (Charoenchai et al. 1997; Masoud and Jespersen 2006; Maturano et al. 2012). There is a significant role of enzymes produced by both *Saccharomyces* and non-*Saccharomyces* yeasts during wine fermentation (Maturano et al. 2012; Strauss et al. 2001), which has a considerable impact on the final product. The production of hydrolytic enzymes in fermented foods by indigenous yeast needs to be understood and managed for the improvement of the beneficial factors in fermented food products.

3.5 Flavour, Aroma and Texture

Flavour, aroma and texture are the important properties that account for the popularity and acceptance of any fermented food product. Flavour and aroma enhancement during fermentation by yeast takes place either by biotransformation of fruit components into flavour and aroma impacting compounds or by the *de nova* synthesis of compounds such as ethanol, glycerol, acetaldehyde, organic acids, esters, fatty acids and higher alcohols, which are responsible for aroma and flavour development (Styger et al. 2011). Several aroma compounds such as organic acids, esters and carbonyl compounds are also the metabolic products of fermentative conversion of carbohydrate by yeast (Jayaram et al. 2013). Changes in flavour and aroma in wines can even differ depending on the strain of *S. cerevisiae* used for fermentation as production of acetic acid esters, fatty acid ethyl esters and a higher alcohol varies even at strain level (Herjavec et al. 2003; Perez-Coello et al. 1999; Swiegers et al. 2009). Apart from alcohol production, *S. cerevisiae* is most intensively studied in fermented foods and beverages for the production of aroma components such as organic acids, esters and carbonyl compounds (Torner et al. 1992; Janssens et al. 1992). Among different products, *Saccharomyces* yeasts have been shown to have a greater impact on the favour of fermented products in brewing industry (Hammond 1993).

Yeast has also been shown to have a greater impact on soya sauce as more than 300 types of flavour compounds has been reported in Japanese soya sauce and most of them are produced by yeasts (Nunomura and Sasaki 1992). The major flavour compound in Japanese soya sauce includes ethanol, glycerol, 4-hydroxy-5-methyl-3(2H)-furanone (HMMF), 4-hydroxy-2(or 5)-ethyl-5(or 2)-methyl-3(2H)-furanone (HEMF), isobutyl alcohol, 4-hydroxy-2,5-dimethyl-3(2H)-furanone (HDMF), isoamyl alcohol, 2-phenylethanol methanol, 4-ethylguaiacol (4-EG) and γ-butyllactone 4-ethylphenol (4-EP). HDMF and HEMF are flavour active compounds that produce a strong caramel-like odour has also been isolated from wide range of fermented product *shoyu*, *miso*, beer and cheese (Slaughter 1999).

The metabolites produced by yeast during fermentation contribute to the flavour and aroma of the baked products. During spontaneous fermentation of maize dough, totally 76 compounds were identified, which included 19 alcohols, 21 carbonyls, 12 acids, 17 esters, a furan and an alkene compound (Annan et al. 2003). The production of alcohols and some esters compounds also coincided with higher count of yeasts during fermentation. The esters in fermented dough were also reported during spontaneous fermentation of Danish rye sourdoughs and beer fermentation as a result of activities by the yeast involved in the fermentation process (Hammond 1993; Hansen et al. 1989). As yeast plays a major role in the flavour and texture parameters of the finished bread by affecting the rate of substrate breakdown and microbial acidification, it is important to characterize the yeasts responsible for such activities. Yeast also affects the textural and sensory properties of fermented food products. Yeast present during *dhokla* fermentaion raises the volume of the batter and imparts sponginess to the final product (Kanekar and Joshi 1993). Fermentation of *Idli* batter with yeast suggests that yeast fermentation resulted in product with good texture and sensory qualities (Aidoo et al. 2006).

Red *sufu*, a *Monascus* fermented soybean product contains high amounts of alcohols, esters, and organic acids, which may be formed during fermentation of angkak rice by *Monascus* spp. The esters produced during fermentation give red *sufu* a characteristic fruity aroma (Han et al. 2001). Growth of yeast is also essential for the development of a typical aroma profile and texture of certain fermented milk products (Alvarez-Martin et al. 2008). Yeasts are also believed to play a significant role in various fermented milk products such as *Kefir, Koumis, Viili, Kurut, Laban, Longfil* and *Amasi* (Rai et al. 2015). Yeast also influences the flavour components and accelerates ripening process in several cheese products (Alvarez-Martin et al. 2008; Rai et al. 2015). The major yeast species reported to be associated during cheese ripening process are *Candida zeylanoides* (Fadda et al. 2010), *Trichosporon cutaneum* (Corbo et al. 2001), *Debaryomyces hansenii* (Padilla et al. 2014) and *Geotrichum candidum* (Tornadijo et al. 1998), which can have an impact on cheese flavour. Production of acetaldehyde has also been reported to provide a typical flavour of yoghurt and acceptable up to a level of 37 mg kg^{-1} (Tamine and Robinson 1999). In case of fermented milk products some yeast results in good appearance and pleasent flavour but others give rise to off-flavours, unpleasant odours and excessive gas (Alvarez-Martin et al. 2008). It is also suggested that yeasts make a positive contribution to the development of flavour in cured hams and fermented sausages (Mauriello et al. 2004; Tamang and Fleet 2009).

3.6 Improvement of Nutritive Values

Fermentation is a very well known technology for the improvement of nutritive values of a food and increasing the digestibility by the hydrolysis of the larger biomolecules into a utilisable form. Some of the yeast associated with fermented foods are known for the production of hydrolytic enzymes for improving the digestibility of the product and enhancing vitamins levels during fermentation. During sourdough fermentation yeast influence the nutritional quality of bread by improving the bio accessibility and level of phenols, sterols, vitamins, solubilise the fibre, enhance bioavailability of minerals and reduction in starch digestibility (Poutanen et al. 2009).

Pearl millet flour fermented with *Saccharomyces cerevisiae* enhanced the in vitro protein digestibility, whereas flour fermented by *Saccharomyces cerevisiae* subsp. *diastaticus*, enhanced the digestibility of starch (Khetarpaul and Chauhan 1990). Many researchers have shown that yeast fermentation increases the folate content in baked product prepared from wheat (Kariluoto et al. 2004) as well as rye (Katina et al. 2007; Kariluoto et al. 2004, 2006; Liukkonen et al. 2003). The effect of yeast in improving the folate content was found to be more than lactic acid bacteria during sourdough fermentation (Kariluoto et al. 2006).

3.7 Reduction of Toxins and Anti-nutritional Factors

Fermentation is an economical approach which has been used to reduce the toxins and antinutritional factors present in the raw material (Egounlety and Aworh 2003; Rai and Anu-Appaiah 2014). Yeast particularly *Saccharomyces cerevisiae*, has been suggested to reduce mycotoxins during ethanol fermentation (Shetty and Jespersen 2006; Bata and Lasztity 1999). *Phaffia rhodozyma* and *Xanthophyllomyces dendrorhous* has also been reported to degrade mycotoxin (Moslehi-Jenabian et al. 2010). Yeasts have been reported as one of the useful microorganisms for the production of phytase, which is responsible for degradation of phytic acid, an antinutritional factor (Greppi et al. 2015). Degradation of phytate by yeast during food fermentation improves the bioavailability of divalent metals such as iron, calcium, magnesium and zinc (Moslehi-Jenabian et al. 2010). It was also suggested that high-phytase producing yeast strains can be a potential phytase carriers for improving minerals bioavailability in the gastrointestinal tract.

Commercial baker's yeasts were shown to possess high phytase activity (Turk et al. 1996), which can result in improvement of minerals bioavailability. Phytate content was found to reduce in three breads (whole wheat, rye and white) by doubling the yeast content and extending the fermentation time (Harland and Harland 1980). Calcium absorption from leavened bread was found to increase after phytate hydrolysis by yeast during fermentation (Weaver et al. 1991). The common phytase producing yeast includes *Saccharomyces cerevisiae, Debaryomyces castellii, Saccharomyces kluyveri, Kluyveromyces lactis, Schwanniomyces castellii, Arxula adeninivorans, Pichia anomala, Pichia rhodanensis, Pichia spartinae, Rhodotorula gracilis, Torulaspora delbrueckii* and *Candida krusei* (Moslehi-Jenabian et al. 2010). Apart from phytate hydrolysis, yeast has also been reported to reduce oxalic acid content during wine fermentation. *Saccharomyces cerevisiae* and *Hansiniospora* sp. have been associated with more than 90% reduction in oxalic acid content during fermentation of *Garcinia* must for the production of alcoholic *Garcinia* beverage (Rai et al. 2010; Rai and Anu-Appaiah 2014).

4 Interaction of Yeast and Associated Microbes During Fermentation

Food fermentation takes place with diverse microorganisms that plays significant role at different stages of fermentation to obtain a stable and acceptable fermented product. Fermented foods are complex microbial ecosystem constituting of diverse groups of bacteria, yeasts and filamentous fungi, where they interact and their balance population contribute to the final quality of the fermented product. Interaction of yeast in fermented foods with associated microorganisms can be in various ways such as (i) utilisation of food components and production of metabolites, which is essential for the growth of the main starter culture, (ii) inhibition or elimination of pathogenic microorganisms and other microorganisms, which affect the quality of the food in a negative manner, (iii) inhibition of the starter culture and lead to spoilage of the food, (iv) enhancement of the fermentation process by supporting the starter culture, and (v) conversion of the metabolite produced by the starter culture into functional bioactive metabolites with potential health benefits.

In fermented foods yeast interacts with other microorganisms in a beneficial manner by supporting the starter culture in the formation of the final product (Jakobsen and Narvhus 1996; Viljoen 2001) or by inhibiting and eliminating the undesirable microorganisms responsible for deteriorating the product quality (Fleet 2003; Viljoen 2001). The pathogenic bacteria and spoilage causing microorganisms in fermented foods and beverages are reduced by yeast by the production of proteinaceous toxin referred as mycocin, organic acids, antibiotic factors and hydrogen peroxide, which prevents the growth of unwanted microorganisms (Chen et al. 2015; Viljoen 2006). Yeasts can improve the growth of lactic acid bacteria associated during food fermentation and also produce lactic acid for the reduction of pH (Tsapatsaris and Kotzekidou 2004; Viljoen 2001). As lactic acid bacteria needs a complex medium for their optimal growth, yeast actively synthesises substances such as amino acids, purines and vitamins and utilisable sugars by breakdown complex carbohydrates, which is essential for the optimal growth of lactic acid bacteria (Ruiz-Barba and Jimenez-Diaz 1995; Viljoen 2006). Lactic acid bacteria have also been reported to create a selective environment by producing organic acids and reducing the pH of the medium and favouring growth of yeast (Aidoo et al. 2006; Viljoen 2001).

On the other hand, the interactions which affect the quality of the product in a negative manner is by the growth of spoilage causing microorganisms (Seiler and Busse 1990; Viljoen and Greyling 1995; Jakobsen and Narvhus 1996), by retarding the growth of starter cultures itself and producing metabolites, which results in off-flavors, discoloration or slime formation in the product (Rohm et al. 1992; Tudor and Board 1993; Jakobsen and Narvhus 1996). Apart from yeast-lactic acid bacteria interaction, yeast-filamentous fungi interaction also considerably affects the quality of fermented products specially in wine (Fleet 2003). Growth of filamentous fungi such as *Aspergillus* spp., *Botrytis cinerea* and *Penicillium* spp. have shown to produce metabolites that retard the growth of fermentative yeasts (Reed and

Nagodawithana 1988; Doneche 1993; Fleet 2007). Another interesting interaction during alcoholic fermentation involves the deactivation of metabolites produced by fungus by yeasts (Fleet 2003).

5 Health Benefits of Yeast Fermented Products

In the recent years, application of yeast is gaining more popularity in improving the bioactive components responsible for health benefits. The components responsible for health benefits in yeast or fermented foods are either directly produced by specific yeast species or produced by the changes in the food components mediated by the enzymes produced the associated yeast species. In this section health benefits of yeast fermented products are classified as (i) yeast and yeast metabolites, and (ii) yeast mediated transformed metabolites.

5.1 Yeast and Yeast Metabolites

Yeast cells and some of the metabolites produced by yeast exhibits several potential health benefits (Hatoum et al. 2012). Yeast can also find its application as a potential probiotic and can be delivered in the form of fermented foods as selected yeast species such as *Torulaspora delbrueckii*, *D. hansenii*, (Psani and Kotzekidou 2006), *Kluyveromyces marxianus*, *K. lodderae* (Kumura et al. 2004), *Yarrowia lipolytica* and *K. lactis* (Chen et al. 2010) have been found to exhibit strong antimicrobial effect against pathogenic bacteria and the ability to resist in the gastrointestinal tract. The metabolites produced by yeast during fermentation include health promoting components such as niacin, pantothenic acid, biotin and riboflavin (Fleet 2003; Jakobsen and Narvhus 1996).

Monascus fermented products are used as medicinal purpose as several metabolites produced during fermentation such as γ-aminobutyric acid (GABA), polysaccharide, lovastatin and ergosterol have been known to have certain beneficial physiological functions (Shi and Pan 2011). An important bioactive metabolite, monacolin K is also produced during *Monascus* fermentation, which has the ability to inhibit HMG-CoA reductase and lower plasma cholesterol levels (Manzoni and Rollini 2002). Functional *Monascus* components and fermented *Monascus* products, such as Red yeast rice, fermented soybean have attracted many researchers and gained much attention in recent years (Lim et al. 2010; Shi and Pan 2011). Pyo and Seong (2009) have also developed a medicinal soy yogurt having a high level of GABA. *Saccharomyces cerevisiae* is also known for the production of GABA, which acts as an antihypertensive molecule. Recently, GABA producing *Saccharomyces cerevisiae* strain has been isolated from Korean traditional Bokbunja wine (Song and Baik 2014). GABA was also found to be present as a non protein amino acid fraction of red yeast, *Rhodotorula glutinis* (Krishnaswamy and

Giri 1953). Application of yeast producing specific health promoting metabolites as a co-starter in fermented food can have an added advantage by having the properties of the bioactive components.

5.2 *Yeast Mediated Transformed Metabolites*

Yeast interaction with food components by their hydrolysis and transformation results in different type of food metabolites, which possess specific health benefits. These metabolites mainly include bioactive peptides, free polyphenols and oligosaccharides (Rai et al. 2015). In recent years, yeast is also being studied to play a significant role in production of bioactive peptides in protein rich fermented foods (Rai et al. 2015). Bioactive peptides are sequence of amino acids, which exhibit several physiological effects depending on their size, amino acids sequence and composition (Rai and Jeyaram 2015). Thus the type and activity of the peptides formed during fermentation of protein rich foods depends not only at genus or species of starter but may vary among different strain of a same species (Chaves-López et al. 2012; Sanjukta et al. 2015). As yeasts are present in several protein rich fermented foods they have a significant role in the production of bioactive peptides (Chaves-López et al. 2014; Li et al. 2015).

In the last decade, yeast protease has been reported for the production of bioactive peptides in fermented products (Addis et al. 2001; Chaves-López et al. 2012; Didelot et al. 2006; Li et al. 2015). Yeast has been used as a co-starter for the production of fermented milk products and such combination has resulted in product having peptides with angiotensin I converting enzyme (ACE) inbibitory activity (Bai et al. 2010; Rai et al. 2015). Sour milk, a milk product fermented by co-culturing *Saccharomyces cerevieseae* and *Lactobacillus helveticus* exhibits antihypertensive effect in spontaneously hypertensive rat model (Nakamura et al. 1995). In a recent study on Colombian *Kumis* (low alcoholic traditional fermented cow milk product), higher yeast count has been related to production of ACE-I peptides (Chaves-López et al. 2012). Milk fermented with *Kluyveromyces marxianus* KL26A and *P. kudriavzevii* KL84A resulted in production of peptides with higher ACE-I activity among the yeasts isolates (Chaves-López et al. 2012). Recently, a combination of proteolytic co-culture of *Pichia kudriavzevii* KL84A with lactic acid bacteria resulted in product with high ACE-I activity (Chaves-López et al. 2014). Studies have also shown that yeast not only affect the product quality during cheese ripening but also affect the biological activities of the product (Padilla et al. 2014; Fleet 2007). In a recent report, Li et al. (2015) have purified and identified novel peptides with angiotensin I-converting enzyme inhibitory effect from milk fermented with the yeast *Kluyveromyces marxianus* and optimized the process conditions of milk fermentation. Research findings on impact of yeast on production of bioactive peptides are less in comparison to lactic acid bacteria, but there are higher possibilities of production of novel bioactive peptides on hydrolysis by yeast proteases.

Fermented soybean products are consumed as a source of digested proteins and antioxidants (Sanjukta et al. 2015). *Monascus* fermented soybean products have been developed to provide a single product having combination of *Monascus* metabolites and fermented soybean properties. Soybean fermented with *Monascus* was found to possess enhanced antioxidant activity in comparison to unfermented soybean (Lee et al. 2008). *Monascus* fermented soybean extracts were also found to possess additional lipid lowering effect in hyperlipidemic rats (Pyo and Seong 2009).

Therapeutic values of fermented alcoholic beverages have been attributed to both polyphenols and alcohol in the final product (Rani and Appaiah 2015). Many fermented alcoholic beverages have been studied for their polyphenol content and antioxidant activities (Heinonen et al. 1998; Rani and Appaiah 2015; Vuorinen et al. 2000; Rai et al. 2010). In a study, totally 44 different berry and fruit wines were assessed for polyphenol content and antioxidant activity (Heinonen et al. 1998). The results showed that the total phenolic contents (expressed as gallic acid equivalents) ranged between 91 and 1820 mg/L, and wines exhibited reasonably good antioxidant activity. Apart from antioxidant activities several wines have been reported to posses antimicrobial activity, antimutagenic activity, inhibition to platelet aggregation, anti-inflammatory, beneficial for bone and lower the risk of rip fracture, prevent cardiovascular diseases and reduced incidence of heart attacks (Rani and Appaiah 2015; Keli et al. 1994; Truelsen et al. 1998; Soleas et al. 1997; Ganry et al. 2000).

In traditional Chinese medicine, rice wine is also used for medical purpose and claimed to possess several beneficial effects in prevention of cardiovascular diseases, cancer and aging (Chen et al. 2002; Que et al. 2006; Wang 1998). In their study, Que et al. (2006), showed that that rice wines exhibited free radical scavenging activity, total antioxidant activity and reducing power potential. Among five different rice wines, *Nuomi* exhibited highest antioxidant activity and high phenolic acid content, whereas *Foshou* showed the lowest antioxidant activity with a low phenolic acid content. In another study, *Haria*, Indian rice based fermented alcoholic showed free radical scavenging activity (Ghosh et al. 2015).

6 Future Prospects and Conclusions

Yeast is present in wide range of fermented products and play important roles in many fermented foods and beverages in maintaining the quality and functional properties of the fermented product. Yeast also produces several important metabolites, which are present in traditionally fermented products and these can be exploited more effectively for industrial application in food industry. In some of the fermented foods the role of individual group of yeast is still need to be studied. A deeper understanding of their technological roles even in the established products, such as wine, cheese, bread and beer will allow development of novel products with improved quality. A better knowledge and more exclusive research in microbial interaction in fermented foods will aid in controlling the production of beneficial and undesirable. Desired metabolites can be achieved in fermented foods by using selected yeast species or a group of cultures as a starter. In case of protein rich fermented foods application of yeast as a co-starter can also lead in the production of novel bioactive peptides. Apart from beneficial attributes the challenge in selecting a yeast species for fermentation lies in its ability to result in a product having good sensory quality in combination with nutritional and health benefits.

References

Addis, E., Fleet, G. H., Cox, J. M., Kolak, D. and Leung, T. 2001. *Inter. J. Food Microbiol.* 69: 25–36.

Aidoo, K.E., Smith, J.E. and Wood, B.J.B. 1994. In: The Genus Aspergillus: From Taxonomy and Genetics to Industrial Application (Powell KA, Renwick A & Peberdy JF, eds), Plenum Press, New York, NY, pp. 155–169.

Aidoo, K.E., Nout, M.J.R. and Sarkar, P.K. 2006. *FEMS Yeast Res.* 6: 30–39.

Akabanda, F., Owusu-Kwarteng, J., Tano-Debrah, K., Glover, R. L., Nielsen, D. S. and Jespersen, L. 2013. *Food Microbiol.* 34: 277–283.

Akdogan, H. and Ozilgen, M. 1992. *Enz. Microb. Technol.* 14: 141–143.

Alvarez-Martin, P., Florez, A. B., Hernández-Barranco, A. and Mayo, B. 2008. *Food Control*, 19, 62–70.

Annan, N.T., Poll, L., Plahar, W.A. and Jakobsen, M. 2003. *European Food Res. Technol.* 217: 53–60.

Arroyo-López, F. N., Querol, A., Bautista-Gallego, J. And Garrido-Fernández, A. 2008. *Int. J. Food Microbiol.* 128: 189–196.

Aryuman, P., Lertsiri, S., Visessanguan, W., Niamsiri, N., Bhumiratana, A., and Assavanig, A. 2015. *Int. J. Food Microbiol.* 192: 7–12.

Atanassova, M., Fernández-Otero, C., Rodríguez-Alonso, P., Fernández-No, I. C., Garabal, J. I., and Centeno, J. A. 2015. *Food Microbiol.* doi:10.1016/j.fm.2015.09.012.

Bai, M., Qing, M., Guo, Z., Zhang, Y., Chen, X., Bao, Q., Zhang, H. and Sun, T.S. 2010. *Canadian J Microbiol.* 56: 707–714.

Bata, A., and Lasztity, R. 1999. *Trends Food Sci. Technol.* 10: 223–228.

Berenguer, M., Vegara, S., Barrajón, E., Saura, D., Valero, M. and Martí, N. 2016. *Food Chem.* 190: 848–855.

Bilinski, C. A., Emery, C. J., Zygora, P. S., Russell, I. and Stewart, G. G. 1988. *Critical Reviews in Biotechnology*, 8, 111–120.

Blanco, P., Sieiro, C. and Villa, T.G. 1999. *FEMS Microbiol. Letters* 175: 1–9.

Blandino, A., Al-Aseeri, M. E., Pandiella, S. S., Cantero, D. and Webb, C. (2003). *Food Res. Inter.* 36: 527–543.

Blasco, L., Veiga-Crespo, P., Poza, M. and Villa, T.G. 2006. *World J. Microbiol. Biotechnol.* 22: 1229–1233.

Boraam, F., Faid, M., Larpent, J. P. and Breton, A. 1993. *Sciences des aliments* 13: 501–509.

Boutrou, R., Kerriou, L., and Gassi, J. Y. 2006. *Inter. Dairy J.* 16: 775–783.

Bovo, B., Andrighetto, C., Carlot, M., Corich, V., Lombardi, A. and Giacomini, A. 2009. *Int. J. Food Microbiol.* 129: 221–228.

Budak, S. O., Figge, M. J., Houbraken, J. and de Vries, R. P. 2015. *Int. Dairy J.* doi:10.1016/j.idairyj.2015.09.011.

Callon, C., Retureau, E., Didienne, R., and Montel, M. C. 2014. *Inter. J. Food Microbiol.* 174: 98–109.

Capece, A., and P. Romano. 2009. *Int. J. Food Microbiol.* 132: 180–184.

Capece, A., Salzano, G. and Romano, P. 2003. *Inter. J. Food Microbiol. 84:* 33–39.
Carrau, F. M., Medina, K., Farina, L., Boido, E., Henschke, P. A., and Dellacassa, E. 2008. *FEMS Yeast Res.* 8: 1196–1207.
Ceugniez, A., Drider, D., Jacques, P. and Coucheney, F. 2015. *Food Microbiol. 52:* 177–184.
Charoenchai, C., Fleet, G.H., Henschke, P.A. and Todd, B.E.N. 1997. *Australian J. Grape Wine Res.* 3: 2–8.
Chaves-López, C., Tofalo, R., Serio, A., Paparella, A., Sacchetti, G. and Suzzi, G. 2012. *Int. J. Food Microbiol.* 159: 39–46.
Chaves-López, C., Serio, A.R., Paparella, A., Martuscelli, M., Corsetti, A., Tafalo, R. and Suzzi, G. 2014. *Food Microbiol.* 42: 117–121.
Chen, C., Yin, Z. J., and Xu, S. 2002. *Liquor Making* 29: 55–56.
Chen, L.S., Ma, Y., Maubois, J.L., He, S.H., Chen, L.J. and Li, H.M. 2010. *Dairy Sci. Technol.* 90: 537–548.
Chen, Y., Aorigele, C., Wang, C., Simujide, H. and Yang, S. 2015. *J. Food Nutr. Res.* 3: 52–56.
Chiotellis, E., and Campbell, G. M. 2003a. *Food and Bioproducts Processing* 81: 194–206.
Chiotellis, E. and Campbell, G. M. 2003b. *Food and Bioproducts Processing* 81: 207–216.
Conterno, L. and Delfini, C. 1994. *Journal of Wine Research 5*: 113–126.
Coote, N. and Kirsop, B. H. 1974. *J. Inst. Brew.* 80: 474–482.
Corbo, M. R., Lanciotti, R., Albenzio, M. and Sinigaglia, M. 2001. *Int. J. Food Microbiol.* 69: 147–152.
De Wit, M., Osthoff, G., Viljoen, B. C. and Hugo, A. 2005. *Enz. Microbial Technol.* 37: 606–616.
Desiye, A. and Abegaz, K. 2013. *Adv. Res. Biol. Sci.* 1: 36–44.
Didelot, S., Bordenave-Juchereau, S., Rosenfeld, E., Piot, J.M. and Sannier, F. 2006. *J Dairy Res.* 73: 163–170.
Dizy, M. and Bisson, L.F. 2000. *American J. Enol. Viti.* 51: 155–167.
do-Amaral-Santos, C. C. A., de Almeida, E. G., de Melo, G. V. P. and Schwan, R. F. 2012. *Int. J. Food Microbiol.* 156: 112–121.
Doneche, B., 1993. In: Fleet, G.H. (Ed.), Wine Microbiology and Biotechnology. Harwood Academic Publishers, Chur, Switzerland, pp. 327–353.
Dung, N. T. P., Rombouts, F. M. and Nout, M. J. R. 2006. *Food Microbiol.* 23: 331–340.
Dung, N. T. P., Rombouts, F. M. and Nout, M. J. R. 2007. *LWT-Food Sci. Technol.* 40: 130–135.
Ebine, H. 1989. In: Industrialization of Indigenous Fermented Foods (Steinkraus KH, ed), Marcel Dekker, New York, NY, pp. 89–126.
Egounlety, M. and Aworh, O.C. 2003. *J. Food Engineerg* 56: 249–254.
Fadda, M. E., Mossa, V., Pisano, M. B., Deplano, M., and Cosentino, S. 2004. *Int. J. Food Microbiol.* 95: 51–59.
Fadda, M.E., Viale, S., Deplano, M., Pisano, M.B. and Cosentino, S. 2010. *Int. J. Food Microbiol.* 136: 376–380.
Farag, A.A., Aly, M.E. and El-Alfy, M.B. 1992. *Die Nahrung* 36: 1–7.
Fernández, M., Ubeda, J. F. and Briones, A. I. 2000. *Int. J. Food Microbiol.* 59: 29–36.
Fernandez-Pachon, M. S., Medina, S., Herrero-Martín, G., Cerillo, I., Berná, G., Escudero-López, B., Ferreres, F., Martin, F., Garcia-Parrilla, M.C. and Gil-Izquierdo, A. 2014. *J. Pineal Res.* 56: 31–38.
Ferreira, A. D. and Viljoen, B. C. 2003. *Int. J. Food Microbiol.* 86: 131–140.
Fleet, G. H. 2003. *Int. J. Food Microbiol., 86*: 11–22.
Fleet, G.H. 2007. *Curr. Opin. Biotechnol.* 18: 170–175.
Fox, P.F. and Law, J. 1991. *Food Biotechnol.* 5: 239–262.
Gadaga, T. H., Mutukumira, A. N. and Narvhus, J. A. 2001. *Int. J. Food Microbiol.* 70: 11–19.
Ganry, O., Baudoin, C. and Fardellone, P. 2000. *American J. Epidemiol.* 151: 773–780.
Gardini, F., Tofalo, R., Belletti, N., Iucci, L., Suzzi, G., Torriani, S., Guerzoni, M.E. and Lanciotti, R. 2006. *Food Microbiol.* 23: 641–648.
Garofalo, C., Russo, P., Beneduce, L., Massa, S., Spano, G. and Capozzi, V. 2015. *Annals Microbiol.* 1–8.

Ghosh K., Ray M., Adak A., Dey P., Halder S.K., Das A., Jana A., Parua S., Das Mohapatra P.K., Pati B.R. and Mondal K.C. 2015. *Food Chem.* 168: 196–202.

Gkatzionis, K., Yunita, D., Linforth, R. S., Dickinson, M., and Dodd, C. E. 2014. *Inter. J. Food Microbiol.* 177: 109–116.

Gobbetti, M. 1998. *Trends Food Sci. Technol.* 9: 267–274.

Gobbetti, M., Morea, M., Baruzzi, F., Corbo, M. R., Matarante, A., Considine, T. And Fox, P. F. 2002. *International Dairy Journal*, 12: 511–523.

Greppi, A., Krych, Ł., Costantini, A., Rantsiou, K., Hounhouigan, D. J., Arneborg, and Jespersen, L. (2015). *International journal of food microbiology, 205*: 81–89.

Häggman, M. and Salovaara, H. 2008. *Food Res. Inter.* 41: 266–273.

Hamad, S. H., Dieng, M. C., Ehrmann, M. A. and Vogel, R. F. 1997. *J. App. Microbiol.* 83: 764–770.

Hammes, W. P., Brandt, M. J., Francis, K. L., Rosenheim, J., Seitter, M. F. and Vogelmann, S. A. 2005. *Trends Food Sci. Technol.* 16: 4–11.

Hammond, J.R.M. 1993. In: The Yeasts. Yeast Technology, Vol 5, 2nd edn. (Rose, A.H. and Harrison, J.S., Eds.), Academic Press, London. pp. 7–67.

Han, B.Z., Rombouts, F.M. and Nout, M.J.R. 2001. *Int. J. Food Microbiol.* 65: 1–10.

Hansen, A.Z. Lund, B. and Lewis, M.J. 1989. *Lebens. Wiss. Technol.* 22: 145–149.

Hansen, T. K. and Jakobsen, M. 2001. *Int. J. Food Microbiol.* 69: 59–68.

Harland, B.F. and Harland, J. 1980. *Cereal Chem.* 57: 226–229.

Hatoum, R., Labrie, S. and Fliss, I. 2012. *Frontiers Food Microbiol.* 3: 421.

Heinonen, I. M., Lehtonen, P. J. and Hopia, A. I. 1998. *J. Agric. Food Chem.* 46: 25–31.

Herjavec, S., Podgorski, V., Redzepovic, S. and Mirosevic, N. 2003. *Food Technol. Biotechnol.* 44: 77–81.

Hernández-Lopez, M. J., Pallotti, C., Andreu, P., Aguilera, J., Prieto, J. A. and Randez-Gil, F. 2007. *Int J. Food Microbiol.* 116: 103–110.

Herrero, M., Cuesta, I., García, L. A., and Díaz, M. 1999. *J. Institute Brewing* 105: 191–196.

Hidalgo, C., Torija, M. J., Mas, A. and Mateo, E. 2013. *Food Microbiol.* 34: 88–94.

Infantes, M. and Schmidt, J. L. 1992. *Sciences des Aliments* 12: 271–287.

Jakobsen, M. and Narvhus, J. 1996. *Int. Dairy J.* 6: 755–768.

Janssens, L., De Pooter, H.L., Schamp, N.M. and Vandamme, E.J. 1992. *Process Biochem.* 27: 195–215.

Jayaram, V. B., Cuyvers, S., Lagrain, B., Verstrepen, K. J., Delcour, J. A. and Courtin, C. M. 2013. *Food Chem.* 136: 301–308.

Jayaram, V.B., Cuyvers, S., Verstrepen, K.J., Delcour, J.A. and Courtin, C.M. 2014a. *Food Chem.* 151: 421–428.

Jayaram, V.B., Rezaei MN, Cuyvers, S., Verstrepen, K.J., Delcour, J.A. and Courtin, C.M. 2014b. *J. Agric. Food Chem.* 62: 9326–9335.

Jayaram, V. B., Cuyvers, S., Verstrepen, K. J., Delcour, J. A. and Courtin, C. M. 2014c. *Food Chemistry, 151*: 421–428.

Jeyaram, K. 2002. *Studies on The Selection Of Suitable Yeast Cultures And Standardization Of Fermentation Conditions For Wine Making From Banana, Papaya And Grapes*. PhD Thesis. Tamil Nadu Agricultural University; Coimbatore.

Jeyaram, K., Singh, W.M., Capece, A. and Romano, P. 2008. *Int. J. Food Microbiol.* 124: 115–125.

Jeyaram, K., Tamang, J. P., Capece, A. and Romano, P. 2011. *Antonie van Leeuwenhoek* 100: 569–578.

Joshi, V.K., Sandhu, D.K. and Thakur, N.S. 1999. In: Biotechnology: Food Fermentation (Joshi VK & Pandey A, eds), Educational Publishers, Ernakulam, India, pp. 647–744.

Kanekar, P. Joshi, N. 1993. *Indian J. Microbiol.* 33: 111–117.

Kariluoto, S., Vahteristo, L., Salovaara, H., Katina, K., Liukkonen, K.H. and Piironen, V., 2004. *Cereal Chem.* 81: 134–139.

Kariluoto, S., Aittamaa, M., Korhola, M., Salovaara, H., Vahteristo, L. and Piironen, V. 2006. *Int. J. Food Microbiol.* 106: 137–143.

Katina, K., Laitila, A., Juvonen, R., Liukkonen, K.H., Kariluoto, S., Piironen, V., Landberg, R., Åman, P. and Poutanen, K. 2007. *Food Microbiol.* 24: 175–186.
Keli, S. O., M. G. L. Hertog, E. J. M. Feskens, and D. Kromhout. 1994. *Archives Internal Med.*154: 637–642.
Khetarpaul, N. and Chauhan, B. M. 1990. *Plant Foods Human Nutr.* 40: 167–173.
Kozaki, M. and Uchimura, T. 1990. *J. Brew. Soc. Japan* 85: 818–824.
Krishnaswamy, P. R. and Giri, K. V. 1953. *Curr. Sci.* 22: 143–144.
Kumura, H., Tanoue, Y., Tsukahara, M., Tanaka,T. and Shimazaki, K. 2004. *J. Dairy Sci.* 87: 4050–4056.
Lagace, L.S. and Bisson, L.F. 1990. *American J. Enol. Viti.* 41: 147–155.
Laureys, D. and De Vuyst, L. 2014. *Appl. Environ. Microbiol.* 80: 2564–2572.
Lee, Y. L., Yang, J. H., and Mau, J. L. 2008. *Food chemistry*, *106*: 1128–1137.
Lhomme, E., Urien, C., Legrand, J., Dousset, X., Onno, B. and Sicard, D. 2016. *Food Microbiol.* 53: 41–50.
Li, X., Lim, S. L., Yu, B., Curran, P. and Liu, S.Q. 2013. *South African J. Enol Viti.* 34: 119–128.
Li, J.-Y., Zhao, H.-W. and Huang, W.-D. 2015. *Australian Journal of Grape and Wine Research*, 21: 404–410. doi: 10.1111/ajgw.12140.
Liao, Y.E., Miller, R.A. and Hoseney, R.C. 1998. *Cereal Chem.* 75: 612–616.
Lim, J. Y., Kim, J. J., Lee, D. S., Kim, G. H., Shim, J. Y., Lee, I. and Imm, J. Y. 2010. *Food Chem.* 120: 255–260.
Liukkonen, K.-H., Katina, K., Wilhelmson, A., Myllyma¨ki, O., Lampi, A.-M., Kariluoto, S., Piironen, V., Heinonen, S.-M., Nurmi, T., Adlercreutz, H., Peltoketo, A., Pihlava, J.-M., Hietaniemi, V. and Poutanen, K. 2003. *Proc. Nutr. Soc.* 62: 117–122.
Lopandic, K., Zelger, S., Banszky, L. K., Eliskases-Lechner, F. and Prillinger, H. 2006. *Food Microbiol.* 23: 341–350.
Lore, T.A., Samuel K. M. and John, W. 2005. *LWT-Food Sci. Technol.* 38: 125–130.
Louis, E. J. 2011. *Fungal Biol. Rev.* 25: 136–142.
Luangsakul, N., Keeratipibul, S., Jindamorakot, S. and Tanasupawat, S. 2009. *LWT-Food Sci. Technol.* 42: 1404–1412.
Lv, X. C., Huang, X. L., Zhang, W., Rao, P. F. and Ni, L. 2013. *Food Control* 34: 183–190.
Mäntynen, V. H., Korhola, M., Gudmundsson, H., Turakainen, H., Alfredsson, G. A., Salovaara, H. and Lindström, K. 1999. *Systematic Appl. Microbiol.* 22: 87–96.
Manzoni, M. and Rollini, M. 2002. Appl Microbiol Biotechnol, *58*: 555–564.
Marsh, A. J., Hill, C., Ross, R. P. and Cotter, P. D. 2014. *Trends Food Sci. Technol.* 38: 113–124.
Masoud, W. and Jespersen, L. 2006. *Int. J. Food Microbiol.* 110: 291–296.
Maturano, Y.P., Rodríguez, A.L.A., Toro, M.E., Nally, M.C., Vallejo, M., Castellanos de Figueroa L.I., Combina, M. and Vazquez, F. 2012. *Int. J. Food Microbiol.* 155: 43–50.
Mauriello, G., Casaburi, A., Blaiotta, G. and Villani, F. 2004. *Meat Sci.* 67: 149–158.
Mayser, P., Fromme, S., Leitzmann, G. and Gründer, K. 1995. *Mycoses* 38: 289–295.
McKay, A.M. 1990. *Let. App. Microbiol.* 11: 41–44.
Meroth, C. B., Hammes, W. P. and Hertel, C. 2004. *Systematic Appl. Microbiol.* 27: 151–159.
Minervini, F., Celano, G., Lattanzi, A., Tedone, L., De Mastro, G., Gobbetti, M. and De Angelis, M. 2015. *App.Environ. Microbiol.* 81: 6736–6748.
Monnet, C., Bleicher, A., Neuhaus, K., Sarthou, A. S., Leclercq-Perlat, M. N. and Irlinger, F. 2010. *Food Microbial.* 27: 302–310.
Moroni, A.V., Arendt, E.K., Morrissey, J.P. and Dal Bello, F., 2010. *Int. J. Food Microbiol.* 142: 142–148.
Moslehi-Jenabian, S., Lindegaard Pedersen, L. and Jespersen, L. 2010. *Nutrients* 2: 449–473.
Mu, Z., Yang, X. and Yuan, H. 2012. *Food Microbiol.* 31: 301–308.
Nakamura, Y., Yamamoto, N., Sakai, K., Okubo, A., Yamazaki, S. and Takano, T. 1995. *J. Dairy Sci.* 78: 777–783.
Narvhus, J. A. and Gadaga, T. H. 2003. *Int. J. Food Microbiol.* 86: 51–60.
Nelson, G. and Young, T. W. 1986. *Journal of the Institute of Brewing*, *92*: 599–603.

Newberry, M.P., Phan-Thien, N., Larroque, O.R., Tanner, R.I. and Larsen, N.G. 2002. *Cereal Chem.* 79: 874–879.
Nout, M.J.R. and Aidoo, K.E. 2002. In: Osiewacz, H.D. (Ed.), The Mycota, Vol. X Industrial applications. Springer-Verlag, Berlin, Heidelberg, New York, pp. 23–47.
Nunez-Guerrero, M. E., Paez-Lerma, J. B., Rutiaga-Quiñones, O. M., Gonzalez-Herrera, S. M., and Soto-Cruz, N. O. 2016. *Food Microbiol.* 54: 91–97.
Nunomura, N. and Sasaki, M. 1992. Japanese soy sauce flavour with emphasis on off-flavours. In: Off-flavours in Foods and Beverages (Charalambous G, ed), Elsevier, Amsterdam, The Netherlands, pp. 287–312.
Nuraida, L., Wacher, M. C. and Owens, J. D. 1995. *World J. Microbiol. Biotechnol.* 11: 567–571.
Nyambane, B., Thari, W. M., Wangoh, J. and Njage, P. M. 2014. *Food Sci. Nutr.* 2: 692–699.
Padilla, B., Manzanares, P. and Belloch, C. 2014. *Food Microbiol.* 38: 160–166.
Palaniveloo, K. and Vairappan, C.S. 2013. *J Tropical Biol. Conserv.* 10: 31–41.
Pangallo, D., Sakova, N., Koreňová, J., Puškárová, A., Kraková, L., Valík, L. and Kuchta, T. 2014. *Int. J. Food Microbiol.* 170: 38–43.
Paramithiotis, S., Müller, M. R., Ehrmann, M. A., Tsakalidou, E., Seiler, H., Vogel, R. and Kalantzopoulos, G. 2000. *Systematic App. Microbiol.* 23: 156–164.
Pepe, O., Villani, F., Oliviero, D., Greco, T., and Coppola, S. 2003. *Int. J. Food Microbiol.* 84: 319–326.
Pereira-Dias, S., Potes, M. E., Marinho, A., Malfeito-Ferreira, M. and Loureiro, V. 2000. *Int. J. Food Microbiol.* 60: 55–63.
Perez-Coello, M.S., Briones Pérez, A.I., Ubeda Iranzo, J.F. and Martin Alvarez, P.J. 1999. *Food Microbiol.* 16: 563–573.
Poutanen, K., Flander, L. and Katina, K. 2009. *Food Microbiol.* 26: 693–699.
Price, E. J., Linforth, R. S., Dodd, C. E., Phillips, C. A., Hewson, L., Hort, J., and Gkatzionis, K. 2014. *Food Chem.* 145: 464–472.
Psani, M. and Kotzekidou, P. 2006. *World J. Microbiol. Biotechnol.* 22: 1329–1336.
Pyo, Y. H., and Seong, K. S. 2009. *J Agric Food Chem*, *57*: 8617–8622.
Que, F., Mao, L. C. and Zhu, C. G. 2006. *LWT-Food Sci. Technol.* 39: 111–117.
Rai, A. K. and Anu-Appaiah, K. A. 2014. *Food Biosci.* 5: 101–107.
Rai, A.K. and Jeyaram, K. 2015. In: JP Tamang (Eds.). Health Benefits of Fermented Foods and Beverages. CRC Press, Taylor and Francis Group of USA. pp. 455–474.
Rai, A. K., Maya, P. and Anu Appaiah K. A. 2010. *Int. J. Food Sci. Technol.* 45: 1330–1336.
Rai, A. K., Bhagat M., and Anu-Appaiah K. A. 2012. In: Recent Advances in Microbiology (Vol. 1), S. P. Tiwari, ed. Nova Science Publishers, New York. pp. 421–440.
Rai, A.K., Sanjukta, S. and Jeyaram, K. 2015. *Critical Rev. Food Sci. Nutr.* 10.1080/10408398.2015.1068736.
Randez-Gil, F., Sanz, P., Prieto, J.A., 1999. *Trends Biotechnol.* 17: 237–244.
Rani, U.M. and Appaiah, A.K.A. 2015. In: JP Tamang (Eds.). Health Benefits of Fermented Foods and Beverages. CRC Press, Taylor and Francis Group of USA. pp. 489–513.
Reed, G. and Nagodawithana, T.W. 1988. *Am. J. Enol. Vitic.* 39: 83–90.
Rezaei, M.N., Aslankoohi, E., Verstrepen, K.J., Courtin, C.M. 2015. *Int. J. Food Microbiol.* 204: 24–32.
Rhee, S.J., Lee, C.Y.J, Kim, K.K. and Lee, C.H. 2003. *Food Sci. Biotechnol.* 12: 242–247.
Rohm, H., Eliskases-Lechner, F. and Bräuer, M. 1992. *J. Appl.* Bacteriol, *72*: 370–376.
Romano, P., Ricciardi, A., Salzano, G. and Suzzi, G. 2001. *Int. J. Food Microbiol.* 69: 45–51.
Romi, W., Keisam, S., Ahmed, G. and Jeyaram, K. 2014. *BMC Microbiol.* 14: 52.
Romi, W., Ahmed, G. and Jeyaram, K. 2015. *Mol. Ecol.* 24: 3372–3389.
Rosenquist, H. and Hansen, A. 2000. *Food Microbiol.* 17: 241–250.
Salovaara, H. and Savolainen, J. 1984. *Acta Aliment. Pol.* 10: 241–246.
Sandhu, D.K. and Soni, S.K. 1989. *J. Food Sci. Technol.* 26: 21–25.
Sandhu, D.K., Soni, S.K. and Vilkhu, K.S. 1986. In: Yeast Biotechnology (Vashishat RK & Tauro P, eds), Haryana Agricultural University, Hisar, India, pp. 142–148.

Sanjukta, S., Rai, A.K., Ali, M.M., Jeyaram, K. and Talukdar, N.C. 2015. *J. Func. Foods* 14: 650–658.
Schifferdecker, A. J., Dashko, S., Ishchuk, O. P. and Piškur, J. 2014. *Yeast*, 31: 323–332.
Seiler, H. and Busse, M. 1990. *International journal of food microbiology*, *11*: 289–303.
Shetty, P. H. and Jespersen, L. 2006. *Trends Food Sci. Technol.* 17: 48–55.
Shi, Y. C. and Pan, T. M. 2011. *App. Microbiol. Biotechnol.* 90: 1207–1217.
Shori, A.B. 2012. *J. Saudi Soc. Agric. Sci.* 11: 79–88.
Shurpalekar, S.R. 1986. In: Legume-based Fermented Foods (Reddy NR, Pierson MD & Salunkhe DK, eds), CRC Press, Boca Raton, FL, pp. 191–217.
Sicard, D. and Legras, J. L. 2011. *Comptes rendus biologies* 334: 229–236.
Slaughter, J.C. 1999. *Biol. Rev.* 74: 259–276.
Soleas, G. J., Diamandis, E. P. and Goldberg, D. M. 1997. *J. Clin. Lab. Anal.* 11: 287–313.
Song, N.E. and Baik, S.H. 2014. doi:10.1016/j.jbiotec.2014.07.285.
Soni, S. K., Sandhu, D. K., Vilkhu, K. S. and Kamra, N. 1986. *Food Microbiol.* 3: 45–53.
Spicher, G. and Schröder, R. 1978. *Z. Lebensm. Unters. Forsch.* 167: 342–354.
Strauss, M.L.A., Jolly, N.P., Lambrechts, M.G. and van Resemburg, P. 2001. J. App. Microbiol. 91: 182–190.
Styger, G., Prior, B., Bauer, F. F. 2011. *J. Ind. Microbiol. Biotechnol.* 38: 1145–1159.
Suárez-Lepe, J. A. and Morata, A. 2012. *Trends Food Sci. Technol.* 23: 39–50.
Swiegers, J.H., Willmott, R.L., Siebert, T.E., Lattey, K., Bramley, B.R., Francis, J.L., King, E.S. and Pretorius, I.S. 2009. *Food Microbiol.* 26: 204–211.
Tamang, J.P. and Fleet, G.H. 2009. In: T., Kunze Satyanarayana (Ed.), Yeast Biotechnology: Diversity and Applications, Springer, Netherlands. pp. 169–198.
Tamang, J.P, and Thapa, S. 2006. *Food Biotechnol.* 20: 251–261.
Tamine, A. Y. and Robinson, R. K. 1999. *Yoghurt science and technology*, 11–109.
Thapa, S. and Tamang, J. P. 2004. *Food Microbiol.* 21: 617–622.
Tofalo, R., Fasoli, G., Schirone, M., Perpetuini, G., Pepe, A., Corsetti, A. and Suzzi, G. 2014. *Int. J. Food Microbiol.* 187: 41–49.
Tornadijo, M. E., Fresno, J. M., Sarmiento, R. M. and Carballo, J. 1998. *Le Lait* 78: 647–659.
Torner, M.J., Martinez-Anaya, M.A., Antun, B. and Benedito de Barber, C. 1992. *Int. J. Food Microbiol.* 15: 145–152.
Truelsen, T., Gronbaek M., Schnohr, P. and G. Boysen. 1998. *Stroke* 29: 2467–2472.
Tsapatsaris, S. and Kotzekidou, P. 2004. *Int. J. Food Microbiol.* 95: 157–168.
Tudor, E. A., and Board, R. G. 1993. *The yeasts. Volume 5: yeast technology.*, (Ed. 2), 435–516.
Turk, M., Carlsson, N. and Sandberg, A.S. 1996. *J. Cereal Sci.* 23: 257–264.
Valmorri, S., Tofalo, R., Settanni, L., Corsetti, A. and Suzzi, G. 2010. *Ant. Van Leeuwenhoek* 97: 119–129.
Verheyen, C., Albrecht, A., Elgeti, D., Jekle, M. and Becker, T. 2015. *Food Res. Int.* 76: 860–866.
Viljoen, BC. 2001. *Int. J. Food Microbiol.* 69: 37–44.
Viljoen, B. 2006. In Yeasts in Food and Beverages, eds. A. Querol and G. Fleet, Springer, Berlin, pp. 83–110.
Viljoen, B. C. and Greyling, T. 1995. *Int. J Food. Microbiol*, *28*: 79–88.
Vogelmann, S. A., Seitter, M., Singer, U., Brandt, M. J. and Hertel, C. 2009. *Int. J. Food Microbiol.* 130: 205–212.
Vrancken, G., De Vuyst, L., Van der Meulen, R., Huys, G., Vandamme, P. and Daniel, H.M. 2010. *FEMS Yeast Res.* 10: 471–481.
Vuorinen, H., K. Mtt, and R. Trrnen. 2000. *J Agric. Food Chem.* 48: 2675–2680.
Wacher, C., Cañas, A., Bárzana, E., Lappe, P., Ulloa, M. and Owens, J. D. 2000. *Food Microbiol.* 17: 251–256.
Wah, T. T., Walaisri, S., Assavanig, A., Niamsiri, N. and Lertsiri, S. 2013. *Int. J. Food Microbiol.* 160: 282–289.
Wang, J. G. 1998. *China Brewing* 98: 34–35.
Weaver, CM., Heaney, R.P., Martin, B.R. and Fritzsimmons, M.L. 1991. *J. Nutr.* 121: 1769–1775.
Westall, S. and Filtenborg, O. 1998. *Food Microbiol.* 15: 215–222.

Wojtatowicz, M., Chrzanowska, J., Juszczyk, P., Skiba, A. and Gdula, A. 2001. *Int. J. Food Microbiol.* 69: 135–140.
Wolfe, B. E., Button, J. E., Santarelli, M. and Dutton, R. J. 2014. *Cell* 158: 422–433.
Yang, Y., Shevchenko, A., Knaust, A., Abuduresule, I., Li, W., Hu, X., Wang, C. Shevchenko, A. 2014. *J. Archaeological Sci.* 45: 178–186.
Yoshida, S. and Yokoyama, A. J. 2012. *Biosci. Bioeng.* 113: 556–561.
Zhang, J., Liu, W., Sun, Z., Bao, Q., Wang, F., Yu, J. and Zhang, H. 2011. *Food Cont.* 22: 767–774.
Zhu, Y. and Tramper, J. 2013. *Biotechnol. Adv.* 31: 1448–1457.

Probiotic Yeasts in Human Welfare

V. Choudhary, A. Vohra, A. Madan and Tulasi Satyanarayana

Abstract Species of *Saccharomyces* particularly *S. boulardii* are gaining popularity as promising probiotic organisms. This organism acts by releasing bacteriostatic or bactericidal substances, which inhibit pathogenic effects of bacterial toxins, has anti-secretory action, and shows trophic, immune-stimulatory and anti-inflammatory response. Evidence showing the anti-diarrhoeal activity of lyophilized *S. boulardii* in various diarrhoeal diseases is well documented. Numerous randomized controlled trials have shown that *S. boulardii* prevents acute gastroenteritis, enteral nutrition-related diarrhoea, traveller's diarrhoea and decrease the symptoms related with *Helicobacter pylori* treatment. *S. boulardii* effectively prevents Antibiotic Associated Diarrhoea (AAD), treating irritable bowel syndrome and recurrent *Clostridium difficile* disease. This yeast has shown promising results when used for the treatment of human immunodeficiency virus-related diarrhoea, giardiasis and Crohn's disease. Yeasts are superior probiotics over bacteria as they are naturally resistant to conventional antibacterial antibiotics. *S. boulardii* is, thus, commercially available as a safe and well-tolerated probiotic. This chapter discusses various aspects of probiotic yeasts and their role in human well-being.

Keywords Probiotic · *Saccharomyces boulardii* · Diarrhoea · Probiotic yeast · Human welfare

V. Choudhary · A. Vohra (✉) · A. Madan
Department of Microbiology, Institute of Home Economics,
University of Delhi, F-4 Hauz Khas Enclave, New Delhi 110 016, India
e-mail: vohra_ashi@yahoo.co.in

T. Satyanarayana
Division of Biological Sciences and Engineering, Netaji Subhas Institute of Technology,
Azad Hind Fauz Marg, Sector 3, Dwarka, Delhi 110078, India
e-mail: tsnarayana@gmail.com

T. Satyanarayana and G. Kunze (eds.), *Yeast Diversity in Human Welfare*, DOI 10.1007/978-981-10-2621-8_5

1 Introduction

Gut microbiota refers to the total population of microorganisms including bacteria as well as fungi, archaea, viruses and protozoa inhabiting human gastrointestinal tract (Sekirov et al. 2010; Jandhyala et al. 2015). These microorganisms play a significant functional role in maintaining our intestinal microbial balance and human health as a whole. The important functions of gut microorganisms include stimulation of the immune system, prevention of colonization by pathogenic microbes, maintenance of structural integrity of the gut mucosa, drug metabolism, and further, aids in digestion (Jandhyala et al. 2015; Guarner and Malagelada 2003a, b).

Balanced gastrointestinal microflora comprises more number of beneficial or health promoting bacteria than the harmful ones. The composition of this microflora is host specific and it constantly evolves throughout an individual's lifetime. There is mounting evidence that clearly demonstrates effects of excessive use of antibiotics, diet and stress on normal gut microflora, thereby threatening human health. An imbalance in gut microflora could result in mild infections to severe life threatening diseases such as constipation, gastrointestinal tract infections, (Kennedy et al. 2014), inflammatory bowel disease (Ferreira et al. 2014), irritable bowel syndrome, cardiovascular disease, food allergies (Berin and Sampson 2013), antibiotic-induced diarrhoea (Barnes and Yeh 2015), and colorectal cancer (Erejuwa et al. 2014). Development of high level of resistance to commonly used antibiotics makes the treatment difficult and forces to search alternative disease control strategies such as using microorganisms as probiotics.

The word probiotic comes from greek word 'probios' meaning 'for life'. By definition, probiotics are living microorganisms which are beneficial for the host in adequate quantities. Generally, probiotic microorganisms are *Lactobacillus* and *Bifidobacterium*; recently strains of *Bacillus*, *Pediococcus* and yeast such as *Saccharomyces* are considered as promising candidates. They help in protecting us from pathogens as well as strengthening the host's immune system. A probiotic microorganism must fulfill the following parameters: safe to use, resistant to pancreatic secretions and acidic conditions prevailing in stomach, capable of adhering to epithelial cells of host, antimicrobial activity, antibiotic resistance, inhibits adhesion of pathogenic bacteria to probiotics, tolerates the presence of food additives and remains stable in the food matrix. Probiotic organisms can be consumed in the form of dairy as well as non-dairy products. It is important to include these organisms in the regular diet to establish a positive balance of beneficial microbial populations in the intestinal flora especially after antibiotic therapy, which destroys the normal gastrointestinal microbial flora. Probiotics are dietary supplements containing useful micro-organisms such as bacteria or yeast, which help in fighting pathogens and thus provides health benefits to humans (Agheyisi 2014). This chapter is aimed at reviewing the developments in the utility of probiotic yeasts in treatment of various human diseases, health benefits and future trends.

2 Probiotic Microorganisms

Different strains of any microbial species vary in their probiotic potential. These diverse strains are always unique and may differ in probiotic effects such as in their ability to adhere to host cells, specific immune activation and actions on a healthy and infected mucosal milieu. The widely used probiotics for humans include *Lactobacillus* spp. and *Bifidobacterium* spp., members of normal gut microflora (McFarland 2015). Also, non pathogenic *E. coli* strains and some *Saccharomyces* species are considered as probiotics. Yeasts constitute <0.1% of gut microflora. Yeast cells are ten times bigger than bacteria and thus they act as significant steric hindrance to the commensal bacteria.

2.1 Bifidobacteria

Tissier isolated Bifidobacteria from intestinal microflora present in breast-fed infants for the first time in 1899. The morphology of this bacteria was bifid (Y-shaped), therefore, named as *Bacillus bifidus*. Bifidobacteria are categorized as gram-positive rod shaped bacteria which are non-motile, catalase-negative and non-spore forming (Sgorbati 1995). The age of the individual and diet determines overall distribution of Bifidobacteria in the human gut.

Most common intestinal *Bifidobacterium* species are *Bifidobacterium adolescentis, B. breve, B. bifidum, B. catenulatum, B. angulatum, B. dentium, B. longum* and *B. pseudocatenulatum* (Biavati et al. 2006). *Bifidobacterium* spp. become predominant in the newborn shortly after birth as the infant is exposed to microflora of the mother and environment. The number of Bifidobacteria keeps on reducing with the increasing age of a person, ultimately making it the third most abundant genus in the adult gut microflora (Finegold et al. 1983).

2.2 Lactobacillus

Lactobacilli were first isolated by Moro in 1990 as *Bacillus acidophilus*. *Lactobacillus* is Gram-positive, non-motile rods or coccobacilli which does not forms spores (Hammes and Vogel 1995). These are either facultative anaerobes or microaerophilic. Lactobacilli belong to lactic acid bacteria, capable of fermenting glucose to lactic acid in either homo-fermentative or hetero-fermentative pathway. Lactobacilli are predominant in small intestine and helps in the initial steps in digestion of food. These bacteria have positive health promoting effects on gastrointestinal and genitourinary tracts. Thus, tend to improve the general overall health of the individual. The most important *Lactobacillus* species includes:

L. acidophilus: It is abundant in the small intestine and colon because of its microaerophilic nature. It plays important roles in promoting phagocytosis, prevents the growth of opportunistic pathogens like *Candida albicans* in vagina, develops robust immune system against intestinal bacteria and viruses, reduces diarrhoeal incidence in children and *H. pylori* infections in the human.

L. rhamnosus: It is present in intestinal and vaginal tract. It is able to tolerate gastric acidity and bile conditions. It is able to translocate across intestinal tract very easily. But it remains adhered to intestinal mucosa for a prolonged period within the gut.

2.3 *Saccharomyces*

The principal probiotic yeasts are *Saccharomyces boulardii* (Sb) and *S. cerevisiae* or Brewer's yeast. Another name for *S. boulardii* is *Saccharomyces cerevisiae* Hansen CBS 5926. It is a nonpathogenic yeast widely used for treating antibiotic associated diarrhoea and recurrent *Clostridium difficile* infection. *Saccharomyces cerevisiae*, also called as Brewer's yeast, is used for baking and beer fermentation. *Saccharomyces cerevisiae* stimulates the host immune system and provides good immunity against infections as it is a source of proteins, B-complex vitamins, and minerals like iron, zinc, magnesium, potassium, selenium, chromium; makes unknown growth factors; boosts the digestive system; source of various digestive enzymes, and degrades anti-nutritional factor, phytic acid in human food and animal feeds (Yamada and Sgarbieri 2005).

French microbiologist Henri Boulard discovered *Saccharomyces boulardii* while looking for a new strain of yeast that can withstand high temperature in Indochina for wine production. During the cholera outbreak, he observed that locals who were drinking tea showed no symptoms of cholera. This tea was developed from the skins of the fruit lychee (*Litchi chinensis*) and mangosteen (*Garcinia mangostana*). He named the active agent as "*S. boulardii*" (McFarland 2010). This yeast is being used for prevention and treatment of various gastrointestinal problems that occur by the administration of antimicrobial agents in many parts of the world. Various properties of *Saccharomyces boulardii* make it a suitable probiotic (Czerucka et al. 2007) include:

1. its ability to pass through the GI tract because it is tolerant to variations in pH, temperature, bile salts, pancreatic juices and GI enzymes,
2. it helps in maintaining and restoring the normal intestinal flora,
3. its nonpathogenic and localized in the gastrointestinal tract,
4. its optimum growth at 37 °C, and
5. its ability to restrict a variety of microbial pathogens.

Saccharomyces boulardii has been commonly used in the treatment of various health complications. It is a eukaryotic organism and, thus, is different from

Table 1 Genetic and physiological differences between *Saccharomyces boulardii* and *Saccharomyces cerevisiae* (Czerucka et al. 2007)

Property	*S. cerevisiae* var *boulardii*	*S. cerevisiae*
Optimum temperature	~37 °C	~30 °C
Survives at low pH	Yes	No
Galactose as carbon source	No	Yes
Sporogenous	No, asporogenous	Sporogenous
Ty1/2 elements	Absent	Present
Genetic stability	Chromosome IX trisomy	Stable strains having variable ploidy

bacterial probiotics, which are prokaryotic in nature. The most important differences are the presence of different physiological structures in yeast cells, their large size, not acquiring antibiotic-resistance genes and not affected by antibacterial antibiotics. The classification of *Saccharomyces* strains has been debated, but the recent typing methods identified *S. cerevisiae* var. *boulardii* as a strain different from other *S. cerevisiae* strains. The differences between the two are listed in Table 1.

The potential of *S. cerevisiae* var. *boulardii* as a probiotic has been well documented because it tolerates low pH and bile conditions. It also provides protection against bacterial infections by reducing the intestinal pro-inflammatory response. This probiotic yeast has received Qualified Presumption of Safety (QPS) status from The European Food Safety Authority (EFSA). Presently, it is the only probiotic with clinical importance.

3 Impact of Antibiotics on Yeast

There could be transfer of genes (horizontal or vertical) between different bacterial species which makes the pathogenic bacteria resistant to antibiotics. The genetic exchange occurs in gastrointestinal tract between resident gut flora and transient bacterial probiotics. Recent research suggests that commensals such as lactic acid bacteria may contain genes resistant to antibiotics. The genes responsible for imparting resistance to common antibiotics like tetracycline, vancomycin and erythromycin are well studied in *Lactobacillus* spp. used as probiotics such as *L. lactis*, and others (Egervarn et al. 2009). The genetic exchange of drug resistance genes from these bacteria to the pathogens is the main problem associated with the use of *Lactobacillus* and bacterial probiotics. The natural ability of yeast to remain resistant to bacterial antibiotics increases their effectiveness as probiotic for antibiotic-treated patients. No genetic exchange can occur between bacteria and yeast, thereby making the yeast a better and safer probiotic.

4 Clinical Studies Involving Probiotic Yeast

4.1 Traveller's Diarrhoea

This health problem is very common among those travelling to developing countries, where the probability of acquiring enteric-infection is quite frequent. Enterotoxicogenic strains of *E. coli*, *Salmonella* and *Shigella* are responsible for 80% of the diarrhoea cases (Diemert 2006; Sanders and Tribble 2001). In a study involving 3,000 Austrian tourists going to various countries, researchers found that *S. boulardii* reduced the incidences of traveler's diarrhoea. The diarrhoeal incidence was 40, 34 and 29% in patients given placebo, *S. boulardii* 150 mg daily and 500 mg daily, respectively. The efficiency of the yeast varies with the dosage given and its preparation. In a meta-analysis of 12 different studies, combination of the probiotics (*S. cerevisiae* var. *boulardii* and *Bifidobacterium bifidum* plus *Lactobacillus acidophilus*) prevented traveller's diarrhoea (McFarland 2007).

4.2 Antibiotic Associated Diarrhoea (AAD)

Antibiotic therapy commonly leads to this condition due to disruption of normal bacteria present in digestive tract and colonization of harmful bacteria causing inflammation of the intestinal mucosa (Coté and Buchman 2006). The opportunistic pathogen infecting persons with AAD are *C. difficile* (Rohde et al. 2009), *Staphylococcus aureus*, *Clostridium perfringens*, *Klebsiella oxytoca*, *E. coli*, *Candida* and *Salmonella* sp. (Beaugerie and Petit 2004; Asha et al. 2006). Numerous investigations have reported the use of *S. boulardii* in antibiotic associated diarrhoea in adults and children and its beneficial effects (Moré and Swidsinski 2015; Shan et al. 2013; Hatoum et al. 2012; McFarland 2010; D'Souza et al. 2002; Kotowska et al. 2005; Can et al. 2006; Bravo et al. 2008).

Adam et al. (1977) reported that only 4.5% of the patients receiving *S. boulardii* (200 mg) for a week showed lesser AAD symptoms as compared to the control. Diarrhoea was seen in only 9.5% cases in probiotic supplemented individuals as compared to 21.8% in the control group (Surawicz et al. 1990). In another investigation, patients on β-lactam antibiotic treatment, AAD was seen in 7.2% people who took *S. boulardii* (1 g) for three days in comparison with 14.6% in placebo (Mcfarland et al. 1994). Can et al. (2006) showed that AAD was seen in only 1.4% of the *S. boulardii* treated patients versus 9.0% in placebo. In the study carried out by Surawicz et al. (2000), *S. boulardii* was shown to prevent diarrhoea in 7% patients (n = 193), when the probiotic (1 g/day) was given at the time of starting the antibiotic treatment and 3 days after completion of the antibiotic course. Meta-analysis of 21 randomized controlled trials (4780 individuals), where probiotic treatment was given to patients on antibiotic therapy, a remarkable decline (18.7–8.5%) in the occurrence of AAD was observed. *S. boulardii*, in children, decreased the risk from 20.9 to 8.8%, whereas in adults, it was from 17.4 to 8.2% (Szajewska and Mrukowicz 2015). Meta-analysis done by McFarland (2010) also focused on the prevention of AAD symptoms using of *S. boulardii* as therapeutic probiotic.

Treatment module in which broad spectrum antibiotics were given to the children, 11–40% diarrhoea cases were seen (Turck et al. 2003). First randomized trial where *S. boulardii* was able to prevent AAD was done on 269 children (Kotowska et al. 2005). Recently a meta-analysis explaining the data of 11 RCTs further confirmed its effectiveness as probiotic for the therapy of AAD (Szajewska and Kołodziej 2015).

AAD in hospitalized patients (20–25% cases) and community patients (10% cases) was caused by the infection of *Clostridium difficile* (Beaugerie et al. 2004). McFarland et al. (1990) carried out randomized, placebo-controlled trial on 124 patients to demonstrate the effectiveness of *S. boulardii* and placebo having drugs like metronidazole or vancomycin. In this report, *C. difficile* infection was documented in 64 individuals and its recurrence in 60 patients. There was 50% reduction in the recurrence of *C. difficile* infection after administration of *S. boulardii*. Similar observations are recorded for patients taking high-dose of vancomycin (Surawicz et al. 2000). Thus, on the basis of these trials and another meta-analysis involving six RCTs where different probiotics were used (McFarland 2006), *S. boulardii* was the sole effective probiotic that prevented recurrent *C. difficile* associated diarrhoea.

4.3 Helicobacter pylori *Diarrhoea*

Helicobacter pylori colonizes gastric mucosal lining and causes peptic ulcers. For its treatment, patients are given antibiotic (amoxicillin, clarithromycin and omeprazole or lansoprazole) to suppress the bacterial growth. However, these antibiotics affect the normal gut microflora also.

Multiple clinical trials suggest that *S. boulardii* only reduces the negative impacts of the standard triple therapy rather than completely eradicating *H. pylori* infections. To access the impact of *S. boulardii* in *H. pylori* infection, randomized controlled trials were carried out on children (Hurduc et al. 2009; Gotteland et al. 2005) and on adult population (Cindoruk et al. 2007; Cremonini et al. 2002). In one of the trials, use of *S. boulardii* reduced *H. pylori* infection in 12% children (Gotteland et al. 2005). Another study on 145 children, the eradication rate was 87.7% and side effects were greatly diminished in the *S. boulardii* group (Hurduc et al. 2009).

Cremonini et al. (2002) investigated the probiotic potential of *S. boulardii*, well established bacterial probiotics (*Lactobacillus rhamnosus* GG or a mixture of *L. acidophilus* and *Bifidobacterium lactis*) and placebo in triple therapy for people carrying *H. pylori* infections with no symptoms. Eradication rates of *H. pylori* infection were same in all groups during second week, but antibiotic-associated diarrhoea was less (5%) in probiotic groups in comparison to placebo (30%).

Cindoruk et al. (2007) correlated eradication rate as well as reduction rate of *H. pylori* infection to standard triple therapy. There was hardly any appreciable change in *H. pylori* eradication (71% in Sb vs. 60% in placebo) after six weeks but reduction in epigastric distress in 14.5% patients compared with placebo (43.5%) (Cindoruk et al. 2007). A recent meta-analysis of 11 RCTs (2200 total, 330 children), showed 80% eradication in *S. boulardii* group compared to 71% in the control. Also, the risk of *H. pylori* triple therapy related adverse effect was reduced in *S. boulardii*, mainly diarrhoea and nausea (Szajewska et al. 2015). These studies show that *H. pylori* is not eliminated by *S. boulardii*, but the yeast effectively reduces the adverse effects of the standard therapy.

4.4 Cold and Flu

Brewer's yeast has been used to prevent colds and seasonal influenza (flu) in a large, randomized trial on 116 adults. In the 12 weeks trial, participants received either 500 mg of a brewer's yeast (Epicor) or placebo. Low incidence of cold and flu was seen in people who received Epicor. These people became ill for a shorter time period as compared to placebo (Moyad et al. 2008). Further research is called for routinely using brewer's yeast for the prevention of colds and flu.

4.5 Inflammatory Bowel Diseases (IBD)

Crohn's disease and ulcerative colitis are the main inflammatory bowel diseases. Its characteristic feature is chronic inflammation in the mucosa of intestinal cells (Kaser et al. 2010).

4.5.1 Crohn's Disease

The distinctive feature of Crohn's disease is disruption of mucosal integrity. It mainly affects small intestine and colon. Symptoms include diarrhoea, abdominal pain and GI bleeding. Being a sporadic disease of no defined etiology, it becomes challenging for clinical trials, presents multiple disease outcome, requires long term treatment with regular follow-ups. Three RCTs have analysed the role played by *S. boulardii* in treating Crohn's disease patients (Guslandi et al. 2000; Plein and Hotz 1993).

Plein and Hotz (1993) performed a randomized double blind controlled study involving 20 Crohn's disease patients. These people were treated with either 750 mg *S. boulardii* for 7 weeks or placebo. Maintenance therapy was given to the patients throughout the trial period. In patients receiving *S. boulardii*, reduction in frequency of bowel movements was seen in comparison with the placebo group. In second study involving 31 Crohn's disease patients, *S. boulardii* was given for 3 months and the patients were given their maintenance medicines like mesalamine, azathioprine or others. Intestinal permeability was highly reduced upon *S. boulardii* treatment as compared to placebo, thus reducing the probability of translocation of bacteria. Guslandi et al. (2000) carried out a study on 32 Crohn's disease patients in remission for ≥ 3 months; only 6% patients relapsed as compared to control group (38%). No ill effects were observed while conducting the trial. Thus, *S. boulardii* can be given as a safe and reliable therapy for this condition.

4.5.2 Ulcerative Colitis

S. boulardii has been used for treatment and prevention of relapses in ulcerative colitis patients (Varankovich et al. 2015). Adults experiencing less severe ulcerative colitis were given mesalazine and *S. boulardii* for a period of four weeks. Almost 68% patients showed positive response to this treatment (Guslandi et al. 2003). As the treatment was given only for a short period of time and follow up was not properly done, these promising results need to be confirmed.

Guslandi (2010) carried out another pilot study, where 6 patients having ulcerative colitis were given a combination of *S. boulardii* with rifaximin for 3 months. This treatment effectively prevented early symptoms of ulcerative colitis. Further, additional studies with large number of patients and treatment with probiotic agent for longer duration are required (Floch et al. 2008). Moreover, placebo-controlled studies need to be done for confirming the role played by probiotic yeast in ulcerative colitis treatment.

4.6 Irritable Bowel Syndrome

Saccharomyces boulardii showed promising results for treatment of patients having diarrhoea-predominant IBS (IBS-D). A randomized double blind, placebo-controlled clinical trial, 179 adults who received 500 mg *S. boulardii*, showed considerable reduction in abdominal pain or discomfort without changes in frequency and consistency of stools (Pineton de Chambrun et al. 2015). In another study, where Sb was used either alone or in combination therapy with mesalamine, prominent reduction in IBS-D symptoms was observed (Bafutto et al. 2013). There had been significant reduction in the number and improvement in the stool consistency ($P < 0.05$), when 250 mg of *S. boulardii* was given twice daily (Czerucka et al. 2007).

4.7 Aids

HIV-associated diarrhoea is a major complication, which is difficult to control by standard therapy. In a random double blind study involving 35 patients, *S. boulardii* was given (3 g each day) for a week effectively controlled diarrhoea. Saint Marc et al. (1995) showed that after *S. boulardii* treatment for one week, no diarrhoea symptoms were observed in 61% cases versus 12% in placebo. Another controlled, double blind work on 35 AIDS adult patients presenting chronic diarrhoea for more than 24 days, showed that 10 out of 18 patients that were given *S. boulardii* (1.5 g, twice a day) did not show any symptoms of diarrhoea, while it was effective in only 1 out of 17 patients ($P < 0.001$) in the placebo group (Penna et al. 2000).

4.8 Giardiasis

The main characteristic feature of giardiasis is prolonged diarrhoea, severe weight loss, weakness and abdominal pain. It occurs commonly among people who consume apparently clean but contaminated water during outdoor activities. In a placebo-controlled double blind study on 65 adults in turkey, *S. boulardii* in combination with metronidazole for treatment of giardiasis infection was assessed. The two groups of patients were—group1 (30 patients) was given metronidazole (750 mg) with *S. boulardii* capsules (250 mg), while Group 2 (35 patients) received metronidazole (750 mg) without probiotic yeast capsules as placebo for 10 days. After examining the patients two weeks later, cysts of the parasite *Giardia lambia* were detected in 17.1% i.e. 6 cases of placebo. No cyst was seen in case of *S. boulardii*. These findings indicate that *S. boulardii* effectively treats giardiasis infection, in combination with metronidazole (Kelesidis et al. 2012; Besirbellioglu et al. 2006). This probiotic has been found to be effective in the treatment of other parasitic infections such as acute amebic colitis (Mansour-Ghanaei et al. 2003; Dinleyici et al. 2011, 2009).

4.9 Skin Health

S. boulardii also may be beneficial for making the skin healthy. In a randomized, controlled double-blind trial on 139 patients having acne, the probiotic yeast was assessed and compared with placebo for five months. The therapy was found to be effective in 74.3% of the patients, whereas it was 21.7% in placebo. Acne was reduced in more than 80% individuals receiving *S. boulardii*, while in the placebo group it was only 26% (Weber et al. 1989). Thus, *S. boulardii* can be exploited as an alternative therapy for acne treatment.

5 Safety Issues Associated with Use of *Saccharomyces boulardii* as Probiotic

Saccharomyces boulardii is generally available in lyophilized form as capsule or heat-dried formulations. Lyophilized powder stays stable at room temperature for long periods (one year or more) whereas heat dried form is unstable at room temperature and needs refrigerated conditions. Also, the lyophilized form is more convenient to package and easily portable (Graff et al. 2008; Schwenzer 1998). The four potential risks of using live organism as therapeutics include: their property of transferring antibiotic-resistance genes, their movement from the intestine to other regions of body, continued presence in intestinal mucosa and creating further adverse conditions.

S. boulardii has not become resistant to antimicrobials, unlike other well established probiotic bacteria *Enterococcus faecium* and *Lactobacillus rhamnosus*, which are highly resistant to antibiotics (Temmerman et al. 2001; McFarland 2010). Evidence from animal studies showed decreased translocation upon *S. boulardii* treatment (Karen et al. 2010). Pharmacokinetic analysis suggests that *S. boulardii* does not remain in the body 3–5 d after stopping the oral regimen, so persistence is not a problem with this probiotic yeast (McFarland 2015; Klein et al. 1993). *S. boulardii* was first used as probiotic in 1950s, and since then, it has been analysed in multiple clinical studies worldwide. These numerous trials documented the safe use of *S. boulardii* for treating gastro-intestinal disorders like antibiotic-associated diarrhoea, IBD, IBS, HIV-related diarrhoea and giardia infections without any adverse effects. Thus, *Saccharomyces boulardii* has proven as a 'probiotic drug' (Czerucka et al. 2007).

Several sporadic incidences of *S. boulardii* associated fungaemia have been reported in patients having severe general or intestinal diseases and on central venous catheters (Hennequin et al. 2000). Fungemia, was not due to direct inhalation of *S. boulardii* probiotics, but might be due to contaminated environmental fomites (Hennequin et al. 2000; Herek et al. 2004). It is a challenging task to determine whether the fungemia is due to *S. cerevisiae* (non-boulardii strains) or *S. boulardii* cases because of lack of available advanced yeast identification assays (Enache-Angoulvant and Hennequin 2005). Till date, almost 100 cases of fungemia associated with *S. boulardii* have been reported (Vandenplas et al. 2009). Once diagnosed, *S. boulardii* fungemia can be effectively treated using antifungal drug, however, in some cases, failure of fluconazole treatment has been reported (Kelesidis and Pothoulakis 2012; Burkhardt et al. 2005). *S. boulardii* as probiotic is not recommended to people having yeast allergies, diabetes, in immunocompromised or critically ill condition (Hwang et al. 2009; Whelan and Myers 2010; Riquelme et al. 2003). Few persons may experience general side effects like bloating or gastric problem upon ingesting probiotics. A doctor should always be consulted for taking any supplement or medication. A comprehensive list of commercially available probiotics is presented in Table 2.

Table 2 Commercially available probiotics containing *Saccharomyces boulardii* (McFarland 2010)

Probiotic	Manufacturer	Colony forming unit (CFU) in billion per mg or ml	Microbial strains	Stability at room temp
Nexabiotic	Bioprosper labs (US)	30	*S. boulardii* and 22 other strains [*Streptococcus thermophilus*, *Lactobacillus* spp., *Bacillus* spp., *Bifidobacterium* spp.]	NO
Syntol AMD	Arthur Andrew Medical (US)	13.6	*S. boulardii* and 7 other strains (*Lactobacillus* spp., *Bacillus* spp., *Bifidobacterium bifidum*)	YES
Jarrows S. boulardii + MOS	Jarrow formulas (US) and Gnosis (Italy)	1.5	*S. boulardii* + MOS (MannanOligoSaccharides)	NO
Florastor®	Biocodex (France)	5/250 mg	*S. boulardii* lyophilised (lyo)	YES
Perenterol®	Medice Arzneimittel Pütter GmbH & Co. (Germany) Sanofi-aventis Pharmaceuticals Ltd. (Turkey)			
Reflor® Ultra-Levure	Biocodex (US)			
Protecflor®	Lallemand health solutions (France)	1/2 ml	*S. boulardii* and (*Lactobacillus helveticus* rosell-52, *Lactobacillus rhamnosus* rosell-11, *Bifidobacterium longum* rosell-175)	NOT STATED
Erce Flora	Sanofi-aventis Pharmaceuticals Ltd. (Belgium)			
MitoMix ® (cats, dogs and small animals)	Imagilin Technology (USA)	2.3 per capsule	*S. boulardii* and *Pediococcus acidilactici*	YES
NutriLots™ (humans)				
Pro-Bio Defense™	Kirkman group (Spain)	20 per capsule	*S. boulardii* + 7 other strains (*Lactobacillus* spp., *Bifidobacterium lactis*, *Streptococcus thermophilus*)	NO
ABX Support™	Klaire's lab (US)	5/300 mg	*S. boulardii* and 3 other strains (*L. rhamnosus*, *Bifidobacterium bifidum*, *Bifidobacterium breve*)	NO
Kombucha fermented tea	Millennium products, Inc (US)	1 per 16 oz.	*S. boulardii* and two other strains (*Lactobacillus* bacterium and blue-green algae)	NO

6 Mechanism of Action of Probiotic Yeast

Multiple studies show that *S. cerevisiae* var. *boulardii* acts as probiotic in a variety of ways as described below:

6.1 Effect on Enteric Bacterial Pathogens

S. boulardii interferes with intestinal pathogens either directly or indirectly. The probiotic yeast can directly stop the growth of various microbial pathogens (e.g. *Yersinia enterocoliticum*, *Candida albicans*, *Salmonella typhimurium*, *Aeromonas hemolysin*) (Zbinden et al. 2011; Ducluzeau and Bensaada 1982; Altwegg et al. 1995). Different mechanisms by which *S. cerevisiae* var. *boulardii* acts against various enteric pathogens are described below.

6.1.1 Preventing Adherence and Movement of Bacteria in the Epithelial Cells of Intestine

The cell wall of this probiotic yeast binds to enteric bacteria such as enterohaemorrhagic *Salmonella typhimurium* and *Escherichia coli* (Gedek 1999). The probiotic yeast reduced the adhesion of bacteria *C. difficile* to modified kidney epithelial cells (Vero cells) by modifying the cell surface receptors involved in adhesion process by steric hindrance or proteolytic activity (Tasteyre et al. 2002). Additionally, it was shown to decrease *E. coli* attachment to mesenteric lymph node of pigs (Lessard et al. 2009). *S. boulardii* has shown beneficial effects on infected animal models without affecting the bacterial adherence (Czerucka et al. 2000; Mumy et al. 2008; Rodrigues et al. 2000).

6.1.2 Producing Bacterial Toxin Neutralising Factors

S. boulardii may exert its anti-toxin effect by either blocking receptor sites of the toxin (Pothoulakis et al. 1993), acting as a bait for the toxin (Brandao et al. 1998) or by directly destroying the toxin (Castagliuolo et al. 1996).

C. difficile toxins (A and B) are cleaved by serine protease (54 kDa) of *S. boulardii* (Castagliuolo et al. 1996), thus, the binding of *C. difficile* toxins to brush border membrane receptor is inhibited. This finding is further confirmed when oral *S. cerevisiae* var. *boulardii* or its supernatant decreased intestinal secretion and permeability induced by toxin A (Pothoulakis et al. 1993; Castagliuolo et al. 1999). *S. boulardii* synthesizes a large protein (120 kDa) which inactivates cholera toxin. This protein lowers cAMP levels in the intestinal cells by competing with the chloride ions stimulated by *Vibrio cholerae* toxins (Czerucka and Rampal 1999;

Vidon et al. 1986). *S. boulardii* also secretes a protein phosphatase, which makes the *E. coli* 055B5 inactive by dephosphorylating its lipopolysaccharide (Buts et al. 2006).

6.2 Alterations in the Pro-inflammatory Response

S. cerevisiae var. *boulardii* blocked the activation of signaling molecules such as nuclear factor-kappa B (NF-κB) and mitogen-activated protein kinase (MAPK) in in vitro studies carried out on mammalian cell lines. These molecules reduced the expression of inflammation-associated cytokines for example, interferon gamma (IFN-γ), tumor necrosis factor alpha (TNF-α) and interleukin 8 (IL-8) (Mumy et al. 2008; Dalmasso et al. 2006a, b; Dahan et al. 2003). This probiotic strain produces a small 10 kDa protein that inhibits the formation of pro-inflammatory cytokines in human colonocytes and enhances mitogen-activated protein kinases 1 and 2 (ERK1/2) activity in human as well as mouse cells (Chen et al. 2006). Sougioultzis et al. (2006) demonstrated that *S. boulardii* produced a small soluble factor (<1 kDa) which inhibited activation of NF-κBand expression of IL-8 in monocytes and epithelial cells of intestine.

6.3 Anti-inflammatory Effect

The anti-inflammatory effects are important in many gastrointestinal conditions such as *C. difficile*, gut inflammation and in general irritable bowel disease (Chen et al. 2006; Pothoulakis 2009). *S. boulardii* shows its anti-inflammatory action by interfering with the NFkB and MAP kinase pathways in intestinal host cells and thus exerts its anti-inflammatory effect. The probiotic yeast increased the levels of peroxisome proliferator-activated receptor-gamma (PPAR-γ) in human colonocytes (Lee et al. 2005). This receptor is expressed by cells such as intestinal epithelial cells, dendritic cells, T and B cells, and can regulate inflammation (Su et al. 1999; Dubuquoy et al. 2006). In studies involving mice model of IBD, *S. boulardii* has been found to inhibit the inflammation of colon because it decreases IFN-γ production and increases the distribution of T-cells in mesenteric lymph nodes. Additionally, *S. boulardii* modifies adhesion of endothelial cells to leucocytes (Dalmasso et al. 2006a, b). Inflammatory response occurs due to increased level of nitric oxide (NO) and inducible nitric oxide synthase (iNOS) (Dijkstra et al. 1998; Moslehi-Jenabian et al. 2010). *S. boulardii* inhibited citrulline (marker of NO production) and iNOS activity in rats suffering from diarrhoea (Girard et al. 2005).

6.4 Trophic Effects on Intestinal Mucosa

Numerous investigations show that this probiotic yeast exhibits its trophic effect on intestinal mucosa by various mechanisms such as secreting spermidine and spermine or other brush border enzymes aiding in the enterocytes maturation (Buts et al. 1994; Jahn et al. 1996), reducing mucositis (Sezer et al. 2009), restoring fluid transport pathways (Breves et al. 2000; Schroeder et al. 2004), stimulating protein synthesis and ATP generation (Buts et al. 2008). *S. boulardii* cells synthesize and secrete polyamines (spermine, spermidine and putrescine) in rat ileum which enhances the expression of sucrase and maltase in the intestine (Buts et al. 1994). Proliferation of small intestine enterocytes requires polyamines. The release of polyamines might be involved in fast recovery of diarrhoea patient.

Upon administrating this probiotic to human volunteers or experimental animals (rats), an increased activity of brush border membrane enzymes like alkaline phosphatase, maltase-glucoamylase, sucrase-isomaltase and α-glucosidase and lactase was reported. These enzymes accelerated the process of nutrient absorption (Buts et al. 1986; Jahn et al. 1996). The disaccharides are converted to monosaccharides which are taken to bloodstream through enterocytes (Zaouche et al. 2000). In brush border epithelial cells of intestine, this yeast increased disaccharide, improved D-glucose absorption and enhanced the sodium/glucose cotransporter-1 (SGLT-1) expression (Buts et al. 1999). This increased expression of SGLT-1 has a significant impact on diarrhoea treatment and sucrase-isomaltase deficiency, because it improves the rate of water and electrolytes reabsorption. In cholera infection, *S. boulardii* reduces hypersecretion of water and electrolytes (Czerucka and Rampal 1999).

Another important trophic effect of probiotic yeast is alteration in the amounts of luminal short-chain fatty acids (SCFAs). Anaerobic bacteria produce SCFAs that help in the water and electrolyte absorption in the colon (Bowling et al. 1993). Upon giving oral *S. cerevisiae* var. *boulardii* to these patients, the concentrations of fecal SCFAs increased up to 9 days after stopping the treatment (Schneider et al. 2005). This yeast is, therefore, important in prevention of enteral nutrition-induced diarrhoea.

6.5 Increased Immune Response

The host cell immunity, both innate and adaptive, are known to be stimulated by *S. cerevisiae* var. *boulardii*. Orally administrating *S. cerevisiae* var. *boulardii* to healthy individuals activated the reticuloendothelial and complement system (Caetano et al. 1986; Kelesidis et al. 2012). *S. boulardii* is known to increase secretory IgA levels in the intestine of animal models (Ozkan et al. 2007). IgA is very effective in protecting the host gastrointestinal and respiratory tracts against invading pathogens. Qamar et al. (2001) reported increase in total sIgA and antitoxin sIgA levels when *S. cerevisiae* var. *boulardii* was given to *C. difficile* toxin A treated mice. It has also resulted in higher levels of serum IgG against toxins A and B of *C. difficile* (Kyne et al. 2001). *S. boulardii* may also stimulate pro-inflammatory cytokine production by interfering with signaling pathway mediated by NF-κB (Fidan et al. 2009; Thomas et al. 2009). A study by Chen et al. (2006) showed that the yeast stimulated IL-8 production in in vitro mouse models as it halts the activation of ERK1/2 and MAP kinases, and cell necrosis. This probiotic reduces inflammation by trapping T helper cells into mesenteric lymph nodes (Dalmasso et al. 2006a, b).

7 Beneficial Effects of Probiotic Yeast on Human Health

7.1 Amelioration in Digestion

Probiotic yeasts help in the release of multiple molecules into the lumen of intestine. The vitamins and enzymes aid in digestion, thus, alleviating symptoms of intestinal malabsorption. This leads to increased secretion of short chain fatty acids (SCFA). Upon absorption, SCFAs contribute towards total energy available for use by the host (Rombeau et al. 1990; Rolfe 2000).

7.2 Stimulation of Immune System

S. boulardii can stimulate both innate (first line of defense) and adaptive (antibodies) immune response against pathogens as suggested by studies conducted on animal models. It exerts immune stimulatory effect by increasing levels of secretory IgA, immunoglobulins and pro-inflammatory cytokines, activates macrophages and NK cells (Czerucka et al. 2007; Buts et al. 1999). β-Glucans of yeast cell wall activates the immune system (Barreto-Bergter et al. 2014; Samuelsen et al. 2014). Glucans binds to the cell surface receptors like complement receptors, Toll-like receptors and dectin-1, dendritic cell receptors (Vannucci et al. 2013; Goodridge et al. 2011). Thus, immune stimulation enables the host to destroy the intestinal pathogens at an early stage.

7.3 Inhibition of Enteropathogenic Bacteria

S. boulardii is known to lessen the adherence of enteropathogens such as *E. coli* to epithelial cells (Gedek 1999) as observed in case of mesenteric lymph node in pigs (Lessard et al. 2009). The yeast may show its anti-toxin effect either by blocking receptor sites of the toxin (Pothoulakis et al. 1993) or acting as a bait for the toxin (Brandao et al. 1998) or secrete enzymes and proteins that directly make the toxin inactive (Castagliuolo et al. 1996).

7.4 Prevention and Treatment of Gastro-intestinal Disorders

Saccharomyces boulardii has been shown to be effective in the prevention and treatment of gastrointestinal tract infections, constipation, irritable bowel syndrome (Kennedy et al. 2014), inflammatory bowel disease (Ferreira et al. 2014), food allergies (Berin and Sampson 2013), antibiotic-induced diarrhoea (Barnes and Yeh 2015; Kurugöl and Koturoglu 2005; Htwe et al. 2008), colorectal cancer (Erejuwa et al. 2014) and cardiovascular disease in adults and children. Nowadays, probiotic yeasts are given either as fermented products or in lyophilized powders to severe diarrhoea patients (Hatoum et al. 2012).

Table 3 lists the applications of *S. boulardii* as a probiotic.

8 Future Perspectives

Probiotic therapy has been proven to be quite beneficial for the treatment of a number of diseased conditions, despite main problems associated with their dosage and viability, absence of proper standardization protocols in the industry and safety issues. The food industry has marketed live organisms as food supplements, but there is scarcity of data available to support such claims. Hence, there is an urgent need to carry out further research with well designed placebo-controlled studies to determine the exact benefits of the probiotics, identify and characterize various probiotics strains, the strain specific mechanism of action, optimum dosage required for the best results and their viability and biosafety assessment. Industry-oriented research focusing on enhancing the half-life and survival in the GIT, adherence capability and appropriate procedures for production, handling and packaging of

Table 3 Applications of *Saccharomyces boulardii* as a probiotic

Acute diseases	Chronic diseases
Antibiotic-associated diarrhoea	(1) Crohn's disease
Traveller's diarrhoea	(2) Ulcerative colitis
Helicobacter pylori infection	(3) Irritable bowel syndrome
Clostridium difficile infection	(4) Human immunodeficiency virus (HIV)-associated diarrhea
Amebic colitis	(a) Giardiasis

probiotics are required before these products come into the public domain (market). With time, novel food products having probiotics will be made available in the market in the form of energy bars, juices, cereals and specific medicinal foods. Gene technology can effectively help in exploring novel potential strains of the probiotic yeasts other than *S. boulardii*. Only a strain of *S. boulardii* has been used in all the randomized control trials assessing its probiotic potential, but a probiotic mixture may prove as a superior probiotic than a single strain preparation in humans. A mixture of *L. acidophilus, L. rhamnosus, Bifidobacterium* and *S. boulardii* has shown promising results in pre-clinical studies in rats for treating *E. coli* diarrhoea. Currently researchers all over the world are working on the probiotic potential of *S. boulardii.*

9 Conclusions

The human gut microbiota is very complex and is important for maintaining homeostasis. However, it often gets unbalanced as we use antibiotics. This unbalanced condition can be prevented by using appropriate probiotics. But the activity of probiotics is strain specific and these probiotic strains with their human health benefits should be well characterized. Although prokaryotes have always been at the center of probiotic research in the recent times, the scientific and commercial setups are exploring the potential of eukaryotic probiotics. Yeasts are superior probiotics as compared to the traditional lactic acid bacteria as yeasts are resistant to antibiotics. The mode of action and effectiveness of S. boulardii, as documented in numerous studies on animal models and clinical trials, proves its use as a therapeutic probiotic. The probiotic potential of other strains of *Saccharomyces* needs to be thoroughly evaluated by clinical studies. *S. boulardii* has been found to be significantly effective for the prevention of several human ailments such as inflammatory bowel diseases, *Clostridium difficile* infection antibiotic associated diarrhoea, *Helicobacter pylori* diarrhea, Traveller's diarrhoea, and Irritable bowel syndrome. Diarrhoea is a major threat in developing countries, especially in children, being responsible for alarmingly high rate of mortality. In contrast, in the developed parts of the world, although it is not responsible for the death of children, its burden is having a huge economic impact on the society. Probiotic yeast exerts its beneficial effect by multiple mechanisms such as releasing antibacterial substances in vivo that makes the bacterial toxins ineffective, trophic effects, immune-stimulatory and anti-inflammatory effects. In recent clinical trials, *S. boulardii* has shown promising results in the treatment of chronic diseases (Crohn's disease, HIV-related diarrhoea) and preventing the relapse of *C. difficile* infections. The fields of potential applications of *S. boulardii* as probiotic are growing day by day.

References

Adam, J., Barret, C., Barret-Bellet, A., Benedetti, E., Calendini, A., Darchen, P., Galibert, J.M., Guerci, P., Guiot, G., Haechler, M., Junqua, R., Le Matelot, J., Legrand, J.P., Mallez, B., Massonneau, A., Mestas, J.J., Michelet, F.X., Papel, B., Petit, P., Pfeiffer, G., Pidoux, A., Portalis, H., Prado, F., Rameau, J. and Valence, P. 1977. Gazette Medicale de France. 84: 2072–2078.

Agheyisi, R. 2014. The probiotics market: Ingredients, supplements, foods, Report code: FOD035C, BCC Research, Wellesley, MA, USA. (http://www.bccresearch.com/report/FOD035B.html).

Altwegg, M., Schnack, J. and Zbinden, R. 1995. Med. Microbiol. Lett. 4: 417–425.

Asha, N.J., Tompkins, D. and Wilcox, M.H. 2006. J. Clin. Microbiol. 44: 2785–2791.

Bafutto, M., Almeida, J.R., Leite, N.V., Costa, M.B., Oliveira, E.C. and Resende-Filho, J. 2013. Arq. Gastroenterol. 4: 304–309.

Barnes, D. and Yeh, A.M. 2015. Nutr. Clin. Pract. 30: 747–759.

Barreto-Bergter, E. and Figueiredo, R.T. 2014. Front Cell Infect. Microbiol. 4: 145.

Beaugerie, L. and Petit, J.C. 2004. Best Pract. Res. Clin. Gastroenterol. 18: 337–352.

Berin, M.C. and Sampson, H.A. 2013. Curr. Biol. 9: R389–400.

Besirbellioglu, B.A., Ulcay, A., Can, M., Erdem, H., Tanyuksel, M., Avci, I.Y., Araz, E. and Pahsa, A. 2006. Scand. J. Infect. Dis. 38: 479–481.

Biavati, B. and Mattarelli, P. 2006. In: The Prokaryotes (eds. M. Dworkin, S. Falkow, E. Rosenberg, K.H. Schleifer and E. Stackebrandt), Springer, New York, USA, pp. 322–382.

Bowling, T.E., Raimundo, A.H., Grimble, G.K. and Silk, D.B. 1993. Lancet. 342: 1266–1268.

Brandao, R.L., Castro, I.M., Bambirra, E.A., Amaral, S.C., Fietto, L.G., Tropia, M.J., Neves, M.J., Dos Santos, R.G., Gomes, N.C. and Nicoli, J.R. 1998. Appl. Environ. Microbiol. 64: 564–568.

Bravo, M. V., Bunout, D., Leiva, L., De La Maza, M. P., Barrera, G., De La Maza, J. and Hirsch, S. 2008. Rev. Med. Chil. 136: 981–988.

Breves, G., Walter, C., Burmester, M. and Schröder, B. 2000. J. Anim. Physiol. 84: 9–20.

Burkhardt, O., Kohnlein, T., Pletz, M. and Welte, T. 2005. Scand. J. Infect. Dis. 37: 69–72.

Buts, J.P., Bernasconi, P., Van Craynest, M.P., Maldague, P. and De, M.R. 1986. Pediatr. Res. 20: 192–196.

Buts, J.P., De Keyser, N. and De Raedemaeker, L. 1994. Pediatr. Res. 36: 522–527.

Buts, J.P., De, K.N., Marandi, S., Hermans, D., Sokal, E.M., Chae, Y.H., Lambotte, L.,Chanteux, H. and Tulkens, P.M. 1999. Gut. 1999 45: 89–96.

Buts, J.P., Dekeyser, N., Stilmant, C., Delem, E., Smets, F. and Sokal, E. 2006. Pediatr. Res. 60: 24–29.

Buts, J.P., Stilmant, C., Bernasconi, P., Neirinck, C. and De Keyser, N. 2008. Scand. J. Gastroenterol. 43: 1489–1496.

Caetano, J.A., Parames, M.T., Babo, M.J., Santos, A., Ferreira, A.B., Freitas, A.A., Coelho, M.R. and Mateus, A.M., 1986. Int. J. Immunopharmacol. 8: 245–259.

Can, M., Besirbellioglu, B. A., Avci, I. Y., Beker, C. M. and Pahsa, A. 2006. Med. Sci. Monit. 12: PI19–PI22.

Castagliuolo, I., LaMont, J.T., Nikulasson, S.T. and Pothoulakis, C. 1996. Infect. Immun. 64: 5225–5232.

Castagliuolo, I., Riegler, M.F., Valenick, L., LaMont, J.T., Pothoulakis, C. 1999. Infect. Immun. 67: 302–307.

Chen, X., Kokkotou, E.G., Mustafa, N., Bhaskar, K.R., Sougioultzis, S., O'Brien, M., Pothoulakis, C. and Kelly, C.P. 2006. J. Biol. Chem. 281: 24449–24454.

Cindoruk, M., Erkan, G., Karakan, T., Dursun, A., Unal, S. 2007. Helicobacter. 12: 309–316.

Coté, G. A. and Buchman, A. L. 2006. Expert Opin. Drug Saf. 5: 361–372.

Cremonini, F., Di Caro, S., Covino, M., Armuzzi, A., Gabrielli, M., Santarelli, Nista, E.C., Cammarota, G., Gasbarrini, G. and Gasbarrini, A. 2002. Am. J. Gastroenterol. 97: 2744–2749.

Czerucka, D., and Rampal, P. 1999. Dig. Dis. Sci. 44: 2359–2368.

Czerucka, D., Dahan, S., Mograbi, B., Rossi, B. and Rampal, P. 2000. Infect. Immun. 68: 5998–6004.

Czerucka, D., Piche, T. and Rampal, P. 2007. Aliment Pharmacol. Ther. 26 :767–778.

D'Souza, A. L., Rajkumar, C., Cooke, J. and Bulpitt, C. J. 2002. BMJ. 324: 1361–1366.

Dahan, S., Dalmasso, G., Imbert, V., Peyron, J.F., Rampal, P. and Czerucka, D. 2003. Infect. Immun. 71: 766–773.

Dalmasso, G., Cottrez, F., Imbert, V., Lagadec, P., Peyron, J.F., Rampal, P., Czerucka, D., Groux, H., Foussat, A. and Brun, V. 2006a. Gastroenterol. 131: 1812–1825.

Dalmasso, G., Loubat, A., Dahan, S., Calle, G., Rampal, P. and Czerucka, D. 2006b. Res. Microbiol. 157: 456–465.

Diemert, D.J. 2006. Clin. Microbiol. Rev. 3: 583–594.

Dijkstra, G., Moshage, H., van Dullemen, H.M., de Jager-Krikken, A., Tiebosch, A.T., Kleibeuker, J.H., Jansen, P.L. and Van, G.H. 1998. J. Pathol. 186: 416–421.

Dinleyici, E.C., Eren, M., Yargic, Z.A., Dogan, N. and Vandenplas, Y. 2009. Am. J. Trop. Med. Hyg. 80: 953–955.

Dinleyici, E.C., Eren, M., Dogan, N., Reyhanioglu, S., Yargic, Z.A. and Vandenplas, Y. 2011. Parasitol. Res. 108: 541–545.

Dubuquoy, L., Rousseaux, C., Thuru, X., Peyrin Biroulet, L., Romano, O., Chavatte, P., Chamaillard, M. and Desreumaux, P. Gut. 2006. 55: 1341–1349.

Ducluzeau, R., and Bensaada, M. 1982. Ann. Microbiol. 3: 491–501.

Egervarn, M., Roos, S. and Lindmark, H. 2009. J. Appl. Microbiol. 107: 1658–1668.

Enache-Angoulvant, A., and Hennequin, C. 2005. Clin. Infect. Dis. 41: 1559–1568.

Erejuwa, O.O., Sulaiman, S.A. and Ab Wahab, M.A. 2014. Int. J. Mol. Sci. 15: 4158–4188.

Ferreira, C.M., Vieira, A.T., Vinolo, M.A.R., Oliveira, F.A., Curi, R. and Martins, F.D.S. 2014. J. Immunol. Res. 2014: 10.1155/2014/689492.

Fidan, I., Kalkanci, A., Yesilyurt, E., Yalcin, B., Erdal, B., Kustimur, S. and Imir, T. 2009. Mycoses. 52: 29–34.

Finegold, S.M., Sutter, V.L. and Mathisen, G.E. 1983. In: Human Intestinal Microflora in Health and Disease, D.J. Hentges (Ed.), Academic Press, New York, USA, pp. 3–31.

Floch, M.H., Walker, W.A., Guandalini, S., Hibberd, P., Gorbach, S., Surawicz, C., Sanders, M. E., Garcia-Tsao, G., Quigley, E.M., Isolauri, E., Fedorak, R.N. and Dieleman, L.A. 2008. J. Clin. Gastroenterol. 42: S104-S108.

Gedek. B.R. 1999. Mycoses. 42: 261–264.

Girard, P., Pansart, Y. and Gillardin, J.M. 2005. Nitric. Oxide. 13: 163–169.

Graff, S., Chaumeil, J.C., Boy, P., Lai-Kuen, R. and Charrueau C. 2008. Biol. Pharm. Bull. 31: 266–272.

Goodridge, H.S., Reyes, C.N., Becker, C.A., Katsumoto, T.R., Ma, J., Wolf, A.J., Bose, N., Chan, A.S., Magee, A.S., Danielson, M.E., Weiss, A., Vasilakos, J.P. and Underhill, D.M. 2011. Nature. 7344: 471–475.

Gotteland, M., Poliak, L., Cruchet, S. and Brunser, O. 2005. Acta Paediatr. 94: 1747–1751.

Guarner, F. and Malagelada, J.R. 2003a. Lancet. 9356: 512–519.

Guarner, F. and Malagelada, J.R. 2003b. Lancet. 361: 512–519.

Guslandi, M. 2010. J. Clin. Gastroenterol. 44: 385.

Guslandi, M., Mezzi, G., Sorghi, M. and Testoni, P.A. 2000. Dig. Dis. Sci. 45: 1462–1464.

Guslandi, M., Giollo, P. and Testoni, P.A. 2003. Eur. J. Gastroenterol. Hepatol. 15: 697–698.

Hammes, W.P. and Vogel R.F. 1995. In: The Lactic Acid Bacteria (eds. B.J.B. Wood and W.H. Holzapfel), Chapman and Hall, London, UK, pp. 19–54.

Hatoum, R., Labrie, S. and Fliss, I. 2012. Front. Microbiol. 3: 421.

Hennequin, C., Kauffmann-Lacroix, C., Jobert, A., Viard, J.P., Ricour, C., Jacquemin, J.L. and Berche, P. 2000. Eur. J. Clin. Microbiol. Infect. Dis. 19: 16–20.

Herek, O., Kara I.G. and Kaleli I. 2004. Surg. Today. 34: 256–260.

Htwe K., Yee K. S., Tin M.and Vandenplas Y. 2008. Am. J. Trop. Med. Hyg. 78: 214–216.

Hurduc, V., Plesca, D., Dragomir, D., Sajin, M. and Vandenplas, Y. 2009. Acta Paediatr. 98: 127–131.

Hwang, J.B., Kang, K.J., Kang, Y.N. and Kim, A.S. 2009. Ann. Allergy Asthma Immunol. 103: 87–88.
Jahn, H.U., Ullrich, R., Schneider, T., Liehr, R.M., Schieferdecker, H.L., Holst, H. and Zeitz, M. 1996. Digestion. 57: 95–104.
Jandhyala, S.M., Talukdar, R., Subramanyam, C., Vuyyuru, H., Sasikala, M. and Reddy, D.N. 2015. World J. Gastroenterol. 29: 8787–8803.
Klein, S.M., Elmer, G.W., McFarland, L.V., Surawicz, C.M. and Levy, R.H. 1993. Pharm. Res. 10: 1615–161.
Karen, M., Yuksel, O., Akyürek, N., Ofluoğlu, E., Cağlar, K., Sahin, T.T., Paşaoğlu, H., Memiş, L., Akyürek, N. and Bostanci, H. 2010. J. Surg. Res. 160: 139–144.
Kaser, A., Zeissig, S. and Blumberg, R.S. 2010. Annu. Rev. Immunol. 28: 573–621.
Kelesidis, T. and Pothoulakis, C. 2012. Therap. Adv. Gastroenterol. 2: 111–125.
Kennedy, C., Brooks-Young, P., Gray, C.B., Larkin, P., Connolly, M., Wilde-Larsson, B., Larsson, M., Smith, T., Chater, S. 2014. BMJ Support Palliat. Care. 3: 263–270.
Klein, S.M., Elmer, G.W., McFarland, L.V., Surawicz, C.M. and Levy, R.H. 1993. Pharm. Res. 10: 1615–1619.
Kotowska, M., Albrecht, P. and Szajewska, H. 2005. Aliment. Pharmacol. Ther. 21: 583–590.
Kurugöl Z. and Koturoglu G. (2005). Acta Paediatr. 94: 44–47.
Kyne, L., Warny, M., Qamar, A. and Kelly, C.P. 2001. Lancet. 357: 189–193.
Lee, S.K., Kim, H.J., Chi, S.G., Jang, J.Y., Nam, K.D., Kim, N.H., Joo, K.R., Dong, S.H., Kim, B. H., Chang, Y.W., Lee, J.I. and Chang, R. 2005. Korean J. Gastroenterol. 45: 328–334.
Lessard, M., Dupuis, M., Gagnon, N., Nadeau, E., Matte, J.J., Goulet, J. and Fairbrother, J.M. 2009. J. Anim Sci. 87: 922–934.
Mansour-Ghanaei, F., Dehbashi, N., Yazdanparast, K. and Shafaghi, A. 2003. World J. Gastroenterol. 9: 1832–1833.
McFarland, L.V. 2006. Am. J. Gastroenterol. 101: 812–822.
McFarland, L.V. 2007. Travel. Med. Infect. Dis. 5: 97–105.
McFarland, L.V. 2010. BMJ open,.8: e005047.
McFarland, L.V. 2015. Clini. Infect. Dis. S2: S85-S90.
McFarland, L.V., Surawicz, C.W. and Stamm, W.E. 1990. J. Infect. Dis. 162: 678–684.
McFarland, L. V., Surawicz, C. M., Greenberg, R. N., Fekety, R., Elmer, G.W., Moyer, K. A., Melcher, S.A., Bowen, K.E., Cox, J.L., Noorani, Z., Harrington, G., Rubin, M. and Greenwald, D. 1994. J. Am. Med. Assoc. 271: 1913–1918.
Moré, M.I., and Swidsinski, A. 2015. Clin. Exp. Gastroenterol.11: 237–55.
Moslehi-Jenabian, S., Pedersen, L.L. and Jespersen, L., 2010. Nutrients. 4: 449–473.
Moyad, M.A., Robinson, L.E., Zawada, E.T. Jr, Kittelsrud, J.M., Chen, D.G., Reeves, S.G. and Weaver, S.E. 2008. Urol. Nurs. 1: 50–55.
Mumy, K.L., Chen, X., Kelly, C.P. and McCormick, B.A. 2008. Am. J. Physiol. Gastrointest. Liver. Physiol. 294: G599–G609.
Ozkan, T.B., Sahin, E., Erdemir, G. and Budak, F. 2007. J. Int. Med. Res. 35: 201–212.
Penna, F.J., Filho, L.A., Calçado, A.C., Junior, H.R. and Nicolli, J.R. 2000. J. Pediatr. (Rio J) 76: S209–217.
Pineton de Chambrun, G., Neut, C., Chau, A., Cazaubiel, M., Pelerin, F., Justen, P. and Desreumaux, P. 2015. Dig. Liver Dis. 47: 119–124.
Plein, K. and Hotz, J. 1993. J. Gastroenterol. 31: 129–134.
Pothoulakis, C. 2009. Aliment. Pharmacol. Ther. 8: 826–833.
Pothoulakis, C., Kelly, C.P., Joshi, M.A., Gao, N., O'Keane, C.J., Castagliuolo, I. and Lamont, J. T. 1993. Gastroenterol. 104: 1108–1115.
Qamar, A., Aboudola, S., Warny, M., Michetti, P., Pothoulakis, C., LaMont, J.T. and Kelly, C. P. 2001. Infect. Immun. 69: 2762–2765.
Riquelme, A. J., Calvo, M. A., Guzmán, A. M., Depix, M. S., García, P., Pérez, C. and Labarca, J. A. (2003). J. clini. Gastroenterol. 36: 41–43.
Rodrigues, A.C., Cara, D.C., Fretez, S.H., Cunha, F.Q., Vieira, E.C., Nicoli, J.R. and Vieira, L.Q. 2000. J. Appl. Microbiol. 89: 404–414.

Rohde, C.L., Bartolini, V. and Jones, N. 2009. Nutr. Clin. Pract. 24: 33–40.
Rolfe, R.D. 2000. J. Nutr. 130: 396S–402S.
Rombeau, J. L., Kripke, S.A. and Settle, R.G. 1990. In: Dietary Fiber: Chemistry, Physiology, and Health Effects (ed. D. Kritchesvsky), Plenum Press, New York, USA, pp. 317–337.
Saintmarc, T., Blehaut, H., Musial, C. and Touraine, J.L. 1995. Sem. Hôsp. Paris. 71: 735–741. Salminen, S., Isolauri, E. and Onnela, T. 1995. Chemotherapy. 41: 5–15.
Samuelsen, A.B., Schrezenmeir, J. and Knutsen, S.H. 2014. Mol. Nutr. Food Res. 1:183–193.
Sanders, J.W. and Tribble, D.R. 2001. Curr. Gastroenterol. Rep. 3: 304–314.
Schneider, S.M., Le Gall, P., Girard-Pipau, E., Piche, T., Pompei, A., Nano, J.L., Hebuterne, X., Schneider S.M., Girard-Pipau F., Filippi J., Hebuterne X., Moyse D., Hinojosa G.C., Pompei A. and Rampal P. 2005. World J. Gastroenterol. 11: 6165–6169.
Schroeder, B., Winckler, C., Failing, K. and Breves, G. 2004. Dig. Dis. Sci. 49: 1311–1317.
Schwenzer, V.B. 1998. Deutsche Apotheker Zeitung. 138: 75–77.
Sekirov, I., Russell, S.L., Antunes, L.C. and Finlay, B.B. 2010. Physiol. Rev. 3: 859–904.
Sezer, A., Usta, U. and Cicin, I. 2009. Med. Oncol. 26: 350–357.
Sgorbati, B., Biavati, B., Palenzona, D. 1995. In: The Lactic Acid Bacteria (eds. B.J.B. Wood and W.H. Holzapfel) Chapman and Hall, London, UK, pp. 279–306.
Shan, L.S., Hou, P., Wang, Z.J., Liu, F.R., Chen, N., Shu, L.H., Zhang, H., Han, X.H., Han, X.X., Cai, X.X., Shang, Y.X. and Vandenplas, Y. 2013. Benef. Microbes. 4: 329–334.
Sougioultzis, S., Simeonidis, S., Bhaskar, K.R., Chen, X., Anton, P.M., Keates, S., Pothoulakis, C. and Kelly, C.P. 2006. Biochem. Biophys. Res. Commun. 343: 69–76.
Su, C.G., Wen, X., Bailey, S.T., Jiang, W., Rangwala, S.M., Keilbaugh, S.A., Flanigan, A., Murthy, S., Lazar, M.A. and Wu, G.D. 1999. J. Clin. Invest. 104: 383–389.
Surawicz, C. M., Elmer, G. W., Speelman, P., McFarland, L. V., Chinn, J. and Van Belle, G. 1990. Z. Gastroenterol. 28: 259–260.
Surawicz, C. M., McFarland, L. V., Greenberg, R. N., Rubin, M., Fekety, R., Mulligan, M. E., Garcia, R.J., Brandmarker, S., Bowen, K., Borjal, D. and Elmer, G.W. 2000. Clin. Infect. Dis. 4: 1012–1017.
Szajewska, H and Kołodziej, M. 2015. Aliment. Pharmacol. Ther. 7: 793–801.
Szajewska, H. and Kolodziej M. 2015 Aliment Pharmacol. Ther. 42 : 1149-1157.
Szajewska, H, Horvath, A. and Kołodziej, M. 2015. Aliment. Pharmacol. Ther. 12: 1237–1245.
Tasteyre, A., Barc, M.C., Karjalainen, T., Bourlioux, P. and Collignon, A. 2002. Microb Pathog. 2002. 32: 219–225.
Temmerman, R., Pot, B., Huys, G. and Swings, J. 2001. Meded Rijksuniv Gent Fak Landbouwkd Toegep Biol. Wet. 66: 535, 537–542.
Thomas, S., Przesdzing, I., Metzke, D., Schmitz, J., Radbruch, A. and Baumgart, D.C. 2009. Clin. Exp. Immunol. 156: 78–87.
Turck, D., Bernet, J.P., Marx, J. Kempf, H., Giard, P., Walbaum, O., Lacombe, A., Rembert, F., Toursel, F., Bernasconi, P., Gottrand, F., McFarland, L.V. and Bloch, K. 2003. Gastroenterol. Nutr. 37: 22–26.
Vandenplas, Y., Brunser, O. and Szajewska, H. 2009. Eur. J. Pediatr. 168: 253–265.
Vannucci, L., Krizan, J., Sima, P., Stakheev, D., Caja, F., Rajsiglova, V., Horak, V. and Saieh, M., 2013. Int. J. Oncol. 2: 357–364.
Varankovich N.V., Nickerson, M.T. and Korber, D.R. 2015. Front Microbiol. 6: 685.
Vidon, N., Huchet, B. and Rambaud, J.C. 1986. Gastroenterol. Clin. Biol. 10: 1–4.
Weber, G., Adamczyk, A. and Freytag, S. 1989. Fortschr. Med. 26: 563–566.
Whelan K. and Myers C.E. 2010. Am. J. Clin. Nutr. 91: 687–703.
Yamada, E.A. and Sgarbieri, V.C. 2005. J. Agric. Food Chem. 10: 3931–3936.
Zaouche, A., Loukil, C., De Lagausie P., Peuchmaur, M., Macry, J., Fitoussi, F., Bernasconi, P., Bingen, E. and Cezard, J.P. 2000. Scand. J. Gastroenterol. 2: 160–165.
Zbinden, R., Gönczi, E.E. and Altwegg, M. 2011. Microb. Ecol. Health Dis. 3: 1651–2235.

Yeast Biofilms in the Context of Human Health and Disease

Jayant Shankar Raut, Sonali Kashinath Doke and Sankunny Mohan Karuppayil

Abstract Microbial biofilms play important roles in ecology, industry and most importantly in the human health. Extensive research is being done to study their involvement in chronic infectious diseases. Yeasts, members of the Kingdom fungi, are no exception and flourish as biofilms in their natural habitats. Yeasts either exist as a part of human microbiota or reside in close proximity environment and may turn pathogenic to cause superficial or systemic infections. The majority of these infections involve growth in biofilm form. Particularly, a large population of immunocompromised individuals and patients using prosthetic devices are susceptible to biofilm related infections. *Candida*, *Cryptococcus* and *Histoplasma* are the major yeast species responsible for high morbidity and mortality associated with mycoses. Interestingly, these yeasts colonize host tissues or medical devices to form biofilms which are highly resistant to antifungal drugs. Also, biofilms may act as a reservoir for recurrent infections and consequently complicate the antifungal therapy. Efforts are being done to characterize biofilms as an important virulence factor in fungi. This review, with a special emphasis on *Candida albicans*, discusses biofilm formation and associated drug resistance. Also, the involvement of yeast biofilms in human diseases and the therapeutic strategies are briefly reviewed.

Keywords Antifungal · Biofilm · *Candida* · Drug resistance · Infection · Prostheses · Virulence · Yeast

---

J.S. Raut
UGC Center for Advanced Study, University Institute of Pharmaceutical Sciences (UIPS), Panjab University, Chandigarh 160 014, India

S.K. Doke
School of Pharmacy, SRTM University, Nanded 431 606, Maharashtra, India

S.M. Karuppayil (✉)
DST-FIST and UGC-SAP Sponsored School of Life Sciences, SRTM University (NAAC Acredited with a Grade), Nanded 431 606, Maharashtra, India
e-mail: prof.karuppayil@gmail.com

T. Satyanarayana and G. Kunze (eds.), *Yeast Diversity in Human Welfare*, DOI 10.1007/978-981-10-2621-8_6

1 Introduction

Approximately 1.2 billion people are estimated to suffer from fungal diseases worldwide (Denning and Bromley 2015). Most of these (around 1 billion) are cutaneous/superficial infections involving the skin, nail, hair and oral or urino-genital mucosa. Although cured by antifungal therapy, superficial infections have a substantial effect on the quality of life (FIT 2012). Remaining are the systemic/invasive infections, which are hard to diagnose, difficult to treat, and hence a threat to the patients. Up to two million people die of severe fungal infections every year. Various species of yeast and filamentous fungi are known to be human pathogens, and more are being reported to be associated with mycoses. However, species belonging to the genera *Aspergillus*, *Candida*, *Cryptococcus*, *Pneumocystis*, and *Histoplasma* are the major fungal killers (Brown et al. 2012a). It would be interesting to note that three of the major culprits are yeasts. Few of them are dimorphic in nature and can switch between yeast and hyphal morphologies.

Over the last 30 years, incidences of fungal infections have greatly increased parallel with advances in medical technology and an increase in the population of immunocompromised patients. Prolonged lives of old age people, increase in the number of ICU patients, prolonged stay in ICU, HIV/AIDS-infected, organ transplant, cancer, and neonatal patients, people undergoing surgeries; all this coincided with a rise in the use of broad-spectrum antibiotics, prolonged chemotherapy, immunosuppression therapy, indwelling catheters and medical devices, culminating into rising of opportunistic fungal pathogens (Diekema et al. 2012; Pfaller et al. 2012). More importantly, there has been an increase in the fungal infections related to the biofilm growth form (Mathe and Van Dijck 2013).

Biofilm is a community of microorganisms characterized by cells which are irreversibly attached to a surface and embedded in a matrix of extracellular polymeric substances. Cells in a biofilm (sessile cells) exhibit altered phenotype compared to the free-living (planktonic) cells, due to surface induced gene expression (Donlan and Costerton 2002). Living as a community is a survival strategy which provides several ecological benefits to microorganisms; like, escape from host immune defense, protection from environmental stress, better acquisition of nutrients, metabolic cooperation, and persistence in unfavorable niches (West et al. 2007). Hence, the majority of the microorganisms, in their natural habitats, live as aggregated communities attached to a surface (Donlan and Costerton 2002). Unfortunately, human pathogens adapt this strategy to colonize host tissues or indwelling prostheses, survive the attack of the defensive immune system, resist the antimicrobial agents/antibiotics and flourish to cause infections (Costerton et al. 1999). It is believed that in humans, 80% of all microbial infections are biofilm-related. Efforts are being done to study the involvement of biofilms in chronic infectious diseases, medical device associated infections and the drug resistance associated with them (Harriott et al. 2010; Fox and Nobile 2012).

This is also true for yeasts, which can colonize human host leading to recalcitrant infections. Yeast infections on moist surfaces, mucosal tissues and prosthetic devices inside the body involve biofilm formation (Cuellar-Cruz et al. 2012). For example, oropharyngeal candidiasis, vaginitis, and native valve endocarditis infections are associated with *Candida* biofilms. Similarly, *Candida* biofilms are the third leading cause of catheter-related fungemia (Ganguly and Mitchell 2011; Desai et al. 2014). Overall, biofilm formation is a crucial step in yeast infections (d'Enfert 2009). Most of the knowledge on the fungal biofilms has been obtained from studies on yeasts. *Candida albicans* is capable of formation of highly structured biofilms and has emerged as a model system to study pathogenic biofilms. Various in vitro and in vivo studies on *C. albicans* have contributed significant information on biofilm formation and associated characteristics (Shinde et al. 2012b; Nett and Andes 2015). In this chapter, we discuss biofilm formation in yeasts, particularly pathogenic yeasts and its consequences on the human health. Biofilm-related drug resistance, underlying mechanisms and various strategies to overcome biofilm associated infections are also discussed, with a special emphasis on *C. albicans* biofilm.

2 Yeasts as Human Pathogens

Unicellular fungi, yeasts as they are commonly known, belong to the kingdom fungi which occupy a diverse range of environments with an estimated 1.5 million species (Hawksworth 2001). Only a fraction of these are yeasts, but they exhibit recognizable effects on the human life. Selective yeasts are applicable in human welfare. A well-known example is commercial use of different strains of budding yeast, *Saccharomyces cerevisiae*. It is used to ferment sugars in the production of alcoholic beverages and also routinely applied in bakery industry and food industry. *S. cerevisiae* is often taken as a vitamin supplement because it is a rich source of B vitamins, niacin, and folic acid. It is also popular as a model organism in cell and molecular biology and hence indirectly contributes to advances in human medicine (d'Enfert 2009; Alexandre 2013). However, few of the yeasts are human pathogens and exert substantial effects on human health.

Recent reports on host-fungal interactions have revealed that fungi are an integral part of the human microbiome, and must be playing an important role in defining commensal microbial communities (Huffnagle and Noverr 2013). The interactions among commensal yeasts or bacteria and yeast and their consequences to the host are being explored actively (Klotz et al. 2007; Underhill and Iliev 2014). For example, the interplay between *C. albicans* and bacterial pathogens such as *Pseudomonas aeruginosa*, *Acinetobacter baumanii*, and *Salmonella enteric* in the gastrointestinal tract has been demonstrated either in vitro or in invertebrate models of infection (Davis-Hanna et al. 2008). Few bacterial pathogens may limit the infectivity of *C. albicans* through the secretion of small molecules such as homoserine lactones; while, secretion of quorum sensing molecules like farnesol by *Candida* may influence bacterial colonization (Hogan 2006).

Table 1 Most significant fungal infections and their estimated worldwide burden

Infection	Causative fungal pathogen	Predominant morphology (yeast/filamentous)	Estimated infections worldwide/year	Associated mortality (%)
Cryptococcosis	*Cryptococcus neoformans*; *C. gattii*	Yeast	>10,00,000	20–70
Candidiasis	*Candida albicans* and non-albicans *Candida* species (NACS)	Yeast	>4,00,000	46–75
Pneumocystis	*Pneumocystis jirovecii*	–	>4,00,000	20–80
Aspergillosis	*Aspergillus fumigatus*	Filamentous	>2,00,000	30–95
Histoplasmosis	*Histoplasma capsulatum*	Yeast	25,000	28–50

Even though very few of the fungi are pathogenic to insects, amphibians, plants, animals and humans; they have a pronounced effect on the global biota (Fisher et al. 2012; Gundacker and Baddley 2015). Fungi are supposed to cause billions of infections every year and estimated to kill around two million people Worldwide (Table 1) (Brown et al. 2012a). Unfortunately, their influence on the human health is still under-recognized. For example, the fungal infections have not been mentioned by World Health Organization in their program. Almost everybody experience superficial fungal infection at least once in a lifetime, the majority of which are cured easily in healthy individuals. However, millions of immunocompromised individuals contract life-threatening invasive infections which are much harder to be cured. In many cases, the rate of mortality often exceeds as high as 50%, with total deaths exceeding that of associated with TB and malaria (Brown et al. 2012a).

About 300 fungal species are well recognized to be associated with human diseases and infections, only a few of these are yeasts and only a minor of the latter are human pathogens; but, exhibit the substantial effect on the human health and disease. Approximately 90% of the deaths related to fungal infections involve species belonging to five genera i.e. *Aspergillus*, *Candida*, *Cryptococcus, Histoplasma* and *Pneumocystis* (Table 1) (Brown et al. 2012b). Three of this i.e. *Candida* spp., *Cryptococcus* spp., and *Histoplasma* spp. are yeasts or dimorphic fungi predominantly existing in yeast morphological form.

Yeasts can be found in different natural habitats such as plants, soil, water, animals and importantly humans. Two important yeasts, *Cryptococcus neoformans* and *Histoplasma caspulatum*, are naturally found in the soil and other environmental niches. They frequently get access to the human host and may reside inside the body for a long time without causing any harm. However, being opportunistic, they take advantage of a weak immune system and proliferate in the human host to cause infections (d'Enfert 2009). Few members such as *C. albicans* and non-albicans *Candida* species (NACS) are commensal and grow on the skin surfaces, mucous membranes, oropharyngeal, urinogenital and gastrointestinal tracts as normal microbiota; but, may turn pathogenic to invade tissues and proliferate to cause disease in immunocompromised patients.

The opportunistic behavior of pathogenic yeasts like *Candida* and *Cryptococcus* is responsible for a sharp rise in the infections to a population of immunocompromised individuals such as HIV-infected people, cancer patients, diabetics, patients under long term antibiotic treatment or chemotherapy, people undergoing organ transplantation or heart surgeries, hospitalized patients using catheters, bone implants and other prostheses (Raut and Karuppayil 2016; Polvi et al. 2015). It has been estimated that in the United States the cost burden of fungal infections, the majority of which are due to yeasts, may be as high as $2.6 billion per year (Wilson et al. 2002).

3 Biofilm Formation in Yeasts

Biofilm formation represents an important intrinsic property of most of the microorganisms. Fungi are no exception to this and are capable of biofilm formation in vitro and in vivo. Fungal species, particularly those which are involved in human disease are being studied in detail for their biofilm forming abilities (Table 2) (Desai et al. 2014). Most common fungal pathogens belong to phyla Ascomycota or Basidiomycota. Major pathogenic yeasts belong to the phylum Ascomycota (including species from the genera *Candida* and *Histoplasma*) and the Basidiomycota (include the genera *Cryptococcus* and *Trichosporon*). Biofilm formation and biofilm mediated pathogenesis have been studied in only a few of these pathogens (Desai et al. 2014; Fox et al. 2015).

3.1 *Candida*

Candida species are the most common fungal pathogens responsible for the superficial and life-threatening systemic infections. Candidiasis is prevalent in immunocompromised patients, people undergoing chemotherapy, invasive clinical procedures, major trauma and prolonged stay in intensive care units. Advanced medical procedures such as the use of catheters, neonatal intensive care, gut surgeries, or organ transplantation are predisposing factors to disseminated *Candida* infections (Calderone and Clancy 2012). *Candida* species are the fourth most common cause of nosocomial (hospital-acquired) bloodstream infections and the third major reason for catheter-related infections (Pfaller and Diekema 2007). The estimated annual global incidence of *Candida* bloodstream infections is approximately 400,000 cases, per year, with very high mortality rates of 30–40% (Brown et al. 2012a). Mucosal *Candida* infections of the oral and genital tracts are very common. For example, 50–75% of women suffer from at least one episode of vulvovaginal candidiasis and 5–8% (75 million) experience at least four episodes annually (Sobel 2007). Also, there are at least 10 million cases of oral thrush in HIV/AIDS patients, cancer patients and other immunocompromised patients (Pfaller and Diekema 2007). More than 20 species of *Candida* have been found to be involved in human disease. However, *C. albicans* predominates and is the

Table 2 Biofilm formation by important yeast pathogens and involvement in human infections

Pathogenic yeast	Biofilm formation		Biofilm associated infections	Drug resistance	References
	In vitro	In vivo			
Candida albicans; non-albicans *Candida* species (NACS) like, *glabrata; parapsilosis; dubliniensis; krusei; tropicalis*	Yes	Yes	Superficial and systemic candidiasis; mucosal infections; invasive tissue infections; candidemia; colonization of catheters, endotracheal tubes, cardiac devices, implants, voice prostheses, joint prostheses and bone implants	Yes	Chandra et al. (2001a), Donlan and Costerton (2002), Kojic and Darouiche (2004), Kaur et al. (2005), Kumar and Menon (2006), Ramage et al. (2006), Al-Fattani and Douglas (2006), Shinde et al. (2012b) and Raut et al. (2013b)
Cryptococcus neoformans; C. gattii	Yes	Yes	Meningoencephalitis and pulmonary infections; device related infections such as cardiac valves, peritoneal dialysis equipments, ventriculoatrial shunt	Yes	Braun et al. (1994), Banerjee et al. (1997), Martinez and Casadevall (2007) and Robertson et al. (2012)
Histoplasma capsulatum	Yes	Yes	Histoplasmosis; human respiratory system (lung) infections; device associated infections	Yes	Pitangui et al. (2012), Pierce et al. (2013) and Brilhante et al. (2015)
Trichosporon asahii	Yes	Yes	Mainly device associated infections of dialysis grafts, breast implants	Yes	Reddy et al. (2002), Krzossok et al. (2004) and Bonaventura et al. (2006)

most common species associated with human infections. Medically important species other than C. albicaus i.e. NACS include *Candida glabrata*, *Candida parapsilosis*, *Candida tropicalis*, *Candida dubliniensis*, *Candida krusei*, *Candida rugosa* and *Candida lusitaniae* (Calderone and Clancy 2012).

Candida can cause a wide spectrum of clinical manifestations and attack nearly every organ in the human body. It is the most common yeast species isolated from blood and mucosal surfaces. How it remains at mucosal surfaces in the presence of adaptive immunity is still not known. It has been speculated that ability to produce immunomodulatory compounds (such as oxylipins) and to adhere various surfaces leading to biofilm formation helps *Candida* to persist various niches for prolonged periods (Huffnagle and Noverr 2013). In general, *Candida* species can colonize various mucosal surfaces at oral and nasal cavities, gastrointestinal and urogenital tract and develop a community structure (Samaranayake et al. 2009; Sardi et al. 2013). *Candida* readily adheres to prosthetic devices implanted in a patient and form biofilms leading to a device associated infections (Kojic and Darouiche 2004). Formation of biofilm is being thoroughly investigated in *C. albicans* and various in vitro and in vivo studies have contributed to our understanding of biofilm mode of growth (Nett and Andes 2015). Obviously, it has been considered as a model to study the formation of fungal biofilms and various manifestations associated with it.

Candida albicans biofilm is not just an aggregation of cells; but, is a heterogeneous structure developed by cells interacting as a community (Nickerson et al. 2006). The behavior of individual cells in a biofilm is regulated by diffusible molecules. A signal transduction process which involves the production, release and response to signalling molecules secreted by the microbial cells themselves, is called quorum sensing. The diffusible molecules involved in quorum sensing are called autoinducers or quorum sensing molecules (QSMs) (Hornby et al. 2001). QSMs accumulate in the medium as a microbial population grows and may convey the physiological changes to individual cells in response to the density of population. *C. albicans* biofilm development is also regulated through quorum sensing; however, the molecular mechanisms behind it are not fully understood.

Farnesol, a QSM in *C. albicans*, is a sesquiterpene continuously produced during growth of *C. albicans* cells. Extracellular farnesol accumulates in the culture to exert various physiological effects on individual cells as well as community growth. For example, at a cell density of $>10^6$ cells/ml, it reaches to a threshold concentration and prevents yeast to hyphal morphogenesis. Exogenously added farnesol (2–250 μM) prevents morphogenesis induced by various inducers (Mosel et al. 2005; Rathod et al. 2013). Tyrosol is another QSM which acts opposite to farnesol and is known to enhance germ tube formation during growth. At low cell densities, it reduces lag phase in the growth and promotes the formation of hyphae (Chen et al. 2004). In addition, few more molecules like nerolidol, isoamyl alcohol, dodecanol, ethanol and acetaldehyde were detected in *Candida* cultures and known to act as morphogenetic signalling molecules to inhibit filamentation (Chauhan et al. 2011a, b).

Formation of biofilm by *C. albicans* takes place through three distinct stages such as early, intermediate and maturation phases (Chandra et al. 2001a). Early phase extends over 0–6 h approximately. It consists of adhesion of blastospores/yeast form cells to a surface (0–2 h), the formation of micro-colonies and dimorphic transition to give rise hyphal forms (3–6 h). Intermediate phase (6–18 h) is characterized by cellular growth, an increase in cell density, the formation of multiple layers of cells, and elongation of filaments to form a mesh-like network of yeast, hyphae and pseudohyphae. In the last maturation phase (18–48 h), multiple layers of cells start depositing extracellular polymeric matrix (EPM). A dense network of filamentous and yeast cells embedded in EPM gives a three-dimensional, heterogeneous community structure. At the end of the maturation phase controlled dispersion of planktonic cells from community takes place (Raut 2014).

Adhesion of blastospores (yeast cells) to a solid surface is of prime importance in biofilm formation. During initial attachment which is reversible, nonspecific interactions like *van der waals* forces and electrostatic forces between cells and abiotic surface are involved (Klotz 1990). Cell surface hydrophobicity and hydrophobic interactions play important role in this stage (Panagoda et al. 2001; Raut et al. 2010). In later stages, anchoring of the cells takes place by means of specific cell surface molecules called adhesins (Verstrepen and Klis 2006). Binding through adhesins is irreversible i.e. if there are no strong physical/chemical forces acting, the cells cannot be removed easily. Proteins and mannoproteins present in *Candida* cell wall are involved in binding to host tissue surface as well as abiotic surfaces. Ability to bind to abiotic surfaces is important in device-related infections. The peptide portion of cell surface mannoproteins, particularly the exposed hydrophobic domains may be involved in binding to plastic materials (Chaffin et al. 1998). The Agglutinin like sequence genes (*ALS*) are known to code for adhesins in *C. albicans*. Out of the eight different proteins encoded by *ALS* gene family, Als3p protein shows stronger adhesive properties and is involved in adhesion to plastic. *BCR1* gene is found to act as a transcription factor and regulate expression of Als3p surface protein (Nobile and Mitchell 2006). Increased expression of drug efflux pumps in response to contact and adhesion of *C. albicans* to the surface plays an important role in antifungal resistance in biofilms (Kumamoto and Vinces 2005).

Formation of hyphae is another important event in *C. albicans* biofilm development and maturation. The presence of multiple layers of filamentous growth is important in typical *Candida* biofilm structure. Mutants of *C. albicans* that are unable to form hyphae were observed to form only a basal layer of biofilm (Nickerson et al. 2006). Biofilms formed in vitro and in vivo were detected to secrete farnesol and tyrosol, which suggests that *C. albicans* biofilms are regulated by various QSMs. Tyrosol is shown to enhance filamentation in early and intermediate biofilms, while farnesol is found dominant in maturation phases to overcome tyrosol activity and inhibit mycelial growth (Chen et al. 2004).

Specific genes are expressed during biofilm development and maturation (Garcia-Sanchez et al. 2004). Northern blot analysis of sessile and planktonic cells of *C. albicans* showed differential gene expression. Particularly, *ALS* gene family and genes belonging to drug efflux pump proteins, *CDR1 and CDR2* are found to be over-expressed (Chandra et al. 2001b). Analysis of 1850 different genes showed that 325 genes are differentially expressed in biofilm phenotype compared to that of planktonic. Two hundred fourteen of 325 genes were shown to be over expressed (Garcia-Sanchez et al. 2004). These genes were from various functional categories of metabolism, cell cycle, DNA processing, protein synthesis, cell signalling and transport. Among all, 34 genes involved in protein synthesis were up-regulated significantly. Genes for synthesis of aromatic amino acid and sulfur amino acids were overexpressed indicating their importance in the biofilm mode of growth. Expression of a set of genes for lipid synthesis like ergosterol, sphingolipids and phospholipids was increased. Genes that control cell wall synthesis and organization and genes involved in adhesion were up-regulated significantly. Also, the hyphal regulatory genes were found to be differentially regulated indicating the importance of filamentous forms in normal biofilm development (Garcia-Sanchez et al. 2004).

Biofilm-associated infections of *C. albicans* range from superficial oral thrush, denture stomatitis, ophthalmic infections, and wound/burn infections, to severe candidaemia and colonization of internal tissues and organs (Ramage et al. 2006). Many non-albicans *Candida* species are also involved in clinical biofilm-related infections. For example, *C. glabrata, C. dubliniensis, C. parapsilosis, C. tropicalis,* and *C. krusei,* have been observed to cause biofilm-associated infections (Ramage et al. 2006; Silva et al. 2011). Interestingly, few studies have reported that *C. albicans* biofilm production is significantly less frequent than non-albicans *Candida* spp. (Ramage et al. 2014). *C. dubiliensis* is capable of formation of a complex biofilm structure, consisting of blastospores, pseudohyphae, and hyphae, as seen in typical *C. albicans* biofilm (Ramage et al. 2001). Same is the case with *C. tropicalis*, which is able to form heterogeneous biofilms consisting of hyphae and filamentous forms, enclosed in a matrix layer (Bizerra et al. 2008). However, the EPM content shows the presence of hexosamine, small amounts of protein, phosphorous, and more uronic acid than that in *C. albicans* biofilms (Al-Fattani and Douglas 2006). *C. glabrata* readily form biofilms on biotic as well as abiotic surfaces in the host body, although in vitro biofilms show reduced thickness. Another characteristic feature of *C. glabrata* biofilm is that there is no morphogenetic switching of cells to give rise filamentous growth. Instead, cells adhered to a surface form layered clusters of blastospores. This community is covered by the EPM which exhibit comparatively higher concentration of carbohydrate and protein than biofilm formed by other NACS (Silva et al. 2009; Kucharíkova et al. 2011).

Similarly, *C. parapsilosis* also develop into a biofilm devoid of hyphae formation, hence the three-dimensional community possesses only layers of clustered yeast form cells. EPM of these biofilms is prominently rich in carbohydrates, while protein content is comparatively less (Silva et al. 2009). Strain-dependent variation in biofilm formation has been observed in *C. parapsilosis* isolates (Lattif et al. 2010; Silva et al. 2011). Although these two NACS do not form true hyphae, their biofilms may show the presence of elongated yeast cells resembling pseudohyphae (Ramage et al. 2014). *C. parapsilosis* and *C. glabrata* are responsible for 13 and 24% of *Candida* bloodstream infections, respectively. *C. parapsilosis* is most commonly infects neonates, transplant patients, and patients receiving parenteral nutrition. *C. glabrata* is known to form biofilms on voice prostheses. Such infections are clinically important as they hamper normal work of the device, restrict airflow and impede normal activities like speech, swallowing and respiration (Fanning and Mitchell 2012).

The gastrointestinal tract of 30–80% of healthy individuals is colonized by *Candida* and at many instances may enhance the inflammation (Kumamoto 2011). Colonization of percutaneous endoscopy gastronomy tubes by *C. albicans* and *C. tropicalis* may contribute to the degradation of the polyurethane to cause diarrhoea and sepsis (Trevisani et al. 2005). Colonization of mucosal layers of urinary/vaginal tract leading to vulvovaginal candidiasis is common. *Candida* biofilms on urethral stents are associated with pyelonephritis, cystitis and prostatitis (Sobel 2011). It has been reported that pathogenic fungal species play a role in wound infections and in combat trauma cases. Moreover, molecular analysis of chronic wound infections, including ulcers, non-healing surgical wounds and venous leg ulcers, showed that *Candida* spp. were the most abundant fungal pathogens (Branski et al. 2009; Paolino et al. 2012; Ramage et al. 2014).

3.2 Saccharomyces

Baker's yeast, *S. cerevisiae*, is known to form biofilms on solid surfaces and is being developed as an in vitro model system for biofilms. These biofilms are characterized by a thin matrix of budding yeast cells and elongated pseudohyphal cells (Reynolds and Fink 2001). *S. cerevisiae* can undergo a transition from budding yeast form to a filamentous multicellular community (Bastidas and Heitman 2009). Haploid cells show invasive growth to form biofilms on semisolid agar medium upon carbon starvation (Reynolds and Fink 2001); while diploid cells form pseudohyphae in response to nitrogen starvation (Gimeno et al. 1992). The filamentous transition is regulated by cyclic adenosine monophosphate (cAMP)—protein kinase A (PKA) pathway and a mitogen-activated protein kinase (MAPK) pathway (Ryan et al. 2012). Downstream of these pathways is *FLO11* gene which encodes a cell-surface protein involved in haploid invasive growth, biofilm formation, and diploid pseudohyphal growth (Guo et al. 2000). The expression of *FLO11* is controlled by numerous transcriptional regulators which are being investigated.

Biofilm formation in *S. cerevisiae* takes place through specific steps in which cell-to-cell interactions and cell-to-surface interaction occur simultaneously to result in adhesion and colonization of cells (Bojsen et al. 2012). It has the ability to adhere biotic and abiotic surfaces such as polystyrene, silicone, polypropylene, and polyvinylchloride. Besides its role in colonization of semisolid and solid agar, Flo11p directly plays a significant role in the adhesion to solid surfaces and is responsible for hydrophobic properties of the cell wall (Reynolds and Fink 2001). The same protein (Flo11p) also contributes to the yeast biofilm formation at the air-liquid interphase. For example, 'flor' observed in some alcoholic beverages is nothing but *S. cerevisiae* biofilm on air-liquid interface. It is useful in the aerobic growth of yeast and synthesis of specific metabolites in the production of sherry wines (Vallejo et al. 2013). Generally, *S. cerevisiae* is not observed to be involved in human infections; however, on rare occasions, it is reported as a member in mixed-species biofilm infections on catheters in ICU patients. Hence, it is speculated that *S. cerevisiae* is capable of biofilm formation in vivo and may be associated with infections in severely immunocompromised patients (Fox et al. 2015).

3.3 *Cryptococcus*

Cryptococcus species rank high among the prominent fungal pathogens of the humans. Cryptococcosis is mainly caused by *C. neoformans* and a closely related species *C. gattii*. These species which mainly resides in soil and avian habitats has a worldwide distribution and are frequently involved in meningoencephalitis and severe pulmonary infections (Gullo et al. 2013). Ability to form biofilms may play an important role in the survival of *C. neoformans*, in its environmental niche (Pierce et al. 2013). *Cryptococcus* biofilms have a well organized structure consisting of yeast cells. Layers of yeast cells are surrounded by the matrix material which mainly contains glucuronoxylomannan and galactoxylomannan and various sugars such as xylose, mannose, and glucose (Martinez and Casadevall 2007). Exposure to *Cryptococcus* mainly occurs by inhalation of airborne organisms into the lungs. It can cause local as well as systemic infections and mainly invades the central nervous system to cause meningoencephalitis. Both the species can form biofilms which is a threat not only to immunocompromised but also immunocompetent individuals (Alvarez et al. 2008; Fox et al. 2015).

Quantitative or qualitative defects in cellular immune functions, particularly in CD4+ lymphocytes due to AIDS, immunosuppressive medications, and solid organ transplantation are the major risk factors for cryptococcal infection; while biofilm formation on host tissues further complicates it (Park et al. 2009). Similarly, *C. neoformans* is reported to form biofilms on prostheses such as ventricular shunts, cardiac valves and peritoneal dialysis equipments to cause device related infections (Ramage et al. 2009a). The biofilm growth of *Cryptococcus* is very well tolerant to the attack of immune cells and various antifungal drugs. The estimated yearly global burden of Cryptococcal meningitis is around 1 million cases, with more than 620,000 deaths in sub-Saharan Africa. Mortality rates associated with *Cryptococcus* infections in AIDS patients are estimated to be 15–20% in the United States and 55–70% in Latin America and sub-Saharan Africa, despite the availability of the treatment (Brown et al. 2012a).

3.4 *Histoplasma*

Histoplasma capsulatum, the causative agent of histoplasmosis is an opportunist which infects the human respiratory system, primarily in immunocompromised patients. *H. capsulatum* var. capsulatum is a dimorphic fungus and exists as a filamentous form in the environment and predominates as a yeast-form in vivo (McKinsey and McKinsey 2011). Pitangui et al. (2012) have reported a dense community of yeast-form cells in vitro and suggested that same architecture may be prevalent in vivo, depicting biofilm formation abilities of *Histoplasma* (Pitangui et al. 2012). Such a biofilm growth may be responsible for clinical infections of

Histoplasma and exhibit resistance to antifungal drugs (Pierce et al. 2013; Brilhante et al. 2015).

Conidia (spores) of *Histoplasma* when inhaled, germinate in the lungs to give budding yeast form which has infective abilities. Cells ingested by pulmonary macrophages can survive and multiply within phagolysosomes to turn pathogenic under favorable conditions (Nucci and Marr 2005). Histoplasmosis may range from localized tissue infection to a lethal disseminated infection. Dissemination of the cells to various tissues results in damage to multiple organs and proves fatal to severely immunocompromised individuals (Kauffman 2007; McKinsey and McKinsey 2011).

3.5 *Trichosporon*

Infections caused by other yeasts, such as *Trichosporon* species, are also on the rise (Kontoyiannis et al. 2004). *Trichosporon asahii* is an emerging fungal pathogen and majorly infects patients with suppressed immune status. For example, disseminated *Trichosporon* infections are mainly observed in organ transplant patients (Ramage et al. 2009a). Trichosporonosis have been observed associated with implanted medical devices and is supposed to colonize there as biofilm growth forms. Biofilms formed are typical complex structures consisting of yeast and hyphal cells. This network is embedded in protective EPM (Bonaventura et al. 2006). *Trichosporon* biofilms are mainly found associated with dialysis graft and breast implants (Reddy et al. 2002; Krzossok et al. 2004).

4 Biofilm as a Virulence Factor in Pathogenic Yeasts

Cellular aggregation and surface colonization by fungi, particularly yeasts, was reported as early as in 1938 (Vallejo et al. 2013). The intrinsic ability of microorganisms to group and form communities is widely distributed in nature. It is supposed to play crucial roles in reproduction, colonization, pathogenesis, and survival under environmental stress (Costerton et al. 1999). Primary colonization of yeasts in the human host is through the acquisition of maternal flora in the perinatal period and later human contact, like in the case of *C. albicans*; or it is through interaction/exposure with surrounding environment; for example, *Cryptococcus* infection (Alvarez et al. 2008). Once a fungal cell reaches the mucosal surface or blood stream, it colonizes a tissue to survive there either as a commensal or a pathogen. The commensal association doesn't cause any damage to the host unless the immune status or the microbiota of the host is disturbed (Casadevall and Pirofski 2007).

Relatively little is known about the molecular requirements for commensalism of yeasts, as no reliable animal models are available which mimics in vivo conditions of the human host (Miceli et al. 2011). Similarly, details about mutualistic/beneficial fungal colonization and its relationship with the human host are not known. Only one example is of *Saccharomyces cerevisiae var. boulardii*, which is considered a beneficial fungus. It is well-described as a probiotic for the relief of gastroenteritis (Dinleyici et al. 2012). However, few reports have described that it can grow on indwelling catheters and form biofilms to cause fungemia. This may happen when the catheters are contaminated through accidental aerosolization of probiotic preparation intended to be given to the patients (Cassone et al. 2003).

Usually, yeasts follow strategies like, persistence in macrophages, commensalism with other microorganisms or colonization and formation of the community known as 'biofilms'; which allow them to survive and flourish in the host (d'Enfert 2009). Also, yeasts can readily adhere to abiotic surfaces of indwelling medical devices. Colonization of mucosal layers, tissues and prostheses, may result in subsequent biofilm formation; which leads to either asymptomatic persistence of the pathogen or extensive association and overgrowth culminating in an infection (Casadevall and Pirofski 2007). Yeast infections associated with biofilm growth have been observed in oral soft tissues, teeth, skin, wounds, the middle ear, the gastrointestinal tract, the urogenital tract, airway/lung tissue, heart valves, the eyes, dental implants, urinary tract prostheses, the peritoneal membrane and peritoneal dialysis catheters, indwelling catheters for hemodialysis and for chronic administration of chemotherapeutic agents, cardiac implants such as pacemakers, prosthetic heart valves, ventricular assist devices, and synthetic vascular grafts and stents, internal fixation devices, and percutaneous sutures, and tracheal and ventilator tubing, penile implants, hip and joint prostheses (Kojic and Darouiche 2004; Desai et al. 2014).

Yeast biofilms on human skin have been linked to the development of many dermatologic conditions or diseases (Kong and Segre 2012; Nusbaum et al. 2012). For example, *Candida* is reported to be involved in the development of atopic dermatitis (AD) (Zhang et al. 2011). Particularly, *C. albicans*, *Cryptococcus diffluans*, and *Cryptococcus liquifaciens* are the yeast species which have been found to colonize skin of AD patients (Sonesson et al. 2013). Quantification of microbial flora has revealed that fungi contribute to >50% of the microbial burden at the majority of wounds. *Candida* biofilms have been associated with the delayed healing of chronic wounds (Leake et al. 2009). Although microbial communities in the oral cavity are dominated by bacteria, considerable fungal organisms are also detected which may have significant effects on oral microbiota and overall health. *Candida* and *Cryptococcus* are the yeasts most frequently colonizing the oral cavity; and the species majoraly present are *C. albicans*, *C. parapsilosis*, *C. tropicalis*, and *C. neoformans* (Ghannoum et al. 2010).

The lungs harbor a low level of microflora and little is known about the fungal burden of the lungs. However, the presence of yeast like fungi, *Pneumocystis* spp. is most frequently observed, which proliferates to cause pneumonia in immunocompromised patients (Huffnagle and Noverr 2013). Similarly, limited information is available on the fungal communities of the gastrointestinal tract. It harbors low pH tolerant yeasts such as *Candida* species. *C. albicans* have been isolated from the sites of gastric ulceration in addition to *Helicobacter* and *Lactobacillus* bacteria. It is now being realized that the fungal/yeast biofilms may play a decisive role in overall health, especially in patients where normal bacterial biota is disturbed. Various prostheses, notably different type of catheters, are readily colonized by yeasts leading to biofilm formation (Kojic and Darouiche 2004). Strikingly, yeasts (mainly *C. albicans*) are the third leading cause of catheter-related infections (Crump and Collignon 2000). *C. neoformans* frequently form biofilms on ventricular shunts, cardiac valves and peritoneal dialysis equipment (Ramage et al. 2009b). The presence of indwelling prostheses is considered as a risk factor for the development of *C. glabrata* infections. It readily forms biofilm on venous catheters, prosthetic joints and peritoneal dialysis systems. Also, *C. parapsilosis* is found to colonize indwelling catheters in neonates, prosthetic knees in old age people, hip joint and breast implants (Ramage et al. 2014). Other biofilm forming yeasts involved in device-related clinical infections include species of *Histoplasma, Cryptococcus and Trichosporon* (Table 2) (Fanning and Mitchell 2012; Pierce et al. 2013).

Biofilm-related infections are difficult to treat and hard to eradicate, hence considered as a clinical threat. Besides the reasons like poor diagnosis, lack of effective therapy, and the emergence of resistant strains, biofilm formation is also the main reason behind high mortality and morbidity related to fungal infections. Biofilm formed by pathogenic yeasts display elevated resistance to most of the antifungal drugs available for the treatment and hampers the normal treatment procedures. It has been reported that biofilms formed by members of genus *Candida*, *Cryptococcus*, *Histoplasma*, and *Trichosporon* show reduced susceptibility to various antifungal agents compared to their planktonic growth (Pettit et al. 2010; Ramage et al. 2012; Zhang et al. 2012; Brilhante et al. 2015). Moreover, this community structure can very well withstand host immune defense (Fanning and Mitchell 2012). Further, severity increases as biofilms may act as reservoirs to keep releasing the cells which cause repeated infections when antibiotic therapy is discontinued or immune system is compromised.

Colonization of prostheses like catheters, heart valves, pacemakers and another bio-medical-assist devices in patients can compromise the normal function of the device or even may lead to its failure (Srinivasan et al. 2014). It was found that involvement of biofilm forming strain in nosocomial infections increased the risk of death as compared to non-biofilm forming isolates. It has been revealed that *Candida* clinical isolates which are able to form biofilms, have significantly more contribution to hospital mortality, costs of antifungal therapy, and increased the length of hospital stay of the patients (Ramage et al. 2014). Overall, biofilm formation is an important virulence factor in yeasts; and hence need to be studied thoroughly for its role in human health and disease.

5 Drug Resistance and the Mechanisms Involved

5.1 Drug Resistance

Antifungal drugs available for the treatment of candidiasis are mainly confined to four classes of molecules i.e. polyenes, 5-fluoro-cytosine (5-FC), azoles and echinocandins (Nosanchuk 2006). Nucleoside analogue, 5-flucytosine (5-FC) is converted to

5-fluorouracil (5-FU) when enters a fungal cell. 5-FU mimics a pyrimidine analogue to interfere with nucleic acid synthesis and inhibit cell cycle. Although 5-FC was found promising against *Candida* in the initial period, its use was limited by the high prevalence of resistance in *C. albicans* isolates (White et al. 1998). Polyenes represented by two heterocyclic molecules, amphotericin B and nystatin, are amphipathic in nature. Polyenes get intercalated into lipid bilayers, bind to membrane sterols and aggregate, which ultimately causes the formation of pores and leakage of cellular ions resulting in cell death. Polyenes also cause oxidative damage to the *Candida* cells. Although resistance to polyenes is not very common, it may be evident in few mutant populations (*ERG3* mutants) or growth forms like late biofilms where cells have decreased ergosterol content. The main limiting factor for polyenes is the severe toxicity associated with them (Xie et al. 2014).

The azole antifungals include some of the most widely prescribed drugs like fluconazole, active against *Candida* and other yeasts. Initially derived imidazoles, for example, miconazole and ketoconazole, have been replaced with less toxic and more efficient triazoles like fluconazole, itraconazole, and voriconazole. Azole drugs mainly interfere in ergosterol biosynthetic pathway. They inhibit the cytochrome P_{450} enzyme, 14α-lanosterol demethylase encoded by the *ERG11* gene. Depletion of membrane ergosterol affects membrane fluidity and integrity to cause loss of membrane function. Intervention in sterol synthesis may result in the synthesis of alternate toxic sterols resulting in inhibition of *C. albicans* growth (Cannon et al. 2009). Yeasts like *C. krusei* and *Cryptococcus* are resistant to azoles, while the emergence of drug resistance in susceptible yeasts (e.g. *C. albicans*) has been reported from all over the world (Rathod et al. 2012; Ghannoum and Rice 1999; Mishra et al. 2007). Echinocandins such as caspofungin, micafungin and anidulafungin, inhibit the enzyme 1,3-β-glucan synthetase resulting in a reduction of 1,3-β-glucan in the cell wall. Although recent, clinical resistance to echinocandin has been reported; a point mutation in 1,3-β-glucan synthase subunit was found responsible for echinocandin resistance in *C. albicans* (Xie et al. 2014).

A characteristic feature of yeast biofilms is its resistance to most of the available antifungal drugs including the widely prescribed azoles. It has been reported that biofilms formed by members of genus *Candida*, *Cryptococcus*, *Histoplasma* and *Trichosporon* show reduced susceptibility to various antifungal agents compared to their planktonic growth (Bonaventura et al. 2006; Pettit et al. 2010; Pitangui et al. 2012; Ramage et al. 2012). Susceptibility studies have revealed that biofilms formed by *C. albicans* may be up to 2,000 times more resistant to antifungal drugs than that of planktonic cells (Baillie and Douglas 2000; Shinde et al. 2012b). Also, biofilms of NACS show enhanced drug resistance to antifungal (Ramage et al. 2012; Desai et al. 2014; Fox et al. 2015). Resistance to antifungal drugs increases with the development of biofilm structure making mature biofilms totally non-responsive to the drug therapy (Shinde et al. 2012b), thus administration of very high doses of antifungals for a prolonged time is usually required to treat such infections. However, side effects due to toxicity put limitations on the effective use of antifungal drugs against biofilms.

5.2 Mechanisms of Drug Resistance

Based on *C. albicans* biofilm studies, various reasons have been proposed to be responsible for drug resistance associated with yeast biofilms. No single reason could fully explain the antifungal resistance exhibited by biofilms, hence considered as a multifactorial phenomenon. General mechanisms supposed to be responsible include, sequestration of drugs by extracellular polymeric matrix (EPM), enhanced drug efflux, high cell density, changes in metabolic state, the presence of persister cells and activation of the stress-responsive pathway (Mathe and Van Dijck 2013; Taff et al. 2013).

Formation of EPM is an important characteristic of biofilm formation. Individual cells remain embedded in this matrix which is composed of carbohydrates, proteins, and nucleic acids, often secreted by the biofilm cells (Al-Fattani and Douglas 2006; Martins et al. 2012). It was found that reduced drug diffusion may not be a problem in *Candida* biofilms and drugs like fluconazole could diffuse through normally. Instead, specific components of the matrix must be contributing to the resistance (Al-Fattani and Douglas 2004). Interestingly, treatment of biofilms with DNAse was found to enhance the sensitivity of *C. albicans* biofilm to the activity of caspofungin and amphotericin B. Hence, extracellular DNA in association with other components must be providing structural integrity and strength to EPM and contributing to drug resistance (Martins et al. 2012; Rajendran et al. 2013). Similarly, both, biofilm cells and biofilm matrix contain higher levels of β-1,3-glucans in their cell wall, compared to planktonic cells. The glucan was observed to bind four- to five-fold more drug than that of planktonic and contribute sequestering of antifungal azoles and polyenes (Nett et al. 2007; Mitchell et al. 2013). Disruption of β-1,3-glucans by glucanase treatment resulted in increased drug susceptibility of biofilms. Further evidence comes from an observation where low expression of glucan synthase gene was found to enhance the antibiofilm efficacy of amphotericin B, anidulafungin, and flucytosine (Nett et al. 2010). Overall, glucan-mediated binding/sequestering of drugs is an important resistance mechanism in biofilm growth form.

Up-regulation of drug efflux protein after exposure to antifungal drugs is a well-known mechanism of resistance in planktonic cells (Xie et al. 2014). ATP-binding cassette (ABC) transporter superfamily (e.g. *CDR1* and *CDR2*) and the major facilitator (MF) class (e.g. *MDR1*) are two main types of efflux pump proteins in *C. albicans* (Akins 2005; Cowen et al. 2014). Overexpression of these transporter proteins was observed in both, in vitro and in vivo biofilms, even in the absence of drug. Hence, upregulation of efflux pumps seems to be a normal mechanism associated with biofilm development (Ramage et al. 2002). Adhesion of *C. albicans* to a solid surface is sufficient to activate expression of the genes encoding the efflux pumps (Mateus et al. 2004). The efflux of drugs entering the cells remains active in mature biofilms too and continue to be a cause of biofilm-related drug resistance (Nobile et al. 2012).

Biofilm is an aggregated community of the cells attached to a surface, and cell density in that microenvironment is high. In the microplate based in vitro biofilm model for susceptibility testing, cell density ranges between 10^6 and 10^8 cells/ml. If the biofilm community is dispersed the cells with lower density exhibit increased sensitivity. Even, in the planktonic cell assays, it has been observed that increasing cell concentration results in reduced susceptibility to the drugs, fluconazole, ketoconazole, caspofungin and amphotericin B (Perumal et al. 2007; Mathe and Van Dijck 2013). Moreover, there is density-dependent secretion of quorum sensing molecules in biofilms; for example farnesol in *C. albicans* (Hornby et al. 2001). The presence of molecules like farnesol influences the overall gene expression of individual cells and may contribute to lower drug susceptibility (Cao et al. 2005; Garcia-Sanchez et al. 2004).

Reduced rate of metabolic activity of bacterial cells could contribute to the low drug sensitivity. In the bacterial biofilms, there is a limitation of nutrients so the cells may exhibit lower growth rates resulting in resistance to drugs which are effective against actively growing (like planktonic) cells (Martinez and Rojo 2011). But this may not be true for fungi; for example, biofilms were found equally resistant to amphotericin B, over a range of growth rates. Similarly, limitation to important nutrients like glucose or elements like iron did not cause changes in *Candida* biofilm susceptibility to amphotericin B (Baillie and Douglas 1998a, b). However, the role of altered metabolism in fungal resistance is not well investigated. Persister cells are a subset of cells which are phenotypically dormant and highly tolerant to the antimicrobial drugs. Bacterial biofilms harbor a notable (1%) percentage of persister cells which contribute to overall antibiotic resistance (Lewis 2010).

Persister cells have been observed in *C. albicans* biofilms too, and are highly resistant to antifungal agents (Khot et al. 2006). These are supposed to be phenotypic variants of the wild type exclusively present in biofilms and which gives rise to subpopulations of cells to form a new biofilm. Persisters act as a reservoir to initiate a new biofilm cycle and their drug tolerant nature is an important reason for the failure of antifungal treatment in clinical settings (LaFleur et al. 2006). Furthermore, *C. albicans* persister cells are exclusively recovered from biofilms and not from planktonic populations, regardless of their growth phase, and require attachment to a substrate to initiate the dormant phenotype. Biofilms of *C. krusei* and *C. parapsilosis* have been observed to harbor persisters and may be contributing to tolerance to drugs like amphotericin B (Al-Dhaheri and Douglas 2008). The molecular mechanisms underlying the drug refractory characteristic of fungal persisters is not investigated in detail.

Adhesion to a surface after initial contact is first important step in biofilm formation. The reversible attachment to a substrate results in activation of various signalling pathways. For example, the protein kinase C (PKC) pathway is an important pathway activated in response to cell wall stress (Kumamoto and Vinces 2005). Activation of such a stress-responsive pathway in fungal cells turns them drug tolerant. Mkc1 is the terminal mitogen-activated protein (MAP) kinase in PKC cascade. Deletion of *MKC1* gene has been shown to form abnormal *C. albicans* biofilms (Kumamoto and Vinces 2005). Interestingly, such a biofilm was found several times more sensitive to the antifungal activity of azoles.

Activation of a heat shock protein Hsp90 also contributes to azole and echinocandin resistance. This is through calcineurin pathway for stress responses (Cowen 2009; Singh et al. 2009). Inhibition of the protein phosphatase i.e. calcineurin or intervention of Hsp90 results in sensitization of *C. albicans* biofilm to various antifungal drugs (Uppuluri et al. 2008; Shinde et al. 2012a). Overall, drug resistance exhibited by yeast biofilms is governed by a complex network of multiple factors.

6 Therapeutic Strategies

6.1 Therapeutics

Current therapeutics against yeast biofilm includes the use of antifungal drugs to achieve inhibition of sessile cells and eradicate biofilm mass from the surface of biomaterials (Ramage et al. 2013). However, prevention of biofilm appears to be the best strategy because the drug resistance comes into picture once biofilms are developed and complicate the treatment. In fact, once biofilm is formed, removal of a colonized device (mainly catheters) is a strategy applied whenever suitable and helps to reduce mortality in device-associated infections (Andes et al. 2012; Cornely et al. 2012). Removal of the medical device may not be always possible, as the surgical procedures involve risk and increased costs. In such scenario, use of antifungal agents is necessary. Antifungal lock therapy (ALT) is one of the initial options for the treatment of catheter-related infections (Walraven and Lee 2013). Usually, polyenes and echinocandins are applied; for example, amphotericin B and its liposomal form are two agents commonly used for ALT purpose (Cornely et al. 2012). Similarly, caspofungin has been used to deal with catheter-related *Candida* biofilms (Ozdemir et al. 2011).

Various in vitro studies have indicated the efficacy of polyenes and echinocandins against *C. albicans* biofilm; hence, it would be useful to treat the biofilm infections in vivo (Kuhn et al. 2002). Although ALT using caspofungin and micafungin have shown high efficacy, it fails to completely eradicate the biofilm growth (Cateau et al. 2011). Liposomal AMB exhibited better antibiofilm activity than echinocandins (Ramage et al. 2013). Animal catheter models studies suggested that azoles are ineffective against biofilm growth, while liposomal AMB significantly reduced *C. albicans* biofilm infection. Similarly, AMB deoxycholate and caspofungin have been observed to achieve 80–100% removal of *C. albicans* colonization from catheters in rabbit models (Shuford et al. 2006). Infections associated with medical devices other than catheters, for example, prosthetic heart valves, knee implants or pacemakers, are hard to deal with; because removal of such a device is not easy and it involves a risk. For example, *Candida* related infective endocarditis is difficult to treat and involves mortality rates around 50%. Such infections can be treated with liposomal amphotericin B or caspofungin (Ellis et al. 2001; Falcone et al. 2009).

The combination of polyenes and azoles has been found useful in the inhibition of wound-related biofilms. *Candida* biofilm infections at wounds and joints can be efficiently treated with a combination of liposomal AMB and voriconazole or posaconazole. Combinatorial therapy is also applied to treat oral fungal biofilms like denture-related stomatitis and oral candidiasis (Rautemaa and Ramage 2011). Despite the available options of biofilm therapy, treatment of biofilm infections remains a challenge. In the majority of cases, complete removal of colonized growth is not achieved and may result in recurrent infections (Ramage et al. 2014; Fox et al. 2015). Hence, there is need to find alternative therapeutic options for the treatment of yeast biofilms.

6.2 Future Strategies

Development of an antifungal agent is difficult as fungi are eukaryotic organisms and share many similarities with the human host (Routh et al. 2011). Hence, to find a cellular mechanism that can be specifically targeted in the fungal cells and use it from the drug discovery point of view is relatively complicated. This becomes a more difficult task when the infections are biofilm-associated and exhibit increased resistance to antifungal agents. Various approaches are being followed to increase the antifungal arsenal.

Rational drug designing is one of the approaches which target a specific protein or biochemical pathway (Srinivasan et al. 2014). For example, identification of the mechanisms behind biofilm dispersal may help for developing a compound that dismantles biofilm community. Similarly, a better understanding of the proteins involved in the transformation of a sessile cell into a persister during biofilm formation would allow devising strategies to reverse their physiology. The combination of such a strategy with available antifungals would successfully remove biofilms, kill the released planktonic cells and prevent recurrence of infections (Fox et al. 2015). A diverse range of genes involved in adherence, morphogenesis, quorum sensing, matrix production, cell wall biosynthesis, and metabolism have been found to play important roles in biofilm formation and regulation (Garcia-Sanchez et al. 2004; Nobile et al. 2012; Desai et al. 2014). Various proteins are differentially expressed between biofilms and planktonic cells. Many of these proteins may be enzymes resulting in a different metabolic state of biofilms. This may be used to target a metabolic pathway important for biofilm growth and can be used as drug targets (Fox et al. 2015).

A systems biology study to target important protein involved in biofilm formation is another approach. For example, Nobile et al. (2012) have identified transcription factors regulating biofilm growth of *C. albicans* (Nobile et al. 2012). The study identified six main regulators of transcription, Bcr1, Tec1, Efg1, Ndt80, Rob1 and Brg1. They are involved in controlling the expression of at least 1000 target genes. Deletion of *ALS1*, *HWP1*, and *CAN2* genes has been found to result in the defective biofilm. It has been proposed that Als1 and Hwp1 which are cell surface proteins involved in adhesion, Can2 and Tpo4 (probably play a role in transport), and Eht1 protein involved in fatty acid synthesis and morphogenesis may be explored as antibiofilm targets (Fox and Nobile 2012; Nobile et al. 2012). A study has identified transcription regulators Bcr1, Ace2, Snf5, and Arg81 important for adhesion to silicone and subsequent biofilm formation (Finkel et al. 2012). Zap1 is another important regulator of extracellular matrix production and also govern the synthesis of β-1,3-glucan and other matrix constituents (Nobile et al. 2009).

However, rational drug designing is time-consuming and involves a lot of money. Many researchers are following an empirical approach for antifungal discovery through screening of synthetic/semi-synthetic chemicals (Srinivasan et al. 2014). Plant extracts, essential oils and their constituent molecules exhibit novel antimicrobial and antifungal properties (Raut et al. 2013a; Raut 2014; Raut and Karuppayil 2014a). Most importantly, phytochemicals have been found to possess inhibitory potential against drug-resistant biofilms of bacterial and fungal pathogens (Raut and Karuppayil 2014b). Efforts are being done to identify molecules with antibiofilm potential through random screening of small molecules of natural origin including phytochemicals. It includes a search for plant actives which can prevent biofilm development as well as those which disrupt mature biofilms (Raut et al. 2012, 2013b, 2014; Raut and Karuppayil 2016).

Phytochemicals or other synthetic molecules can be used in combination with existing drugs so that to potentiate the activity of available antifungal agents. The combinatorial approach may be useful to mitigate the drug-resistance associated with biofilm communities. Drug efflux inhibitors or cell sensitizer molecules may be used to overcome the problem of biofilm mediated resistance (Shinde et al. 2013; Doke et al. 2014). Other miscellaneous approaches include combination of biofilm disruptive agents with drugs so that EPM surrounding the biofilm is disturbed. For example, combinatorial therapy of AMB and CSP with DNase significantly disrupted EPM and sensitized *C. albicans* biofilm to antifungal drugs (Martins et al. 2012).

Use of broad-spectrum antimicrobial metal ions like silver or nanoparticles of silver is another interesting way. It can be used for coating a catheter surface or medical device to prevent adhesion and biofilm formation by yeasts. Silver interferes with DNA replication, denatures proteins, and inhibit oxidative enzymes (Rai et al. 2009). Its combination with antifungal drugs can be very effective. Silver nanoparticles have been shown to inhibit *C. albicans* and *C. glabrata* biofilms at various stages of development. These have been utilised in hydrogels used to treat chronic wounds and also in denture prostheses (Monteiro et al. 2011, 2012). Molecules that interfere with the quorum sensing involved in biofilm formation and regulation are also an attractive alternative for biofilm mitigation (Nickerson et al. 2006; Kalia 2013). Screening of clinical and preclinical non-antifungal drugs, drug compound libraries, and repurposing of them against fungal biofilms is a recent approach being investigated (Routh et al. 2011; Shinde et al. 2013; Pierce and Lopez-Ribot 2013).

7 Conclusions

Biofilms of pathogenic yeasts are increasingly being recognized for their involvement in the human health and disease. Yeast biofilms on tissue surfaces and/or indwelling prostheses and drug-resistant infections associated with these have emerged as a serious threat to a large population of immunocompromised individuals. The available arsenal of antifungal agents is not sufficient to successfully mitigate biofilms; hence, there is an urgent need to search for novel therapeutic agents. Further understanding of the mechanisms involved in biofilm formation and regulation may provide clues to the development of antibiofilm strategies for the prevention and treatment of yeast infections.

Acknowledgements Dr. Raut Jayant S. is thankful to UGC, New Delhi, for the award of Dr. D.S. Kothari Postdoctoral Fellowship. Raut J.S. and Doke S.K. have equal contribution to this book chapter.

References

Akins, R.A. 2005. *Med. Mycol.* 43: 285–318.
Al-Dhaheri, R.S. and Douglas, L.J. 2008. *Antimicrob. Agents Chemother.* 52: 1884–1887.
Alexandre, H. 2013. *Internat. J. Food Microbiol.* 167: 269–275.
Al-Fattani, M.A. Douglas, L.J. 2004. *Antimicrob. Agents Chemother.* 48: 3291–3297.
Al-Fattani, M.A. and Douglas, L.J. 2006. *J. Med. Microbiol.* 55: 999–1008.
Alvarez, M., Saylor, C. and Casadevall, A. 2008. *Cell. Microbiol.* 10: 1622–1633.
Andes, D.R., Safdar, N., Baddley, J.W., Playford, G., Reboli, A.C., Rex, J.H., Sobel, J.D., Pappas, P.G., Kullberg, B.J. and Mycoses Study Group. 2012. *Clin. Inf. Dis.* cis021.
Baillie, G.S. and Douglas, L.J. 1998a. *Antimicrob. Agents Chemother.* 42: 1900–1905.
Baillie, G.S. and Douglas, L.J. 1998b. *Antimicrob. Agents Chemother.* 42: 2146–2149.
Baillie, G.S. and Douglas, L.J. 2000. *J. Antimicrob. Chemother.* 46: 397–403.
Banerjee, U., Gupta, K. and Venugopal, P. 1997. *J. Med. Vet. Mycol.* 35: 139–141.
Bastidas, R.J. and Heitman, J. 2009. *Proc. Nat. Acad. Sci.* 106: 351–352.
Bizerra, F.C., Nakamura, C.V., De Poersch, C., Estivalet Svidzinski, T.I., Borsato Quesada, R.M., Goldenberg, S., Krieger, M.A. and Yamada-Ogatta, S.F. 2008. *FEMS Yeast Res.* 8: 442–450.
Bojsen, R.K., Andersen, K.S. and Regenberg, B. 2012. *FEMS Immunol. Med. Microbiol* 65: 169–182.
Bonaventura, G.D., Pompilio, A., Picciani, C., Iezzi, M., D'Antonio, D. and Piccolomini, R. 2006. *Antimicrob. Agents Chemother.* 50: 3269–3276.
Branski, L.K., Al-Mousawi, A., Rivero, H., Jeschke, M.G., Sanford, A.P., Herndon, D.N. 2009. *Surg. Inf.* 10: 389–397.
Braun, D.K., Janssen, D.A., Marcus, J.R. and Kauffman, C.A. 1994. *Amer. J. Kidney Dis.* 24: 864–867.
Brilhante, R.S.N., de Lima, R.A.C., de Farias Marques, F.J., Silva, N.F., Caetano, E.P., Castelo, D. D.S.C.M., Bandeira, T.J.P.G. and Rocha, M.F.G. 2015. *J. Med. Microbiol.* 64: 394–399.
Brown, G.D., Denning, D.W. and Levitz, S.M. 2012a. *Science* 336: 647–647.
Brown, G.D., Denning, D.W., Gow, N.A., Levitz, S.M., Netea, M.G. and White, T.C. 2012b. *Science Trans. Med.* 4: 165rv13–165rv13.
Calderone, R.A. and Clancy, C.J. (Eds.). 2012. *Candida and candidiasis.* 2 nd ed. ASM Press, Washington DC, USA.

Cannon, R.D., Lamping, E., Holmes, A.R., Niimi, K., Baret, P.V., Keniya, M.V., Tanabe, K., Niimi, M., Goffeau, A. and Monk, B.C. 2009. *Clin. Microbiol. Rev.* 22: 291–321.

Cao, Y.Y., Cao, Y.B., Xu, Z., Ying, K., Li, Y., Xie, Y., Zhu, Z.Y., Chen, W.S. and Jiang, Y.Y. 2005. *Antimicrob. Agents Chemother.* 49: 584–589.

Casadevall, A. and Pirofski, L.A. 2007. *Eukar. Cell* 6: 2169–2174.

Cassone, M., Serra, P., Mondello, F., Girolamo, A., Scafetti, S., Pistella, E. and Venditti, M. 2003. *J. Clin. Microbiol.* 41: 5340–5343.

Cateau, E., Berjeaud, J.M. and Imbert, C. 2011. *Int. J. Antimicrob. Agents* 37: 380–384.

Chaffin, W.L., Lopez-Ribot, J.L., Casanova, M., Gozalbo, D. and Martínez, J.P. 1998. *Microbiol. Mol. Biol. Rev.* 62: 130–180.

Chandra, J., Kuhn, D.M., Mukherjee, P.K., Hoyer, L.L., McCormick, T. and Ghannoum, M.A. 2001a. *J. Bacteriol.* 183: 5385–5394.

Chandra, J., Mukherjee, P.K., Leidich, S.D., Faddoul, F.F., Hoyer, L.L., Douglas, L.J. and Ghannoum, M.A. 2001b. *J. Dent. Res.* 80: 903–908.

Chauhan, N.M., Raut, J.S. and Karuppayil, S.M. 2011a. *Mycoscience* 52: 356–360.

Chauhan, N.M., Raut, J.S. and Karuppayil, S.M. 2011b. *Mycoses* 54: e697-e703.

Chen, H., Fujita, M., Feng, Q., Clardy, J. and Fink, G.R. 2004. *Proceed. Nat. Acad. Sci.* 101: 5048–5052.

Cornely, O.A., Bassetti, M., Calandra, T., Garbino, J., Kullberg, B.J., Lortholary, O., Meersseman, W. and Ullmann, A.J. 2012. *Clin. Microbiol. Inf.* 18: 19–37.

Costerton, J.W., Stewart, P.S. and Greenberg, E.P. 1999. *Science* 284: 1318–1322.

Cowen, L.E. 2009. *PLoS Pathog.* 5: e1000471.

Cowen, L.E., Sanglard, D., Howard, S.J., Rogers, P.D. and Perlin, D.S. 2014. *Cold Spring Harb. Perspect. Med.* 10.

Crump, J.A. and Collignon, P.J. 2000. *Europ. J. Clin. Microbiol. Inf. Dis.* 19: 1–8.

Cuellar-Cruz, M., Vega-Gonzalez, A., Mendoza-Novelo, B., Lopez-Romero, E., Ruiz-Baca, E., Quintanar-Escorza, M.A. and Villagomez-Castro, J.C. 2012. *Eur. J. Clin. Microbiol. Inf. Dis.* 31: 2513–2527.

d'Enfert, C. 2009. *Curr. Opin. Microbiol.* 12: 358–364.

Davis-Hanna, A., Piispanen, A.E., Stateva, L.I. and Hogan, D.A. 2008. *Mol. Microbiol.* 67: 47–62.

Denning, D.W. and Bromley, M.J. 2015. *Science* 347: 1414–1416.

Desai, J.V., Mitchell, A.P. and Andes, D.R. 2014. *Cold Spring Harbor Perspect. Med.* 1: 10.

Diekema, D., Arbefeville, S., Boyken, L., Kroeger, J. and Pfaller, M. 2012. *Diagno. Microbiol. Infect. Dis.* 73: 45–48.

Dinleyici, E.C., Eren, M., Ozen, M., Yargic, Z.A. and Vandenplas, Y. 2012. *Exp. Opin. On Biol. Ther.* 12: 395–410.

Doke, S.K., Raut, J.S., Dhawale, S. and Karuppayil, S.M. 2014. *The J. Gen. Appl. Microbiol.* 60: 163–168.

Donlan, R.M. and Costerton, J.W. 2002. *Clin. Microbiol. Rev.* 15: 167–193.

Ellis, M., Shamoon, A., Gorka, W., Zwaan, F. and Al-Ramadi, B. 2001. *Clin. Inf. Dis.* 32: e87-e89.

Falcone, M., Barzaghi, N., Carosi, G., Grossi, P., Minoli, L., Ravasio, V., Rizzi, M. and Venditti, M. 2009. *Med.* 88: 160–168.

Fanning, S. and Mitchell, A.P. 2012. *PLoS Pathog.* 8: e1002585.

Finkel, J.S., Xu, W., Huang, D., Hill, E.M., Desai, J.V., Woolford, C.A., Nett, J.E. and Mitchell, A.P. 2012. *PLoS Pathog.* 8: e1002525-e1002525.

Fisher, M.C., Henk, D.A., Briggs, C.J., Brownstein, J.S., Madoff, L.C., McCraw, S.L. and Gurr, S. J. 2012. *Nature* 484: 186–194.

FIT, (The Fungal Infection Trust). 2012. *Fungal Research Trust*, 20th Anniversary meeting, London, June 18th 2011, updated December 2012.

Fox, E.P. and Nobile, C.J. 2012. *Transcript.* 3: 315–322.

Fox, E.P., Singh-Babak, S.D., Hartooni, N. and Nobile, C.J. 2015. In: Antifungals: From Genomics to Resistance and the Development of Novel Agents, Caister Academic Press, pp. 71–90.

Ganguly, S. and Mitchell, A. P. 2011. *Curr. Opin. Microbiol.* 14: 380–385.
Garcia-Sanchez, S., Aubert, S., Iraqui, I., Janbon, G., Ghigo, J.M. and d'Enfert, C. 2004. *Eukar. Cell* 3: 536–545.
Ghannoum, M.A. and Rice, L.B. 1999. *Clin. Microbiol. Rev.* 12: 501–517.
Ghannoum, M.A., Jurevic, R.J., Mukherjee, P.K., Cui, F., Sikaroodi, M., Naqvi, A. and Gillevet, P.M. 2010. *PLoS Pathog.* 6: e1000713.
Gimeno, C.J., Ljungdahl, P.O., Styles, C.A. and Fink, G.R. 1992. *Cell* 68: 1077–1090.
Gullo, F.P., Rossi, S.A., de CO Sardi, J., Teodoro, V.L.I., Mendes-Giannini, M.J.S. and Fusco-Almeida, A.M. 2013. *Europ. J. Clin. Microbiol. Inf. Dis.* 32: 1377–1391.
Gundacker, N.D. and Baddley, J.W. 2015. *Curr. Clin. Microbiol. Reports* 2: 76–83.
Guo, B., Styles, C.A., Feng, Q. and Fink, G.R. 2000. *Proc. Nat. Acad. Sci.* 97: 12158–12163.
Harriott, M.M., Lilly, E.A., Rodriguez, T.E., Fidel, P.L. and Noverr, M.C. 2010. *Microbiol.* 156: 3635–3644.
Hawksworth, D.L. 2001. *Mycol. Res.*105: 1422–1432.
Hogan, D.A. 2006. *Eukar. Cell* 5: 613–619.
Hornby, J.M., Jensen, E.C., Lisec, A.D., Tasto, J.J., Jahnke, B., Shoemaker, R., Dussault, P. and Nickerson, K.W. 2001. *Appl. Environ. Microbiol.* 67: 2982–2992.
Huffnagle, G.B. and Noverr, M.C. 2013. *Trends Microbiol.* 21: 334–341.
Kalia, V.C. 2013. *Biotechnol. Adv.* 31: 224–245.
Kauffman, C.A. 2007. *Clin. Microbiol. Rev.* 20: 115–132.
Kaur, R., Domergue, R., Zupancic, M.L. and Cormack, B.P. 2005. Curr. Opin. Microbiol. 8: 378–384.
Khot, P.D., Suci, P.A., Miller, R.L., Nelson, R.D. and Tyler, B.J. 2006. Baillie, G.S. and Douglas, L.J. 1998a. *Antimicrob. Agents Chemother.* 50: 3708–3716.
Klotz, S.A. 1990. *FEMS Microbiol. Lett.* 68: 249–253.
Klotz, S.A., Chasin, B.S., Powell, B., Gaur, N.K., and Lipke, P.N. 2007. *Diagnos. Microbiol. Inf. Dis.* 59: 401–406.
Kojic, E.M. and Darouiche, R.O. 2004. *Clin. Microbiol. Rev.* 17: 255–267.
Kong, H.H. and Segre, J.A. 2012. *J. Invest. Dermatol.* 132: 933–939.
Kontoyiannis, D.P., Torres, H.A., Chagua, M., Hachem, R., Tarrand, J.J., Bodey, G.P. and Raad, I. I. 2004. *Scand. J. Inf. Dis.* 36: 564–569.
Krzossok, S., Birck, R., Henke, S., Hof, H., van der Woude, F.J. and Braun, C. 2004. *Clin. Nephrol.* 62: 66–68.
Kucharíkova, S., Tournu, H., Lagrou, K., Van Dijck, P. and Bujdakova, H. 2011. *J. Med. Microbiol.* 60: 1261–1269.
Kuhn, D.M., George, T., Chandra, J., Mukherjee, P.K. and Ghannoum, M.A. 2002. *Antimicrob. Agents Chemother.* 46: 1773–1780.
Kumamoto, C.A. 2011. *Curr. Opin. Microbiol.* 14: 386–391.
Kumamoto, C.A. and Vinces, M.D. 2005. *Annu. Rev. Microbiol.* 59: 113–133.
Kumar, C.G. and Menon, T. 2006. *Med. Mycol.* 44: 99–101.
LaFleur, M.D., Kumamoto, C.A. and Lewis, K. 2006.
Lattif, A.A., Mukherjee, P.K., Chandra, J., Swindell, K., Lockhart, S.R., Diekema, D.J., Pfaller, M.A. and Ghannoum, M.A. 2010. *Int. J. Med. Microbiol.* 300: 265–270.
Leake, J.L., Dowd, S.E., Wolcott, R.D., Zischkau, A.M. and Sun, Y. 2009. *J. Wound Care* 18: 103–108.
Lewis, K. 2010. *Ann. Rev. Microbiol.* 64: 357–372.
Martinez, L.R. and Casadevall, A. 2007. *Appl. Environ. Microbiol.* 73: 4592–4601.
Martinez, J. L. and Rojo, F. 2011. *FEMS Microbiol. Rev.* 35: 768–789.
Martins, M., Henriques, M., Lopez-Ribot, J.L. and Oliveira, R. 2012. *Mycoses* 55: 80–85.
Mateus, C., Crow, S.A. and Ahearn, D.G. 2004. *Antimicrob. Agents Chemother.* 48: 3358–3366.
Mathe, L. and Van Dijck, P. 2013. *Curr. Gen.* 59: 251–264.
McKinsey, D.S. and McKinsey, J.P. 2011. *Sem. Resp. Crit. Care Medicine* 32: 735–744.
Miceli, M.H., Díaz, J.A. and Lee, S.A. 2011. *The Lancet Inf. Dis.* 11: 142–151.

Mishra, N., Prasad, T., Sharma, N., Payasi, A., Prasad, R., Gupta, D. and Singh, R. 2007. *Acta Microbiol. et Immunol. Hungar.* 54: 201–235.
Mitchell, K.F., Taff, H.T., Cuevas, M.A., Reinicke, E.L., Sanchez, H. and Andes, D.R. 2013. *Antimicrob. Agents Chemother.* 57: 1918–1920.
Monteiro, D.R., Gorup, L.F., Silva, S., Negri, M., de Camargo, E.R., Oliveira, R., Barbosa, D.B. and Henriques, M. 2011. *Biofouling* 27: 711–719.
Monteiro, D.R., Silva, S., Negri, M., Gorup, L.F., De Camargo, E.R., Oliveira, R., Barbosa, D.B. and Henriques, M. 2012. *Lett. Appl. Microbiol.* 54: 383–391.
Mosel, D.D., Dumitru, R., Hornby, J.M., Atkin, A.L. and Nickerson, K.W. 2005. *Appl. Environ. Microbiol.* 71: 4938–4940.
Nett, J.E. and Andes, D. 2015. *Microbiol. Spectrum* 3: E30.
Nett, J., Lincoln, L., Marchillo, K., Massey, R., Holoyda, K., Hoff, B., VanHandel, M. and Andes, D. 2007. *Antimicrob. Agents Chemother.* 51: 510–520.
Nett, J.E., Sanchez, H., Cain, M.T. and Andes, D.R. 2010. *J. Inf. Dis.* 202: 171–175.
Nickerson, K.W., Atkin, A.L. and Hornby, J.M. 2006. *Appl. Environ. Microbiol.* 72: 3805–3813.
Nobile, C.J. and Mitchell, A.P. 2006. *Cell. Microbiol.* 8: 1382–1391.
Nobile, C.J., Nett, J.E., Hernday, A.D., Homann, O.R., Deneault, J.S., Nantel, A., Andes, D.R. Johnson, A.D. and Mitchell, A.P. 2009. *PLoS Biol.* 7: e1000133–1.
Nobile, C.J., Fox, E.P., Nett, J.E., Sorrells, T.R., Mitrovich, Q.M., Hernday, A.D., Tuch, B.B., Andes, D.R. and Johnson, A.D. 2012. *Cell* 148: 126–138.
Nosanchuk, J.D. 2006. *Rec. Pat. on Anti-inf. Drug Disc.* 1: 75–84.
Nucci, M. and Marr, K.A. 2005. *Clin. Inf. Dis.* 41: 521–526.
Nusbaum, A.G., Kirsner, R.S. and Charles, C.A. 2012. *Skin Thear. Lett.* 17: 1–5.
Ozdemir, H., Karbuz, A., Ciftci, E., Dincaslan, H. U., Ince, E., Aysev, D., Yavuz, G. and Dogru, U. 2011. *Mycoses* 54: e647-e649.
Panagoda, G.J., Ellepola, A.N.B. and Samaranayake, L.P. 2001. *Mycoses* 44: 29–35.
Paolino, K.M., Henry, J.A., Hospenthal, D.R., Wortmann, G.W. and Hartzell, J.D. 2012. *Military Med.* 177: 681–685.
Park, B.J., Wannemuehler, K.A., Marston, B.J., Govender, N., Pappas, P.G. and Chiller, T.M. 2009. *Aids* 23: 525–530.
Perumal, P., Mekala, S. and Chaffin, W.L. 2007. *Antimicrob. Agents Chemother.* 51: 2454–2463.
Pettit, R.K., Repp, K.K. and Hazen, K.C. 2010. *Med. Mycol.* 48: 421–426.
Pfaller, M.A. 2012. *American J. Med.* 125: S3–S13.
Pfaller, M.A. and Diekema, D.J. 2007. *Clin. Microbiol. Rev.* 20: 133–163.
Pierce, C.G. and Lopez-Ribot, J.L. 2013. *Exp. Opin. On Drug Dis.* 8: 1117–1126.
Pierce, C.G., Srinivasan, A., Uppuluri, P., Ramasubramanian, A.K. and Lopez-Ribot, J.L. 2013. *Curr. Opin. Pharmacol.* 13: 726–730.
Pitangui, N.S., Sardi, J.C.O., Silva, J.F., Benaducci, T., Moraes da Silva, R.A., Rodriguez-Arellanes, G., Taylor, M.L., Mendes-Giannini, M.J.S. and Fusco-Almeida, A.M. 2012. *Biofouling* 28: 711–718.
Polvi, E.J., Li, X., O'Meara, T.R., Leach, M.D. and Cowen, L.E. 2015. *Cell. Mol. Life Sci.* 72: 2261–2287.
Rai, M., Yadav, A. and Gade, A. 2009. *Biotechnol. Adv.* 27: 76–83.
Rajendran, R., Williams, C., Lappin, D.F., Millington, O., Martins, M. and Ramage, G. 2013. *Eukar. Cell* 12: 420–429.
Ramage, G., Walle, K.V., Wickes, B.L. and Lopez-Ribot, J.L. 2001. *Rev. Iberoam. Micol.* 18: 163–170.
Ramage, G., Bachmann, S., Patterson, T.F., Wickes, B.L. and Lopez-Ribot, J.L. 2002. *J. Antimicrob. Chemother.* 49: 973–980.
Ramage, G., Martínez, J.P. and Lopez-Ribot, J.L. 2006. *FEMS Yeast Res.* 6: 979–986.
Ramage, G., Mowat, E., Jones, B., Williams, C. and Lopez-Ribot, J. 2009a. *Crit. Rev. Microbiol.* 35: 340–355.
Ramage, G., Mowat, E., Williams, C. and Ribot, J.L.L. 2009b. In: Pathogenic Yeasts (eds. H.R. Ashbee and E.M. Bignell), Springer-Verlag, Heidelberg, pp. 121–144.

Ramage, G., Rajendran, R., Sherry, L. and Williams, C. 2012. *Int. J. Microbiol.* doi:10.1155/2012/528521.

Ramage, G., Jose, A., Sherry, L., Lappin, D. F., Jones, B. and Williams, C. 2013. *Antimicrob. Agents Chemother.* 57: 2369–2371.

Ramage, G., Robertson, S.N. and Williams, C. 2014. *Int. J. Antimicrob. Agents* 43: 114–120.

Rathod, V.S., Raut, J.S. and Karuppayil, S.M. 2012. *Asian J. Pharma. Clin. Res.* 5: 170–173.

Rathod, V.S., Raut, J.S., Raut, S.S., More, S.R. and Karuppayil, S.M. 2013. *Brit. Microbiol. Res. J.* 3: 150–157.

Raut, J.S. 2014. *Modulation of Candida albicans biofilm formation by selected molecules of plant origin*, Ph.D. Thesis submitted to S.R.T.M. University, Nanded, Maharashtra, India, January 2014.

Raut, J.S. and Karuppayil, S.M. 2014a. In: *Environment and Sustainable Development,* (Eds. M. H. Fulekar, B. Pathak, R.K. Kale), Springer, India, pp. 59–76.

Raut, J.S. and Karuppayil, S.M. 2014b. *Ind. Crops Prod.* 62: 250–264.

Raut, J.S. and Karuppayil, S.M. 2016. *Curr. Pharma. Des.* 22: 4111– 4134.

Raut, J., Rathod, V. and Karuppayil, S.M. 2010. *Jpn. J. Med. Mycol.* 51: 131–136.

Raut, J.S., Shinde, R.B. and Karuppayil, M.S. 2012. *Afr. J. Microbiol. Res. 6*: 6005–6012.

Raut, J.S., Chauhan, N.M., Shinde, R.B. and Karuppayil, S.M. 2013a. *J. Med. Plant Res.* 7: 777–782.

Raut, J.S., Shinde, R.B., Chauhan, N.M. and Karuppayil, S.M. 2013b. *Biofouling* 29: 87–96.

Raut, J.S., Chauhan, N.M., Shinde, R.B. and Karuppayil, S.M. 2014. *J. Microbiol. Biotechnol.* 24: 1216–1225.

Rautemaa, R. and Ramage, G. 2011. *Crit. Rev. Microbiol.* 37: 328–336.

Reddy, B.T., Torres, H.A. and Kontoyiannis, D.P. 2002. *Scand. J. Inf. Dis.* 34: 143–144.

Reynolds, T.B. and Fink, G.R. 2001. *Science* 291: 878–881.

Robertson, E.J., Wolf, J.M. and Casadevall, A. 2012. *Appl. Env. Microbiol.* 78: 7977–7984.

Routh, M.M., Raut, J.S. and Karuppayil, S.M. 2011. *Chemother.* 57: 372–380.

Ryan, O., Shapiro, R.S., Kurat, C.F., Mayhew, D., Baryshnikova, A., Chin, B., Lin, Z.Y. and Boone, C. 2012. *Science* 337: 1353–1356.

Samaranayake, L.P., Keung-Leung, W. and Jin, L. 2009. *Periodontol. 2000* 49: 39–59.

Sardi, J.C.O., Scorzoni, L., Bernardi, T., Fusco-Almeida, A.M. and Giannini, M.M. 2013. *J. Med. Microbiol.* 62: 10–24.

Shinde, R.B., Chauhan, N.M., Raut, J.S. and Karuppayil, S.M. 2012a. *Ann. Clin. Microbiol. Antimicrob.* 11: 1–6.

Shinde, R.B., Raut, J.S. and Karuppayil, M.S. 2012b. *Mycoscience* 53: 220–226.

Shinde, R.B., Raut, J.S., Chauhan, N.M. and Karuppayil, S.M. 2013. *Braz. J. Inf. Dis.* 17: 395–400.

Shuford, J.A., Rouse, M.S., Piper, K.E., Steckelberg, J.M. and Patel, R. 2006. *J. Inf. Dis.* 194: 710–713.

Silva, S., Henriques, M., Martins, A., Oliveira, R., Williams, D. and Azeredo, J. 2009. *Med. Mycol.* 47: 681–689.

Silva, S., Negri, M., Henriques, M., Oliveira, R., Williams, D.W. and Azeredo, J. 2011. *Trends Microbiol.* 19: 241–247.

Singh, S.D., Robbins, N., Zaas, A.K., Schell, W.A., Perfect, J.R. and Cowen, L.E. 2009. *PLoS Pathog.* 5: e1000532.

Sobel, J.D. 2007. *The Lancet* 369: 1961–1971.

Sobel, J.D. 2011. In: Sexually Transmitted Infections and Sexually Transmitted Diseases, Springer, Berlin Heidelberg, pp. 613–624.

Sonesson, A., Bartosik, J., Christiansen, J., Roscher, I., Nilsson, F., Schmidtchen, A. and Back, O. 2013. *Acta Dermato-venereol.* 93: 340–346.

Srinivasan, A., Lopez-Ribot, J.L. and Ramasubramanian, A.K. 2014. *Drug Disc. Today: Technol.* 11: 65–71.

Taff, H.T., Mitchell, K.F., Edward, J.A. and Andes, D.R. 2013. *Fut. Microbiol.* 8: 1325–1337.

Trevisani, L., Sartori, S., Rossi, M.R., Bovolenta, R., Scoponi, M., Gullini, S.S. and Abbasciano, V. 2005. *Digest. Dis. Sciences* 50: 463–469.
Underhill, D.M. and Iliev, I.D. 2014. *Nat. Rev. Immunol.* 14: 405–416.
Uppuluri, P., Nett, J., Heitman, J. and Andes, D. 2008. Baillie, G.S. and Douglas, L.J. 1998a. *Antimicrob. Agents Chemother.* 52: 1127–1132.
Vallejo, J.A., Sanchez-Perez, A., Martínez, J.P. and Villa, T.G. 2013. *Appl. Microbiol. Biotechnol.* 97: 2305–2318.
Verstrepen, K.J. and Klis, F.M. 2006. *Mol. Microbiol.* 60: 5–15.
Walraven, C.J. and Lee, S.A. 2013. *Antimicrob. Agents Chemother.* 57: 1–8.
West, S.A., Diggle, S.P., Buckling, A., Gardner, A. and Griffin, A.S. 2007. *Ann. Rev. Ecol. Evol. Syst.* 53–77.
White, T.C., Marr, K.A. and Bowden, R.A. 1998. *Clin. Microbiol. Rev.* 11: 382–402.
Wilson, L.S., Reyes, C.M., Stolpman, M., Speckman, J., Allen, K. and Beney, J. 2002. *Value in Health* 5: 26–34.
Xie, J.L., Polvi, E.J., Shekhar-Guturja, T. and Cowen, L.E. 2014. *Fut. Microbiol.* 9: 523–542.
Zhang, E., Tanaka, T., Tajima, M., Tsuboi, R., Nishikawa, A. and Sugita, T. 2011. *Microbiol. Immunol.* 55: 625–632.
Zhang, X., Sun, X., Wang, Z., Zhang, Y. and Hou, W. 2012. *Invest. Ophthalmol. Vis. Sci.* 53: 7774–7778.

Biology of Killer Yeast and Technological Implications

Bijender Kumar Bajaj and Satbir Singh

Abstract The killer phenomenon has been reported among various genera of yeast. Potential ability of certain yeast (killer yeast) to kill the other yeast (sensitive yeast) was first observed in the strains of *Saccharomyces cerevisiae*. The killer yeasts secrete extracellular protein toxin that is fatal for the sensitive yeast, and kills the latter. Killer toxin producer yeasts are immune towards their own toxin but can kill the other sensitive yeasts by employing variety of mechanisms like targeting several cellular components viz. cell wall, plasma membrane, tRNA, DNA etc. The genetic information for production of killer toxin (killer phenotype) is generally present as extra-chromosomal genetic elements like dsRNA or linear DNA, or on the chromosome. The protein toxins produced by several killer yeasts have been thoroughly studied after purification and characterization. Killer toxins encoding genes have been cloned, characterized and expressed in heterologous systems. Significance of yeast-derived killer toxins and/or killer yeasts have been implicated in various areas including food fermentations/yeast-based bioprocesses. Yeast killer phenomenon may play a substantive role in stabilizing the ecosystem. Killer toxins of yeast may have potential for application as biopreservatives, biocontrol agents and as new therapeutic molecules especially against multidrug resistant pathogens. For ethanol-based industries like distilleries, breweries, and wineries, killer yeast can be employed as starter industrial yeast cultures to protect against the wild contaminants. Current article presents recent developments on biological and technological implications of killer yeasts.

Keywords Killer yeast · Yeast killer toxin · Receptor · *Saccharomyces cerevisiae* · Applications

B.K. Bajaj (✉) · S. Singh
School of Biotechnology, University of Jammu, BAWE WALI RAKH, Jammu 180006, India
e-mail: bajajbijenderk@gmail.com

T. Satyanarayana and G. Kunze (eds.), *Yeast Diversity in Human Welfare*, DOI 10.1007/978-981-10-2621-8_7

1 Introduction

The origin of the word 'yeast' has been originated from a Dutch word '*gist*' that means froth/foam produced due to metabolic activities of yeast during brewing (Hatoum et al. 2012). Yeasts, the eukaryotic microorganisms are extensively spread in various natural environments viz. the food products, fruits, vegetables, body surfaces of plants, animals and humans and several other ecological niches (Satyanarayana and Kunze 2009; Bajaj et al. 2013). Generally nutrient rich substrates are more frequently and rapidly colonized by yeasts. Yeasts are integral and very important components of several complex ecosystems especially due to their tremendous capability for potential interactions with other microorganisms of those ecosystems. The most common mode of vegetative growth of yeast is by budding but yeasts may have both sexual and asexual states. The genus *Saccharomyces* of the phylum *Ascomycota* is one of the most studied yeasts. Members of several yeast genera have got enormous biotechnological significance due to their industrial applications. Yeast may be utilized for production of bread, beer, wine, industrial alcohol, vitamins, biocontrol agents and rDNA products. Yeast may have food/feed importance as single cell protein (Satyanarayana and Kunze 2009). Among various species of yeast *Saccharomyces cerevisiae* has got the foremost importance due to its association with human since ancient times for producing bread, beer wine, and due to its industrial usage in recent times (Bajaj and Sharma 2010). Industrial application potential of *S. cerevisiae* has been described by Pretorius et al. (2003). More than 6×10^5 tons of *S. cerevisiae* is used as bakery yeast, and about 8×10^5 tons as single cell protein, annually. Similarly, approximately 3×10^7 tons of wine, and 6×10^7 tons of beer is produced annually using of *S. cerevisiae.* Generally the *S. cerevisiae* vegetative cells are diploid, but can be euploid, tetraploid or polyploids (Bajaj and Tauro 1994; Webster and Weber 2007).

Certain yeasts have the potential ability to produce killer toxins/mycocins which may inhibit growth of other yeast/fungi/bacteria (Bajaj et al. 2013). Such yeasts are called killer yeasts. This attribute of producing killer toxins provides additional advantage to such yeast species to successfully compete with their co-inhabitants. Thus, killer yeast acts selfishly to prosper at the expense of others as they have evolved a mechanism (killer toxin) to successfully compete/kill other yeasts/microbes inhabitating the same ecological niches (Bomblies 2014). Killer phenotype in yeast was first reported by Makower and Bevan in 1963. They reported that some strains of *S. cerevisiae* have ability to secrete proteinaceous mycocins (killer toxins) which may be lethal to other yeasts that are susceptible. Later on the killer phenomenon was found to be widely distributed among yeast strains (Marquina et al. 2002; Bajaj et al. 2003). Killer activity has been reported in almost 100 yeast species belonging to 20 or more genera. The secreted killer toxin differs among various killer yeast genera, species and strains with respect to structural genes, molecular size, biosynthesis, processing and maturity, and mechanism of killing (Marquina et al. 2002).

The genetic determinants of killer phenomenon are represented either by dsRNA, dsDNA or nuclear genes. The chemical nature of killer toxin is generally either protein or glycoprotein of low molecular weight. The toxin binds to the specific receptor on the sensitive yeast cells and kills them (Hatoum et al. 2012). The killer strains themselves are immune to their own toxin, but may be susceptible to the toxins secreted by other killer yeasts. The components of yeast cell wall viz. mannoprotein, β-1,6-glucans and chitin serve as primary binding sites or cell wall receptors for killer toxins. The killing mechanisms of action of killer toxins from various yeast spp. may show differences. The yeast killer toxins K1 and K28 from *S. cerevisiae* are most intensively investigated variants of killer toxins. The primary receptors for killer toxins K1 and K28 are β-1, 6-D-glucan and α-1, 6-mannoprotein of the cell wall of the target cell. After binding these toxins initiate different apoptotic pathways for killing the target cell (Reiter et al. 2005).

During recent years application potential of killer yeasts/and their toxins has been realized in several industries especially for fermentation industries involving production of fermented foods and/or variety of other industrial products. Considering that cell wall components generally serve as primary binding sites for killer toxins, intensive investigations are on for potential development of novel killer toxins based antifungal chemotherapeutic agents for future medical applications (Marquina et al. 2002; Bajaj et al. 2013). As soon as it was realized that wild contaminating yeasts during wine making may not only lead to slow, sluggish or stuck fermentation but may deleteriously influence the sensory properties of wine, serious efforts were undertaken to develop industrial strains (brewery, wine making) that possessed killer phenotype. (Bajaj and Sharma 2010). By the application of starter culture equipped with killer property several of the common yeast contaminants (*Pichia*, *Candida*, *Hansenispora* etc.) of ethanol fermentation were avoided. One of the major limitations of using killer yeast as starter culture is restricted activity of yeast killer toxin against closely related yeast genera only. However, several yeasts especially non-*Saccharomyces* have been reported to produce killer toxins which exhibit activity against a wide range of microorganisms particularly the pathogenic ones (Keszthelyi et al. 2006).

Considering that killer systems of yeasts are relatively recently discovered as compared to those of bacterial and other fungal systems. Much of the understanding about the synthesis of killer toxins, and action mechanisms of various killer yeasts of different genera/species is yet to be unraveled at molecular level with precise details. Furthermore, vast application potential of killer yeasts/toxins is still under exploited and requires immense research inputs for fully harnessing their application for antimicrobial chemotherapy, bio-based food preservation and biocontrol approaches, and of course in food/fermentation industries. The current chapter presents fundamentals of killer yeast/toxins, and their applications in diverse areas.

2 Yeast Warfare

Killer yeasts are considered selfish as they can kill the other closely related yeast in the near vicinity in the ecosystem. It has been established that certain stresses internal, environmental (exposure to UV) or others even the aging may induce a process of apoptosis/self destruction in yeast (LeBrasseur 2005). The killer yeasts possess toxin-producing viral sequences which are actually responsible for destruction of other yeast which are susceptible. The killing mechanisms of yeast killer toxins include altering the permeability properties of plasma membrane by making some holes in it or by inhibiting other cellular vital processes like DNA synthesis (Reiter et al. 2005). Since the killer yeast are immune to their own toxin, but the toxin is generally lethal for other yeasts or fungi that are susceptible. So in the ecosystem the killer yeasts may get preferential survival advantage, and dominate over other yeast/fungi.

The killer phenomenon or antiyeast activity is the ability to kill other susceptible yeast while keeping them immune against the toxin. Thus killer yeasts get competitive survival advantage over others in an ecosystem. Bevan and Makower first described the killer phenomenon in yeast *Saccharomyces cerevisiae*, and with regard to this killer attribute, yeasts are termed as killer yeast (K, which produce toxins that kills/destroys other yeasts), sensitive (S, which got killed by the killer toxin) and neutral (N, which neither produce toxin not get killed by the killer toxin of some other yeast). In the natural environment/ecological niches killer toxin producing yeasts get competitive advantages, and the killer toxin may help eliminating the competitors that may otherwise compete for limited available nutrients.

2.1 *Killer Yeasts/Toxins*

Killer toxins/mycocins are proteins or glycoproteins of low molecular mass which are synthesized and secreted by some yeasts. The sensitive yeasts of same/closely related genera are killed by the killer toxins secreted by the killer yeasts (Coelho et al. 2009; Hatoum et al. 2012). The killing mechanism of killer toxins may seem analogous to that of bacteriocins. However, most of the studies on killer toxins of yeasts reports ‘yeast-against-yeast antagonism’ but scarce reports are available on antibacterial characteristics of yeast killer toxins (Bajaj et al. 2013). The killer toxins show great diversity among various genera, species or strains. The differences are attributed to toxin encoding genes, processing for production of mature toxin, size, composition and molecular weight of toxin, and finally the mechanisms of action of the toxins (Marquina et al. 2002).

The killer phenomenon has been reported to be quite wide-spread among numerous yeast genera (Buzdar et al. 2011; Table 1). The species of *Candida*,

Table 1 Killer and sensitive yeast strains

Killer strain	Sensitive strain	Reference
S. cerevisiae *P. kluyveri*	*Dekkera anomala*	Dabhole and Joishy (2005)
P. anomala	*Saccharomyces cerevisiae* *Candida albicans* *Rhodotorula sloffiae*	Baeza et al. (2008)
C. guilliermondii *P. ohmeri*	*Penicillium expansum*	Coelho et al. (2009)
Williopsis saturnus	14 food spoilage yeast genera	Goretti et al. (2009)
P. anomala	*Brettanomyces/Dekkera* sp.	Ingeniis et al. (2009)
Tetrapisispora phaffii	*Saccharomyces cerevisiae*	Comitini et al. (2009)
Kluyveromyces siamensis	*Metschnikowia bicuspidate*	Buzdar et al. (2011)
S. cerevisiae *Wickerhamomyces anomalus*	*Penicillium digitatum*	Platania et al. (2012)
C. glabrata	*Saccharomyces cerevisiae*	Arroyo-Helguera et al. (2012)
Wickerhamomyces anomalus	Different yeast genera	Guo et al. (2013)
P. kudriavzevii	*Saccharomyces cerevisiae*	Bajaj et al. (2013)
P. membranifaciens	Spoilage yeasts and fungi	Santos et al. (2013)
C. parapsilosis	*Saccharomyces cerevisiae*	Robledo-Leal et al. (2014)
S. cerevisiae	*Saccharomyces cerevisiae*	Ullivarri et al. (2014)
P. kluyveri	Food and beverage spoilage yeasts	Labbani et al. (2015)

Saccharomyces, *Hansenula*, *Cryptococcus*, *Ustilago*, *Debaryomyces*, *Torulopsis*, *Zygosaccharomyces*, *Hanseniaspora*, *Kluyveromyces*, *Williopsis*, *Pichia*, *Metschnikowia* etc. have been reported to possess killer toxin producing ability. Killer yeast/their toxins may have potential applications in various industrial sectors like food, pharmaceutical, agricultural, and fermentation (Magliani et al. 2004; Lopes and Sangorrin 2010) due to the antagonistic capability against undesired microorganisms like human, animal or plant pathogens, food spoilage ones etc. (Chi et al. 2010). Killer toxin ability has been reported in many species of terrestrial yeasts as well as yeasts of marine origin (Chi et al. 2010; Guo et al. 2013).

2.2 *Killer Activity Assay*

For ascertaining the killer phenomenon in yeast, the first step is to determine activity of the toxin. The killer-sensitive interaction depends on the type of killer and sensitive (susceptible) strains used and the appropriateness of assay conditions

(Maturano et al. 2012). Various assay systems have been used to measure killer toxin activity (Baz and Shetaia 2005) are given below:

- The agar diffusion well bio-assay
- Rhodamine B method
- Bromocresol purple method
- ATP bio-luminescence measurement method
- Flowcytometry method
- 2,3,5,6-tetramethyl-1,4-benzoquinone method

The methylene blue agar diffusion assay has been a commonly used method for assay of killer activity. In this method the sensitive yeast cells are spread plated on methylene blue agar. Then well are cut on the agar plates, and poured with cell-free cultural filtrate from a presumed log phase grown killer yeast cell broth. The appearance zone of inhibition encircled by blue margins is suggestive of killer activity of yeast under study. The sensitive or neutral yeast would show no such inhibition zone (Bajaj et al. 2013).

3 Incidence of Killer Yeast in Nature

The prevalence of killer yeast has been investigated in natural environments including alcoholic, food fermentation units, hospitals, and other ecological niches like plant/animal/human sources. Killer yeasts have been reported to be widely spread in nature, and have been isolated from medical/pathological/clinical laboratory settings, plant and animal sources and from ethanol based industries viz. wineries, breweries and distilleries (Caramalac et al. 2007; Bajaj et al. 2013). Incidence of killer yeasts in natural plant associated environments was studied (Starmer et al. 1987), and it was observed cactus-stems/fruits and slime fluxes host killer yeasts. Killer phenotype was ascertained by analyzing toxin action against sensitive yeasts. Deteriorating cactus fruits had higher killer yeast proportion than the dead tissue, and the tree fluxes.

The yeast isolates from flowers' nectar and trees slime fluxes were investigated for killer/sensitive phenotypes (Mushtaq et al. 2010). *Pichia strasburgensis* was earmarked as the maximally sensitive strain as it got inhibited by numerous others, while the highest killing activity was observed in *Sporidiobolus ruineniae*, *Bullera pseudoalba*, and *Pichia anomala*, in the yeast isolates from trees slime fluxes. The analysis of yeast isolates from nectar of flowers showed that the highest killing potential was in *Bullera megalospora* and *Bensingtonia miscanthi*; several of the isolates viz. *Williopsis californica*, *Cryptococcus laurentii*, *Pichia jadinii*, *Candida valdiviana*, *Saccharomyces kluyveri* exhibited maximum sensitive potential. The killer/sensitive character was observed to be a strain dependent function. Bajaj et al. (2003) studied the occurrence of yeast killer phenomenon in various natural sources, and studied the standing of the industrial yeast (distillery and brewery yeast with respect to killer, sensitive or neutral phenotype) for killer/sensitive phenotype. Among a total of 210 yeast isolates from molasses, thirteen were found to be of killer phenotype, and belong to different genera.

Among more than one thousand isolates of several *Candida* species 52 were found to possess killer character (Robledo-Leal et al. 2012). These *Candida* species were isolated from patients of a Mexican hospital (various body parts/sources like nails, vagina, blood, semen, closed cavities etc.) and few of them have medical importance. The killer yeast incidence was more in closed cavities. Although adequate frequency of killer character was observed in *Candida utilis* the highest number was represented by *Candida glabrata*. The killer yeast may potentially inhibit/suppress lodging of other yeasts which are devoid of killer character.

Examination of yeast isolates from leafcutter ant (*Atta sexdens*) showed the presence of killer character in several yeasts viz. *Aureobasidium*, *Rhodotorula*, *Tremella* and *Trichosporon*. The analysis of killer activity was carried out by employing reference sensitive yeasts (Carreiro et al. 2002), and quite high frequency (78%) of killer phenomenon was observed. Similarly Baeza et al. (2008) reported reasonably high frequency (51%) of killer character in yeasts that were isolated from various clinical sources, yeast based industries, and other ecological niches. The yeast isolates belonged to genera *S. cerevisiae* and *Pichia anomala*. The test for killer character was conducted over different temperature and pH by using reference sensitive strains. The yeast isolate *P. anomala* exhibited activity against several other yeast species especially against *C. albicans*. Contrary to these studies Bajaj et al. (2013) showed relatively low frequency (18.4%) of killer yeast. Among 65 yeast isolates of different origin like fermented foods, soil samples, spoiled vegetables and/or fruits, only 12 exhibited killer activity against standard sensitive yeasts but remarkably three isolates showed antagonism against pathogenic and/or food spoilage bacteria as well. The isolate RY55 (*P. kudriavzevii* RY55) was earmarked to possess maximum activity against several such bacterial species like *Staphylococcus aureus*, *Pseudomonas* spp., *Klebsiella* sp. and *Enterococcus faecalis*.

4 Molecular Basis of Toxicity

Several symbols have been used to denote phenotype of killer yeasts. The phenotype 'K^+R^+' is used for killer yeast strains that generate an extracellular protein toxin which kills other yeast strains that are sensitive, indicated as 'K^-R^-'. There also exist neutral yeasts (K^-R^+) that are resistant to killer toxin but do not produce it. The basis of 'killer phenotype' in *Saccharomyces* spp. is the presence of dsRNA viruses in the cytoplasm. The virus belongs to the family Totiviridae, which is a class of mycoviruses (Schmitt and Breinig 2002). Most of these viruses are non-infectious and not generate any symptoms in their typical hosts. These viruses are generally called as 'virus-like particles' (VLPs). VLPs are responsible for generation of low molecular weight protein toxins (Sinclair 2014). Cytoplasmically inherited M-dsRNA satellite virus, also named as ScV-M1, ScV-M2 or ScV-M28, with the simultaneous presence of L-A helper virus, is responsible for killer phenotype of *S. cerevisiae*. These VLPs are typically responsible for stable maintenance of killer phenotype. L-A virus, however, is an autonomously replicating mycovirus which do not confer any demonstrable phenotype upon its host (yeast) cell. Furthermore, the occurrence of M-dsRNA satellite virus is not mandatory for L-A virus to successfully and stably maintain itself. L-A is the member of the family Totiviridae (Marquina et al. 2002).

The constitution of each particle of the L-A virus is a linear single copy of dsRNA (4.6 kb). Two open reading frames (ORFs) are present in the coding L-A (+) ssRNA strand with overlap of 130 nucleotides which code for the Gag protein. The Gag protein is the main protein of capsid that is needed for structural formulation of the virus particle, and for encapsulation. Contrary to L-A, either of the 3 M-dsRNA genomes have a single ORF that encodes a precursor protein called preprotoxin protein. The preprotoxin protein represents the precursor (crude/unrefined) form of the fully developed/functional killer toxin that is excreted by yeast (Schmitt and Breinig 2002). Thus, the yeast cells which either lack dsRNA or have only L-A, are sensitive cells and devoid of killer property; however yeast cells containing ScV-M1, ScV-M2 or ScV-M28, along with L-A are killers but immune to their own toxins. The cells which segregate as virus free become sensitive and get destroyed. Mating of sensitive yeast strains with killer strains results in destruction of the former as mating involve cytoplasmic mixing of the numerous ScV-M killer particles. Thus inheritance appears to be non-Mendelian type in the following pattern of meiosis (Wickner et al. 2013).

4.1 Secretion and Processing of Toxin

The yeast toxins K1 and K28 are the most extensively investigated toxins of *S. cerevisiae*. Both these toxins are synthesized from precursor's preprotoxins. The preprotoxins are subjected to post-translational modification (through endoplasmic reticulum, golgi bodies and secretory vesicles), and finally secreted in the form of a mature heterodimer (α/β). The N-glycosylated γ segment (which is not a part of mature toxin) is flanked by α and β subunits of the toxin (Marquina et al. 2002; Fig. 1).

Processing of killer toxin occurs in following steps:

- **Cleavage of pre-region of toxin:** With the help of N-terminal hydrophobic secretion signal the preprotoxin enters the endoplasmic reticulum. After reaching endoplasmic reticulum lumen, pre-region of toxins is removed by signal peptidase.

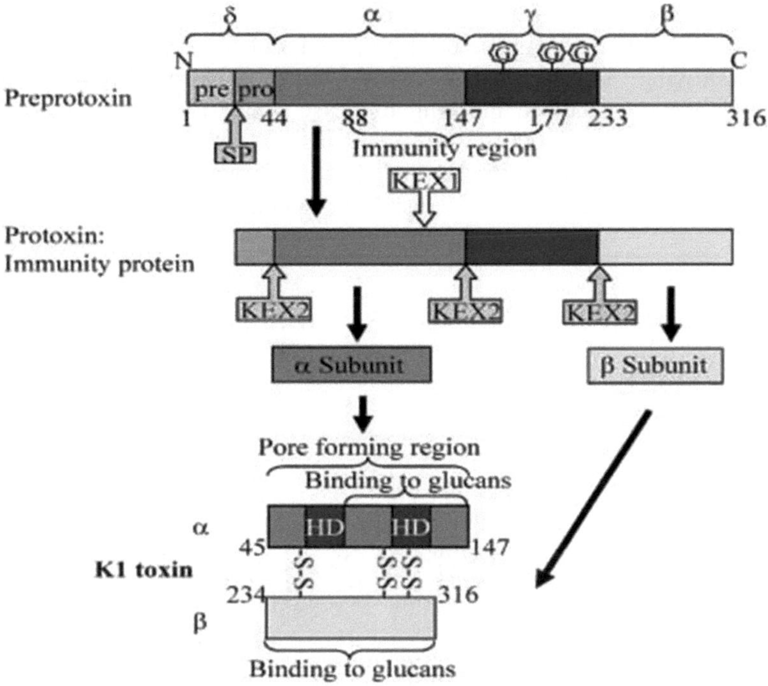

Fig. 1 Processing of K1 killer toxin and structure of mature toxin (Marquina et al. 2002)

- **Conversion of protoxin to mature toxin:** Two different proteases i.e. Kex2p and Kex1p execute the conversion of protoxin to mature toxin. Kex1p represents a serine carboxypeptidase while Kex2p is an endoprotease. This modification process takes place in the lumen of golgi apparatus The N-glycosylated γ region is removed by the action of Kex2p/Kex1p. Finally the carboxy termini of α and β chains are trimmed, and a mature toxin of 21 kDa (α/β) is formed.

4.2 Mode of Action

Although the antagonistic attribute of bacteria against other microorganisms has been well investigated and established. But antimicrobial features of yeast/fungi represent a relatively recent discovery. Several yeast/fungal species are also known which secrete compounds that have potential antimicrobial action against numerous other microorganisms. Study of in-depth molecular mechanisms of the antimicrobial principles of yeast/fungal species may help developing potentially novel chemotherapeutic agents for treatment of fungal infections in humans/animals keeping in view the fast emergence of drug resistance among pathogens (Liu et al. 2013). Several mechanisms have been proposed which are used by the toxins for killing/destroying sensitive yeast cells which include distorting of the cell-membrane pH gradient, blocking of DNA synthesis, and caspase-mediated apoptosis (Breinig et al. 2006).

The action mechanism of the K1 and K28 killer system has been studied well (Fig. 2), however, other toxins have also been elucidated for their modes of action.

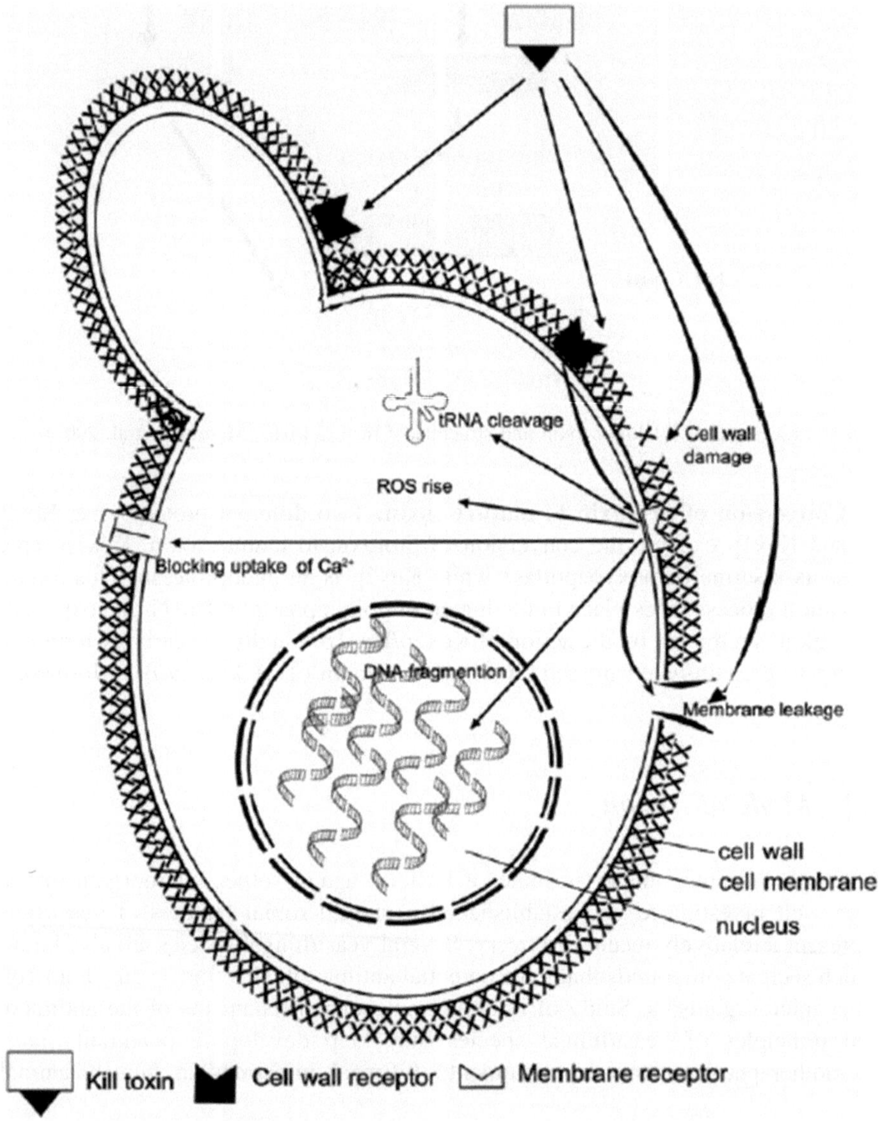

Fig. 2 Mode of action of different killer toxins (Liu et al. 2013)

K1 killer toxin kills the sensitive cells by using two-step mechanism. The adherence of the toxin protein to the appropriate receptor in the cell wall represents the first step of the mechanism. The cell wall component β-1,6-D-glucan serve as the receptor for binding of the toxin protein. This step does not require energy (energy independent). These receptors are present in adequate number on the surface of sensitive yeast cells and on the strains of the killer phenotype. Second step is energy dependent step and involves interaction of the toxin with receptors on cell membrane. The interaction of toxin with cell membrane destroys the typical permeability function of the cell membrane. Due to loss of unique semi-permeability function of the cell membrane there is loss of protons and potassium ions (Marquina et al. 2002; Magliani et al. 2008) ultimately leading to disruption of potential gradient across cell membrane and finally cell death.

The β-subunit has been believed to be involved in binding to the cell wall receptor, while the α-component might have accounted for the lethal effect on the membrane. The interaction and transfer of toxin across the membrane are energy-dependent processes. Obviously the action of killer toxin damages cell membrane which in turn loses its permeability properties, and results in collapse of the proton gradient. This causes distinct pH-reduction inside the cell which in turn disturbs the vital metabolic processes including release of potassium ions and ATP to the medium and eventual death of the cell (Marquina et al. 2002). The K1 killer toxin is believed to disturb the potassium homeostasis in sensitive cells by activating TOK1 channel. Evidence in this regards has been presented in which mutants with TOK1 deletion have been shown to possess resistance against K1 toxin (Ahmed et al. 1999). Killing mechanism of killer toxin K28 differs from that of K1. Killer toxin K28 binds to α-1,3-mannoprotein a high molecular mass receptor, and enters the sensitive cell by means of endocytosis, and finally reaches the nucleus using the retrograde path. DNA synthesis is inhibited due to action of killer toxin K28 which in turn results in arrest of cell cycle (at G_1/S phase) and caspase-mediated apoptosis, ultimately resulting in lethal action on cells and death of cells (Schmitt and Breinig 2002; Breinig et al. 2006). Other mechanisms of killer toxin action include chitinase, anticodon nuclease activity (Jablonowski and Schaffrath 2007; Klassen et al. 2011). However, Melvydas et al. (2007) showed that *S. cerevisiae* strain isolated from Lithuanian apple wine had a novel 'X' factor which has potential capability of inhibiting the killing activity of K1, K2 and K28 toxins.

4.3 Immunity to Killer Toxin

The killer yeast that has potential to secrete the killer toxin is immune to its own toxin, but may be susceptible to the toxins produced by other yeasts. The mechanisms by which the killer yeast is immune towards their own toxins have not yet been deciphered with absolute certainty. It has been reported that the yeasts secreting the K28 toxin have the ability to re-internalize the secreted mature toxin. This internalized toxin is transported via retrograde transport through the secretion pathway. In the cytosol, this toxin complexes with unprocessed preprotoxin encoded by the M-dsRNA killer virus. Following which the β-subunit is selectively polyubiquitinated and the whole complex is rapidly (mature and unprocessed toxins) degraded by proteasomes, thus inactivating the toxin action against the toxin-producing host. The speculated immunity mechanisms of K1 involves: a competition between toxin precursor and the mature toxin that might result in saturation and/or elimination of plasma membrane receptors; or the γ-component of the toxin precursor might be functioning as a potential defense system of the host against damage by the hydrophobic α-component of toxin; or the toxin receptor (Kre1p) may interact with the K1 protoxin during secretion which may cause diversion of the receptor-protoxin complex to the vacuole for degradation (Magliani et al. 2008).

5 Characteristics of Yeast Killer Toxins

5.1 Molecular Mass of the Toxin

Killer yeast toxins are generally proteins or glycoproteins of low molecular weight. Killer toxins K1 and K28 from *S. cerevisiae* have the molecular mass of 21 kDa (Schmitt and Breinig 2002). A newly isolated killer toxin (30 kDa) from *P. membranifaciens* exhibited antimicrobial activity against several food spoilage fungi and yeast (Santos et al. 2009). Similarly, a high molecular weight toxin (54 kDa) from *Pichia kluyveri* demonstrated antiyeast/fungal activity against numerous food/beverage spoilage yeast and fungal genera (Labbani et al. 2015). *Wickerhamomyces anomalus* an isolate from marine sources showed broad spectrum killer activity. The purified killer toxin had molecular weight of 47 kDa (Guo et al. 2013). In the yeast *Pichia kudriavzevii* (*P. kudriavzevii* RY55) the killer phenomenon was documented for the first time by Bajaj et al. (2013). The secreted toxin of killer yeast *P. kudriavzevii* RY55 was subjected to purification, and purity (homogeneity) of the toxin protein was ascertained by HPLC. The toxin protein of *P. kudriavzevii* RY55 possessed molecular weight of 39.8 kDa as examined by SDS-PAGE. The toxin had possessed killer activity not only against sensitive yeast but also against several pathogenic bacteria of human health significance (Bajaj et al. 2013).

5.2 Optimum pH and Temperature of Killer Toxin

Toxins from different yeast genera generally show optimal activity at acidic pH and temperature below 40 °C (Table 2). Wang et al. (2012) purified and characterized killer toxin from marine-derived yeast *Williopsis saturnus* which was active against pathogenic yeast *Metschnikowia bicuspidate*. The toxin showed optimal activity at pH 3.0–3.5 and temperature of 16 °C.

Halotolerant killer yeast *Candida nodaensis* produced a toxin that had activity and stability over wide pH and temperature for quite long time periods (Silva et al. 2008). Purified preparation of toxin of killer yeast *P. kudriavzevii* RY55 showed activity at 20–37 °C but has optimum temperature of 30 °C for activity (Bajaj et al. 2013). Killer toxin exhibited thorough activity/stability at low temperatures (4–20 °C). But at elevated temperatures activity and stability decreased (Bajaj et al. 2013).

Killer toxin being a protein or glycoprotein may lose activity at very high temperatures and at highly acidic/alkaline pH. Analysis of constructed killer fusants showed that the killer toxin activity was maximum in pH range of 4.0–5.5, and temperature range of 20–35 °C (Bajaj and Sharma 2010). A marine killer yeast *Wickerhamomyces anomalus* YF07b toxin exhibited maximum activity at pH and temperature of 3.5 and 16 °C, respectively, against several yeasts (Guo et al. 2013). Two *S. cerevisiae* killer strains (Cf8 and M12) were investigated by Ullivarri et al. (2014). The production of killer toxins, and killer activity of both the toxins showed temperature optimum of 15–25 °C, and pH optimum of 3.5. These conditions are generally the same as used during the process of wine-making.

5.3 Killer Activity and Halotolerance

Certain reports published in the literature indicated existence of a relationship between killer activity of yeast cells and the salt concentration. Killer toxins produce the ion-permeable channels, and interrupt the ionic equilibrium of the plasma membrane especially under high salt concentration environments. This may lead to increased mortality due to killer toxin (Guo et al. 2013). *Wickerhamomyces anomalus* killer toxin required 4.0% NaCl for its optimal killer activity. Wang et al. (2007) optimized the cultural conditions for production of killer toxin from *P. anomala* YF07b strain. The toxin production medium contained NaCl (2.0%), had pH at 4.5 and incubation temperature at 20 °C. The most suitable assay/medium conditions for activity of killer toxin against *Metschnikowia bicuspidata* WCY, a pathogenic yeast, were: medium supplemented with NaCl at 6.0%, at 15 °C and pH 4.5. Suzuki et al. (2000) showed that *Pichia farinosa* KK1, a halotolerant yeast produced a toxin (SMKT) that required NaCl (2 M) for exerting highest inhibitory activity towards *Saccharomyces cerevisiae* and *Zygosaccharomyces rouxii*. The name of toxin was given accordingly *i.e.* SMKT—salt-mediated killer toxin. Aguiar and Lucas (2000) studied the correlation between

Table 2 Characteristics of killer toxins from different yeast genera

Organism	Optimal conditions	Comments	Reference
S. cerevisiae	pH 4.1–4.5, temperature 22–25 °C	Molecular mass 18–20 kDa	Soares and Sato (2000)
K. phaffii	pH 3–5, temperature <40 °C	Biopreservative application of the toxin for wine production	Ciani and Fatichenti (2001)
P. anomala	Toxin stability at pH 3–5.5 and temperature up to 37 °C	Molecular mass 47 kDa, pI 3.4 and 3.7	Izgu et al. (2006)
P. anomala	pH 4.5, temperature 15 °C	Toxin production was maximum in 2% NaCl	Wang et al. (2007)
C. nodaensis	Toxin inactivated above 40–50 °C and high pH	Toxin was extremely halotolerant	Silva et al. (2008)
P. membranifaciens	pH 4.5, temperature 20 °C	Molecular mass 30 kDa, pI 3.7	Santos et al. (2009)
K. siamensis	pH 4, temperature 25 °C	Toxin effective against pathogenic yeast *Metschnikowia bicuspidate*	Buzdar et al. (2011)
C. glabrata	pH 4–7, temperature 37 °C	Toxin causes DNA fragmentation in sensitive yeast	Arroyo-Helguera et al. (2012)
Williopsis saturnus	pH 3–3.5, temperature 16 °C	Molecular mass 11 kDa, 10% NaCl is optimum for killer activity	Wang et al. (2012)
Wickerhamomyces anomalus	pH 3.5, temperature 16 °C	Molecular mass 47 kDa, high salt concentration promotes toxin activity	Guo et al. (2013)
P. kudriavzevii	pH 5.0, temperature 30 °C	Molecular mass 39.8 kDa, First report of killer phenomenon in species, Antagonism against bacterial pathogens	Bajaj et al. (2013)
S. cerevisiae	pH 3.5, temperature 15–25 °C	Toxin production induced in presence of nitrogen source	Ullivarii et al. (2014)
P. kluyveri	–	Molecular mass 54 kDa	Labbani et al. (2015)

killer/sensitivity and salt-stress tolerance phenotypes, and categorized the yeasts. Some of the killer yeast especially the halotolerant ones exhibited enhanced killer potential and spectrum with increased salt concentration during killer activity assay.

Table 3 Toxins from different organism with their receptors

Receptor	Toxin (Organism)	Reference
β-1,3 D-glucan, β-1,6-D-glucan	(K1 and K2) *S. cerevisiae*	Schmitt and Breinig (2002)
	W. saturnus	Guyard et al. (2002)
	K. phaffii	Comitini et al. (2004)
	Wickerhamomyces anomalus	Muccilli et al. (2013)
Protein	*Z. bailii*	Santos et al. (2000)
	(K28) *S. cerevisiae*	Schmitt and Breinig (2002)
Chitin	*K. lactis*	Santos et al. (2000)
	P. acacia	Klassen et al. (2004)

But the killer activities of other yeasts which had relatively low halotolerance remained either unaffected or decreased slightly in presence of salt.

5.4 Receptors for Killer Toxin

Two kinds of receptors have been reported to play a role in toxicity: these include cell wall receptors (primary) and plasma membrane (secondary) receptors (Magliani et al. 2008). The involvement of different receptors in toxicity of yeast killer toxins is shown in Table 3.

6 Applications of Killer Yeast/Toxin

Killer yeast and their toxins have got several potential applications for various industrial processes. In the bioprocess/fermentation industries and in food industries the killer yeast/toxin can suitably be exploited for combating wild yeast contaminants. During production of ethanol or alcoholic beverages (beer, wine etc.) the process conditions are generally not absolutely sterile, and there is likelihood of contamination by undesired wild yeast that may lead to sluggish/stuck fermentation. The contaminating wild yeasts exert competition for the nutrients with industrial yeast starter resulting in low product yields. However, should the wild contaminant happen to be a killer yeast it may altogether destroy/kill the industrial yeast strain, thus, causing process failure. Besides, the contaminating wild yeast may contribute towards spoilage of products like wine/beer due to production of undesired flavours/aromatic compounds. Therefore, it is highly desirable that the starter culture must possess killer phenotype, and must be immune to killer toxins of other killer yeasts. Thus, the killer yeast/toxins can be used for preventing yeast/bacterial contaminants in fermentation, as biopreservatives for protection against food spoilage microorganisms, as biocontrol in agriculture for inhibiting/killing plant pathogens. Furthermore, killer toxins have been considered as potentially novel chemotherapeutic agents due to their antimicrobial activity. In addition, killer yeasts have been used as model systems in fundamental research for investigating the mechanisms of regulation of eukaryotic polypeptide processing, secretion and receptor binding. Application of killer yeast/toxins in diverse areas is described below.

6.1 Killer Yeast in Wineries/Breweries

Production of wine by the fermentation of grape juice is a complex process involving growth and biochemical activity of yeasts. The fermentation is executed either by natural flora of grapes or by the starter cultures. Natural wine fermentation is executed under non-sterile process conditions, and involves sequential development of a large number of microorganisms. Several *Saccharomyces* and non-*Saccharomyces* yeasts are the first dominant group and yeasts belonging to genera such as *Candida*, *Debaryomyces*, *Hanseniaspora*, *Hansenula*, *Kloeckera*, *Metschnikowia*, *Pichia*, *Schizosaccharomyces*, *Torulaspora* and *Zygosaccharomyces* are mostly found during the initial three days of fermentation (Esteve-Zarzoso et al. 1998). Non-*Saccharomyces* yeasts in the winemaking environment are confined to the early stages of fermentation as they have weak fermenting capacity and rate, poor resistance to SO_2 and feeble ethanol tolerance (Ciani et al. 2010). However, some non-*Saccharomyces* yeast species viz. *Kloeckera apiculata*, *Lachancea thermotolerans*, *Torulaspora delbrueckii*, *Schizosaccharomyces pombe*, and *Starmerella bombicola* have been reported to survive until the end of fermentation when co-inoculated with *S. cerevisiae* (Ciani et al. 2010; Comitini et al. 2011). During alcoholic fermentation yeasts utilize grape juice sugar and transform it into ethanol, acetaldehyde, fatty acids and amino acids which contribute towards flavour component of wine. Additionally the enzyme complement of yeasts may transform neutral compounds of grape juice into flavor-active compounds such as aldehydes, ketones, organic acids, alcohols, esters, polyols, volatile sulphur compounds etc. Also yeast autolysis may release nutrients and metabolites which help in the extraction of secondary flavour metabolites (Fleet 2003).

During later stages of natural wine fermentation, the non-*Saccharomyces* population declines, and the highly fermentative and ethanol tolerant *Saccharomyces* species dominate and completes alcoholic fermentation (Manzanares et al. 2011). Lastly, lactic acid bacteria convert grape malate to lactate through malolactic fermentation. The main lactic acid bacterium found in wine is *Oenococcus oeni* that is able to proliferate at the low pH values (3.2–3.9) commonly found in grape must (Bisson 2004).

During the process of ageing of wines uncontrolled growth/biochemical activities of certain yeasts, lactic and acetic acid bacteria (Ciani and Comitini 2011) can cause microbial spoilage of wines. The spoilage effects may be manifested in the form of sediments and gas production in bottled wines, off-odour and off-taste, cloudiness or haziness, formation of film in stored wines, and others at various stages of wine spoilage (Ingeniis 2009). A wide variety of yeast species belonging to several genera have been reported in the spoiled wine. Some of yeasts associated with wine spoilage are for instance *Dekkera/Brettanomyces*, *Saccharomyces*, *Hanseniaspora*, *Metschnikowia*, *Schizosaccharomyces Candida*, *Pichia*, and *Zygosaccharomyces* species. Some of the non-*Saccharomyces* yeast species such as *Candida* spp., *Starmerella bombicola*, *Hanseniaspora uvarum*, *Metschnikowia pulcherrima* and *Wickerhamomyces anomalus*, in axenic fermentations are reported to produce esters, acetoin, acetaldehyde, and volatile acids that may be undesirable above a certain threshold level. Interestingly, however, these negative metabolic activities are not pronounced when these yeasts are co-inoculated with *S. cerevisiae*. Either these activities are not expressed or are modified by the metabolic activity of *S. cerevisiae* (Ciani and Comitini 2011).

Killer phenomenon of yeast may have immense importance in wine making process. Biotransformation of grape must into wine is mainly attributed to the activity of yeast *S. cerevisiae* due to its high fermentation rate as compared to the non-*Saccharomyces* yeasts. However, contamination by killer toxin-producing strains of *S. cerevisiae* may not only delay the fermentation but may cause sluggish/stuck fermentations, and may all together inhibit the starter culture if the latter happens to be the sensitive one. Wines produced from such fermentations may have an undesired organoleptic properties, excess H_2S, off-flavours and volatile, acidity caused by fusel oils, acetaldehyde and lactate, and have lower ethanol yields. However, sluggish or stuck fermentations may occur due to some other factors like oxygen levels, nitrogen deficiency, vitamin (thiamine, biotin and pantothenic acid) deficiency, high initial sugar content/high ethanol concentration, pH/temperature of fermentation, extreme clarification of must, grape solids, and varieties of grape/vineyards conditions, and pesticides. Additionally the type of wine yeast strain, and its metabolic potential may influence the quality of wine (Vadasz et al. 2002; Malherbe et al. 2007). These factors effect individually, and can alter the fermentation profile. However, a synergistic effect is thought to exist among these factors in sluggish fermentations.

The ratio of killer to sensitive cells during fermentation is extremely important, and determines the overall influence of the killer activity (Gutierrez et al. 2001). There are conflicting reports with regards to the killer to sensitive cells ratio that can have an effect on killer activity during fermentation. However, it has been proven that killer toxins are continuously secreted in the absence of the sensitive yeast (Heard and Fleet 1987; Malherbe et al. 2007).

The influence of killer toxins has also been studied in mixed culture fermentations. Killer to sensitive ratios of 1:1, 1:10 and 1:100 lead to complete elimination of the sensitive cells within 24 h (Heard and Fleet 1987; Carrau et al. 1993). Heard and Fleet (1987) reported a rapid disappearance of sensitive strains and dominance of killer strains in mixed culture fermentations. Under different inoculum conditions, high residual sugar was obtained when a 1:100 killer to sensitive ratio was used, and a population of 82% dead cells was observed. Supplementation of ammonium sulphate and Roviferm circumvents the stuck fermentations even at 1:100 killer to sensitive cells ratio. The addition of these nitrogen sources resulted in a residual sugar concentration between 1.0–1.7 g l^{-1} towards the completion of fermentation. The same residual sugar concentrations could be observed when bentonite and activated carbon was added to similar fermentations indicating that exhaustion of nutrients may not allow the cells of killer yeast to end the fermentation (Carrau et al. 1993).

Killer and sensitive mixed culture analysis of *S. cerevisiae* cells under batch process conditions showed that there was a decrease in the viable biomass population when the killer cells accounted for 10% of the population. In addition, in axenic cultures the killer culture was higher than that of the sensitive culture from 10 h of fermentation until the end of fermentation (23 h). The mixed culture (sensitive cells +5–10% killer cells) studies showed that viable biomass decrease was more pronounced at higher proportion of killer cells. Furthermore, substrate utilization (glucose) and product synthesis (ethanol) kinetics was severally affected at higher killer cell concentration (Ramon-Portugal et al. 1998).

The yeast isolates belonging to *Hanseniaspora* spp. that were isolated from samples of grapes, were found to be sensitive against the killer toxin of *Kluyveromyces phaffii* DBVPG 6076 killer yeast toxin (Ciani and Fatichenti 2001). The killer toxin of *K. phaffii* killer yeast was examined for few characteristics like optimum temperature (<40 °C) and pH (pH 3–5) for activity, and its relevance for wine production process. The antagonistic activity of killer toxin of *K. phaffii* was quite as effective as that by sulfur dioxide. Killing activity of toxin against *Hanseniaspora uvarum* was concentration dependent. Killing mechanism was based on existence of receptor for the toxin. The study indicates potential application of toxin of *K. phaffii* killer yeast for preservation of wine. Several yeast strains isolated from the ecological niches involving Turkish wine producing areas were studied (Elmaci et al. 2014) for many technologically important characteristics like rate of growth and fermentation, ability to grow at elevated temperatures, and in presence of high ethanol and SO_2 concentrations, high strength sugar fermentation potential, and killer phenotype etc., however, no isolate exhibited killer character.

Considering the problems associated with killer yeast contamination in wine fermentations, starter industrial yeast cultures have been developed that are equipped with killer phenotype by various approaches (Schmitt and Schernikau 1997; Bajaj and Sharma 2010). Protoplast fusion approach was used for developing industrial killer yeast for winery. The newly constructed killer yeast exhibited all the vitally desired important properties of industrial yeast. Remarkably the newly developed yeast showed ability to produce enhanced ethanol yield, and provided protection against contamination by wild yeasts (Farris et al. 1992). Similarly electrofusion strategy was employed for producing killer yeast, and the newly designed yeast showed tolerance towards high ethanol and sugar concentrations

(Salek and Arnold 1994). Still another approach was used for producing distillery yeast with killer property by Javadekar et al. (1995). The cells of standard killer yeast (*S. cerevisiae* NCIM 3578) were transformed into protoplasts, and UV killed, and then allowed to fuse with industrial yeast *Saccharomyces cerevisiae* that had high ethanol producing and flocculation potential. The protoplast fusion was effected by polyethylene glycol (PEG) 6000. The resultant fusants possessed killer character and other desired characteristics of industrial yeast like fast fermentation of molasses, and high activity of major enzymes which are instrumental for ethanol production.

Recombinant DNA approach was employed in which K1 killer yeast (wine yeast *Saccharomyces cerevisiae*) was further transformed with killer genes from two other killer yeasts i.e. K2 and K28 (Schmitt and Schernikau 1997). The resultant killer yeast had strong power of three different killer yeasts ($K_1/K_2/K_{28}$), and showed a wide range killing potential. Such killer yeasts with broad spectrum killing ability may have potential application for wineries/distilleries/breweries for protection against contamination by wild yeast/killer yeasts.

An industrial yeast *Saccharomyces cerevisiae* HAU-1 possessed features that would suit its application as bakery/distillery/winery yeast. Though *S. cerevisiae* HAU-1 had several of the industrially important fermentation features like ability to grow rapidly and ferment at high sugar/ethanol concentrations even at elevated temperatures but this yeast was of sensitive phenotype like many of the other Indian industrial yeast strains (Bajaj et al. 2003). The industrial *S. cerevisiae* HAU-1 was transformed with killer character by using protoplast fusion approach (Bajaj and Sharma 2010). The protoplast of standard killer yeast *S. cerevisiae* MTCC 475 and that of *S. cerevisiae* HAU-1 were fused, and fusants that had killer potential in addition to all the process suitable characteristics were selected. It was further established that the killer property in the newly developed yeast remained unaltered in the processes involving recycling of biomass, and under fermentation of molasses. Recently, Branco et al. (2014) reported that *S. cerevisiae* has potential to produce certain peptides during ethanol fermentation that have antimicrobial activity and may be important for providing protection against a wide range of contaminating yeasts/bacteria in wine fermentation. Nature of peptides was elucidated with mass spectrometry, and it was shown that peptides belonged to parts of *S. cerevisiae* protein glyceraldehyde 3-phosphate dehydrogenase. Such peptides might be quite important, and may provide protection against contaminating wild yeast which may spoil the wines. Thus these peptides may serve as natural preservatives for wine.

Similar to wine industry, killer yeast/toxins have implications in brewing industry as well. The killer phenotype of starter culture may inhibit the undesired yeast contaminants that may not only reduce ethanol yield, but may contribute towards producing off-flavors/undesired products in the beer. The key step of beer making is preparation of malted cereal i.e. barley or rye. Germination of cereals is done (malt) and followed by kilning and roasting. The germination causes activation of starch degrading enzymes amylases (α and β amylases) that transform cereal starch into simple sugars. Malt is subjected to milling for breaking cereal grains, and followed by mixing with hot water (mashing). During this step, enzymes act on starch to convert it into simple monosaccharides/disaccharides or oligosaccharides (dextrins). The mash is subjected to lautering in which the grain part is separated from the liquid/wort. The wort is subjected to boiling in kettle with hops. During this process some additional components like herbals, sugars etc. may also be included in the boiling wort. The flavour, aroma and pleasant bitter taste of beer are due to hops. The hopped wort is subjected to fermentation by yeast which transforms the sugars into ethanol, carbon dioxide and other components. After 2–3 weeks of fermentation, the freshly produced beer is subjected to aging for a few weeks/months in which beer is incubated at low temperatures and yeast is removed. After aging finally the beer is subjected to filtration for removal of particles/remaining yeast, and now the 'ready to serve' beer is called as 'bright beer.'

Brewing is based on pure culture fermentation that is executed by specific starter culture in contrast to wine making process that may involve more than one yeast or bacteria. Similar to wine making process killer yeast contamination at any stage will be devastating for the brewing process. As the contaminating killer yeast can destroy the starter culture or/and may produce off flavors, and cause yield losses. Bajaj et al. (2003) reported that all the brewery yeast strains used in India were of sensitive nature. An industrial yeast *S. cerevisiae* that has suitable characteristic features as brewery or distillery yeast was introduced with killer character by protoplast fusion with standard killer strain. Newly constructed killer yeast possessed all desirable fermentation features (Bajaj and Sharma 2010). *Ale* and *lager* strains of brewery yeasts were transformed with killer trait (cytoplasmically inherited) from a laboratory killer yeast *Saccharomyces cerevisiae* (Hammond and Eckersley 1984). Killer character transfer efficacy varied among brewery yeasts. Beer quality and fermentation properties of newly developed killer brewery strains were very similar that of parent brewery yeast. Newly developed killer yeast produced good quality beer, and possessed killer trait due to which it was able to kill the sensitive contaminants. Electroinjection approach was employed for transferring the dsRNA isolated from super-killer strain into brewery yeast *S. cerevisiae* and *S. carlsbergensis* that were of killer-sensitive nature (Salek et al. 1992). Thermo-tolerant and sensitive strains, and the mutant strains were transformed with killer character. The killer yeast developed by electrotransformation exhibited high ethanol yields and strong killer character for prolonged time periods.

6.2 Killer Yeast in Biotyping

Precise identification, discrimination or differentiation of pathogens is mandatory for epidemiology associated studies. Biotyping methods need to be developed that are sensitive, reproducible, easy, and economical. Furthermore, the methods/techniques that are developed should have application spectrum to a broad/wide range of distinct pathogens. The killer phenomenon of yeast was utilized for differentiation of strains of *Candida albicans* for the first time. This system was found to be very effective epidemiological tool for earmarking especially the fungal associated nosocomial infections cases (Polonelli et al. 1983). Following this killer yeast based system was developed for discrimination/differentiation of a variety of yeasts. Later, the killer system was applied for the differentiation of other opportunistic molds (*Cryptococcus neoformans*, *Aspergillus niger* etc.) and yeasts like *Candida glabrata*, and other *Candida* spp., *Saccharomyces cerevisiae*, and *Pseudallescheria boydii* (Buzzini et al. 2004). Following these successful investigations, killer phenomenon-based system was used for identification and/or discrimination or differentiation of other pathogens like *Nocardia* spp., and other Gram positive and Gram negative bacterial pathogens, and even mycobacteria (Izgu and Altinbay 1997).

The discrimination of dissimilar *Candida* species and/or other pathogenic microorganisms can be executed by using the toxins from earmarked killer yeasts having a broad range antimicrobial potential (Polonelli and Conti 2009). This approach has successfully been utilized for discrimination/accurate identification of fungal pathogens from clinical/nosocomial sources. This approach of biotyping (differentiating/discriminating) pathogens is considered typically reliable, cost-effective, and unique especially for the laboratory-establishments which have meager resources to sustain advanced molecular methods for accurate identification.

Staphylococcus epidermidis strains originated from Brazilian hospitals and clinical sources were differentiated based on the antagonistic action of eleven selected killer yeasts (Fuentefria et al. 2008). Based upon the antimicrobial action of selected killer yeasts sensitive *S. epidermidis* strains and coagulase-positive *Staphylococcus* strains were differentiated with absolute certainty and reliability. Thus, the killer yeast based biotyping represents a legitimate, straightforward and low-cost system for differentiation/discrimination of pathogenic microorganisms.

Molecular tools though have gained immense importance in recent years for biotyping of yeast and other microorganisms. But the approach based on pattern of killer/sensitive phenotype of yeast may play a vital role in supplementation of molecular identification data for yeast. In fact the real potential of this system *i.e.* killer/sensitive pattern for discriminating yeasts has not been fully investigated with regard to methods/mechanism and for ecological connotation (Buzzini et al. 2007). The strains of *Candida dubliniensis* from *Candida albicans* were appreciably discriminated using a novel approach that was based upon susceptibility against killer toxins (Scheid et al. 2010). Similarly antimycotic potential of killer yeasts was exploited for categorization of fungal pathogens obtained from various environmental niches and clinical sources (Stopiglia et al. 2014). The study indicated that killer yeast/toxins may be successfully used for differentiation/discrimination of fungal pathogens.

6.3 Killer Toxin in Medicine

The antagonistic action of killer yeast/toxins against pathogenic fungi/bacteria may help developing specific/novel chemotherapeutic agents. Killer toxin of yeast *Filobasidium capsuligenum* showed strong antagonistic action against pathogenic (opportunistic) fungus *Cryptococcus neoformans* (Keszthelyi et al. 2006), and may have the potential for therapeutic applications for cryptococcosis. *C. neoformans* obtained from various environmental/clinical sources were categorized into 8 bio-types based upon their sensitivity patterns. Further investigation of mechanistic details showed that killer action was due to disruption of cell membrane functioning due to toxin protein that behaved like ionophore, and the cell wall hosted the receptor for binding of killer protein toxin. Interestingly, the killer mechanism has no association at all with cell cycle and/or cell wall biosynthesis. Similarly Bracesco et al. (2006) observed a strong antifungal action of K1 and K2 killer toxins of killer yeast *Saccharomyces cerevisiae*. Killer yeast *Saccharomyces cerevisiae* K1 and *S. cerevisiae* K2 secreted killer toxin that exerted toxic action against *Candida albicans* and *Trichophyton mentagrophytes*. The killer toxin protein showed activity and stability over wide range of temperatures and pH.

Wickerhamomyces anomalus YF07b, a marine killer yeast produced toxin that had broad spectrum killing activity against several yeasts including the *Candida albicans*, *C. tropicalis*, *Metschnikowia bicuspidate* WCY, *Kluyveromyces aestuartii* and *Saccharomyces cerevisiae* (Guo et al. 2013). The toxin was purified and studied for killing mechanism. The basis of killing mechanism of the sensitive cells was interruptions of various cellular functions especially by alterations in the permeability of cytoplasmic membrane. Killer yeast *Williopsis saturnus* var. *mrakii* produced a killer toxin, named as HM-1 toxin (Selvakumar and Miyamoto (2006). Antiidiotype antibodies were produced against HM-1 toxin (single chain variable fragment). These antiidiotype antibodies demonstrated antifungal activity against *Candida* spp. The antibodies were characterized based on minimum inhibitory concentration (MIC).

Although most of the diseases in several marine animals is caused by a variety of (marine) viruses and bacteria, however, recently some pathogenic marine yeast have been reported that cause disease in some marine animals. A pathogenic yeast *Metschnikowia bicuspidata* WCY is responsible for causing (milky) disease in crab, and accounts for heavy economic loss. Similarly, yeast *Torulopsis mogii* may cause disease in shrimp. Killer yeast toxin-based new/novel therapeutic agents may be developed for controlling human/animal/plant pathogens (Peng et al. 2010). *Williopsis saturnus* WC91-2, a marine yeast produced a killer toxin that had inhibitory activity against *Metschnikowia bicuspidata* WCY, the yeast pathogen responsible for causing disease in crab. Killer toxins from such killer yeasts may be used as potential biocontrol agents for providing protection against pathogens. The killer yeast *Williopsis saturnus* WC91-2 from marine source had ability to kill pathogenic strains of yeast (Wang et al. 2012). Killer toxin was produced, purified and characterized. The killer toxin encoding gene had an ORF of 378 bp. The gene was cloned and expressed, and protein toxin was characterized for N-terminal amino acid sequence. The toxin protein was constituted of 125 amino acids, and had molecular weight of 11.6 kDa.

Antimicrobial potential of yeast derived metabolites has been relatively less investigated as compared to that of bacterial antibiotics, bacteriophages and bacteriocins. The killer toxins of killer yeasts may be exploited for development of new antimycotic agents against pathogenic fungi (Stopiglia et al. 2014). The *Sporothrix schenckii* is causative agent for sporotrichosis, in humans/animals. Killer yeasts species from several genera viz. *Candida*, *Kluyveromyces*, *Trichosporon*, and *Kazachstania* showed excellent antimicrobial potential investigated against many species of *Sporothrix* isolated from clinical/environmental sources. Thus, killer toxin may be developed as potential chemotherapeutic agents. Bajaj et al. (2013) established that a newly earmarked killer species *Pichia* i.e. *P. kudriavzevii* RY55 produced a toxin that had inhibitory activity against several bacterial pathogens (*Pseudomonas aeruginosa*, *P. alcaligenes E. coli*, *Enterococcus faecalis*, *Staphylococcus aureus*, *Klebsiella* sp.), thus, indicating the chemotherapeutic potential of the killer toxin.

6.4 Killer Toxin as Biopreservative

Although role of yeasts is well known for various food fermentations e.g. alcoholic beverages, bread, wine etc. However, some yeast species are implicated as potential food/beverage spoilage agents, and are responsible not only for financial loss but may pose health risks to the consumer (Stratford 2006; Santos et al. 2009). Killer yeast toxins generally show antagonistic activity against other yeasts and fungi that are closely associated with killer toxin producing yeast. However, should the killer toxins of yeasts exhibit a broad spectrum antagonistic potential against several microorganisms (bacteria, yeast, fungi) that are either pathogens, or important for spoilage of foods, or are potential contaminants for fermentation/food industries, they may have prospective for application as therapeutics, biocontrol agents or preservatives of food. *Pichia kudriavzevii* killer toxin showed strong antibacterial activity against several pathogenic/food spoilage bacteria (Bajaj et al. 2013).

Spoilage of wine by unwanted yeasts may potentially be protected by the application of killer yeasts as biocontrol agents (Ullivarri et al. 2014). Killer toxins from strains of *Saccharomyces cerevisiae* Cf8 and *S. cerevisiae* M12 were evaluated for their application in enhancing winemaking process. Killer toxins were characterized by biochemical tests and growth inhibition studies. Genes encoding killer toxin were examined. Both toxins exhibited excellent killer potential under wine making conditions. It was remarkable to observe that unwanted contaminating/spoilage yeasts were controlled substantially by killer yeast strains. Furthermore, ethanol and sulphur dioxide enhanced the killer potential of killer yeasts. The study shows that application of killer strains of *Saccharomyces cerevisiae* as starter cultures may help producing wines of good/controlled quality.

Several *Dekkera* species are known to generate phenolic volatiles and spoil the aromatic properties of wine (Stratford 2006). Especially the *D. bruxellensis* is the big nuisance for production of wine (Comitini and Ciani 2011; Santos et al. 2011). Similarly *Zygosaccharomyces bisporus* has also been implicated for spoilage of wine and variety of other food products. *Z. bisporus* generate inedible aromatics and also produce CO_2. Spoilage of several fruits/fruit juices, drinks is attributed to yeasts *Saccharomyces cerevisiae* and *Schizosaccharomyces pombe* (Stratford 2006). In the food/beverage industries several benzoate or sorbate based preservatives are used for controlling the spoilage yeasts, and maintaining quality and shelf life of foods. (Goretti et al. 2009). However, yeast based killer toxin may also have potential for application as preservatives. Killer toxin protein from *Williopsis saturnus* exhibited antagonistic activity against several food spoilage yeasts (Goretti et al. 2009). Contrary to these observations killer toxin of yeast *Williopsis saturnus* var. *mrakii* NCYC 500 was not able to inhibit the *S. pneumonia* (Ochigava et al. 2011). Considering the importance, consumers' popularity and safety of biobased food preservatives, it is important to develop killer yeast based toxins as potential food preservatives. In such case, however, the toxin must have broad spectrum antagonistic capability against variety of food spoilage microorganisms (Bajaj et al. 2013).

The problem of food spoilage by yeasts may be aggravated due to resistance of many types of yeast towards chemical based commercial food preservatives (Papadimitriou et al. 2007). Thus it is imperative on the part of food/beverage industries to safeguard their products against spoilage yeasts. At the same time it is a challenge for the food/beverage researchers/scientists to develop novel preservatives that are acceptable for usage in foods. Another uniquely important aspect that is picking up very fast is consumers' demand and aspirations especially from developed world that there must be minimal or no usage of chemical preservatives in foods; rather the preservative agents must be absolutely safe food-grade compounds and preferably originated from biological sources (Papadimitriou et al. 2007; Goretti et al. 2009). The yeast killer toxins may play an instrumental role for developing such novel biopreservatives which may potentially be utilized for preservation of foods/beverages by providing protection against food spoilage yeasts and other microorganisms like bacteria and fungi (Comitini and Ciani 2011; Bajaj et al. 2013). Certain bacteria (*Bacillus subtilis* and *Lactobacillus plantarum*) that are frequent contaminants of ethanolic fermentation, can be managed by some killer yeasts like *Candida glabrata*, *Pichia anomala* and *Candida* sp. Killer toxin from these yeasts exerted strong antibacterial activity (Meneghin et al. 2010).

7 Conclusion

The toxin producing potential equips the yeasts to have specific advantages for survival in the ecological niches. Of course killer yeast and/or their toxins have substantial application potential in several industrial sectors like food, pharma, health and agriculture. The killer toxin production, killing mechanisms, and immunity of yeast against toxins, show immense variations among various yeast genera. Molecular processes involved in killer toxin synthesis have been understood well only in *S. cerevisiae*. In other yeast genera mechanistic insights of killer phenomenon is yet awaiting investigation. Furthermore, applications of killer toxins have practically been realized in certain areas, but the complete potential of toxins still require intense investigations, especially in the area of development of novel chemotherapeutic agents which have broad spectrum antimicrobial potential against bacterial/fungal pathogens of human/animal health significance. This area has attracted significant research focus in recent years due to swift development of resistance among human/animal pathogens against traditionally used chemotherapeutic agents. In addition, application of killer toxins for developing biopreservatives and biocontrol agents for prevention of food spoilage need more exhaustive exploration. Considering high awareness among consumers regarding the health deleterious/hazardous affects of chemical based preservatives, bio-based preservatives (yeast killer toxins) that are absolutely safe for human consumption have become focal point of intense research. Thus, further research is required to fully exploit the potential of killer yeast/toxins for various applications.

Acknowledgements Dr. Bijender Kumar Bajaj gratefully acknowledges Council of Scientific and Industrial Research (CSIR) and University Grants Commission (UGC) for financially supporting 'Killer Yeast' research in the form of research projects; ERUSMUS-MUNDUS (EU) and VLIR-UOS (Govt. of Belgium) for providing fellowships for 'Research Stays' respectively, at University of Naples, Naples, Italy, and University of Antwerp, Antwerpen, Belgium. Dr. Satbir Singh acknowledges CSIR for JRF/SRF for doctoral research. Authors thank Director, School of Biotechnology, University of Jammu, Jammu, for necessary laboratory facilities.

References

Aguiar, C. and Lucas, C. 2000. *Food Technol. Biotechnol.* 38: 39–46.
Ahmed, A., Sesti, F., Ilan, N., Shih, T.M., Sturley, S.L. and Goldstein, S.A.N. 1999. *Cell* 99: 283–291.
Arroyo-Helguera, O., Alejandro, D.L.P. and Irene, C. 2012. *Braz. J. Microbiol.* 880–887.
Baeza, M.E., Sanhueza, M.A. and Cifuentes, V.H. 2008. *Biol. Res.* 41: 173–182.
Bajaj, B.K. and Sharma, S. 2010. *Braz. J. Microbiol.* 41: 477–485.
Bajaj, B.K. and Tauro, P. 1994. *Biotechnol. Lett.* 16: 631–636.
Bajaj, B.K., Dilbaghi, N. and Sharma, S. 2003. *J. Sci. Ind. Res.* 62: 714–717.
Bajaj, B.K., Raina, S. and Singh, S. 2013. *J. Basic Microbiol.* 53: 645–656.
Baz, A.F. and Shetaia, Y.M. 2005. *Int. J. Agri. Biol.* 7: 1003–1006.
Bisson, L.F. 2004. *Food Biotechnol.* 18: 63–96.
Bomblies, K. 2014. *eLIFE* 3:e03371.
Bracesco, N., Salvo, V.A. and Nunes, E. 2006. *FEMS Microbiol. Lett.* 256: 132–136.
Branco, P., Francisco, D., Chambon, C., Hebraud, M., Arneborg, N., Almeida, M.G., Caldeira, J. and Albergaria, H. 2014. *Appl. Microbiol. Biotechnol.* 98: 843–53.
Breinig, F., Sendzik, T., Eisfeld, K. and Schmitt M.J. 2006. *PNAS* 103: 3810–3815.
Buzdar, M.A., Chi, Z., Wang, Q., Hua, M.X. and Chi, Z.M. 2011. *Appl. Microbiol. Biotechnol.* 91: 1571–1579.

Buzzini, P., Turchaetti, B. and Vaughan-Martini, A.E. 2007. *FEMS Yeast Res.* 7: 749–760.
Buzzini, P., Turchetti, B. and Martini, A. 2004. *J. Appl. Microbiol.* 96: 1194–1201.
Caramalac, D.A., da Silva Ruiz, L., de Batista, G.C., Birman, E.G., Duarte, M., Hahn, R. and Paula, C.R. 2007. *Pediatr. Infect. Dis. J.* 26: 553–557.
Carrau, F.M., Neirotti, E. and Gioia, O. 1993. *J. Ferment. Bioeng.* 76: 67–69.
Carreiro, S.C., Pagnocca, F.C., Bacci, M., Bueno, O.C., Hebling, M.J. and Middelhoven, W. J. 2002. *Folia Microbiol.* 47: 259–62.
Chi, Z., Liu, G., Zhao, S., Li, J. and Peng, Y. 2010. *Appl. Microbiol. Biotechnol.* 86: 1227–1241.
Ciani, M. and Comitini, F. 2011. *Ann. Microbiol.* 61: 25–32.
Ciani, M. and Fatichenti F. 2001, *Appl. Env. Microbiol.* 67: 3058–3063.
Ciani, M., Comitini, F., Mannazzu, I. and Domizio, P. 2010. *FEMS Yeast Res.* 10: 123–133.
Coelho, A.R., Tachi, M., Pagnocca, F.C., Nobrega, G.M.A., Hoffmann, F.L., Harada, K. and Hirooka, E.Y. 2009. *Food Add. Contaminants* 26: 73–81.
Comitini, F., Di Pietro, N., Zacchi, L., Mannazzu, I. and Ciani, M: 2004. *Microbiol.* 150: 2535–2541.
Comitini, F., Gobbi, M., Domizio, P., Romani, C., Lencion, L., Mannazzu, I. and Ciani, M. 2011. *Food Microbiol.* 28: 873–882.
Comitini, F., Mannazzu, I., and Ciani, M. 2009. *Microb. Cell Fact.* 8: 55.
Dabhole, M.P. and Joishy, K.N. 2005. *Indian J. Biotechnol.* 4: 290–292.
Elmaci, S., Ozcelik, F., Tokatli, M. and Cakir, I. 2014. *Antonie Van Leeuwenhoek.* 5: 835–47.
Esteve-Zarzoso, B., Manzanares, P., Ramon, D. and Querol, A. 1998. *Int. Microbiol.* 1: 143–148.
Farris, G.A., Fatichenti, F., Bifulco, I., Berordi, E., Deiana, P., Satta, T. 1992. *Biotechnol. Lett.* 14, 219–222.
Fleet, G.H. 2003. *J. Food Microbiol.* 86: 11–22.
Fuentefria, A.M., Perez, L.R.R., d'Azevedo, P.A., Pazzini, F., Schrank, A., Vainstein, M.H. and Valente, P. 2008. *J. Basic Microbiol.* 48: 25–30.
Goretti, M., Turchetti, B., Buratta, M., Branda, E., Corazzi, L., Vaughan-Martini, A. and Buzzini, P. 2009. *Int. J. Food Microbiol.* 131: 178–182.
Guo, F.J., Ma, Y., Xu, H.M., Wang, X.H. and Chi, Z.M. 2013. *Antonie van Leeuwenhoek* 103: 737–746.
Gutierrez, A.R., Epifanio, S., Garijo, P., Lopez, R. and Santamaría, P. 2001. *Am. J. Viticulture Enol.* 52: 352–356.
Guyard, C., Seguy, N., Cailliez, J.C., Drobecq, H., Polonelli, L., Dei-Cas, E., Mercenier, A. and Menozzi, F.D. 2002. *J. Antimicrob. Chemother.* 49: 961–971.
Hammond, J.R.M. and Eckersley, K.W. 1984. *J. Inst. Brew.* 90: 167–177.
Hatoum, R., Labrie, S. and Fliss, I. 2012. *Frontiers in Microbiol.* 3: 1–12.
Heard, G.M. and Fleet, G.H. 1987. *Appl. Env. Microbiol.* 53: 2171–2174.
Ingeniis, J.D., Raffaelli, N., Ciani, M. and Mannazzu, I. 2009. *Appl. Env. Microbiol.* 75: 1129–1134.
Izgu, F. and Altinbay, D. 1997. *Microbios.* 89: 15–22.
Izgu, F., Altinbay, D. and Acun, T. 2006. *Enzym Microb. Technol.* 39: 669–676.
Jablonowski, D. and Schaffrath, R. 2007. *Biochemical Society Transactions* 35: 1533–1537.
Javadekar, V.S., Sivaraman, H. and Gokhale, D.V. 1995. *J. Ind. Microbiol.* 15: 94–102.
Keszthelyi, A. and Ohkusu, M. 2006. *Mycoses* 49: 176–183.
Klassen, R., Teichert, S. and Meinhardt, F. 2004. *Mol. Microbiol.* 53: 263–273.
Klassen, R., Wemhoff, S., Krause, J. and Meinhardt, F. 2011. *Mol. Genet. Genomics* 285: 185–195.
Labbani, F.Z.K., Turchetti, B., Bennamoun, L., Dakhmouche, S., Roberti, R., Corazzi, L., Meraihi, Z. and Buzzini, P. 2015. *Antonie van Leeuwenhoek* 107: 961–970.
LeBrasseur, N. 2005. *J. Cell Biol.* 168: 346–347.
Liu, G., Chi, Z., Wang, G., Wang, Z., Li, Y. and Chi, Z. 2013. *Crit. Rev. Biotechnol.*
Lopes, C.A. and Sangorrín, M.P. 2010. *Revista Argentina de Microbiología* 42: 298–306.
Magliani, W., Conti, S., Salati, A., Vaccari, S., Ravanetti, L., Maffei, D.L. and Polonelli, L. 2004. *FEMS Yeast Res.* 5: 11–18.

Magliani, W., Conti, S., Travassos, L.R. and Polonelli, L. 2008. *FEMS Microbiol. Lett.* 288: 1–8.
Malherbe, S., Bauer, F.F., du Toit, M. 2007. *South Afr. J. Enol. Viticult.* 28: 169–186.
Manzanares, P., Valles, S. and Viana, F. 2011. In: Molecular wine microbiology. (eds C.A.V. Santiago, R. Munoz and R.G. Garcia), Academic Press, Valencia, Spain, pp. 85–110.
Marquina, D., Santos, A. and Peinado, J.M. 2002. *Int. Microbiol.* 5: 65–71.
Maturano, Y.P., Nally, M.C., Toro, M.E., de Figueroa, L.I.C., Combina, M. and Vazquez, F. 2012. *World J. Microbiol. Biotechnol.* 28: 3135–3142.
Melvydas, V., Serviene, E., Cernishova, O. and Petkuniene, G. 2007. *Biologija* 53: 32–35.
Meneghin, M.C., Reis, V.R. and Antonini, S.R. 2010. *Braz. Arch. Biol. Technol.* 53: 1043–1050.
Muccilli, S., Wemhoff, S., Restuccia, C. and Meinhardt, F. 2013. *Yeast* 30: 33–43.
Mushtaq, M., Nahar, S. and Hashmi, M.H. 2010. *Pak. J. Bot.* 42: 4313–4327.
Ochigava, I., Collier, P.J., Walker, G.M. and Hakenbeck, R. 2011. *Antonie van Leeuwenhoek* 99: 559–566.
Papadimitriou, M.N., Resende, C., Kuchler, K. and Brul, S. 2007. *Int. J. Food Microbiol.* 113: 173–179.
Peng, Y., Chi, Z., Wang, X. and Li, J. 2010. *Mar. Biotechnol.* 12: 479–85.
Platania, C., Restuccia, C., Muccilli, S. and Cirvilleri, G. 2012. *Food Microbiol.* 30: 219–225.
Polonelli, L. and Conti, S. 2009. *Methods Mol. Biol.* 499: 97–115.
Polonelli, L., Archibusacci, C., Sestito, M. and Morace, G. 1983. *J. Clin. Microbiol.* 17: 774–780.
Pretorius, I.S., duToit, M. and van Rensburg, P. 2003. *Food Technol. Biotechnol.* 41: 3–10.
Ramon-Portugal, F., Delia, M.L., Strehaiano, P. and Riba, J.P. 1998. *World J. Microbiol. Biotechnol.* 14: 83–87.
Reiter, J., Herker, E., Madeo, F. and Schmitt, M.J. 2005. *J. Cell Biol.* 168: 353–358.
Robledo-Leal, E., Elizondo-Zertuche, M., Villarreal-Treviño, L., Treviño-Rangel, R.D., García-Maldonado, N., Adame-Rodríguez, J.M. and González, G.M. 2014. *Folia Microbiol.* 59: 503–506.
Robledo-Leal, E., Villarreal-Trevino, L. and Gonzalez, G.M. 2012. *Trop. Biomed.* 29: 297–300.
Salek, A., Schnettler, R. and Zimmermann, U. 1992. *FEMS Microbiol. Lett.* 75: 103–109.
Salek, A.T. and Arnold, W.M. 1994. *Chem. Mikrobiol. Technol.* 16: 165–183.
Santos, A., Alonso, A., Belda, I. and Marquina, D. 2013. *Fungal Gen. Biol.* 50: 44–54.
Santos, A., Marquina, D., Leal, J.A. and Peinado, J.M. 2000. *Appl. Environ. Microbiol.* 66: 1809–1813.
Santos, A., Mauro, M.S., Bravo, E. and Marquina, D. 2009. *Microbiol.* 155: 624–634.
Santos, A., Navascues, E., Bravo, E. and Marquina, D. 2011. *Int. J. Food Microbiol.* 145: 147–154.
Satyanarayana, T. and Kunze, G. 2009. Yeast biotechnology: diversity and applications. Springer, Berlin, pp. 489–715.
Scheid, L.A., Mario, D.A., Heins-Vaccari, E.M., Santuario, J.M. and Alves, S.H. 2010. *Rev. Inst. Med. Trop. Sao Paulo.* 52: 161–2.
Schmitt M.J. and Breinig, F. 2002. *FEMS Microbiol. Reviews* 26: 257–276.
Schmitt, M.J. and Schernikau, G. 1997. *Food Technol. Biotechnol.* 35: 281–285.
Selvakumar, D. and Miyamoto, M. 2006. *Antimicrob. Agents Chem.* 50: 3090–3097.
Silva, S., Calado, S., Lucas, C. and Aguiar, C. 2008. *Microbiol. Res.* 163: 243–251.
Sinclair, R.M. 2014. *Frontiers in Microbiol.* 5: 1–5.
Soares, G.A.M. and Sato, H.H. 2000. *Braz. J. Microbiol.* 31: 291–297.
Starmer, W.T., Ganter, P.F., Aberdeen, V., Lachance, M.A. and Phaff, H.J. 1987. *Can. J. Microbiol.* 33: 783–96.
Stopiglia, C.D., Heidrich, D., Sorrentino, J.M., Vieira, F.J., Landell, M.F., Valente, P. and Scroferneker, M.L. 2014. *J. Basic Microbiol.* 54: 578–84.
Stratford, M. 2006. In: The yeast handbook: yeast in food and beverages. (eds A. Querol and G.H. Fleet), Springer-Verlag Berlin, pp 335–379.
Suzuki, C., Kawano, M., Kashiwagi, T., Arata, Y., Kawasumi, T. and Kashiwagi, Y. 2000. *Protein Eng.* 13: 73–76.
Ullivarri, M.F., Mendoza, L.M. and Raya, R.R. 2014. *Antonie van Leeuwenhoek* 106: 865–878.

Vadasz, A.S., Franken, D.B., Govender, B.L., Jagganath, D.B., Govender, P., Ariatti, M., Pretorius, I.S. and Gupthar, A.S. 2002. *South Afr. J. Enol. Viticulture* 23: 39–47.

Wang, X., Chi, Z., Yue, L. and Li, J. 2007. *Curr. Microbiol.* 55: 396–401.

Wang, X.X., Chi, Z., Peng, Y., Wang, X.H., Ru, S.G. and Chi, Z.M. 2012. *Microbiol. Res.* 167: 558–63.

Webster, J. and Weber, R. 2007. Introduction to fungi. Cambridge University Press, New York, pp 226–247.

Wickner, R.B., Fujimura, T. and Esteban, R. 2013. *Adv. Virus Res.* 86: 1–36.

Yeast Genetics as a Powerful Tool to Study Human Diseases

Preeti Dabas, Deepak Kumar and Nimisha Sharma

Abstract Yeasts have proven to be an invaluable model organism to explore the fundamental cellular processes and pathways conserved across eukaryotic organisms. The wide array of available genetic and genomic tools, coupled with experimental tractability, make them ideal organisms for genetic research. A significant contribution to our systemic understanding of many human diseases has been made by studies carried out in yeasts. High-throughput genetic screens have been used to identify human disease genes, as well as to dissect the molecular pathways that regulate the function of disease related proteins. Recent developments in the areas of chemical-genetics and chemical genomics have further highlighted the importance of yeasts in the simultaneous analysis of a large number of drugs, as well as facilitating the identification of their mechanism of action. In this chapter, we describe the various genetic tools that have been used by yeast researchers to increase our understanding of the basic human biology and provide insights into the molecular mechanisms underlying various human diseases.

Keywords Yeast · Null mutants · Overexpression · Complementation · Chemical-genetics · Genetic interaction · Two-hybrid

1 Introduction

Yeast has provided an excellent model system by virtue of it being one of the most experimentally and genetically tractable model organism (Botstein et al. 1997; Forsburg 2007; Botstein and Fink 2011), availability of its complete genome sequence (Goffeau et al. 1996; Wood et al. 2002), and development of many genome-scale resources, such as collection of genome-wide deletion mutants (Winzeler et al. 1999; Kim et al. 2010), overexpression libraries (Jones et al. 2008), collection of genes required for cell cycle and cell shape (Hayles et al. 2013),

P. Dabas · D. Kumar · N. Sharma (✉)
School of Biotechnology, G.G.S. Indraprastha University, 110078 Dwarka, Delhi, India
e-mail: nimishaphd@gmail.com

T. Satyanarayana and G. Kunze (eds.), *Yeast Diversity in Human Welfare*, DOI 10.1007/978-981-10-2621-8_8

genome-wide binary protein interaction networks (Auerbach et al. 2005) and GFP-fusion libraries (Huh et al. 2003). A large number of transcriptomic, proteomic and metabolomic studies have been carried out in yeasts and comprehensive datasets have also been generated from these studies. Furthermore, extensive non-coding transcription has also been observed in both *Saccharomyces cerevisiae* and *Schizosaccharomyces pombe*, and since dysregulation of non-coding RNAs underlies several complex human diseases, these yeasts offer enormous potential for functional characterization of these non-coding RNAs. Table 1 lists the various genome-scale resources that are available for yeasts.

It is well established that many of the fundamental molecular and cellular processes including DNA replication and repair, transcription, RNA processing, protein folding and degradation, intracellular protein trafficking, cell cycle control and mitochondrial function are conserved between yeasts and mammalian cells. Furthermore, similar mechanisms have been shown to control these processes in both unicellular and multicellular eukaryotes and thus, a detailed understanding of these processes has important implications in unravelling the molecular mechanisms underlying different human diseases. Several model organisms have been used to study these processes and yeast has played a significant role in expanding our knowledge about basic eukaryotic biology and understanding of human disease (Sunnerhagen 2002; Oliver 2002; Barrientos 2003; Mager and Winderickx 2005; Mustacchi et al. 2006; Smith and Snyder 2006; Perocchi et al. 2008; Khurana and Lindquist 2010). In recent years, the application of high throughput genome-level technologies has provided new avenues for the use of yeast genetics as a tool for biomedical research in studying human genes and dissecting their role in disease pathways (Petranovic and Nielsen 2008).

Most of the genetic screens used in yeast were initially developed in *Saccharomyces cerevisiae*, but more recently *Schizosaccharomyces pombe* has emerged as a complementary platform for performing these genetic assays (Sunnerhagen 2002). Approximately 60% of *S. cerevisiae* genes display sequence homology to a human orthologue (Botstein et al. 1997), and approximately 25% of the human disease-associated genes have a close homologue in *S. cerevisiae* (Basset et al. 1996). In case of *S. pombe*, 850 proteins out of a total of 5054 predicted proteins have not been assigned a biological function as yet, and 182 of these uncharacterized proteins are conserved in mammals (Hoffman et al. 2015). Interestingly, genes encoding many of these proteins are orthologues of human disease genes or are a part of the COSMIC (Catalogue of Somatic Mutations In Cancer) database (Forbes et al. 2015). For instance, an orthologue of BLCAP protein that is associated with bladder cancer has been found in *S. pombe* (SPAC652.01) and similarly, *S. pombe* SPAC6910.10c is an orthologue of MMTAG2 which is linked with multiple myeloma. Functional analysis of these proteins in *S. pombe* will offer critical insights into the roles of these proteins in humans (Hoffman et al. 2015). Moreover, since *S. cerevisiae* and *S. pombe* diverged about 350 million years ago and differ in many aspects, like occurrence of introns, cell-cycle organization, heterochromatin, centromere structure as well as complexity of DNA replication origins (Sunnerhagen 2002; Hoffman et al. 2015), a comparison between these two

Table 1 List of available genome-wide yeast resources

Name of the resource	Information available	URL ID
Saccharomyces genome database (SGD)	Genome sequence data, information about gene and protein products	http://www.yeastgenome.org/
Pombase	Information about all coding and non-coding genes, literature curation	http://www.pombase.org/
MIPS comprehensive yeast genome database (CYGD)	Structural and functional information of genome, comparative analysis	http://www.helmholtz-muenchen.de/en/ibis/institute/groups/fungal-microbial-genomics/resources/cygd/index.html
The localization and quantitation atlas of the yeast proteome (LoQAtE)	Localization of proteins is stressed and non-stressed conditions	http://www.weizmann.ac.il/molgen/loqate/
Interactome database	Physical and genetic interactions	http://interactome-cmp.ucsf.edu/
Functional specification (FunSpec)	Functional and localization information about genes	http://funspec.med.utoronto.ca/
Yeast functional genomics database (YFGdb)	Functional genomics, analysis tools	http://yfgdb.princeton.edu/
The promoter database of *Saccharomyces cerevisiae* (SCPD)	Promoter sequence region	http://rulai.cshl.edu/SCPD/
The yeast metabolome database (YMDB)	Metabolites	http://www.ymdb.ca/
Yeast search for transcriptional regulators and consensus tracking (YEASTRACT)	Transcription factors and target genes	http://www.yeastract.com/
The yeast snoRNA database	Small Nucleolar RNAs	http://people.biochem.umass.edu/sfournier/fournierlab/snornadb/main.php
Collection of yeast cells and localization patterns (CYCLoPS)	Localization of proteins	http://cyclops.ccbr.utoronto.ca/index.html
Yeast interactome database	Y2H interactions	http://interactome.dfci.harvard.edu/S_cerevisiae/index.php
Saccharomyces cerevisiae pathway/Genome databases (Yeast Cyc)	Metabolomics	http://yeast.biocyc.org/
Saccharomyces cerevisiae morphological database (SCMD)	Morphology	http://scmd.gi.k.u-tokyo.ac.jp/datamine/
Yeast fitness database	Chemical-genetic interactions	http://chemogenomics.pharmacy.ubc.ca/fitdb/fitdb2.cgi

evolutionarily distant yeasts will be invaluable for identifying those genes and processes that are more widely conserved among eukaryotic organisms. However, both *S. cerevisiae* and *S. pombe* share a common genetic tool set in terms of plasmids, mutant strains, promoter constructs, etc., which allow for an easy identification of novel genes, followed by their functional characterization. Furthermore, approximately 3038 genes have been shown to be conserved among both these yeasts and other eukaryotes, including humans, strengthening the view that conclusions made from experiments in these two yeasts will contribute significantly to our understanding of metazoan cells (Wood et al. 2002).

Various classical genetic screens have been applied by yeast researchers to identify the function of different genes and elucidate pathways that they may be involved in (Appling 1999; Forsburg 2001; Boone et al. 2007). Many modifications and interesting innovations have been made in these screens over the years that offer additional insights into the biological processes across organisms. The aim of this chapter is to review key yeast genetic approaches that have been employed to address fundamental biological questions, as well as in the systematic analysis of human disease genes.

2 Manipulation of Gene Dosage

An effective and simple strategy to probe gene function is by manipulation of the dosage of the gene of interest in yeast cells. Therefore, if a human gene has a yeast orthologue, the function of the yeast gene can be determined by either deleting or overexpressing it, followed by an analysis of the resulting phenotype(s) of these yeast mutants. Screens based on either deletion of a gene or reducing its dosage allow for studying loss-of-function mutations, while overexpression strains would identify gain-of-function mutations. Any yeast gene can be deleted by homologous recombination (Fig. 1) and appropriate yeast vectors are available for overexpression of the gene of interest. Several examples are reported in literature where both these approaches have provided valuable information about the molecular mechanisms underlying various diseases. Strand et al. (1993) discovered the role of human mismatch repair genes, hPMS1, hMLH1 and hMSH2, in hereditary non-polyposis colon cancer by analysis of phenotypes obtained by deletion of the *S. cerevisiae* orthologues of these genes. *SURF-1*, a gene involved in Leigh's disease, was identified because its yeast homolog encoded a product that is targeted to the mitochondria and impairs mitochondrial respiration when mutated (Tiranti et al. 1998). Yeast mutants have also offered insights into the process of ageing, which is mainly categorized into two types in yeast, i.e. replicative life span (RLS) and chronological life span (CLS). RLS is defined as total number of divisions carried out by a single cell before it dies, and is considered to be similar to the ageing process of asymmetrically dividing cells in higher eukaryotes, like stem cells. On the other hand, CLS is the time that the cells can survive once they reach the non-dividing state, and is linked with the ageing of non-dividing cells, e.g. neurons

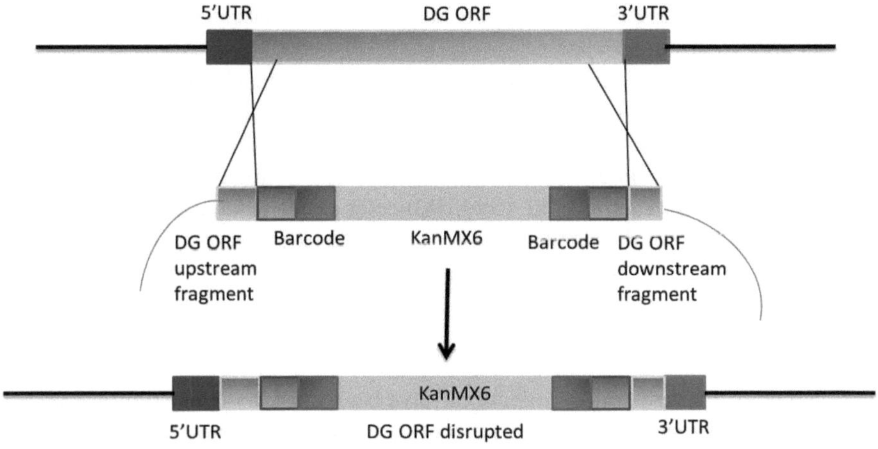

Fig. 1 Gene deletion by homologous recombination. A non-essential desired gene (DG) is deleted using homologous recombination by replacing its ORF with a KanMX6 cassette that is flanked by upstream and downstream unique sequences (BARCODES) as well as upstream and downstream fragments of the desired gene's ORF

in higher eukaryotes (Kaeberlein et al. 2007). Loss-of-function mutations in the human WRN gene that encodes RecQ-family helicases result in Werner's syndrome, which is characterized by premature ageing. The yeast orthologue of WRN gene, SGS1, has both a helicase and telomerase activity, and mutations in the SGS1 gene cause a substantial shortening of the RLS, supporting the view that a shortening of telomeres is important in the regulation of RLS (Johnson et al. 2001). In another example, functional studies carried out in *S. cerevisiae* revealed that loss of HRP1, which is the *S. cerevisiae* orthologue of the HNRPDL, an RNA processing protein important for Limb Girdle Muscular Dystrophy, caused a dramatic reorganization of proteins involved in RNA processing pathways (Vieira et al. 2014).

Complete loss-of-function alleles constitute only a minority of the relevant genetic variation in humans. Therefore, consequences of human genetic variation can be studied by making homologous mutations in their yeast orthologs (Dunham and Fowler 2013). This type of approach was used by Owen et al. (2000) to characterize the function of the *SMN* (Survival Motor Neuron) involved in spinal muscular atrophy (SMA) in *S. pombe*. It was demonstrated that deletion of the *S. pombe SMN1* orthologue caused lethality of yeast cells, but expression of *SMN1* with mutations resembling those in patients with type I SMA resulted in mislocalization of the protein in *S. pombe*. Another study for functional characterization of *MSH2* alleles involved in hereditary colon cancer demonstrated that the different alleles interfered with different aspects of protein function (Martinez et al. 2010). Rak et al. (2007) used this approach to study the NARP (Neuropathy, Ataxia, Retinis Pigmentosa) syndrome, a severe mitochondrial disease caused due to point mutations in the mitochondrially-encoded ATP6 gene. Yeast-NARP models were used to identify genes and drugs that suppressed their respiratory growth phenotype because there was a strong correlation between the severity of mutations seen in patients and yeast phenotypes.

Various genome-wide libraries of deletion mutants have also been constructed in both *S. cerevisiae* and *S. pombe*, providing a powerful tool that has been used in several large scale phenotypic screens to study not only function(s) of individual genes, but genetic interactions, gene-environment interactions as well as chemical-genetic interactions. In 1999, Winzeler et al. deleted 2,026 ORFs in *S. cerevisiae* and 17% of these genes were found to be essential for viability. This was a beginning of deleting all the yeast genes and analyzing loss of function phenotypes associated with them using high-throughput methods. Later, Giaever et al. (2002) deleted 5,916 genes of *S. cerevisiae* that covered almost 96% of the total annotated genes and 18.7% of these genes were found to be essential for viability on rich media. These deletion collections have been used in dozens of novel screens designed to address various questions of human biology and have contributed towards an increased understanding of disease biology. Collectively, both these articles describing the deletion collections have been cited approximately 2500 times since their publication, underscoring the importance of these deletion collections as universal resources (Scherens and Goffeau 2004; Giaver and Nislow 2014). Steinmetz et al. (2002) screened the whole-genome pool of *S. cerevisiae* deletion mutants to discover genes involved in mitochondrial diseases, and identified 466 genes whose deletion caused impairment of mitochondrial respiration. Another screen of the deletion collection carried out by Zhang et al. (2002) with the aim to identify genes controlling cell size revealed 49 genes that dramatically changed cell size and 88% of these genes have human homologs, underlying the extensive level of conservation in the core set of genes regulating cell cycle. A lot of work using the *S. cerevisiae* deletion libraries has laid the foundation for understanding the molecular basis of genome instability and genome maintenance mechanisms in eukaryotes (Measday and Stirling 2015). In case of *S. pombe*, Kim et al. (2010) reported the construction and analysis of 4,836 heterozygous diploid deletion mutants covering 98.4% of the genome. A comparison of orthologous gene pairs in *S. cerevisiae* and *S. pombe* revealed that 83% of these genes shared the same status of being either indispensable (essential genes) or dispensable (non-essential genes) for yeast cell viability. This observation raises the possibility that orthologues conserved among other eukaryotic organisms may also have similar dispensability as yeast genes, facilitating comparisons between different eukaryotic organisms, as well as providing critical insights into their functions. This study also identified genes required for transcription and translation which are haploinsufficient for growth in both these yeasts, suggesting that they may also be important in regulating growth in other eukaryotes. Haploinsufficiency is a dosage effect that results in a mutant phenotype upon deletion of one copy of the gene in a diploid cell since presence of both the functional copies of a gene are required to bring about a wild-type phenotype. In addition, haploproficient genes encoding ribosomal proteins, Rab-GTPase activating protein, TOR pathway proteins, were also identified (Kim et al. 2010). Haploproficient genes are those in which deletion of one copy of the gene in a diploid cell results in a better phenotype than wild type cells. Use of *S. pombe* deletion libraries has led to the identification of novel genes involved in DNA damage response (Deshpande et al. 2009; Pan et al. 2012), understanding regulation of the phosphate signal transduction pathway (Henry et al. 2011), study of TORC1 function and cellular ageing (Rallis et al. 2014). These genome-wide studies have identified several new genes that were earlier not considered to be associated with a specific phenotype. Furthermore, many genes that were previously annotated to only a single function, were found to be involved in multiple cellular processes. During the past decade or so, several methodological improvements and variations of the yeast deletion collections have been generated (Smith et al. 2009; Cooper et al. 2010; Ryan et al. 2012; Mulleder et al. 2012; Gibney et al. 2013; Robinson et al. 2014), expanding the resources available to explore biological questions. In addition, a barcode-tagged insertion mutant library has also been generated in *S. pombe* by Chen et al. (2012) to create viable mutations in both essential and non-essential genes.

Although knockout (null) mutations resulting in a complete loss of function are a highly useful tool to identify and study disease related genes in yeast, but they cannot be used in case of genes that are essential for the viability of yeast cells. Assigning functions to essential genes is relevant not only genetically, but also medically since approximately 40% of these essential genes have human counterparts (Mnaimneh et al. 2004). Thus, such genes can be studied by screening for conditional or heterozygous null mutants. Conditional mutants are those in which the encoded gene product is non-functional under a particular condition (called as restrictive or non-permissive condition), but remains functional in another condition (called as permissive condition). The most common type of conditional mutants are temperature—or thermosensitive (ts) mutants, which are inviable at high temperatures (non-permissive condition), but can grow well with no phenotype at low temperatures (permissive condition). One of the strategies for generating a ts allele of an essential gene involves its fusion with the heat inducible-degron domain that modulates its stability (Kanemaki et al. 2003; Sanchez-Diaz 2004) (Fig. 2a). Campion et al. (2010), using a temperature-degron allele of the *S. pombe* SMN protein, demonstrated that *S. pombe* cells expressing this allele had splicing defects similar to those seen in SMN-deficient metazoan cells. Yet another approach used to control the expression of an essential gene in a conditional manner is to clone the desired gene under a regulatable promoter (Fig. 2b). Tetracycline and thiamine repressible promoters are some examples of regulatable promoters widely used in yeast. Mnaimneh et al. (2004) created a collection where over two-thirds of all the essential genes of yeast were expressed under the control of the tetracycline (TetO7) promoter. In an alternative strategy for studying essential genes, diploid heterozygous null mutants can be created. These mutants having a single functional copy of a gene in a diploid organism can exhibit haploinsufficient or haploproficient phenotypes. The screening of a library of heterozygous null mutants for haploinsufficient or haploproficient phenotypes is a high throughput strategy to determine the effects of quantitative changes in the concentration of gene product on phenotypes. Deutschbauer et al. (2005) used a haploinsufficieny screen to identify proteins required for optimal growth rate and the identified proteins belonged to

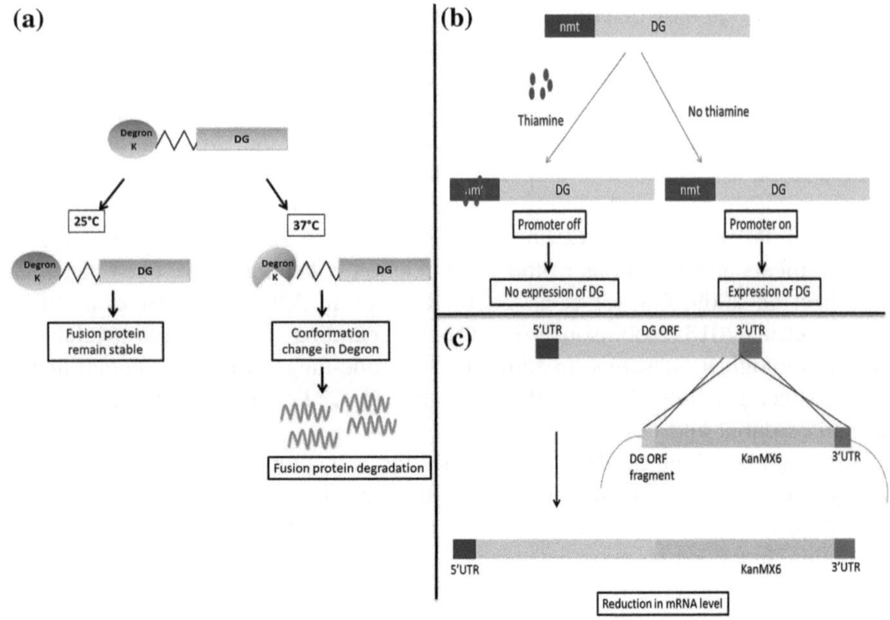

Fig. 2 Approaches to study essential genes. **a** Heat induced degron system: A degron is fused to the amino terminus of the desired gene. At permissive temperature (25 °C), the fusion protein is stable, while at restrictive temperature (37 °C), the degron undergoes a conformational change resulting in degradation of the desired protein. **b** Regulatable promoter: The *nmt* promoter is a commonly used yeast promoter regulated by thiamine. Presence of thiamine shuts down the promoter as a result of which the desired gene is not expressed. In the absence of thiamine, the promoter is active leading to the expression of the desired gene. **c** Damp collection: The 3'UTR of the essential gene of interest is replaced by KanMX6 cassette using homologous recombination. Disruption of 3'UTR renders mRNA unstable and the mRNA levels reduce up to tenfold. *DG*: *Desired gene*; *K*: *Lysine*; *nmt*: *Thiamine regulatable promoter*; *UTR*: *Untranslated region*

those involved in metabolic processes and protein production. Clare and Oliver (2013) have provided evidence that many of the *S. cerevisiae* haploproficient genes, especially those involved in maintenance of genome integrity are orthologues of human cancer genes. In 2008, Breslow et al. constructed a library of hypomorphic alleles (reduced gene function) using the decreased abundance by mRNA perturbation (DAmP) approach covering 842 (approximately 82%) essential genes (Fig. 2c). In this approach, the 3'UTR of a gene is disrupted using an antibiotic resistance cassette, which in turn destabilizes the transcript and reduces mRNA levels 4–10 fold. Proteins are produced under endogenous promoters and remain unmodified. This eliminates the possibility of complications due to misregulated activities. DAmP collection has facilitated high precision functional analysis of yeast genes.

Another alternative yet powerful tool to elucidate gene function is by overexpression of the gene of interest since overexpression can result in hyper—or neomorphic effects often due to misregulation (Prelich et al. 2012). This strategy becomes even more meaningful in case of study of redundant or paralogous genes, where no obvious knockout phenotype may be observed after deletion of the desired gene. Also, hyperactivation and overexpression of genes has been observed in many human diseases, underscoring the relevance and importance of understanding the mechanism by which overexpression causes perturbations in biological processes. Therefore, overexpression studies in yeast are a useful tool to identify novel functions of genes. However, one disadvantage of overexpression studies can be that overexpression of some proteins may prove to be deleterious to the cell. Therefore, overexpression in such cases may be carried out by using regulatable promoters. Infact, Sopko et al. (2006) generated a collection of *S. cerevisiae* strains covering 85% of the total yeast genes, and each strain in the collection conditionally overexpresses a unique yeast gene. This study showed that overexpression of the WHI4 gene in *S. cerevisiae* caused aggregates of large unbudded *S. cerevisiae* cells, identifying its role in regulating progression through the G1 phase of the cell cycle. On the other hand, deletion of WHI4 in *S. cerevisiae* did not show any phenotype. Baruffini et al. (2012) demonstrated that the yeast mitochondrial mutability because of pathological mutations in DNA polymerase gamma can be decreased by overexpression of the DNA polymerase zeta. Several variations of the overexpression approach have been described in the subsequent sections where it has been used to identify different components of a pathway and new drug targets.

3 Identification of Gene/Protein Interactions

Systematic screens have been designed to define genetic interaction networks in yeast since genes involved in the same pathway or biological process are likely to share similar genetic interaction profiles (Tong et al. 2004). Furthermore, since many of the cellular processes implicated in human diseases are conserved between yeast and humans, it is expected that the interaction networks may also be conserved between them. Therefore, information gained from these networks in yeast can be used to reconstruct related pathways in human cells. These screens have also contributed towards our understanding of many human diseases by identifying new disease-related genes acting in the same pathway, or in different pathways affecting the same process of interest. It can be expected that the gene(s) interacting with a disease-causing gene might regulate the disease phenotype and thus may represent a potential therapeutic target. Thus, if a human disease gene has an orthologue in yeast, then two common approaches can be used to identify new genes/proteins that interact either physically and/or functionally with the yeast ortholog, as part of a conserved eukaryotic pathway or cellular process, offering insights into our understanding of the molecular basis of the disease. The first strategy relies on 'suppression analysis', i.e. rescuing or suppressing the phenotype caused by a mutation in the yeast ortholog of the human gene, either by a second mutation on the same gene (intragenic suppression) or on a different gene (intergenic/extragenic suppression) or by increased gene dosage (multicopy suppression) (Fig. 3). In one

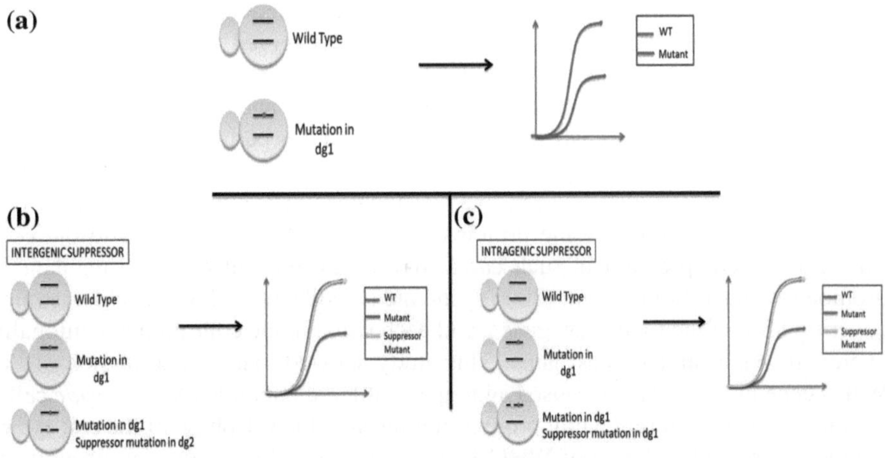

Fig. 3 Suppression analysis. Mutation in the desired gene (dg1) shows growth defect (**a**). This growth defect can be suppressed by **b** a mutation in a different gene, dg2 (intergenic suppression) or **c** a second mutation in dg1 itself (intragenic suppression)

of the early classic examples of the use of multicopy suppressor screen, Suc1 was identified as a suppressor of the Cdc2–33 ts-phenotype (Hayles et al. 1986a). Cdc2 is a protein kinase regulating cell cycle progression in *S. pombe*, and Suc1 directly contacts and regulates Cdc2 (Hayles et al. 1986b). In another study, Treich et al. (1998) identified *RSC6* as a multicopy suppressor of *swh3* conditional ts-mutant. Both *RSC6* and *SWH3* encoded proteins that were constituents of the yeast RSC complex, which is a SWI/SNF-related multiprotein complex important in chromatin remodeling.

The second strategy involves 'synthetic enhancement', i.e. exacerbating the phenotype caused by a mutation in the yeast ortholog of the human gene by a mutation in a second gene, while individual mutation in either of these two genes does not result in any visible phenotype. However, when the combination of mutation in two genes results in cell death, while individual mutation is viable, the phenomenon is called 'synthetic lethality'. A pure synthetic interaction between disease causing genes, where the mutant genes individually display no phenotype, but the combination of two mutant variants results in a disease, is called a 'digenic disease'. For instance, there are no symptoms associated with a mutation in either the *ROM1* gene that encodes retinal outer segment membrane protein 1, or *RDS* (Retinal Degeneration Slow) gene. However, a combination of these two mutations causes retinitis pigmentosa (Kajiwara et al. 1994). The digenic interactions seen for a particular gene can also extend to multiple interacting partners (Badano et al. 2002).

A significant methodological advancement that has helped in uncovering various genetic interactions has been the development of synthetic genetic array (SGA) methodology in both *S. cerevisiae* and *S. pombe* (Baryshnikova et al. 2010).

In a simple SGA screen, a haploid strain carrying a mutation in the gene of interest is crossed to an array of viable yeast deletion mutants, and the resulting double mutants are subsequently scored for genetic interactions based on cellular fitness derived from colony-size measurements. The first SGA screen with eight query genes identified 291 interactions among 204 genes (Tong et al. 2001). In a subsequent SGA screen, 132 query genes were used to detect 4000 interactions among 1000 genes with roles in cytoskeletal organization, DNA metabolism, microtubule-based chromosome segregation and cell-wall biosynthesis (Tong et al. 2004). In a subsequent study, a genome-wide genetic interaction map was generated by studying 5.4 million gene-gene pairs for synthetic genetic interactions, resulting in quantitative genetic interaction profiles for approximately 75% of all genes in *S. cerevisiae*. It was also shown that an unbiased mapping of the complete genetic interaction network serves as a key for analysis of the chemical-genetic interactions and drug-target identification (Costanzo et al. 2010). Bian et al. (2014) used the SGA approach to screen the *S. cerevisiae* deletion library to identify genes whose deletion resulted in synthetic lethality in *mad2* overexpressing *S. cerevisiae* cells, since Mad2 is overexpressed in many cancer cells. Their screen identified a gene encoding protein phosphatase 2 (PP2A), indicating that PP2A can serve as a therapeutic target in Mad2-overexpressing tumours. SGA approach has also been utilized for high-resolution genetic mapping of suppressor mutations (SGA Mapping, SGAM, Jorgensen et al. 2002) and can be used for analysis of multigenic traits. In addition to SGA, dSLAM (diploid-based Synthetic Lethality Analysis on Microarrays) technology has also been developed, where synthetic-lethal genetic interactions are identified and quantified by hybridizing genomic DNA of double mutants to DNA microarrays containing all the deletion tags (Ooi et al. 2003; Pan et al. 2004). It has been used to study network of genes involved in maintaining DNA integrity (Ooi et al. 2003; Pan et al. 2006), and identify members of the DNA helicase interaction network (Ooi et al. 2003). All these techniques relying on synthetic genetic interactions usually suggest compensatory or parallel gene action. Another set of screens has been designed that allow detection of 'alleviating interactions', where the double mutant phenotype is less severe than expected (i.e. growth rate of the double mutant is greater than that expected from the growth rates of the single mutants). These screens often indicate concerted or serial gene action within the same pathway (Komili and Roth 2007).

Genetic interactions as discussed above help in identifying gene products that work in functionally related pathways and do not always imply a physical interaction between the two gene products. Therefore, alternative methods need to be used to confirm physical association between the gene products. One such powerful approach is the yeast two-hybrid system, which has been widely exploited to investigate protein–protein interactions in an in vivo environment (Bruckner et al. 2009). The basic principle underlying the original yeast two-hybrid screen is reconstitution of a transcription factor as a result of the interaction between the two proteins being assayed, which is detected by activation of the reporter gene(s) under the control of this transcription factor (Fields and Song 1989). The reporter genes commonly used in these assays generate either a colorimetric (*LacZ*) or fluorescent

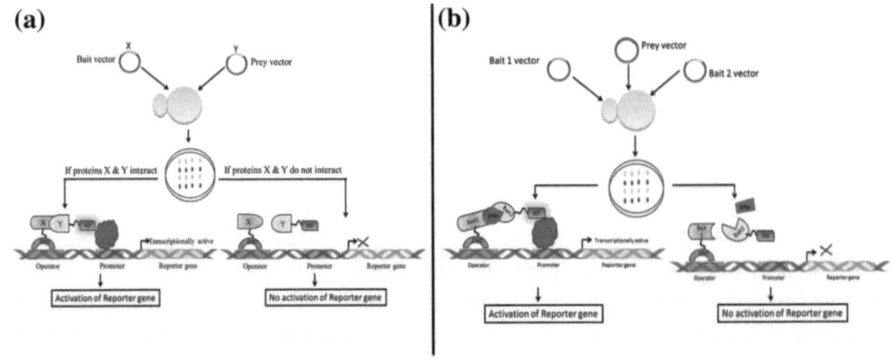

Fig. 4 Protein–protein interaction assays. **a** Classical yeast two hybrid. **b** Yeast three hybrid

(GFP) readout or allow growth on selective media (HIS3). As shown in Fig. 4a, a protein X is expressed as a fusion to the DNA binding domain (DBD) of the transcription factor (generally known as 'bait'). The bait protein can bind to an Upstream Activating Sequence (UAS, a DNA sequence present upstream of the reporter gene), but it cannot activate the reporter gene because it lacks an activation domain. A second protein Y is expressed as a fusion to the activation domain (AD) of the transcription factor (called as 'prey'). The prey is capable of activation of the reporter gene but usually does not do so because it cannot bind to the UAS. However, if bait and prey are expressed together in yeast and the two proteins X and Y interact, then a functional transcription factor is reconstituted that will bind to the UAS causing activation of the reporter gene (Fig. 4a). The yeast two-hybrid approach has been commonly used to generate small interactions networks that focused on protein complexes or components of a defined pathway. Lenk et al. (2011) employed yeast two-hybrid assay to understand how mutation in Fig. 4 causes Charcot-Marie-Tooth disease.

Over the years, several modifications have been made in this strategy to test for interaction between proteins that may not interact in the nucleus, for example membrane proteins or secretory pathways, or 'bait' proteins that cause auto-activation, etc. Some of these variations include the split ubiquitin screen to test interaction between membrane proteins, RNA polymerase III based yeast two hybrid assay for auto-activating bait proteins, reverse two-hybrid screens, where protein–protein interactions are deleterious to the cell and three hybrid system, to detect interactions between small molecules and protein (Bruckner et al. 2009). The yeast three hybrid approach (Fig. 4b) was used to identify the protein interaction partners of the cancer drug, methotrexate, by screening a mammalian cDNA library (Henthorn et al. 2002). It has also been used by Becker et al. (2004) to identify known and novel targets of small molecule cyclin-dependent kinase inhibitors. Many of these candidate targets were also confirmed by biochemical methods.

Optimization and automation of the yeast two-hybrid technique has led to its use in systematic detection of protein–protein interactions on a proteome-wide scale.

Such high throughput yeast two hybrid screens can be carried out by following either the library screening approach, or the matrix screening approach (Zhong et al. 2003). In the library screening approach, a single (or multiple) bait protein(s) is mated with a library of prey proteins, i.e. a collection of strains expressing different prey proteins. Subsequently, an interaction pair is identified based on the expression of the reporter gene (s), and the identity of the interacting prey protein is determined by isolating and sequencing the DNA cloned in the activation domain vector. In the matrix approach, a single bait protein is mated with an array of a pool of defined preys rather than with a random pool of open reading frames as present in a library. The outcome of this approach is a matrix in which the bait carrying strain has been directly screened against every strain in the activation domain array. The first large-scale comprehensive yeast two-hybrid screenings in *S. cerevisiae* was conducted by two independent groups using all yeast ORFs as baits in both matrix and library format (Ito et al. 2000; Uetz et al. 2000). Ito et al. (2000) identified 175 interactions out of which 163 had not been reported earlier. In a subsequent more exhaustive yeast two-hybrid screening by the same group, a total of 841 interactions were identified. Uetz et al. (2000) identified 281 interactions employing the matrix approach, and a total of 692 interactions in the exhaustive library screening approach. Several examples exist in literature where yeast two-hybrid screens have been employed to generate interaction networks for human proteins involved in different pathways, and a few examples have been described below. In 2005, two independent groups reported the first generation of human binary interactome using the yeast two-hybrid approach, where approximately 2700 interactions were identified (Rual et al. 2005; Stelzl et al. 2005). In the second phase of the human interactome project, approximately 14,000 binary interactions were identified by Rolland et al. (2014). A protein–protein interaction network was generated by Lim et al. (2006) with about 50 different proteins involved in 23 different kinds of inherited ataxias, which demonstrated that apparently unrelated ataxia proteins assemble into a highly connected interaction network that can provide clues to the underlying pathology of different cerebellar ataxias. More recently, Grose et al. (2015) generated the first HspB2 (cardiac restricted Heat Shock Protein B2) cardiac interactome.

Collectively, all these examples illustrate the power of using high-throughput systematic screens to map novel functional and physical relationships between genes and their encoded products, as well as characterization of very diverse cellular processes.

4 Cross-Species Complementation and Heterologous Expression

Study of yeast orthologues of human genes has generated a wealth of information regarding the function of human genes as discussed in the above sections. But sometimes the impact of a non-synonymous substitution in the orthologous yeast genes may not accurately predict a similar effect in the context of the human protein (Marini et al. 2010). Moreover, an added advantage of a human gene/protein being functional in yeast is that all the yeast genetics and molecular biology tools can be used to study it. This direct study of human genes can be accomplished by either cross-species complementation or heterologous expression (Osborn and Miller 2007). Many human disorders are caused by a loss of function of the protein encoded by the disease gene. In such cases, if the human disease gene is conserved and its functional homolog exists in yeast, cross-species complementation studies can be carried out to test the ability of the human disease gene to rescue an orthologous loss-of-function mutation in yeast. In this approach, specific yeast deletion strains are used to determine if the corresponding human gene can partially or fully replace the function of its yeast ortholog, provided the yeast deletion strain has a measurable phenotype. Moreover, different mutated forms of the human gene can also be expressed directly in the corresponding yeast deletion strain and the significance of these mutated alleles in disease can also be examined. To make it easier for researchers to identify those genes that are 'swappable' between yeast and humans, functional complementation data has been assembled in the *Saccharomyces* Genome Database (SGD) as well as *S. pombe* database (Table 1). These functional complementation studies have resulted in identification of human orthologues, deciphering the functions of the human genes as well as characterizing gene variants (Osborn and Fowler 2007; Dunham and Fowler 2013). Human-yeast complementation pairs have been systematically tested in different studies. Zhang et al. (2003) screened a human cDNA library for rescue of lethality caused by inducible loss-of-function of 25 essential *S. cerevisiae* genes, and identified six essential genes that could be rescued by a human orthologue. More recently, Kachroo et al. (2015) demonstrated that 176 out of a total of 414 essential genes tested in the study could be replaced by their human orthologue. Hamza et al. (2015) screened 621 essential *S. cerevisiae* deletion mutants for functional complementation by all potential human homologs, and further extended the cross-species complementation to 35 tumor-specific mutations in genes associated with chromosome stability. Several studies have also employed yeast functional complementation assays to test candidate genes for several human diseases, including cancer, mitochondrial disorders and neurodegenerative diseases. For example, Friedreich Ataxia results from a GAA trinucleotide repeat expansion in the frataxin gene causing reduced expression. Studies on its yeast orthologue *YFH1* showed that deletion of this gene impaired oxidative phosphorylation and increased oxidative stress in yeast, suggesting that Friedreich Ataxia is caused by mitochondrial dysfunction. Moreover, its human orthologue, FXN, rescued the phenotypes shown by the yeast mutant (Pandolfo 1999). In another example, identification of the SOD1 orthologue in yeast and functional complementation of its yeast mutants by the human orthologue provided a way to study amyotrophic lateral sclerosis in yeasts (Gunther et al. 2004). Batten disease provides another classic example where work in yeast has provided crucial insights into the disease mechanism (Pearce et al. 1999). Patients with Batten disease have mutations in the *CLN3* gene. Deletion of its yeast homologue, *BTN1*, showed resistance to the chloramphenicol breakdown product, ANP, and expression of human CLN3 in these mutants re-established the ANP sensitivity. Another example is provided by complementation of the *S. cerevisiae cys4* null mutant by its human ortholog, Cystathionine beta Synthase, which enabled 84 gene variants from patients with homocystinuria to be tested for functionality and cofactor dependence (Mayfield et al. 2012). In a recently developed high throughput yeast complementation approach, regulatable promoters are used to independently control the expression level of the yeast and human genes, maximizing complementation (Perocchi et al. 2008). Another approach of creating experimental *S. cerevisiae*-based 'phenomic' models has been developed that can be used to discover gene interaction networks underlying expression of disease phenotypes (Hartman et al. 2015). Using this approach, Louie et al. (2012) screened different *S. cerevisiae* deletion libraries for identifying modifier genes of the *S. cerevisiae* homolog of *CFTR* gene with a mutation of the F508 amino acid that is responsible for cystic fibrosis.

In cases where a gene or pathway is completely absent in yeast, it may be possible to 'humanize' yeast cells by heterologous expression of a human gene from a yeast promoter (Tarnowski et al. 2012). However for using a humanized yeast model for diseases, it is important that the expression of the human disease gene results in a specific cellular phenotype in yeast that can be studied. For instance, the tumor suppressor gene, *p53*, has no ortholog in *S. cerevisiae*, but its overexpression in *S. cerevisiae* causes a growth defect. This phenotype was used to find hyperactive variants of *p53* among alleles created by random mutagenesis (Kato et al. 2003). Yeast-based *p53* transactivation assays have also been used to study the cancer-associated *p53* mutations (Flaman et al. 1995). 'Humanized' yeasts have also been used as models for studies on age-associated neurodegenerative disorders, like Parkinson's disease and Huntington disease. In these disorders although the disease genes are present only in vertebrates, but pathogenicity is believed to be caused by a gain of function of the disease mutant protein (Ocampo and Barrientos 2008). Parkinson's disease is mainly caused by the accumulation of α-synuclein in neuronal cells. Although there is no orthologue present for this gene in yeasts, its heterologous expression showed similar effects as in diseased condition such as formation of protein inclusions and cytotoxicity (Outeiro and Lindquist 2003). Recognition of genes involved in ER-to Golgi transport as suppressors of cytotoxicity caused by synuclein identified a new mechanism of action of this protein different from other neurodegenerative diseases (Spradling et al. 2006). Along with synuclein, Synphilin-1 was also associated with Parkinson's disease, and interaction between synuclein and synphilin-1 was confirmed by the yeast two hybrid technique (Engelender et al. 1999). Yeasts also have no genes for caspases or many other genes involved in apoptosis in multicellular organisms. But the human pro-apoptotic Bax and Bak proteins kill yeast cells by a mechanism similar to apoptosis (Matsuyama et al. 1998; Xu et al. 1998). Human anti-apoptotic Bcl2 protein blocks Bax-induced death in yeast. Bax results in the release of cytochrome c from mitochondria that can be blocked by Bcl2 in both yeast and human cells. F_0F_1-ATPase mitochondrial proton pump was identified as being necessary for Bax-induced cell death by generating yeast mutants resistant to Bax killing (Matsuyama et al. 1998). In case of

Huntington disease, a human mutant Huntingtin protein (Htt) with an expanded polyglutamine-domain was expressed in *S. cerevisiae*, where it aggregated and caused cytotoxicity as in the neuronal cells of patients. Subsequently, it was observed that the mutant Htt protein was not toxic in the *S. cerevisiae* mutants lacking kynurenine 3-monooxygenase, a highly conserved enzyme of the tryptophan degradation pathway, providing a putative therapeutic target for the treatment of the disease (Giorgini et al. 2005). Zwilling et al. (2011) used chemical inhibitors of this enzyme in animal models to ameliorate neurodegeneration. In another example, Yonemura et al. (2011) reconstituted the human secretase associated with Alzheimer's disease in yeast and used it for high throughput selection of its inhibitors. *S. pombe* overexpressing human tankyrase I gene was used to identify a series of flavones as inhibitors of tankyrase I, which is an important target for cancer therapy (Yashiroda et al. 2010). Many mammalian genes that do not have orthologs in yeast, like those involved in angiogenesis, have been found to have yeast 'phenologs' that display an orthologous phenotype. These phenologs can help to identify associations between genes and phenotypes across yeast and humans (Mc Gary et al. 2010).

5 Chemical-Genetic Screens and Chemical Genomics

S. cerevisiae and *S. pombe* have also been used as cellular platforms for conducting high throughput screens with the aim to determine the cellular target of a compound (drug) and to elucidate its mechanism of action. If the cellular target of a given compound is known, then genetic screens can be designed to delineate the steps/processes upstream or downstream of the compound-target interaction that govern cell sensitivity. If the target and/or the mechanism of action of the compound is not known, yeast genetic assays can provide important clues provided the parental yeast strain being used in these assays is sensitive to the compound, or genetic approaches can be used to first create a sensitive phenotype that can be subsequently used to determine the mechanism of action of the given compound (Bjornsti et al. 2002). Two classic genetic approaches have been used to determine the cellular target and mechanism of action of a drug. In the first approach, high copy number suppressors of the drug sensitive phenotype of the parental yeast strain are identified by transforming a multi-copy plasmid DNA library into the drug-sensitive parental yeast strain, and selecting colonies resistant to that specific drug. The plasmid from these resistant colonies is then isolated and sequenced to identify the gene conferring resistance to the drug (Fig. 5). The basis of the screen is that increasing the dosage of the wild-type gene responsible for causing drug-toxicity will lead to suppression of drug sensitivity and thus identify target genes of the drug (Rine et al. 1983). For instance, if the target of a particular compound is an enzyme and the compound causes toxicity by inhibiting the activity of this enzyme, then, overexpressing the enzyme will alleviate drug-induced toxicity. Luesch et al. (2005) used the same strategy and found Pkc1 as a drug target of PAP compound, which was also confirmed by other biochemical assays. In the

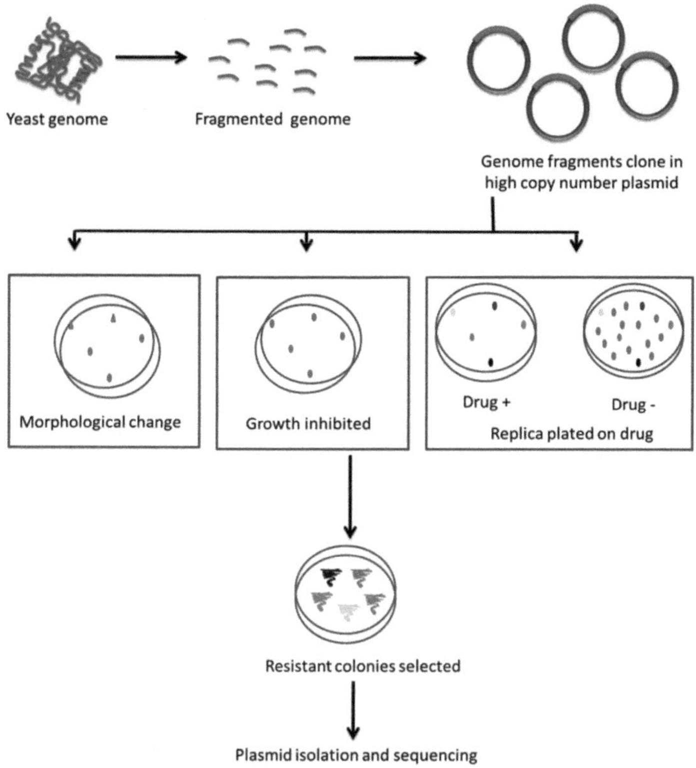

Fig. 5 Overexpression approach. Overexpression libraries are screened for suppression of different phenotypes associated with a mutation or sensitivity towards a drug, to identify genes involved in disease pathways

second approach, drug-sensitive haploid yeast cells are randomly mutagenized and drug-resistant colonies are selected (Lawrence 2002). The underlying principle is that mutation-induced changes in gene function increase cellular resistance to drug-induced effects. This approach was used by Heitman et al. (1991) to identify Tor1 and Tor2 as targets of rapamycin.

More recently, genomic approaches have been employed to screen mutant strains in different yeast libraries (deletion library, haploinsufficient libraries or overexpression libraries) for sensitivity to different compounds. The goal of this area of research, referred to as 'chemical genomics', is to detect functional relationships between particular genes and chemical compounds and identify pathways and target proteins of the given compound through systematic analysis of all genes present in a genome (Andrusiak et al. 2012). The main concept underlying these approaches is that bioactive compounds can act as mimetics of genetic mutations, implying that inactivation of a particular protein by a compound is equivalent to a perturbation caused due to a mutation in the gene encoding the protein, and thus a mutation can act as a 'drug' (Hughes et al. 2000). A 'chemical-genetic' profile can be generated for each compound by treating the complete set of deletion mutants with the compound and documenting the drug-sensitivity of all the mutants. Since gene deletions that make cells sensitive to a specific compound can identify pathways which buffer the cell against the toxic effects of that compound, a clustering of all mutants exhibiting similar sensitivity to a given compound can provide clues about the compound's mechanism of action. Moreover, the molecular target of the compound can also be predicted by comparing the chemical-genetic profile of the compound to the genetic interaction profile of the deletion mutants because as explained above deletion of a gene encoding the target of an inhibitory compound should cause cellular effects resembling inhibition of the target by the treatment with the compound (Parsons et al. 2004, 2006).

In a pioneering study by Parsons et al. (2004), 12 inhibitory compounds were screened against the 4700 non-essential gene mutants of *S. cerevisiae* to create compound-specific chemical-genetic interaction profiles. In addition, genetic interaction profiles for genes encoding targets of these compounds were generated by carrying out a synthetic genetic array analysis with query mutations in the target genes. Subsequently, integration of the chemical-genetic profiles with the compendium of genetic interaction profiles provided a link between bioactive compounds and cellular target pathways or proteins. In another similar study, chemical genetic profile of 82 different compounds and natural product extracts was generated. Among these, tamoxifen, a breast cancer drug, and amiodarone, an antifungal agent, were tested and it was shown that they disrupted the calcium homeostasis in yeast cells. Moreover, based on the same chemical genetic profiling, phosphatidylserine was found to be the cellular target of Papuamide B, a known cytotoxic agent with anti-HIV activity (Parsons et al. 2006). The fission yeast heterozygous gene deletion library has also recently been used for generating the chemical genetic profiles in the presence of 47 drugs and also for comparing it with the profiles of budding yeast. This analysis led to the identification of 9 evolutionarily conserved drug targets that are also present in human cells (Han et al. 2013). These assays are also known as HOP assays (Homozygous Profiling) because the gene is completely deleted. The main advantage offered by the HOP assays is that they can provide information for compounds that do not have a direct protein target.

Heterozygous gene deletion libraries have been screened for 'drug-induced haploinsufficiency', in which the heterozygous deletion mutants show altered sensitivity to a given compound due to reduced dosage of the target gene (known as 'HIP assays', HaploInsufficiency Profiling assays, Roberge 2008). Giaever et al. (1999) used a pool of 233 heterozygous deletion mutants to identify targets of tunicamycin. Baetz et al. (2004) used drug induced haploinsufficiency screen and Kemmer et al. (2009) used chemical-genetic synthetic lethality screen for discovering molecular targets of dihydromotuporamine C (dhMotC), a potent inhibitor of invasion of human carcinoma cell line. In 2004, Lum et al. used the genome-wide collection of *S. cerevisiae* heterozygous deletion mutants as a tool to identify the targets of 78 clinically or agriculturally relevant compounds, and revealed Erg7p, which is also known as lanosterol synthase and has a role in ergosterol biosynthesis, as a molecular target of molsidomine, a potent vasodilator. They also validated the putative target by overexpressing *Erg7* in a wild type strain, which in turn showed resistance to molsidomine. In addition, they discovered that the cell growth inhibitor, 5-Fluorouracil, primarily targets the proteins involved in rRNA processing exosome. These targets were also confirmed using standard molecular biology and biochemical methods. In another study by Giaever et al. (2004), a set of 5916 heterozygous deletion strains were examined for sensitivity to 10 different compounds. Interestingly, they observed that a similar subset of heterozygous deletion mutants exhibited sensitivity to three therapeutically distinct compounds (a muscle relaxant, an antifungal agent and an anesthetic), but all with a common core chemical structure, raising the possibility that analysis of a collection of profiles may reveal novel structure-activity relationships. The approach of screening deletion library collections against a compound has identified L-carnitine transporter, Agp2p, as a new putative bleomycin transporter in yeast, suggesting membrane transport as an important determinant of bleomycin resistance (Aouida et al. 2004). In an independent study, Rpn4p transcription factor was demonstrated to compensate for proteasome inhibition by PS-341, a drug being explored for its anti-cancer potential (Fleming et al. 2002). An extensive chemical-genomic analysis carried out by testing the effect of environmental or chemical stress on the whole-genome heterozygous and homozygous deletion collections of *S. cerevisiae* offers an excellent resource for prediction of drug targets and drug synergy (Hillenmeyer et al. 2008). More recently work by Clare and Oliver (2013) provided evidence that heterozygous deletion of a set of *S. cerevisiae* haploproficient genes is sufficient to cause deleterious phenotypes that are commonly associated with cancer in mammalian cells. Since many of these genes are orthologs of mammalian cancer genes, it suggests that copy number variation of these cancer gene orthologs may be sufficient to induce tumourigenesis in human cells. Moreover, their screen of anti-cancer compounds conducted against a set of *S. cerevisiae* mutants heterozygous for haploproficient genes involved in DNA damage response pathway revealed that the response to a range of anti-cancer drugs is strongly dependent on gene dosage. HIP assays can result in the direct identification of drug targets, although effect of the drug could be masked owing to redundancy of protein activities.

Taken together, the collection of genome-wide yeast mutants along with development of automated screening techniques allows for rapid, high-throughput analysis of large numbers of compounds with a high target pathway resolution.

6 Conclusions and Future Perspective

Work in various model organisms continues to contribute to the expansion of basic biological knowledge and understanding of the important mechanisms and pathways underlying human diseases. Yeast has served as a valuable model organism owing to its experimental tractability and high degree of conservation of gene function and cellular processes with higher eukaryotic cells. However, the real power of yeast lies in the availability of a vast array of yeast-based molecular genetic tools, allowing for the study of individual gene function, mapping of gene/protein interactions and generation of protein networks in cells. Moreover, yeast has played an important role in the development and advancement of genomic technologies that have not only added to its relevance as a model organism, but have also paved the way in developing similar tools in other model organisms.

The long history of yeast research has provided evidence of the role played by yeast in the identification and characterization of important components involved in the development of several human diseases, including mitochondrial diseases, ageing, cancer and many neurodegenerative diseases. The list of yeast orthologues of human disease genes and human-yeast complementation pairs, as well as sets of conserved gene-protein interactions between yeast and humans has been steadily increasing over the years, expanding the connection between these two organisms. Thus, the versatile genetic flexibility of yeast has and will continue to benefit study of human diseases. Yeast is also increasingly being used for modelling human diseases and screening of compounds for therapeutic intervention. However, success of these approaches will depend upon obtaining a yeast phenotype that is close to the known human disease condition and is easy to study by yeast-based functional assays.

It is evident that no single model organism or single technology can be used to study genes or pathways involved in the development of a disease. Therefore, cross-genomic studies across multiple model organisms will be pivotal in making comparisons of genes, their expression patterns, functions, localization and their interaction partners. Furthermore, an integrated approach involving large-scale genetic, transcriptomic, proteomic, metabolomic, phenotypic and interactomic studies is required for a better understanding of the molecular mechanism(s) underlying a particular disease and in facilitating the process of drug discovery. A critical aspect of these integrated approaches will be the application of computational biology tools to search and analyse data obtained from these different studies across organisms such that new links can be made between mutant phenotypes in model organisms and human disease phenotypes. The ultimate goal of the model organism-based research, including yeast, will be the ability to transfer information and technical knowledge obtained from these organisms to humans to aid in developing strategies for improved disease diagnosis and individual-based therapies.

Acknowledgements The authors thank Dr. Aparna Sapra for critical reading of the manuscript.

References

Andrusiak, K., Piotrowski, J. S. and Boone, C. 2012. *Bioorg. Med. Chem.* 20: 1952–1960.
Aouida, M., Pagé, N., Leduc, A., Peter, M. and Ramotar, D. 2004. *Cancer Res.* 64: 1102–1109.
Appling, D. R. 1999. *Methods* 19: 338–349.
Auerbach, D. and Stagljar, I. 2005. In: Proteomics and Protein–Protein Interactions: Yeast two-hybrid protein–protein interaction networks. Springer, US, pp. 19–31.
Badano, J. L. and Katsanis, N. 2002. *Nat. Rev. Genet.* 3: 779–789.

Baetz, K., McHardy, L., Gable, K., Tarling, T., Rebérioux, D., Bryan, J., et al. 2004. *PNAS* 101: 4525–4530.

Barrientos, A. 2003. *IUBMB lif* 55: 83–95.

Baruffini, E., Serafini, F., Ferrero, I. and Lodi, T. 2012. *PLoS one* 7: e34322.

Baryshnikova, A., Costanzo, M., Dixon, S., Vizeacoumar, F. J., Myers, C. L., Andrews, B. and Boone, C. 2010. *Method Enzymol.* 470: 145–179.

Bassett, D. E., Boguski, M. S. and Hieter, P. 1996. *Nature* 379: 589–590.

Becker, F., Murthi, K., Smith, C., Come, J., Costa-Roldán, N., Kaufmann, C., et al. 2004. *Chem. Biol.* 11: 211–223.

Bian, Y., Kitagawa, R., Bansal, P. K., Fujii, Y., Stepanov, A. and Kitagawa, K. 2014 *PNAS* 111: 1628–1633.

Bjornsti, M. A. 2002. *Cancer cel* 2: 267–273.

Boone, C., Bussey, H. and Andrews, B. J. 2007. *Nat. Rev. Genet.* 8: 437–449.

Botstein, D. and Fink, G. R. 2011. *Genetics* 189: 695–704.

Botstein, D., Chervitz, S. A. and Cherry, J. M. 1997. *Science* 277:1259–1260.

Breslow, D. K., Cameron, D. M., Collins, S. R., Schuldiner, M., Stewart-Ornstein, J., Newman, H. W., et al. 2008. *Nat. Methods* 5: 711–718.

Brückner, A., Polge, C., Lentze, N., Auerbach, D. and Schlattner, U. 2009. *Int. J Mol. Sci.* 10: 2763–2788.

Campion, Y., Neel, H., Gostan, T., Soret, J. and Bordonné, R. 2010. *EMBO J.* 29: 1817–1829.

Chen, B. R., Hale, D. C., Ciolek, P. J. and Runge, K. W. 2012. *BMC genomics* 13: 161.

Cooper, S. J., Finney, G. L., Brown, S. L., Nelson, S. K., Hesselberth, J., MacCoss, M. J. and Fields, S. 2010. *Genome Res.* 20: 1288–1296.

Costanzo, M., Baryshnikova, A., Bellay, J., Kim, Y., Spear, E. D., Sevier, C. S., et al. 2010. *Science* 327: 425–431.

de Clare, M. and Oliver, S. G. 2013. *BMC Biol.* 11: 24.

Deshpande, G. P., Hayles, J., Hoe, K. L., Kim, D. U., Park, H. O. and Hartsuiker, E. 2009. *DNA repair* 8: 672–679.

Deutschbauer, A. M., Jaramillo, D. F., Proctor, M., Kumm, J., Hillenmeyer, M. E., Davis, R. W., et al. 2005. *Genetics* 169: 1915–1925.

Dunham, M. J. and Fowler, D. M. 2013. *Curr. Opin. Genet. Dev.* 23: 658–664.

Engelender, S., Kaminsky, Z., Guo, X., Sharp, A. H., Amaravi, R. K., Kleiderlein, J. J., et al. A. 1999. *Nat. Genet.* 22: 110–114.

Fields, S. and Song, O. K. 1989. *Nature* 340:245–246.

Flaman, J. M., Frebourg, T., Moreau, V., Charbonnier, F., Martin, C., Chappuis, P., et al. 1995. *PNAS* 92: 3963–3967.

Fleming, J. A., Lightcap, E. S., Sadis, S., Thoroddsen, V., Bulawa, C. E. and Blackman, R. K. 2002. *PNAS* 99: 1461–1466.

Forbes, S. A., Beare, D., Gunasekaran, P., Leung, K., Bindal, N., Boutselakis, H., et al. 2015, *Nucleic Acids Res* 43: D805–D811.

Forsburg, S. L. 2001. *Nat. Rev. Gene.* 2: 659–668.

Forsburg, S. L. 2007. *Gravitational Space Biol. Bull.* 18: 3–9.

Giaever, G. and Nislow, C. 2014. *Genetics* 197: 451–465.

Giaever, G., Shoemaker, D. D., Jones, T. W., Liang, H., Winzeler, E. A., Astromoff, A. and Davis, R. W. 1999. *Nat. Genet.* 21: 278–283.

Giaever, G., Chu, A. M., Ni, L., Connelly, C., Riles, L., Veronneau, S., et al. 2002. N*ature* 418: 387–391.

Giaever, G., Flaherty, P., Kumm, J., Proctor, M., Nislow, C., Jaramillo, D. F., et al. 2004. PNAS 101: 793–798.

Gibney, P. A., Lu, C., Caudy, A. A., Hess, D. C. and Botstein, D. 2013. *PNAS* 110: e4393-e4402.

Giorgini, F., Guidetti, P., Nguyen, Q., Bennett, S. C. and Muchowski, P. J. 2005. *Nat. Genet.* 37: 526–531.

Goffeau, A., Barrell, B. G., Bussey, H., Davis, R. W., Dujon, B., Feldmann, H., et al. 1996. *Science* 274: 546–567.

Grose, J. H., Langston, K., Wang, X., Squires, S., Mustafi, S. B., Hayes, W., et al. 2015. *PLoS one* 10: e0133994.
Gunther, M. R., VanGilder, R., Fang, J. and Beattie, D. S. 2004. *Arch. Biochem. Biophys.* 431: 207–214.
Hamza, A., Tammpere, E., Kofoed, M., Keong, C., Chiang, J., Giaever, G., et al. 2015. *Genetics* 201: 1263–1274.
Han, S., Lee, M., Chang, H., Nam, M., Park, H. O., Kwak, Y. S., et al. 2013. *Biochem. Biophys. Res. Commun.* 436: 613–618.
Hartman J.L. 4th., Stisher, C., Outlaw, D. A., Guo, J., Shah, N. A., Tian, D., et al. 2015. *Genes* 6: 24–45.
Hayles, J., Aves, S. and Nurse, P. 1986a. *EMBO J.* 5: 3373.
Hayles, J., Beach, D., Durkacz, B. and Nurse, P. 1986b. *Mol. Gen. Genet.* 202: 291–293.
Hayles, J., Wood, V., Jeffery, L., Hoe, K. L., Kim, D. U., Park, H. O., et al. 2013. *Open Biol.* 3: 130053.
Heitman, J., Movva, N. R and Hall, M. N. 1991. *Science* 253: 905–909.
Henry, T. C., Power, J. E., Kerwin, C. L., Mohammed, A., Weissman, J. S., Cameron, D. M. and Wykoff, D. D. 2011. *Eukaryot. Cell* 10: 198–206.
Henthorn, D. C., Jaxa-Chamiec, A. A. and Meldrum, E. 2002. *Biochem. Pharmacol.* 63: 1619–1628.
Hillenmeyer, M. E., Fung, E., Wildenhain, J., Pierce, S. E., Hoon, S., Lee, W., et al. 2008. *Science* 320: 362–365.
Hoffman, C. S., Wood, V. and Fantes, P. A. 2015. *Genetics* 201: 403–423.
Hughes, T. R., Marton, M. J., Jones, A. R., Roberts, C. J., Stoughton, R., Armour, C. D., et al. 2000. *Cell* 102: 109–126.
Huh, W. K., Falvo, J. V., Gerke, L. C., Carroll, A. S., Howson, R. W., Weissman, J. S. and O'Shea, E. K. 2003. *Nature* 425: 686–691.
Ito, T., Tashiro, K., Muta, S., Ozawa, R., Chiba, T., Nishizawa, M., et al. 2000. *PNAS* 97: 1143–1147.
Johnson, F. B., Marciniak, R. A., McVey, M., Stewart, S. A., Hahn, W. C. and Guarente, L. 2001. *EMBO J.* 20: 905–913.
Jones, G. M., Stalker, J., Humphray, S., West, A., Cox, T., Rogers, J., et al. 2008. *Nat. Methods* 5: 239–241.
Jorgensen, P., Nelson, B., Robinson, M. D., Chen, Y., Andrews, B., Tyers, M. and Boone, C. 2002. *Genetics* 162: 1091–1099.
Kachroo, A. H., Laurent, J. M., Yellman, C. M., Meyer, A. G., Wilke, C. O. and Marcotte, E. M. 2015. *Science* 348: 921–925.
Kaeberlein, M., Burtner, C. R. and Kennedy, B. K. 2007. *PLoS Genet.* 3: e84.
Kajiwara, K., Berson, E. L. and Dryja, T. P. 1994. *Science* 264: 1604–1608.
Kanemaki, M., Sanchez-Diaz, A., Gambus, A. and Labib, K. 2003. *Nature* 423: 720–725.
Kato, S., Han, S. Y., Liu, W., Otsuka, K., Shibata, H., Kanamaru, R. and Ishioka, C. 2003. *PNAS* 100: 8424–8429.
Kemmer, D., McHardy, L. M., Hoon, S., Rebérioux, D., Giaever, G., Nislow, C., et al. 2009. *BMC Microbiol.* 9: 9.
Khurana, V. and Lindquist, S. 2010. *Nat. Rev. Neurosci.* 11: 436–449.
Kim, D. U., Hayles, J., Kim, D., Wood, V., Park, H. O., Won, M., et al. 2010. *Nat. Biotechnol.* 28: 617–623.
Komili, S. and Roth, F. P. 2007. *Genes Dev.* 21: 137–142.
Lawrence, C. W. 2002. *Methods Enzymol.* 350:189–199.
Lenk, G. M., Ferguson, C. J., Chow, C. Y., Jin, N., Jones, J. M., Grant, A. E., et al. 2011. *PLoS Genet* 7: e1002104.
Lim, J., Hao, T., Shaw, C., Patel, A. J., Szabó, G., Rual, J. F., et al. 2006. *Cell* 125: 801–814.
Louie, R. J., Guo, J., Rodgers, J. W., White, R., Shah, N., Pagant, S., et al. 2012. *Genome Med.* 4: 103.
Lum, P. Y., Armour, C. D., Stepaniants, S. B., Cavet, G., Wolf, M. K., Butler, J. S., et al. 2004. *Cell* 116: 121–137.

Mager, W. H. and Winderickx, J. 2005. *Trends Pharmacol. Sci.* 26: 265–273.
Marini, N. J., Thomas, P. D. and Rine, J. 2010. *PLoS Genet.* 6: e1000968.
Martinez, S. L. and Kolodner, R. D. 2010. *PNAS* 107: 5070–5075.
Matsuyama, S., Xu, Q., Velours, J. and Reed, J. C. 1998. *Mol cell.* 1: 327–336.
Mayfield, J. A., Davies, M. W., Dimster-Denk, D., Pleskac, N., McCarthy, S., Boydston, E. A., et al. 2012. *Genetics* 190: 1309–1323.
McGary, K. L., Park, T. J., Woods, J. O., Cha, H. J., Wallingford, J. B. and Marcotte, E. M. 2010. *PNAS* 107: 6544–6549.
Measday, V. and Stirling, P. C. 2015. *Brief Funct. Genomics* pii: elv033.
Mnaimneh, S., Davierwala, A. P., Haynes, J., Moffat, J., Peng, W. T., Zhang, W., et al. 2004. *Cell* 118: 31–44.
Mülleder, M., Capuano, F., Pir, P., Christen, S., Sauer, U., Oliver, S. G. and Ralser, M. 2012. *Nat. Biotechnol.* 30: 1176–1178.
Mustacchi, R., Hohmann, S. and Nielsen, J. 2006. *Yeast 23*: 227–238.
Ocampo, A. and Barrientos, A. 2008. *BioTechniques 45*: vii-xiv.
Oliver, S. G. 2002. *Philos. Trans. R Soc. Lond. B Biol. Sci.* 357: 17–24.
Ooi, S. L., Shoemaker, D. D. and Boeke, J. D. 2003. *Nat. Genet.* 35: 277–286.
Osborn, M. J. and Miller, J. R. 2007. *Brief. Funct. Genomic. Proteomic.* 6: 104–111.
Outeiro, T. F. and Lindquist, S. 2003. *Science* 302: 1772–1775.
Owen, N., Doe, C. L., Mellor, J. and Davies, K. E. 2000. *Hum. Mol. Genet.* 9: 675–684.
Pan, X., Yuan, D. S., Xiang, D., Wang, X., Sookhai-Mahadeo, S., Bader, J. S., et al. 2004. *Mol. Cell* 16: 487–496.
Pan, X., Ye, P., Yuan, D. S., Wang, X., Bader, J. S. and Boeke, J. D. 2006. *Cell* 124: 1069–1081.
Pan, X., Lei, B., Zhou, N., Feng, B., Yao, W., Zhao, X., et al. 2012. *BMC genomics* 13: 662.
Pandolfo, M. 1999. *Arch Neurol.* 56: 1201–1208.
Parsons, A. B., Brost, R. L., Ding, H., Li, Z., Zhang, C., Sheikh, B., et al. 2004. *Nat. biotechnol.* 22: 62–69.
Parsons, A. B., Lopez, A., Givoni, I. E., Williams, D. E., Gray, C. A., Porter, J., et al. 2006. *Cell* 126: 611–625.
Pearce, D. A., Ferea, T., Nosel, S. A., Das, B. and Sherman, F. 1999. *Nat. Genet.* 22: 55–58.
Perocchi, F., Mancera, E. and Steinmetz, L. M. 2008. *Mol. Biosyst.* 4: 18–29.
Petranovic, D. and Nielsen, J. 2008. *Trends Biotechnol.* 26: 584–590.
Prelich, G. 2012. *Genetics* 190: 841–854.
Rak, M., Tetaud, E., Duvezin-Caubet, S., Ezkurdia, N., Bietenhader, M., Rytka, J. and di Rago, J. P. 2007. *J Biol. Chem.* 282: 34039–34047.
Rallis, C., López-Maury, L., Georgescu, T., Pancaldi, V. and Bähler, J. 2014. *Biol. Open.* 3: 161–171.
Rine, J., Hansen, W., Hardeman, E. and Davis, R. W. 1983. *PNAS* 80: 6750–6754.
Roberge, M. 2008. *Sci. Signal.* 1: pt5.
Robinson, D. G., Chen, W., Storey, J. D. and Gresham, D. 2014. *G3.* 4: 11–18.
Rolland, T., Taşan, M., Charloteaux, B., Pevzner, S. J., Zhong, Q., Sahni, N., et al. 2014. *Cell* 159: 1212–1226.
Rual, J. F., Venkatesan, K., Hao, T., Hirozane-Kishikawa, T., Dricot, A., Li, N., et al. 2005. *Nature* 437: 1173–1178.
Ryan, O., Shapiro, R. S., Kurat, C. F., Mayhew, D., Baryshnikova, A., Chin, B., et al. 2012. *Science* 337: 1353–1356.
Sanchez-Diaz, A., Kanemaki, M., Marchesi, V. and Labib, K. 2004. *Sci. STKE.* 2004: pl8.
Scherens, B. and Goffeau, A. 2004 *Genome Biol.* 5: 229.
Smith, M. G. and Snyder, M. 2006. *Curr. Protoc. Hum. Genet.* 15–6.
Smith, A. M., Heisler, L. E., Mellor, J., Kaper, F., Thompson, M. J., Chee, M., et al. 2009. *Genome Res.* 19: 1836–1842.
Sopko, R., Papp, B., Oliver, S. G. and Andrews, B. J. 2006. *Cell Cycle* 5: 1397–1402.
Spradling, A., Ganetsky, B., Hieter, P., Johnston, M., Olson, M., Orr-Weaver, T., et al. 2006. *Genetics* 172: 2025–2032.

Steinmetz, L. M., Scharfe, C., Deutschbauer, A. M., Mokranjac, D., Herman, Z. S., Jones, T., et al. 2002. *Nat. Genet.* 31: 400–404.

Stelzl, U., Worm, U., Lalowski, M., Haenig, C., Brembeck, F. H., Goehler, H., et al. 2005. *Cell* 122: 957–968.

Strand, M., Prolla, T. A., Liskay, R. M. and Petes, T. D. 1993. *Nature* 365: 274–276.

Sunnerhagen, P. 2002. *Curr. Genet.* 42: 73–84.

Tarnowski, L. J., Kowalec, P., Milewski, M., Jurek, M., Plochocka, D., Fronk, J. and Kurlandzka, A. 2012. *PLoS one* 7: e38740–e38740.

Tiranti, V., Hoertnagel, K., Carrozzo, R., Galimberti, C., Munaro, M., Granatiero, M., et al. 1998. *Am. J. Hum. Genet.* 63: 1609–1621.

Tong, A. H. Y., Evangelista, M., Parsons, A. B., Xu, H., Bader, G. D., Pagé, N., et al. 2001. *Science* 294: 2364–2368.

Tong, A. H. Y., Lesage, G., Bader, G. D., Ding, H., Xu, H., Xin, X., et al. 2004. *Science* 303: 808–813.

Treich, I., Ho, L. and Carlson, M. 1998. *Nucleic Acids Res.* 26: 3739–3745.

Uetz, P., Giot, L., Cagney, G., Mansfield, T. A., Judson, R. S., Knight, J. R., et al. 2000. *Nature* 403: 623–627.

Vieira, N. M., Naslavsky, M. S., Licinio, L., Kok, F., Schlesinger, D., Vainzof, M., et al. 2014. *Hum. Mol. Genet.* 23: 4103–4110.

Winzeler, E. A., Shoemaker, D. D., Astromoff, A., Liang, H., Anderson, K., Andre, B., et al. 1999. *Science* 285: 901–906.

Wood, V., Gwilliam, R., Rajandream, M. A., Lyne, M., Lyne, R., Stewart, A., et al. 2002. *Nature* 415: 871–880.

Xu, Q. and Reed, J. C. 1998. *Mol. cell.* 1: 337–346.

Yashiroda, Y., Okamoto, R., Hatsugai, K., Takemoto, Y., Goshima, N., Saito, T., et al. 2010. *Biochem. Biophys. Res. Commun.* 394: 569–573.

Yonemura, Y., Futai, E., Yagishita, S., Suo, S., Tomita, T., Iwatsubo, T. and Ishiura, S. 2011. *J. Biol. Chem.* 286: 44569–44575.

Zhang, Z. K., Davies, K. P., Allen, J., Zhu, L., Pestell, R. G., Zagzag, D. and Kalpana, G. V. 2002. *Mol. Cell. Biol.* 22: 5975–5988.

Zhang, N., Osborn, M., Gitsham, P., Yen, K., Miller, J. R. and Oliver, S. G. 2003. *Gene* 303: 121–129.

Zhong, J., Zhang, H., Stanyon, C. A., Tromp, G. and Finley, R. L. 2003. *Genome Res.* 13: 2691–2699.

Zwilling, D., Huang, S. Y., Sathyasaikumar, K. V., Notarangelo, F. M., Guidetti, P., Wu, H. Q., et al. 2011. *Cell* 145: 863–874.

Yeast Expression Systems: Current Status and Future Prospects

Adivitiya, Vikas Kumar Dagar and Yogender Pal Khasa

Abstract The commercialization of new drugs and enzymes based on their novel therapeutic properties and industrial applications has increased remarkably with the advent of recombinant DNA technology. The emerging fields of genomics and proteomics have helped in identifying an increasing number of potentially useful proteins and also triggered the need to develop better and high throughput heterologous expression systems. Proteins requiring post-translational modifications for their biological activity are produced in more complex expression systems. Yeasts are eukaryotic systems which are commonly employed for the secretory production of recombinant proteins because they can grow rapidly to very high cell densities on inexpensive media and offer the advantages of post-translational modifications. *Saccharomyces cerevisiae* is the best studied eukaryotic organism till date and the prevalent yeast species in food, chemical, and pharmaceutical production processes. The recently developed *Pichia pastoris* system is particularly suited for recombinant proteins due to high-level secretory expression and the humanization of glycosylation pattern in its modified expression hosts. Moreover, a stable integration of the expression cassette in multiple numbers in its genome helps in achieving enhanced expression because of gene dosage. Alternate yeast species like *Yarrowia lipolytica*, *Hansenula polymorpha*, and *Kluyveromyces lactis* are also being used. The efficient design of a cultivation medium, fine-tuning of the production strategies and improvement in microbial strains are the hallmarks of any commercial fermentation process. Therefore, in this chapter, we describe the development of suitable yeast expression systems for recombinant products such as enzymes and therapeutically important biomolecules with emphasis on metabolic engineering and bioprocess development.

Keywords Yeast · Recombinant protein · Cell surface display · Metabolic engineering · Bioprocess

Adivitiya · V.K. Dagar · Y.P. Khasa (✉)
Department of Microbiology, University of Delhi South Campus, New Delhi 110021, India
e-mail: yogi110@gmail.com

T. Satyanarayana and G. Kunze (eds.), *Yeast Diversity in Human Welfare*, DOI 10.1007/978-981-10-2621-8_9

1 Introduction

The expression of recombinant proteins using heterologous systems is a critical step towards the commercial production of these molecules at industrial scale. Different hosts such as bacteria (mainly *E. coli*), yeasts and mammalian cells are extensively employed as expression platforms for the production of biopharmaceuticals and industrial enzymes (Demain and Vaishnav 2009). In 1982, Humulin™ (biosynthetic human insulin) became the first recombinant biopharmaceutical produced in *E. coli* to be commercialized for human use. However, the complexities of foreign proteins, particularly of eukaryotic origin, require higher expression systems capable of incorporating many post-translational modifications. Yeasts are well-characterized, single-celled organisms that combine the advantages of prokaryotes and eukaryotes in a single host (Celik and Calik 2012) (Table 1). Yeast-based systems secrete very low amount of endogenous proteins which makes the downstream processing easy and cost effective (Mattanovich et al. 2009). Moreover, yeast cultures are free from pyrogens, pathogens or viral contaminations which in turn reduces the cost of rigorous testing before commercialization. Interestingly, a large number of proteins of mammalian origin which failed to express in *E. coli* have been successfully expressed in yeast systems (Prinz et al. 2004). Being eukaryotic hosts, their protein modification mechanisms are similar to mammalian cells except a high-mannose type glycosylation pattern (Celik and Calik 2012). Such problems have now been overcome by the development of glycoengineered yeast strains capable of producing human origin proteins with authentic N-glycans (Hamilton et al. 2003; Hamilton and Gerngross 2007). The full-length antibody production has been described in *E. coli*, however, only the *P. pastoris* derived mAb provides the essential glycosylation required for its effector functions (Potgieter et al. 2009; Mazor et al. 2010). Several recombinant proteins have been produced using the yeast-based expression systems including human serum albumin (3 g/L) (Fleer et al. 1991), phytase (13.6 g/L) (Mayer et al. 1999), glucose oxidase (9 g/L) (Park et al. 2000b), lipase (0.5 g/L) (Pignede et al. 2000), β-galactosidase (22 g/L) (Katrolia et al. 2011) and streptokinase (4.25 g/L) (Adivitiya et al. 2016).

In 1987, *S. cerevisiae* produced recombinant human insulin (Novolin®) was marketed by Novo Nordisk. *S. cerevisiae* produced insulin showed an enhanced pharmacological efficacy and reduced side effects than that made by the enzymatic conversion of porcine insulin. The first biopharmaceutical from *P. pastoris*, Kalbitor (Dyax Corp), was approved in 2009 for applications in hereditary angioedema. In 2012, *P. pastoris* derived ocriplasmin (Jetrea®, ThromboGenics) was also approved by the FDA for the treatment of symptomatic vitreomacular adhesion. In 2013, a recombinant Hepatitis B vaccine produced in *H. polymorpha* was approved for use in Europe. Currently, the market value of biopharmaceuticals has reached $140 billion with 246 biopharmaceuticals being licensed and approved for use in the US and Europe. The antibody fragments and mAbs are the fastest

Table 1 Comparison of different heterologous expression systems

Host	Doubling time	Vectors	Expression level	Glycosylation		Cost of medium
				N-linked	O-linked	
Bacteria	Rapid (0.3–0.5 h)	Episomal	High (inclusion bodies)	Absent	Absent	Low
Yeast	Rapid (1.5–3 h)	Episomal/integrative	Low-high (secretory)	High-mannose	Present	Low
Insect	Slow (16–72 h)	Bacmid	Low-high (cytosolic)	Complex	Present	High
Mammalian	Slow (14–36 h)	Episomal/integrative	Low-moderate (secretory)	Hybrid-type complex glycans	Present	High

growing class of approved biotherapeutics. A complete list of yeast derived approved biotherapeutics has been given by Walsh (2014).

Yeasts do not produce toxic secondary metabolites and hence are the hosts of choice for production of enzymes for food industries. In 1984, lactase from *K. lactis* was approved by FDA for the production of lactose-free dairy products for the lactose intolerant population while chymosin was approved in 1992 (van Ooyen et al. 2006). The recombinant phospholipase C protein produced in *P. pastoris* has been permitted as a feed additive (Ahmad et al. 2014).

The availability of complete genome sequences, easy transformation methods, promoters and improvements in expression hosts has helped in the development of many yeast species as industrial host organisms for the production of novel enzymes and biotherapeutics (Buckholz and Gleeson 1991; Porro et al. 2005). Therefore, in this chapter we have described the development of various yeast expression systems for production of commercially important biomolecules with emphasis on latest developments in yeast biotechnology like glycoengineering, protein secretion strategies, cell surface display technology, metabolic engineering, systems biology and bioprocess optimization at large scale.

2 *Saccharomyces cerevisiae* Expression System

S. cerevisiae is the most widely studied eukaryotic model organism for the production of recombinant proteins. It was the first eukaryote to have its genome fully sequenced and was utilized for recombinant protein production due to its GRAS status (Goffeau et al. 1996) (www.yeastgenome.org). *S. cerevisiae* secretes only 0.5% of endogenous proteins helping in the downstream processing of extracellularly targeted heterologous proteins (Romanos et al. 1992). The first recombinant vaccine against Hepatitis B produced intracellularly in *S. cerevisiae* was licensed for human use in 1986 in the USA. Since then, several recombinant proteins with therapeutic and industrial applications have been produced in this expression platform like hirudin (500 mg/L) (Mendoza-Vega et al. 1994), human growth hormone (hGH, 1.3 g/L) (Lee et al. 1999), human serum albumin (200 mg/L) (Kang et al. 2000); glucose oxidase (9 g/L) (Park et al. 2000b) and artemisinic acid (100 mg/L) (Ro et al. 2006).

In the following section, we have briefly discussed various vectors, promoters, selection markers and their favorable attributes towards the successful expression of many recombinant products.

2.1 *Host Strains, Vectors and Selection Markers*

S. cerevisiae expression vectors and host strains are commercially available from Invitrogen (USA). Host cells may be transformed by treating spheroplasts with $CaCl_2$/polyethylene glycol or by use of alkali cations (Cs^+ or Li^+). The transformation efficiency can also be enhanced by pretreatment with dithiothreitol (Meilhoc et al. 1990).

All yeast plasmids used in heterologous expression studies are *E. coli*/yeast shuttle vectors that may be episomal or integrative. Autonomously replicating plasmids may be divided into 3 groups, i.e, YRp (yeast replicating plasmids), YEp (yeast episomal plasmids), and YCp (yeast centromere plasmids). YRp vectors are based on the yeast ARS sequence (autonomous replication sequences) that functions as an origin of replication. They have no partitioning control and are extremely unstable with cells easily losing the plasmid in the absence of selection pressure (Murray and Szostak 1983). The YCp (CEN/ARS) based plasmids, on the other hand possess an origin of replication and yeast centromeric sequences to improve stability and are maintained at 1–2 copies per cell (Murray and Szostak 1983). YEp vectors are based on the native 2μ episomal plasmid of *S. cerevisiae* and are maintained at 10–40 copies per cell (Romanos et al. 1992). It is the most commonly used expression vector which is inherited stably (Futcher and Cox 1983). However, episomal plasmids provide limited copy number control even under selection pressure. The plasmid stability may be enhanced by the use of integrative vectors that integrate into the host genome by homologous recombination and can be maintained in the absence of selection pressure. Integration of heterologous genes provides a straightforward, stable and efficient way to introduce the gene of interest in multiple copies leading to an effective bioprocess for protein production (Da Silva and Srikrishnan 2012). The yeast integrating vectors (YIp) possess yeast DNA sequences to facilitate recombination at homologous sites with a selectable marker and a bacterial origin of replication for propagation in *E. coli* (Chee and Haase 2012). The transformation protocol involves linearization at a unique restriction site within the vector, and a single crossover event results in genomic integration. The use of high concentrations of DNA may lead to multi-copy integrations in tandem (Orr-Weaver and Szostak 1983). The multi-copy integrants can also be produced by integration into the ribosomal DNA (rDNA) cluster on chromosome XII for the genetic modification of industrial strains (Leite et al. 2013). In addition to vector-based integration, polymerase chain reaction (PCR) can also be used to generate 38–50 bp homologous flanks for genomic insertion. Using a similar strategy, a series of shuttle vectors to facilitate metabolic engineering in *S. cerevisiae* has been developed (Fang et al. 2011).

Selection markers may be dominant or auxotrophic. Dominant markers provide an entirely new function to the host while auxotrophic markers can only be used in specific auxotrophic mutants. Auxotrophic markers such as *LEU2*, *TRP1*, *URA3,* and *HIS3* are commonly used in corresponding strains mutant for leucine, tryptophan, uracil and histidine respectively (Sikorski and Hieter 1989). 2μ-plasmid based vectors carrying selection markers with defective promoters such as *LEU2*-d and *URA3*-d were developed to increase plasmid copy numbers in selective medium (Erhart and Hollenberg 1983; Loison et al. 1989). Ro and co-workers (2008) have described the production of 1.06 g/L artemisinic acid in a fed-batch bioreactor using the *LEU2*-d marker in a non-selective medium. Dominant selectable markers increase the host range that can be used for heterologous protein production. The

E. coli Tn903 transposon encodes the G418 resistance marker that provides resistance to the amino glycoside antibiotic G418 (Jimenez and Davies 1980). Other dominant selection markers majorly include hygromycin B and chloramphenicol resistance (Gritz and Davies 1983; Hadfield et al. 1986), copper resistance (*CUP1* gene) (Fogel and Welch 1982), herpes simplex virus thymidine kinase gene (Zealey et al. 1988) and dihydrofolate reductase (Miyajima et al. 1984). Furthermore, a phenomenon of auto selection has also been reported in *S. cerevisiae* where expression of a yeast killer toxin and immunity gene killed the plasmid-free cells upon exposure to the killer toxin (Bussey and Meaden 1985).

2.2 Promoters and Signal Sequences

A variety of promoters exist for *S. cerevisiae* where the glycolytic promoters are the most popular and powerful. These promoters are induced by glucose, however, their induction rate is slow and they are poorly regulated making them unsuitable for industrial use especially for producing toxic proteins (Romanos et al. 1992). Examples include the alcohol dehydrogenase I (*ADHI*), phosphoglycerate kinase (*PGK*) and glyceraldehyde-3-phosphate dehydrogenase (*GAP*). The galactose regulated promoters of genes e.g., *GAL1*, *GAL7,* and *GAL10* are involved in galactose metabolism. They are tightly regulated with their mRNAs being induced more than 1000 fold to give approximately 1% of total mRNA upon galactose addition and strongly repressed through glucose (St. John and Davis 1981). Blazeck and co-workers (2012) developed a hybrid promoter engineering approach involving the combination of core promoters with UAS (upstream activating sequences) elements to enhance transcriptional capacity by more than 2.5 folds. The commercially available pESC vector series (Agilent Technologies, USA) can be used to co-express two different genes in the same plasmid using the bidirectional *GAL1*/*GAL10* promoter cassette inducible by galactose and repressed by glucose (Maury et al. 2008). For the same purpose, Li and co-workers (2008) constructed 8 bi-directional expression vectors carrying a modified inducible *GAL* promoter and a constitutive *GAP* promoter.

Protein secretion is often preferable to circumvent its toxicity and to obtain correctly folded proteins in the culture supernatant. Therefore different signal sequences such as the pre-pro region of the α-mating factor of *S. cerevisiae*, *K. lactis* killer toxin leader sequence, yeast acid phosphatase (*PHO5*) signal sequence and yeast invertase (*SUC2*) signal sequence have been successfully used to express many recombinant proteins including interleukin-1β, human salivary α-amylase, human IFN-α2, human single chain urinary plasminogen activator, human IFN-α1 and human EGF (Romanos et al. 1992).

Despite being a well-studied host for recombinant protein expression, the *S. cerevisiae* expression platform possesses a few drawbacks such as hyperglycosylation, lack of tightly regulated strong promoters, plasmid instability, low secretion levels and protein yields (Gellissen et al. 2005; Adrio and Demain 2014). Thereby, alternate yeast expression systems such as *K. lactis*, *Y. lipolytica*, *H. polymorpha,* and *P. pastoris* have been explored for heterologous protein production.

3 *Kluyveromyces lactis* Expression System

The use of *K. lactis*, the milk yeast, in the food industry is well known. This organism shows a less pronounced preference towards glucose utilization than *S. cerevisiae* and is a crabtree negative host. Since 1960s, this yeast has been used as a food supplement and enjoys a GRAS status by the FDA which permitted its use as food and feed additive and also for therapeutic protein production (van Ooyen et al. 2006). From 1950s it has been used to produce β-galactosidase (trade name Maxilact^{TM}, DSM Food Specialities, Delft, Netherlands) due to its ability to grow on lactose as sole carbon source for the production of lactose-free dairy products for the lactose intolerant population (van Ooyen et al. 2006). Recombinant bovine chymosin secretion from *K. lactis* at the industrial scale is a widely recognized biotechnological achievement (Van den Berg et al. 1990). Furthermore, several reports show that *K. lactis* has a higher protein secretion capacity than *S. cerevisiae* (Muller et al. 1998). Numerous secretory and intracellular proteins have been produced using this system like prochymosin (333 U/mL) (Van den Berg et al. 1990), α-galactosidase (250 mg/L) (Bergkamp et al. 1992), xylanase (130 mg/L) (Walsh and Bergquist 1997), scFv (1.3 mg/L) (Robin et al. 2003), human lysozyme (100 mg/L) (Iwata et al. 2004) and human serum albumin (150 mg/L) (Colussi and Taron 2005).

3.1 Host Strains, Vectors and Selection Markers

K. lactis strains are available from CBS-KNAW Fungal Biodiversity Centre (http://www.cbs.knaw.nl/) and ATCC (http://www.atcc.org/). The *K. lactis* CBS 2359 strain is commonly used for research purposes, while, the commercially available kit (NEB) contains *K. lactis* GG799 strain which is a wild-type haploid isolate having a poorly repressed *LAC4* (β-galactosidase) promoter (van Ooyen et al. 2006). Several hyper-secreting mutant strains of *K. lactis* have been developed by disruption of *MNN10* (mannosyltransferase) and *SEL1* genes (a secretion-lowering gene) (Bartkeviciute and Sasnauskas 2003, 2004). The presence of the hexose transporter gene (*KHT2*) in *K. lactis* JA6 strain allows efficient glucose consumption with high ethanol yields (Weirich et al. 1997). Its isogenic mutant dgr151 has been reported with better secretion efficiency for heterologous proteins like HSA, hIL-1β, and glucoamylase with controlled glycosylation (Donnini et al. 2004). The inactivation of the *K. lactis PMR1* gene (Ca^{2+}-ATPase) has also been reported to enhance secretion of non-hyperglycosylated heterologous proteins (Uccelletti et al. 2004).

Both episomal and integrative vectors are available for *K. lactis*. The episomal vectors are mostly derivatives of the pKD1 plasmid of *Kluyveromyces drosophilarum* whose functional organization is similar to the *S. cerevisiae* 2 μm plasmid and has a copy number of about 60–80 copies per cell which drops to about 20 for plasmids carrying foreign genes (Morlino et al. 1999). Morlino and co-workers (1999) enhanced production of human interleukin-1β four-fold using the pKD1 vector by gene dosage amplification and co-overexpression of a recombinase A gene. Episomal vectors have a high copy number but are difficult to maintain in the absence of selection. A lysozyme-producing *K. lactis* strain retained only 17.3% cells carrying the episomal pKD-1 vector in contrast to more than 91.5% cells maintaining an integrative vector under the same conditions (Iwata et al. 2004). To overcome the structural instability of pKD1 vectors, partial pKD1 plasmids composed of only the pKD1 origin of replication and the cis-acting stability locus have been developed and used for invertase production (Hsieh and Da Silva 1998). Plasmid stability of episomal vectors can be improved by auto selection methods (Hsieh and Da Silva 1998). Using this strategy, a *K. lactis* strain carrying a *PGK* (phosphoglycerate kinase) null mutation was forced to maintain a pKD1 vector carrying the *PGK* gene in addition to the foreign HSA gene (Fleer 1992). The linear DNA killer plasmids of *K. lactis* have also been created, but they are often not used due to their poor stability (van Ooyen et al. 2006).

The *K. lactis LAC4* promoter based integrative vectors have been used to direct integration at the *LAC4* locus with integration efficiency being more than 90% (Swinkels et al. 1993; van Ooyen et al. 2006). However, the integrative vectors have lower copy number (but higher genetic stability) than the episomal vectors and most of the transformants produced by integration at the *LAC4* locus usually contain a single copy of the expression cassette. Only 2–5% of the transformants contain 2–10 integrated vectors in tandem (Swinkels et al. 1993). Bergkamp and co-workers (1992) developed a strategy for multi-copy integration (up to 60 copies) into the rDNA locus and stably maintained them during growth under non-selective conditions. The targeted integration is often very difficult in *K. lactis,* and several strategies have been developed to overcome the same. One of these strategies involves the transfer of T-DNA from the Ti plasmid of *Agrobacterium tumefaciens* to *K. lactis* for enhancing the efficiency of gene targeting from 18 to 93% (Bundock et al. 1999). The deletion of the recombination gene *Ku80* helped to achieve an increase in the targeting efficiency to 97% (Kooistra et al. 2004).

The most common auxotrophic markers for selection include *URA3*, *LEU2* and *TRP1* complementing uracil, leucine and tryptophan auxotrophies in the mutant strains respectively (van Ooyen et al. 2006). The dominant selection markers conferring resistance to geneticin (G418) and hygromycin B have also been used (van Ooyen et al. 2006). The *Aspergillus nidulans* acetamidase (*amdS* gene) that breaks down acetamide to ammonia was also reported as a selection marker where transformants are selected on the basis of acetamide utilization (van Ooyen et al. 2006).

3.2 Promoters and Signal Sequences

The most commonly used promoters in *K. lactis* include the constitutive phosphoglycerate kinase (*PGK*) and the phosphate inducible acid phosphatase (*PHO5*) promoters of *S. cerevisiae* as well as the galactose/lactose inducible *K. lactis* lactase (*LAC4*) promoter (van Ooyen et al. 2006). The *LAC4* promoter has been used for industrial production of bovine prochymosin (Van den Berg et al. 1990). A major drawback of this promoter system is its ability to drive expression in *E. coli*. This can be detrimental during assembly of expression vectors in bacteria especially in the case of toxic proteins. So mutants of P_{LAC4} ($P_{LAC4\text{-}PBI}$) have been created by mutagenesis of DNA sequences resembling the *E. coli* Pribnow box which prevent bacterial protein expression (Colussi and Taron 2005). Toxic proteins like bovine enterokinase and mouse transthyretin have been successfully expressed using this promoter variant (Colussi and Taron 2005). The pKLAC series of integrative shuttle vectors commercially available from NEB contain the strong *K. lactis* $P_{LAC4\text{-}PBI}$ promoter and the *K. lactis* α-MF leader sequence for secretory expression. Another promoter, the ethanol-inducible alcohol dehydrogenase promoter (*ADH4*) has been used to drive production of human serum albumin (Saliola et al. 1999).

Several native and heterologous secretion signals have been used to direct extracellular protein secretion in *K. lactis* including the *S. cerevisiae* α-mating factor pre-pro sequence, *K. lactis* α-mating factor leader peptide, *K. lactis* killer toxin and *K. lactis* acid phosphatase (*PHO5*) secretion signal. A detailed review of the *K. lactis* expression system has been done by van Ooyen et al. (2006).

4 *Yarrowia lipolytica* Expression System

Previously known as *Endomycopsis* or *Saccharomycopsis lipolytica*, this obligate aerobic yeast is the teleomorph of *Candida lipolytica* with its perfect form being discovered in the 1960s by Wickerham (Barth and Gaillardin 1997). It is dimorphic, haploid and heterothallic having 2 mating types A and B (Gellissen et al. 2005). Its fully sequenced genome is available on the Genolevures website (Sherman et al. 2004) (http://www.genolevures.org/yali.html). This yeast possesses a unique ability to utilize hydrophobic substrates like n-alkanes, fats, fatty acids and oils (Fickers et al. 2005). Moreover, it naturally secretes proteases (AEP: alkaline extracellular protease encoded by the *XPR2* gene and AXP: acid extracellular protease), lipases, phosphatases, an RNase and an esterase into the culture medium (Barth and Gaillardin 1997; Madzak et al. 2004). Therefore, tools are developed for its genetic engineering to utilize it as a heterologous host for recombinant protein expression (Madzak 2015). Furthermore, it has also been used for the industrial production of organic acids (mainly citric acid) and single cell protein (Barth and Gaillardin 1997; Madzak 2015). The successful expression of many foreign genes in this expression system have been reported in recent years which include *Trichoderma reesei* endoglucanase I (100 mg/L) (Park et al. 2000a), *Candida antarctica* lipase B (190 mg/L) (Edmond et al. 2010), and *Rhizopus oryzae* lipase (7610 U/L) (Yuzbashev et al. 2012).

4.1 Host Strains, Vectors and Selection Markers

The *Y. lipolytica* strains can be obtained from INRA's CIRM-Levures Yeasts Library (France). Transformation can be carried out using lithium acetate method for integrative vectors and by electroporation or heat shock for replicative vectors. The most widely used strains of *Y. lipolytica* for heterologous expression include E129 and the Po1 series like Po1 d, Po1 f, Po1 g and Po1 h (Madzak 2015). The Po1 d strain of *Y. lipolytica* is best fitted for industrial applications because it is deficient in the extracellular alkaline protease encoded by the *XPR2* gene (Madzak et al. 2004). It has been genetically modified to utilize sucrose since the *S. cerevisiae SUC2* invertase gene has been cloned under the control of *Y. lipolytica XPR2* promoter and signal sequence allowing it to grow on cheap substrates like molasses (Nicaud et al. 1989). Furthermore, it possesses high secretion levels and non-reverting Leu^- and Ura^- auxotrophies (Madzak 2015). Derivatives of Po1 d (Po1 f, g, h) are further deficient in the acidic extracellular protease (*axp1-2 allele*) (Madzak et al. 2004). Lazar and co-workers (2013) developed a Po1 d derivative JMY2593 strain capable of producing 4,519 U/L of extracellular invertase for high-level production of citric acid from sucrose-based media for industrial applications.

Native episomal plasmids like the 2μ circle of *S. cerevisiae* have not been detected in *Y. lipolytica* (Barth and Gaillardin 1997). Hence, plasmids carrying the *Y. lipolytica* ARS (autonomously replicating sequences) and CEN (centromere) replicating origins have been designed (Matsuoka et al. 1993). Foreign DNA can be integrated into *Y. lipolytica* genome by homologous recombination. In 2013, Verbeke and co-workers developed a *KU70* (a gene involved in double-stranded break repair in non-homologous end joining) deficient *Y. lipolytica* strain that promotes better homologous recombination with shorter flanking regions. Further, the LTRs (long terminal repeats called zeta sequences) of *Y. lipolytica* Ylt1 retrotransposon were shown to be a possible site of multiple integrations. These zeta sequences provide a method for non-homologous integration into *Y. lipolytica* strains devoid of Ylt1 retrotransposon (Madzak 2015).

Y. lipolytica is resistant to the most commonly used antibiotics. However, phleomycin and hygromycin B sensitive strains have been isolated leading to the utilization of these antibiotics as dominant selection markers in *Y. lipolytica* (Fickers et al. 2003). But, a high frequency of spontaneous resistance undermined their frequent usage (Barth and Gaillardin 1997). The heterologous expression of the *SUC2* gene has been used as a dominant selectable marker since wild-type strains are not able to utilize sucrose (Nicaud et al. 1989). However, the residual growth of wild-type *Y. lipolytica* on sucrose plates led to false positive selections (Barth and Gaillardin 1997). Therefore, the best choice for screening and selection are the auxotrophic selectable markers including *LEU2* and *URA3* (Nicaud et al. 1989; Barth and Gaillardin 1997). Promoter-defective versions of *URA3* gene (*ura3d4 allele*) have also been developed by Le Dall and co-workers (1994) to increase heterologous protein production where multicopy integration is required to complement auxotrophy for growth recovery. For selection of multi-copy integrants, this defective selection marker strategy is quite attractive (Pignede et al. 2000; Juretzek et al. 2001; Nicaud et al. 2002).

4.2 Promoters and Signal Sequences

Extensive data is available on the *XPR2* promoter (Madzak et al. 1999). This promoter is active above pH 6.0 in media lacking preferred carbon and nitrogen sources and requires high peptone concentration in the production medium for full induction, thereby increasing production costs. Therefore, other promoters for *Y. lipolytica* have been discovered. A hybrid promoter carrying 4 tandem copies of the p*XPR2 UAS1* region upstream of a minimal *LEU2* promoter (reduced to its TATA box) has been designed. This recombinant hybrid promoter called hp4d is highly active without any effect of cultivation conditions like pH, carbon and nitrogen sources (Madzak et al. 2000). This technology has been used in the YLEX kit commercialized by Yeastern Biotech Co. (Taiwan). Several heterologous proteins like laccase (Madzak et al. 2005), lipase (Nicaud et al. 2002) and β-galactosidase (Madzak et al. 2000) have been produced using this promoter system. This concept was further extended by Blazeck and co-workers (2013) where they identified putative UAS elements for the development of a series of hybrid promoters with higher efficiency than the endogenous *Y. lipolytica* promoters. Two strong constitutive promoters from *TEF* (translation elongation factor-1α) and *RPS7* (ribosomal protein S7) genes have also been described (Muller et al. 1998). Several inducible promoters were also described since constitutive expression of toxic proteins can be detrimental to cell growth. These majorly include the metallothionein promoter, p*ICL1* (isocitrate lyase), p*POT1* (3-oxo-acyl-CoA thiolase) and p*POX2* (acyl-CoA oxidases). While p*POT1* and p*POX2* are induced by fatty acids and alkanes and tightly repressed by glucose and glycerol, p*ICL1* gets induced by fatty acids, alkanes as well as ethanol and acetate but does not show strong repression by glucose and glycerol (Juretzek et al. 2000).

For secretory expression of heterologous proteins, the leader sequence of the *XPR2* gene has been used widely (Madzak et al. 2004). The pre-pro region of *Y. lipoytica LIP2* gene and a hybrid of the pre-pro regions of *XPR2* and *LIP2* have also been used as signal sequences in recombinant protein secretion from *Y. lipolytica*

(Nicaud et al. 2002; Pignede et al. 2000). The targeting of heterologous proteins into *Y. lipolytica* peroxisomes can be achieved using peroxisomal targeting signals. The tripeptide AKI (alanine-lysine-isoleucine) and the tripeptide SKL (serine-lysine-leucine) were successfully used for the targeting of a bacterial polyhydroxyalkanoate synthase and green fluorescent protein respectively (Haddouche et al. 2010; Xue et al. 2013). A detailed account of the *Y. lipolytica* expression system has been given in Madzak (2015).

5 *Hansenula polymorpha* Expression System

Originally utilized for single cell protein production, this yeast is a promising host for recombinant protein production because of its unique methanol-assimilating property (van Dijk et al. 2000). Upon growth on methanol, the enzymes of the methanol utilization pathway are induced to constitute 20–30% of the total cell protein indicating the strength of the promoters encoding these genes (Gellissen et al. 1992). Methylotrophic yeasts contain peroxisomes that proliferate in methanol medium and are degraded by autophagy upon transfer to glucose medium. The alcohol oxidase localized in the peroxisome oxidizes methanol to formaldehyde and hydrogen peroxide. Formaldehyde can be either dissimilated into CO_2 and energy or assimilated as cell biomass, while H_2O_2 is detoxified by the peroxisomal catalase (van der Klei et al. 2006).

H. polymorpha also possesses a unique ability of nitrate assimilation as the sole nitrogen source (Celik and Calik 2012). This yeast is particularly intriguing due to its thermotolerant nature. Growth at high temperatures up to 37–43 °C prevents contamination during large scale fermentation and surpasses the need for cooling (van Dijk et al. 2000). *H. polymorpha* strains capable of withstanding temperature up to 50 °C have been constructed by overexpression of Hsp16p and Hsp104p (heat shock proteins) and deleting *ATH1* (acid trehalase) gene (Ishchuk et al. 2009). It can ferment glucose, cellobiose and xylose to ethanol making it a suitable candidate for fermentation of lignocellulosic biomass at elevated temperatures (Ryabova et al. 2003).

The yield of protein expression obtained in this system is higher than *S. cerevisiae* with reduced proteolytic degradation and better secretion efficiency (Johnson 2013). The phenomenon of hyperglycosylation in *H. polymorpha* is very rare, therefore making it a better candidate for heterologous protein expression. Berna Biotech (Switzerland) has developed a *H. polymorpha* expression system for the development of recombinant vaccines including that against Hepatitis B (Gerngross 2004). The high cell density fermentation processes have been optimized to produce gram level of proteins like glucoamylase (1.4 g/L) (Gellissen et al. 1992), hirudin (1.5 g/L) (Weydemann et al. 1995), phytase (13.5 g/L) (Mayer et al. 1999), and staphylokinase (1 g/L) (Moussa et al. 2012).

5.1 Host Strains, Vectors and Selection Markers

The genome sequence of three popular *H. polymorpha* strains (CBS4372, NCYC495 and DL-1) is available (JGI Genome Portal http://genome.jgi.doe.gov/Hanpo2/Hanpo2.home.html) (Ramezani-Rad et al. 2003; Ravin et al. 2013). A highly efficient method for *H. polymorpha* transformation involving pretreatment of mid-logarithmic phase cells with DTT has been described by Faber and co-workers (1994) where a transformation frequency of 1.7×10^6 per μg plasmid DNA was obtained. *H. polymorpha*'s autonomously replicating sequences (HARS) and *S. cerevisiae LEU2* gene that can autonomously replicate in *H. polymorpha* have been used to create episomal plasmids (Roggenkamp et al. 1986; Bogdanova et al. 1995). Several integrative shuttle plasmids have also been developed by utilizing the *S. cerevisiae* α-mating factor signal sequence for secretory expression (Song et al. 2003). Both single and double cross-over events can take place resulting in the integration of the linearized vector into homologous sites in the genome e.g. *MOX* (alcohol oxidase) gene or *AMO* (amine oxidase) gene (van Dijk et al. 2000). Integration into the host genome results in stable transformants that do not require antibiotic selection for growth. Furthermore, recombinant genes can also be expressed by co-targeting different vectors to the *H. polymorpha* rDNA locus (Klabunde et al. 2003).

For the selection of *H. polymorpha* transformants, several auxotrophic and dominant markers have been used. Vectors carrying the homologous (*LEU1*, *URA3*, and *TRP3*) or heterologous (*LEU2* and *URA3* from *S. cerevisiae*, *LEU2* from *Candida albicans*) marker genes can be used to complement auxotrophy in the *leu1.1*, *ura3*, and *trp3* mutant *H. polymorpha* strains (Agaphonov et al. 1994; Bogdanova et al. 1995). Several dominant antibiotic selection markers like geneticin G418, phleomycin or Zeocin are also available (van Dijk et al. 2000). The use of *PUR7* gene (purine biosynthesis) from *H. polymorpha* is also reported as a selectable marker for the expression of *H. polymorpha* amine oxidase (AMO) and catalase (CAT) (Haan et al. 2002). Multi-copy integrant selection can also be carried out using high antibiotic concentration (van Dijk et al. 2000).

5.2 Promoters and Signal Sequences

Strong promoters like P_{MOX} (alcohol oxidase), P_{DHAS} (dihydroxyacetone synthase) and P_{FMD} (formate dehydrogenase) are the most popular promoters used to drive foreign gene expression in *H. polymorpha*. P_{MOX} is induced by methanol, repressed by glucose/ethanol and derepressed by glycerol or xylitol hence allowing a methanol-free process (Gellissen et al. 1992; Egli et al. 1980; Celik and Calik 2012). It has one of the highest productivity values for glucoamylase (1.4 g/L) and phytase (13.5 g/L) (Mayer et al. 1999; Adrio and Demain 2014). The P_{DHAS} and P_{FMD} promoters are strongly induced by methanol and derepressed in a similar manner as

P_{MOX} (van Dijk et al. 2000). The *pAMO* (amine oxidase) promoter is induced by amines and fully repressed by ammonium (Zwart et al. 1983). It is much weaker than P_{MOX} ($\sim$20% of P_{MOX}). Other promoters include those of the nitrate assimilation genes (*YNTI, YNII, YNRI*) that are nitrate-inducible and strongly repressed by ammonium (Avila et al. 1998). In 1998, Phongdara and co-workers identified the *H. polymorpha PHO1* gene promoter (repressible acid phosphatase) that is also used for recombinant protein expression. Promoters of the *PEX* genes (peroxisome biogenesis) are weak and hence not suitable for heterologous expression (van Dijk et al. 2000). Constitutive promoters of *H. polymorpha* are the promoters of genes encoding ATPase (*PMA1*) (Hollenberg and Gellissen 1997; Cox et al. 2000) and transcription elongation factor-1α (*TEF1*/*TEF2*) (Baerends et al. 1997).

Several heterologous and homologous signal sequences have been used for recombinant protein secretion from *H. polymorpha* (van Dijk et al. 2000). The targeting of heterologous proteins to peroxisome provides an added advantage since protein modification enzymes are not present in the peroxisome to modify the protein undesirably. Furthermore, this can be beneficial for the production of toxic proteins or proteins prone to degradation (van Dijk et al. 2000). Targeting may be carried out using peroxisomal targeting signals such as PTS1 and PTS2 (Rachubinski and Subramani 1995; Subramani 1996). PTS1 signal was used for targeting a fusion protein (human insulin-like growth factor and a target protein) to the peroxisome resulting in production levels of >20% of total cell protein (Faber et al. 1996). A detailed review of this expression system has been provided by Van Dijk and co-workers (2000).

6 *Pichia pastoris* Expression System

The *Pichia pastoris* expression platform was developed by the Phillips Petroleum Company and Salk Institute Biotechnology/Industrial Associates Inc. (SIBIA, La Jolla, CA USA) (Cereghino and Cregg 2000; Ahmad et al. 2014). Currently, *Pichia pastoris* is one of the most widely used expression hosts for the production of recombinant proteins for therapeutic and industrial applications. Its high cell density fermentation on inexpensive media is a very attractive approach for heterologous protein production (Potvin et al. 2012; Ahmad et al. 2014). Moreover, the USA's Food and Drug Administration (FDA) has granted it a status of 'generally recognized as safe' (GRAS) (Ahmad et al. 2014). Several recombinant proteins such as human insulin (1.5 g/L) (Wang et al. 2001), synthetic gelatin (3–6 g/L) (Werten et al. 2001) and streptokinase (4.25 g/L) (Adivitiya et al. 2016) have been expressed in *P. pastoris*. Till now, more than 5000 recombinant proteins, 70 commercial products, and 2 approved therapeutics have been expressed in *P. pastoris* (www.pichia.com).

6.1 Host Strains, Vectors and Selectable Markers

A variety of *P. pastoris* expression hosts are available for recombinant protein expression. All strains of *P. pastoris* are derivatives of the Y-11430 wild-type strain from Northern Regional Research Laboratories (NRRL, Peoria, IL). The most frequently used expression hosts are GS115, X-33, SMD1168, PichiaPink® and SuperMan5 (Gleeson et al. 1998; Ahmad et al. 2014). On the basis of methanol utilization, the strains are divided into three categories viz. Mut^{+}, Mut^{S} and Mut^{-} phenotypes. The protease-deficient host strains, i.e., SMD1168 (*his4 pep4*) has been shown to control proteolytic degradation of certain heterologous proteins (Cereghino and Cregg 2000; Ahmad et al. 2014).

Mostly integrative vectors are used to promote homologous recombination at *AOX*1 site to generate stable integrants in the *Pichia* system. All the commercially available (Invitrogen) cloning vectors are *E. coli/P. pastoris* shuttle vectors. The linearized constructs can be integrated into *Pichia* genome via homologous recombination using spheroplast fusion, lithium chloride, polyethylene glycol and electroporation methods (Cereghino and Cregg 2000).

Several auxotrophic markers like *HIS4* (histidinol dehydrogenase), *ARG4* (argininosuccinate lyase), *ADE3* (PR-amidoimidazole succinocarboxamide synthase), *URA3* (orotidine 5′-phosphate decarboxylase) and *URA5* (orotate phosphoribosyl transferase) have been developed for selection of *P. pastoris* transformants (Nett and Gerngross 2003; Spohner et al. 2015). The PichiaPink™ expression kit (Invitrogen) allows identification of multi-copy integrants of an adenine (*ade2*) auxotrophic strain on the basis of a pigmentation phenotype (Spohner et al. 2015). *P. pastoris* dominant selectable drug markers include the *E. coli* Tn903kan^{R} kanamycin/geneticin G418 resistance gene (Scorer et al. 1994), the *Aspergillus terreus* blasticidin-S-deaminase gene for blasticidin S resistance (Vogl et al. 2013) and the *Streptoalloteichus hindustanus Sh ble* gene conferring resistance to Zeocin (Cereghino and Cregg 2000).

6.2 Promoters and Signal Sequences

For the expression of foreign genes, the strong and tightly regulated promoters such as the inducible alcohol oxidase 1 (*AOX1*) and the constitutive glyceraldehyde 3-phosphate dehydrogenase (*GAP*) promoters hold advantages for hyper-expression of heterologous proteins (Ahmad et al. 2014; Calik et al. 2015). Several other inducible promoters such as dihydroxyacetone synthase (*DAS*), formaldehyde dehydrogenase-1 (*FLD1*) and enolase (*ENO1*) as well as the promoters of the constitutive translation elongation factor gene (*TEF1*), GTP-binding protein gene (*YPT1*) and 3-phosphoglycerate kinase gene (*PGK1*) are also used for recombinant protein production (Ahmad et al. 2014; Calik et al. 2015).

For secretory expression, the utilization of various signal sequences such as α-mating factor signal sequence of *S. cerevisiae*, acid phosphatase (PHO1), β-fructofuranosidase (SUC2) and HSA has been reported (Cereghino and Cregg 2000). Further, signal peptides of other genes including the *Pichia acacia* killer toxin (Crawford et al. 2003), K28 yeast virus toxin (Eiden-Plach et al. 2004), class 2 hydrophobins of *Trichoderma reesei* (Kottmeier et al. 2011), and *Rhizopus oryzae* α-amylase (Li et al. 2011b) can also be used to direct secretory expression. The *P. pastoris* endogenous signal peptides of *PIR1* and *PIR2* cell wall proteins were also tested for protein secretion. However, only the *PIR1* signal sequence was able to drive extracellular expression (Khasa et al. 2011). A detailed review of the *Pichia* expression system has been given by Ahmad and co-workers (2014).

7 Engineering of Protein Secretion Strategies in Yeast

Secretory expression of heterologous proteins is often hampered by several bottlenecks such as codon usage, cultivation conditions, properties of the target protein and host-vector combinations. Folding and disulfide bond formation are also reported as major rate-limiting steps during protein production (Hohenblum et al. 2004). The expression of scFvs often leads to the formation of intracellular misfolded aggregates possibly due to its high hydrophobicity (Joosten et al. 2003). Therefore, current strategies of protein secretion are mainly focused on the engineering of protein folding in the ER, engineering of protein trafficking pathways and controlling post-secretory protein degradation by host proteases to improve heterologous protein secretion.

The co-expression of chaperones and foldases has been reported to have improved outcomes in various expression hosts. The overexpression of the chaperone BiP (Hsp70) in *S. cerevisiae* increased secretion of bovine prochymosin by 26-folds (Harmsen et al. 1996). Shusta and co-workers (1998) reported the co-operative effect of BiP and PDI on scFv secretion in *S. cerevisiae*. The co-expression of heterologous *Trichoderma reesei Hac1* (regulator of the unfolded protein response pathway and activator of chaperones) in *S. cerevisiae* enhanced α-amylase secretion by 2.4 folds while there was a 70% increase using the endogenous *S. cerevisiae* Hac1p (Valkonen et al. 2003). *PDI* overexpression helped β-glucosidase secretion in *S. cerevisiae* (Smith et al. 2004). Xu and co-workers (2005) co-expressed chaperones (BiP) and foldases (PDI) to promote correct folding and secretion of scFvs. The co-expression of single or multiple chaperones like Jem1p, Sil1p, Lhs1p, Scj1p significantly improved the secretion of recombinant human albumin, GM-CSF and transferrin in *S. cerevisiae* (Payne et al. 2008). The expression of PDI with recombinant proteins improved their secretory expression in *P. pastoris* (Inan et al. 2006). The co-expression of various *S. cerevisiae* chaperones such as Kar2p, Ssa1p or PDI empirically or in combination YDJ1p/PDI, YDJ1p/Sec63, Kar2p/PDI improved secretion of human G-CSF in *P. pastoris* due to their synergistic effect (Zhang et al. 2006). The Ero1p (oxidative folding machinery) and *PDI1* gene duplication also helped in efficient secretion of recombinant human albumin in *K. lactis* (Gross et al. 2004; Lodi et al. 2005). PDI along with polyubiquitin showed a stimulatory effect on expression and secretion of human serum albumin in *K. lactis* (Bao and Fukuhara 2001). Optimization of signal peptides may also be useful for enhancing protein secretion (Idiris et al. 2010). The *S. cerevisiae* α-mating factor pre-pro leader sequence has been engineered to give 180 fold increase in human IgG1 over wild type (Rakestraw et al. 2009). Vacuolar mis-sorting carried out by the *S. cerevisiae* Vps10p (vacuolar protein sorting receptor) promotes intracellular retention of proteins. Deletion of vps10 aids protein secretion in a few cases (Hong et al. 1996) whereas in other cases deletion of 5 other *vps* genes (*vps4*, 8, 13, 35 and 36) was seen to be beneficial (Zhang et al. 2001). Protein degradation was controlled in *S. cerevisiae*, *C. boidinii*, and *P. pastoris* by the deletion of the vacuolar protease genes *PEP4* and *PRB1* along with other protease genes like *CPY1*, *YPS1*, and *KEX2* (Kang et al. 2000; Komeda et al. 2002; Werten and de Wolf 2005). In *S. cerevisiae,* the deletion of the mitochondrial metalloendoprotease gene *CYM1* helped the efficient secretion of pro-cholecystokinin, growth hormone and pro-B-type natriuretic peptide with reduced proteolytic degradation (Jonson et al. 2004). These examples establish the importance of secretion engineering for recombinant protein production to aid downstream processing and design a cost-effective bioprocess. A detailed review of these strategies has been provided by Idiris et al. (2010).

8 Protein Glycosylation and Glycoengineered Strains

Therapeutically important proteins can be divided into the group of glycosylated and non-glycosylated molecules. In humans, the complex N-glycosylation occurs via a trimannose core ($Man_3GlcNAc_2$) that is extended using N-acetylglucosamine (GlcNAc), galactose and sialic acid. The *E. coli* expression system is commonly used to produce non-glycosylated proteins e.g. insulin which remains unglycosylated in its native state (Hamilton et al. 2003). However, glycosylation is indispensable for the proper folding, activity and pharmacokinetic stability of a vast majority of protein-based biotherapeutics (Mitra et al. 2006). IgG1 antibodies lacking glycosylation at Asn 297 residue in the Fc domain of the heavy chain cannot carry out antibody-dependent cell cytotoxicity (ADCC) due to the reduced affinity of its Fc domain for the receptor on the NK cells (Li et al. 2006). It is due to this reason that mammalian cell lines, especially CHO cells that can mimic human-like glycosylation have been extensively used for biotherapeutic protein production. Such cell lines come with their drawbacks of low volumetric productivity, long fermentation times, high media costs, product heterogeneity and viral contamination issues (Hamilton et al. 2003). Expression systems based on yeasts are useful to overcome all such bottlenecks for the production of protein therapeutics. Yeasts are capable of performing both O- and N-linked glycosylation. However, the yeast glycosylation machinery is of the high mannose type which imparts immunogenicity to the proteins thereby limiting their use (Hamilton et al. 2003). *S. cerevisiae* is known to hyperglycosylate N-linked sites (Gerngross 2004). To overcome such complications during recombinant protein production, yeasts have been engineered to produce 'humanized' glycoproteins. Yeast glycoengineering strategies focus on abolishing such hyper-mannosylation of recombinant proteins via the introduction of human-like sialylated complex glycan pattern (Gerngross 2004). A detailed mechanism of protein glycosylation has been reviewed by Hamilton and Gerngross (2007).

8.1 N-Glycoengineering

The humanization of yeast glycosylation pathway requires abolishing the native glycosylation reactions and replacing them with human glycosylation pathway components in the ER and Golgi (Gerngross 2004; Chiba and Jigami 2007). Furthermore, most therapeutic proteins require sialylation in the final step (Hamilton et al. 2006).

The outer chain elongation protein OCH1p (α-1,6-mannosyltransferase) is the key enzyme involved in yeast glycoengineering. In 1992, *S. cerevisiae och1* deficient mutants were isolated that were unable to elongate the mannose outer chains consequently producing secretory invertase having predominantly core-like oligosaccharides ($Man_{9-10}GlcNAc_2$) (Nagasu et al. 1992). Further progress in glycoengineering was made with the creation of a *S. cerevisiae* double mutant deficient for *OCH1* and *MNN1* (α-1,3-mannosyltransferase) genes that produced majorly the core $Man_8GlcNAc_2$ glycan structure (Nakanishi-Shindo et al. 1993). Later, Chiba and co-workers (1998) introduced the *Aspergillus saitoi* α-1,2-mannosidase gene carrying the C-terminal ER retention signal "HDEL" (His-Asp-Glu-Leu) into *S. cerevisiae* to produce successfully a $Man_5GlcNAc_2$ glycan structure. Using an identical strategy in *P. pastoris*, Callewaert and co-workers (2001) were able to produce influenza virus haemagglutinin and *Trypanosoma cruzi* trans-sialidase with an 85% reduction in α-1, 2-mannose. In 2003, the *P. pastoris* OCH1 deletion mutants were constructed that produced $Man_{8-12}NAc_2$ type glycans. The catalytic domain of heterologous α-1,2-mannosidase genes was introduced into the early secretory pathway of this mutant strain using ER or Golgi retention signals which was then able to produce Man_5NAc_2 type structure. In the next step, this glycan structure was converted to the human-like hybrid and complex glycan structure $GlcNAcMan_5GlcNAc_2$ by localization of the enzyme GnTI (human β-1,2-N-acetylglucosaminyltransferase I) (Choi et al. 2003). Further, Hamilton and co-workers (2003) introduced mannosidase II and GnTII (human β-1,2-N acetylglucosaminyltransferase II), thereby producing the complex glycoprotein with $GlcNAc_2Man_3GlcNAc_2$ glycan chain. Finally in 2006, Hamilton and co-workers developed a *Pichia* system that was able to produce complex sialylated humanized glycoproteins. In 2008, Jacobs and coworkers developed GlycoSwitch® vectors to produce recombinant proteins carrying $Gal_2GlcNAc_2Man_3GlcNAc_2$ type N-glycans in the *Pichia* expression system (Jacobs et al. 2008).

The glycoengineered strains of *H. polymorpha*, *Y. lipolytica* and *K. lactis* have also been developed where the *OCH1* gene disruption is the most common strategy (Kim et al. 2006; Song et al. 2007; Liu et al. 2009a). An *OCH*1 and *ALG*3 (dolichyl-phosphate-mannose dependent α-1,3-mannosyltransferase) double mutant *H. polymorpha* strain was able to secrete proteins with a $Man_3GlcNAc_2$ glycan structure (Oh et al. 2008). Strategies to introduce α-1,2-mannosidases and human GnTI have also been adopted in the yeasts *H. polymorpha and Y. lipolytica* to give rise to proteins carrying hybrid-type glycans (Kim et al. 2006; Oh et al. 2008; Cheon et al. 2012; De Pourcq et al. 2012b). Double mutants for α-1,6-mannosyltransferase and α-1,3-mannosyltransferase have been created in *K. lactis* and *Y. lipolytica* to control hyperglycosylation of recombinant proteins (Liu et al. 2009a; De Pourcq et al. 2012a). In 2011, a *Δoch1* and *Δmpo1* (mannosyl phosphorylation) double mutant in *Y. lipolytica* was constructed with impaired mannosyl phosphorylation and hyper mannosylation activity (Park et al. 2011). Recently, Moon and co-workers (2013) displayed fungal α-1,2-mannosidase on *Y. lipolytica* cell surface to convert $Man_8GlcNAc_2$ to $Man_5GlcNAc_2$.

8.2 O-Glycoengineering

Yeast-type O-glycans can induce immunogenic reactions by interacting with human mannose binding lectins thereby reducing the pharmacokinetic properties of the biotherapeutic molecules (Cukan et al. 2012). In *P. pastoris,* the transfer of mannose residues from dolichol-phospho-mannose to a target protein is the preliminary step in the synthesis of O-linked glycans which is mediated by protein-*O*-mannosyltransferase (*PMT*). In 2013, Nett and co-workers reported 5 *PMT* genes from *P. pastoris*. Through gene knockouts and the use of PMT inhibitors, this research group was able to reduce the degree of O-mannosylation as well as the length of the glycan chain of recombinant glycoproteins. Use of chemical inhibitors for PMT has been shown to reduce O-mannosylation (Orchard et al. 2004; Argyros et al. 2013). The co-expression of α-1,2-mannosidase and protein-*O*-linked-mannose β-1,2-*N*-acetylglucosaminyltransferase I was studied to produce glycans with an N-acetylglucosamine cap that can be extended to produce human-like sialylated glycans (Hamilton et al. 2013). Amano and co-workers (2008) reported the engineering of mucin-type human O-glycosylation in *S. cerevisiae*. In the same year, a mammalian O-glycosylation pathway was engineered in *S. cerevisiae* to produce proteins requiring *O*-fucosylation for their activity (Chigira et al. 2008; Okajima et al. 2008).

9 Yeast Cell Surface Display

The yeast surface display technology has various applications in vaccine and antibody development, screening of libraries, bioremediation, bioconversions and biosorptions (Lee et al. 2003). In this strategy, the target protein is displayed on the cell surface using an anchor protein and a signal sequence. The selection of the anchor protein is of utmost importance and varies with the ultimate application of the display system. Many cell wall proteins have been used as anchor motifs for cell surface display of heterologous proteins. Three major types of anchoring proteins (CWPs) exist in yeast: the GPI-CWPs (glycophosphatidylinositol), the PIR-CWPs (proteins with internal repeats) and the FL/FS proteins (Flo1p system responsible for yeast flocculation) (Tanaka et al. 2012). The commonly used C- and N-terminus fusion strategies have been reviewed elsewhere (Fig. 1) (Tanaka et al. 2012). The most common anchor protein for N-terminal display is *S. cerevisiae* α-agglutinin possessing a GPI attachment signal. Van der Vaart and co-workers (1997) used several cell wall proteins (Cwp1p, Cwp2p, Agα1p, Tip1p, Flo1p, Sed1p, YCR89w,

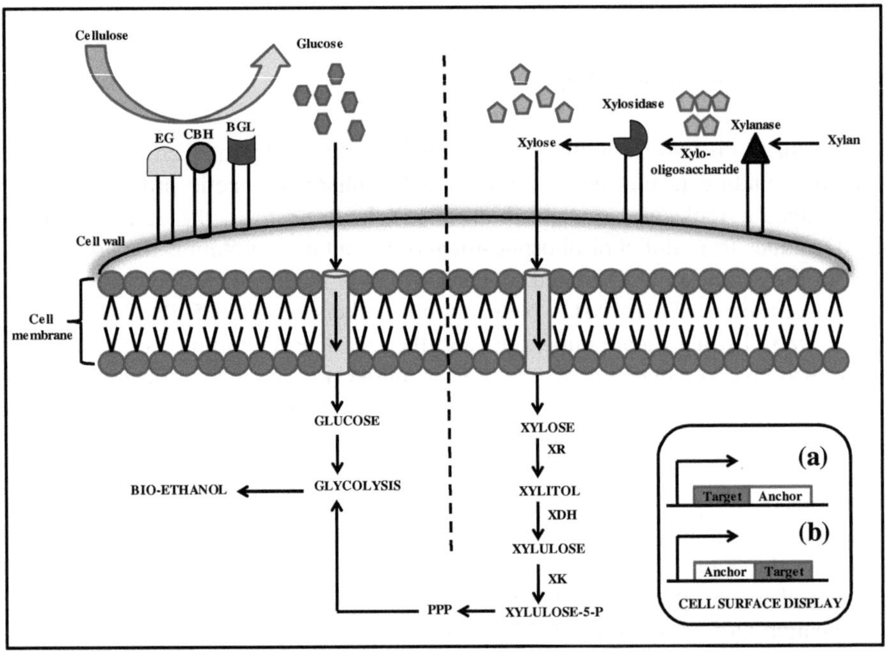

Fig. 1 Figure depicting a recombinant *S. cerevisiae* displaying cellulolytic (*EG* Endoglucanase, *CBH* Cellobiohydrolase and *BGL* Beta-glucosidase) and hemicellulolytic enzymes (Xylanase and Xylosidase) on its surface for bioethanol production. Xylose is connected to the glycolytic pathway via pentose phosphate pathway (PPP) by introduction of enzymes xylose reductase (XR), xylitol dehydrogenase (XDH) and xylulokinase (XK). Inset shows the **a** N-terminal and **b** C-terminal type of anchor fusions for cell surface display

and Tir1p) to display α-galactosidase on *S. cerevisiae* surface of which Cwp2p, Agα1p, and Sed1p showed the highest surface localization capacity.

N-terminal fusions have been used for bio-ethanol production using *S. cerevisiae* strains displaying α-amylase and glucoamylase (Inokuma et al. 2015). Co-display of heterologous xylanolytic enzymes (xylanase and β-xylosidase) on *S. cerevisiae* helped in direct fermentation of xylan to ethanol (Katahira et al. 2004, 2006; Sakamoto et al. 2012). Nakamura and co-workers (2008) demonstrated co-fermentation of xylose and cellobiose by β-glucosidase displaying yeast. This system allows control of the extracellular glucose levels thereby bypassing catabolite repression. The most widely used anchor protein for C-terminal fusions is Aga2 while others include the Pir proteins and the Flo1p system (Tanaka et al. 2012). To improve saccharification and ethanol yield from lignocellulosic biomass, *S. cerevisiae* displaying a functional minicellulosome system was developed using Aga2p (Wen et al. 2010; Tsai et al. 2010).

There are several reports on the construction of N-terminal displays for heavy metal recovery using α-agglutinin as an anchor. Tandem repeats of a yeast metallothionein were displayed on *S. cerevisiae* surface for adsorption of cadmium (Kuroda and Ueda 2006). Similarly, short metal binding NP peptides were efficiently displayed on the *S. cerevisiae* surface using α-agglutinin to enhance affinity and selectivity for Pb^{2+} (Kotrba and Ruml 2010). An α-agglutinin based *S. cerevisiae* display system was created using the *E. coli* ModE protein for selective adsorption of molybdate that could be recovered by papain treatment (Nishitani et al. 2010). The development of *S. cerevisiae* whole cell catalysts include those for fatty acid production from butter using *A. oryzae* cutinase displayed using α-agglutinin (Horii et al. 2010), for enhanced glutathione production from starch using display of amylase on glutathione producing *S. cerevisiae* (Yoshida et al. 2011), for oral vaccine development using display of antigen 380R (Tamaru et al. 2006) and for improving thermostability of *Candida antarctica* lipase B (Kato et al. 2007).

Recently *Pichia* cell surface display system has been developed for some industrially important enzymes and therapeutic proteins (Tanaka et al. 2012). Mergler and co-workers (2004) used *S. cerevisiae* α-agglutinin to display *K. lactis* yellow enzyme (KYE) on *P. pastoris* for its use in bioremediation. The same anchor was used by Wang and co-workers (2007) to display EGFP on the surface of *P. pastoris* cells. The use of *S. cerevisiae* Pir1 protein for *Pichia* cell surface display has also been reported (Wang et al. 2008). Su and co-workers used α-agglutinin and *S. cerevisiae* Sed1 to display *C. antarctica* lipase B on the surface of *P. pastoris* cells (Su et al. 2010a, b). The GPI type TIP1p protein was used in *P. pastoris* to display biologically active human lactoferrin (Jo et al. 2011). In 2011, *P. pastoris* PIR proteins were isolated and used as anchor proteins for EGFP surface display (Khasa et al. 2011). A glycoengineered *P. pastoris* strain capable of producing mammalian type $Man_5GlcNAc_2$ N-linked glycans was used for the surface display of antigen binding (Fab) fragment (Lin et al. 2012). Several other proteins such as *Candia antarctica* lipase B (Liang et al. 2013), monoclonal antibody (Shaheen et al. 2013) and *Bombyx mori* acetylcholinesterase (Dong et al. 2013) were successfully expressed on *Pichia* surface.

In 2002, Kim et al. created a *H. polymorpha* display system using four cell wall proteins (*SED1*, *GAS1*, *TIP1* and *CWP1*) encoding *H. polymorpha* glycosylphosphatidylinositol (GPI)-anchors of which Cwp1p showed the highest anchoring efficiency for *A. niger* glucose oxidase and *B. subtilis* CMCase. *Candida antarctica* lipase B (CalB) displayed on the cell surface of *H. polymorpha* by fusion to a cell wall anchor motif (CwpF) has been functionally improved for its catalytic activity (Kim et al. 2007). The C-terminal domain *of Y. lipolytica* Cwp1p was used to create a GPI-based anchoring system for *Y. lipolytica* (Jaafar and Zueco 2004; Yue et al. 2008). Using this anchor system, the *Aureobasidium pullulans* alkaline protease and *Vibrio* sp. alginate lyase have been displayed on *Y. lipolytica* cell surface (Ni et al. 2009; Liu et al. 2009b). Yuzbasheva and co-workers (2011) displayed lipase on *Y. lipolytica* cell surface using the *S. cerevisiae* Flo1p homolog YALI0C09031p. In 2010, Yu and co-workers developed a *Y. lipolytica* system for the surface display of *Saccharomycopsis fibuligera* acid protease for clotting of skimmed milk, to develop a rennet substitute. Furthermore, Pir proteins have also been used in *Y. lipolytica* to develop cell surface display systems (Duquesne et al. 2014).

10 Metabolic Engineering and Systems Biology

The understanding of the metabolic reactions' network occurring in a cell has led to the development of various systems biology approaches such as transcriptomics (Lashkari et al. 1997); proteomics (Zhu et al. 2001); metabolomics (Villas-Boas et al. 2005; Jewett et al. 2006), fluxomics (Sauer 2006); interactomics (Lee et al. 2002) and locasomics (Huh et al. 2003) etc. Metabolic models help in predicting the outcome of several cellular processes and thereby aid in strain optimization to improve production by altering existing pathways. Knowledge of global regulation phenomena can help in the elimination of various bottlenecks such as product yield, productivity and stress tolerance (Takors et al. 2007; Mukhopadhyay et al. 2008).

Several reports are available for the metabolic engineering of *S. cerevisiae* for biofuel production. Glycerol, the major byproduct formed during ethanol fermentation is essential for reoxidizing NADH generated during the reaction. In 2010, the *S. cerevisiae* host was engineered to bypass glycerol production by expressing the *E. coli mhpF* gene (acetylating NAD-dependent acetaldehyde dehydrogenase) and deleting *GPD1* and *GPD2* genes encoding NAD-dependent glycerol-3-phosphate dehydrogenase (Medina et al. 2010). Using this approach, they were able to reoxidize NADH by converting acetate, a major inhibitor during lignocellulose hydrolysis, to ethanol. Similarly, the redox metabolism in *S. cerevisiae* was engineered using an insilico genome-scale modeling to reduce glycerol production and increase ethanol production on glucose (Bro et al. 2006). Due to the absence of a xylose assimilation pathway, the traditional *S. cerevisiae* cannot ferment xylose. However, it can metabolize xylulose using the pentose phosphate pathway. Efficient xylose utilization is essential for producing biofuels from lignocellulosic biomass (Hahn-Hagerdal et al. 2006). Therefore, xylose metabolic pathways from pentose-fermenting microbes were engineered into *S. cerevisiae* via cloning of xylose isomerase (XI), xylose reductase (XR), xylitol dehydrogenase (XDH) and xylulokinase (XK) to connect xylose assimilation to the endogenous pentose phosphate pathway of *S. cerevisiae* for glucose and xylose co-fermentation (Fig. 1) (Fernandes and Murray 2010). Furthermore, improved industrial strains have also been developed for efficient xylose fermentation in the presence of inhibitors like phenolics and organic acids from cellulosic hydrolysate under low oxygen conditions (Liu 2011; Parawira and Tekere 2011; Sanda et al. 2011; Cai et al. 2012). *S. cerevisiae* was metabolically engineered for fatty acid synthesis, by a gene disruption strategy along with over-expression of a heterologous ATP-citrate lyase, as precursors for the production of biofuels (Tang et al. 2013).

The thermophilic and methylotrophic yeast *Hansenula polymorpha* is also used for simultaneous saccharification and fermentation of xylose at high temperatures. Overexpression of the xylose utilization enzymes like xylose reductase, xylitol dehydrogenase and xylulokinase and a classical selection approach (resistance to the inhibitor of glycolysis, 3-bromopyruvate) in *H. polymorpha* improved ethanol production from 0.6 to 9.8 g/L at 45 °C (Kurylenko et al. 2014). Ishchuk and co-workers (2010) expressed the *S. cerevisiae MPR1* gene (N-acetyltransferase) under the control of the constitutive *GAP* (glyceraldehyde-3-phosphate dehydrogenase) promoter of *H. polymorpha* to produce a strain with improved ethanol tolerance.

Metabolic engineering has allowed the production of novel chemicals, pharmaceuticals and food ingredients using yeasts. Up to 100 mg/L of the antimalarial drug precursor artemisinic acid was produced by engineering the artemisinic acid biosynthetic pathway into *S. cerevisiae* (Ro et al. 2006). In 2011, Wang and co-workers enhanced resveratrol biosynthesis in engineered *S. cerevisiae* by mutating the tyrosine ammonia lyase (*TAL*) gene and overcoming codon bias. Metabolic engineering of *K. lactis* was done to express genes from *A. thaliana* L-galactose pathway to develop a strain producing L-ascorbic acid (Rosa et al. 2013). The introduction of the *Mus musculus* ATP-citrate lyase gene and *S. cerevisiae* acetyl-CoA synthetase gene enhanced the α-ketoglutarate production in *Y. lipolytica* (Zhou et al. 2012). Deletion of *POX3* and introduction of the *POX2* gene (encoding acyl-CoA oxidases having diverse substrate specificities) helped in increasing production of γ-decalactone, a food additive, in *Y. lipolytica* (Guo et al. 2012). A metabolic pathway for lycopene production was also recently engineered in *Y. lipolytica* (Matthaus et al. 2014). *P. pastoris* was metabolically engineered to produce 73.9 mg/L lycopene by the introduction of lycopene pathway enzymes (Bhataya et al. 2009). *Y. lipolytica* cells were able to accumulate ricinoleic acid up to 43% of total lipid content and over 60 mg/g of dry cell weight using a metabolic engineering strategy (Beopoulos et al. 2014). Eilert and co-workers (2013) engineered *H. polymorpha* to secrete 2.8 g/L recombinant 5-hydroxyectoine by inserting 4 genes (*EctA*, *EctB*, *EctC*, and *EctD*) coding for the enzymes of the 5-hydroxyectoine biosynthesis pathway of *Halomonas elongata* into its genome.

Proteomic and transcriptomic approaches have been used to analyze cell stress due to foreign protein expression. Such techniques can be utilized to identify and alleviate bottlenecks to improve protein yields. Proteomic analysis of a chymosin-producing *K. lactis* strain showed stress due to the upregulation of 2 additional proteins, i.e., Hsp26p (heat shock protein) and Sod2p (superoxide dismutase) (van Ooyen et al. 2006). Bonander and co-workers (2009) used transcriptome analysis data to identify genes for optimal membrane protein expression in *S. cerevisiae*. Using this approach, they determined the importance of optimizing *BMS1* transcript levels (involved in ribosome biogenesis) to improve recombinant protein yields.

Researchers have also used the random mutagenesis approach to modify pathways by mutating the existing regulatory proteins and enzymes. This approach, termed as evolutionary engineering has been used for selecting *S. cerevisiae* strains able to grow anaerobically on xylose (Sonderegger and Sauer 2003). Random mutagenesis of an *S. cerevisiae* strain led to the identification of 4 genes regulating the ATPase activity of *KAR2* (hsp70). This genotype was then used for enhancing production of recombinant human albumin, GM-CSF, and transferrin (Payne et al. 2008). Demeke and co-workers (2013) created an industrial strain of *S. cerevisiae* by metabolic and evolutionary engineering (random mutagenesis) that contains 13 genes including enzymes of the pentose phosphate pathway, xylose and arabinose utilization pathways for bioethanol production from lignocellulose. Theerachat and co-workers (2012) report efficient secretion of *Trametes versicolor* laccase from *Y. lipolytica* using the strong constitutive p*TEF* promoter and the native laccase *lcc1* signal peptide combined with directed evolution approach to isolate hyper-producer mutants having four-fold enhanced laccase activity. A mutation in the *PHO13* gene of a xylose-fermenting *S. cerevisiae* strain helped in the improvement of its xylose-fermenting capabilities with a shorter lag time (Kim et al. 2013). Recently, systems biology approaches have helped in the production of several molecules e.g. isoprenoids (Shiba et al. 2007), amorphadiene (Westfall et al. 2012), terpenoids (Dai et al. 2012), vanillin (Brochado and Patil 2013) and opiates (Thodey et al. 2014) indicating the potential of such approaches in revolutionizing yeast biotechnology outcomes.

11 Large Scale Process Development

The bioprocess optimization at fermentor level is necessary for recombinant protein production because the yields correlate principally with the high cell densities (Fickers 2014; Spohner et al. 2015). Bioreactors provide tight control over cultivation conditions for the development of effective feeding and bioprocess strategies (Macauley-Patrick et al. 2005). The efficient and accurate online monitoring of bioprocess parameters such as DO, pH, temperature, methanol, biomass, product and by-product concentration are vital for successful optimization studies. The cultivation parameters may differ with expression host, genotype, and properties of the target protein.

S. cerevisiae is crabtree positive yeast and shows fermentative metabolism in the absence of oxygen or excess of carbon. Hence, most bioprocesses are conducted under glucose limiting conditions to prevent the production of undesired or toxic metabolites (Gorlani et al. 2012). Ferndahl and co-workers (2010) used an engineered strain of *S. cerevisiae* that could carry out aerobic respiration even in the presence of high concentration of glucose to increase recombinant product yields. Shang and co-workers (2006) reported the high cell density fermentation (120 g/L) of *S. cerevisiae* via controlling glucose and ethanol concentration to improve ergosterol yield to 1500 mg/L. Similarly, the mutant *S. cerevisiae* GE-2 strain was used for the simultaneous production of glutathione 2280 mg/L and ergosterol 1510 mg/L at a high cell density of 110 g/L (Shang et al. 2008). The co-expression of cellobiohydrolases (CBH) was optimized in high cell density fermentations where 0.3 g/L of CBH1 and approximately 1 g/L of CBH2 was produced extracellularly (Ilmen et al. 2011). The enhanced cysteine incorporation yield for glutathione was optimized in a repeated fed-batch of *S. cerevisiae* where cysteine feed was regulated via the respiratory quotient (RQ) to obtain a cysteine incorporation yield of 0.40 mol/mol and GSH concentration of 1304.7 mg/L at a cell biomass of 84 g/L (Lorenz et al. 2015). The inducer/repressor-free feeding approach was optimized by the sequential control of biosynthetic pathways for carotenoid production (1156 mg/L) (Xie et al. 2015). The cGMP manufacturing of an animal-free recombinant transferrin was reported where co-expression of protein disulfide isomerase (PDI) increased its secretion by 12-folds (Finnis et al. 2010).

In 2009, Stockmann and co-workers described process development in *H. polymorpha*. The *FMD* promoter (formate dehydrogenase) and the *MOX* promoter of *H. polymorpha* are preferred for industrial use (Gellissen 2000; Hartner and Glieder 2006). Gram level secretory production of hirudin was carried out under the control of the *H. polymorpha MOX* promoter using the *S. cerevisiae* α-mating factor leader sequence (Weydemann et al. 1995). In 1996, Zurek and co-workers produced 350 mg/L of aprotinin under the control of the *MOX* promoter and the α-mating factor pre-pro secretion signal in *H. polymorpha* using a pH/pO_2 controlled feeding strategy. Using the same production approach, 350 mg/L of interferonα-2a was achieved (Muller et al. 2002). Methanol based production has been carried out for HBsAg (Hepatitis B surface antigen) under the control of *FMD* or *MOX* promoter (Gellissen et al. 2005) in 50 L bioreactor. At pilot scale, the use of glucose as a carbon source has improved the phytase production to 13.5 g/L under the control of *FMD* promoter in *H. polymorpha* (Mayer et al. 1999). The limitation of the secondary substrate (phosphate) during batch fermentation directed the metabolic flux toward a 1.87 fold higher GFP production under the control of the *FMD* promoter (Kottmeier et al. 2010). In an engineered strain of *H. polymorpha*, the glucose fed-batch cultivation resulted in a glutathione product yield of 2300 mg/L whereas it was only 250 mg/L in case of methanol-induced culture (Ubiyvovk et al. 2011). In glycerol-limited conditions, 697 mg/L of gamma-linolenic acid production was obtained after optimizing growth conditions in high cell density fed-batch fermentation (Khongto et al. 2010). Youn and co-workers (2010) reported a high-level production of human serum albumin at a level of 5.1 g/L when pure oxygen was supplemented in a 5L bioreactor. The high cell density cultivation conditions were optimized for the expression of *Candida utilis* uricase in *H. polymorpha* where a change in pH from 5.5 to 6.5 during induction phase resulted in 2.1 g/L of its secretory expression (Chen et al. 2008). Recently, the expression of rotavirus VP6 protein was optimized in different expression host such as *E. coli*, *P. pastoris,* and *H. polymorpha*, where, *H. polymorpha* outperformed other systems at bioreactor level (Bredell et al. 2016).

K. lactis batch fermentations performed at pH 6.0 enhanced biomass and *Arxula adeninivorans* glucoamylase production along with its stability (Merico et al. 2004). The creation of hypoxic physiological conditions in *K. lactis* was optimized to produce 180 mg/L of interleukin-1β under the control of pyruvate decarboxylase (*PDC1*) promoter (Camattari et al. 2007). In another fed-batch approach, the use of *GAL7* promoter and *K. lactis* killer toxin signal peptide improved human lysozyme expression from 64.1 U/mL in a flask to 123.6 U/mL in a bioreactor at an optimal temperature of 25 °C and pH near neutrality (Huang and Demirci 2009). The addition of metal ions such as Zn^{2+} and Cu^{2+} to growth medium enhanced the superoxide dismutase activity in *K. lactis* fed-batch culture (Raimondi et al. 2010).

The production of α-ketoglutaric acid was optimized in *Y. lipolytica* (WSH-Z06) fed-batch culture, where pH controlled production phase at 3.0, improved its yields to 66.2 g/L in glycerol medium (Yu et al. 2012). The growth and product formation parameters for citric acid synthesis were optimized in the continuous cultivation of *Y. lipolytica.* A relatively low oxygen with high iron concentration resulted in 120 g/L of citric acid in batch fermentation (Kamzolova et al. 2003). The glycerol-containing waste of biodiesel industry was used to optimize citric acid production in batch, repeated batch, and cell-recycling processes where a highest product concentration of 124.2 g/L was reported (Rymowicz et al. 2010). The multicopy integration of *Rhodotorula araucariae* epoxide hydrolase under hp4d promoter resulted in a product concentration of 1,750 U/g DCW in fed-batch fermentation (Maharajh et al. 2008). Gasmi and co-workers (2011) described human interferon α2b production in *Y. lipolytica* under the control of the oleic acid (OA)-inducible promoter POX2 where 425 mg/L of recombinant protein was produced by continuous feeding of oleic acid. In another approach, the γ-decalactone production was optimized in fed-batch cultivation with a high product yield of 6.8 g/L (Gomes et al. 2012). The simultaneous production of erythritol (80 g/L) and mannitol (27.6 g/L) was also reported using *Y. lipolytica* system where NaCl concentration controlled their production ratios (Tomaszewska et al. 2012). Recently, a novel osmotic pressure control fed-batch strategy was used for *Y. lipolytica* CICC 1675 strain where a maximum erythritol yield of 194.3 g/L was obtained with a lower level of mannitol (Yang et al. 2014). The flux balance analysis approach in fed-batch fermentation was done to improve lycopene production in *Y. lipolytica* where under optimized conditions, a maximum of 242 mg/L of the product was obtained (Nambou et al. 2015). The campesterol accumulation in *Y. lipolytica* was optimized to a level of 453 mg/L using substrate controlled fed-batch fermentation approach (Du et al. 2016).

The bioprocess optimization strategies using *AOX1* promoter in *Pichia* system is divided into two phases, i.e., the biomass accumulation using suitable carbon and nitrogen source, followed by methanol induction (Potvin et al. 2012; Looser et al. 2015). The maximum specific growth rate (μ_{max}) on glucose or glycerol medium varies approximately in the range of 0.16 h^{-1} to 0.29 h^{-1} (Looser et al. 2015). In pre-induction phase, the methanol and glycerol mixed feeding strategies are employed to acclimatize the cells and also to derepress the *AOX1* promoter. Finally, in the induction phase, the methanol is fed in fed-batch mode at a specific growth rate that varies with expression strains or with heterologous protein (Potvin et al. 2012; Looser et al. 2015). Many monoclonal antibodies have been expressed in glycoengineered *P. pastoris* with improved anti-tumor activity (Gong et al. 2013; Gomathinayagam et al. 2015). High cell density cultivation (HCDC) is one of the most frequently used strategies for enhancing the production levels. The HCDC of recombinant *Pichia* having IL-6 gene resulted in a product concentration of 280 mg/L and a total wet cell weight of 470 g/L (Li et al. 2011a). Zhou and co-workers (2014) reported the production of human lysozyme-like 6 (LYZL6) protein in SMD1168 *Pichia* strain with an extracellular lysozyme activity of 2,340 U/mL at a dry cell weight of 116.3 g/L in a 30L bioreactor.

To avoid the methanol toxicity, sorbitol was used as a non-repressing carbon source for *AOX1* promoter (Ramon et al. 2007). The main advantage of mixed feeding of methanol and sorbitol is a reduced oxygen demand without affecting recombinant protein productivity. Jungo and co-workers (2007) reported the use of sorbitol and methanol mix feeding in fed-batch strategy for the production of avidin. The expression of recombinant human growth hormone was improved to the level of 0.64 g/L by co-feeding of sorbitol and methanol in induction phase at a controlled specific growth rate of 0.03 h^{-1} (Calik et al. 2013). Similarly, an enhanced production of *Thermomyces lanuginosus* lipase from 19,500 to 27,000 U/mL was attained when the co-feeding strategy of sorbitol and methanol was implemented in a 3 L fermenter (Fang et al. 2014). The use of mannitol as a co-substrate with methanol enhanced rhEPO production to the level of 0.65 g/L and also reduced the fermentation time due to its fast utilization (Eskitoros and Calik 2014). The production of Fc-fused kringle domain was maximum at the level of 635 mg/L with high productivity (7.2 mg/L/h), under the optimized conditions of fermentation at 5 L scale (Jeong et al. 2014). The fed-batch optimization of an ice-binding protein (rLeIBP) from *Leucosporidium sp.* at pilot-scale (700 L) resulted in a secreted protein concentration of ~300 mg/L (Lee et al. 2013). The glycoengineered GlycoSwitch-Man5 *P. pastoris* strain was used to produce hGM-CSF (~760 mg/L) with a uniform $Man_5GlcNAc_2$ N-glycosylation pattern at a reduced specific growth rate of 0.015 h^{-1} (25% of maximum). The increase in specific growth rate in production phase drastically reduced its production yields (Jacobs et al. 2010). In another approach, the glycoengineered *Pichia* host strain (YGLY8323) capable of producing humanized glycoprotein with terminal galactose was used for monoclonal antibody production up to 1.6 g/L. The process was scaled up from 30 to 1200 L without compromising productivity, and product quality (Ye et al. 2011).

To totally obviate the need of methanol for induction, a strong constitutive promoter like *GAP* is used as an effective alternate. Using this promoter several proteins such as hGM-CSF (250 mg/L), *Candida rugosa* lipase (~14,000 IU/mL) and human angiostatin (176 mg/L) have been successfully expressed (Pal et al. 2006; Zhang et al. 2007; Zhao et al. 2008). Recently, the utilization of *GAP* promoter for bioprocess optimization of various recombinant proteins in *Pichia* system has been reviewed (Calik et al. 2015).

12 Conclusions

Yeasts are attractive hosts for the production of eukaryotic proteins of human origin due to their numerous advantages like ease of genetic manipulation, short generation time and ability to grow in relatively inexpensive media. The advances in recombinant DNA technology, proteomics, genomics, metabolomics, fluxomics and systems biology have helped to produce molecules with desired properties. The use of glycoengineered hosts in combination with optimized fermentation strategies have resulted in optimized industrial processes with reduced production costs. The availability of the versatile yeast cell surface display technology in combination with the powerful metabolic engineering and systems biology tools will help in the development of engineered yeasts with desired properties to broaden their usage as expression platforms for many recombinant therapeutics and value added products.

References

Adivitiya, Dagar, V.K., Devi, N. and Khasa, Y.P. 2016. *Int. J. Biol. Macromol.* 83: 50–60.

Adrio, J.L. and Demain, A.L. 2014. *Biomolecules* 4: 117–139.

Agaphonov, M., Poznyakovski, A.I., Bogdanova, A.I. and Ter-Avanesyan, M.D. 1994. *Yeast* 10: 509–513.

Ahmad, M., Hirz, M., Pichler, H. and Schwab, H. 2014. *Appl. Microbiol. Biotechnol.* 98: 5301–5317.

Amano, K., Chiba, Y., Kasahara, Y., Kato, Y. *et al.* 2008. *Proc. Natl. Acad. Sci. U S A* 105: 3232–3237.

Argyros, R., Nelson, S., Kull, A., Chen, M.T., Stadheim, T.A. and Jiang, B. 2013. *PLoS One* 8: e62229.

Avila, J., González, C., Brito, N. and Siverio, J.M. 1998. *Biochem. J. 335*: 647–652.

Baerends, R.J., Salomons, F.A., Faber, K.N., Kiel, J.A.K.W., Van Der Klei, I.J. and Veenhuis, M. 1997. *Yeast* 13: 1437–1448.

Bao, W.G. and Fukuhara, H. 2001. *Gene* 272: 103–110.

Barth, G. and Gaillardin, C. 1997. *FEMS Microbiol. Rev.* 19: 219–237.

Bartkeviciute, D. and Sasnauskas, K. 2003. *Yeast* 20: 1–11.

Bartkeviciute, D. and Sasnauskas, K. 2004. *FEMS Yeast Res.* 4: 833–840.

Beopoulos, A., Verbeke, J., Bordes, F., Guicherd, M., Bressy, M., Marty, A. and Nicaud, J.M. 2014. *Appl. Microbiol. Biotechnol.* 98: 251–262.

Bergkamp, R.J., Kool, I.M., Geerse, R.H. and Planta, R.J. 1992. *Curr. Genet.* 21: 365–370.

Bhataya, A., Schmidt-Dannert, C. and Lee, P.C. 2009. *Process Biochem.* 44: 1095–1102.
Blazeck, J., Garg, R., Reed, B. and Alper, H.S. 2012. *Biotechnol. Bioeng.* 109: 2884–2895.
Blazeck, J., Reed, B., Garg, R., Gerstner, R., Pan, A., Agarwala, V. and Alper, H.S. 2013. *Appl. Microbiol. Biotechnol.* 97: 3037–3052.
Bogdanova, A.I., Agaphonov, M.O. and Ter-Avanesyan, M.D. 1995. *Yeast* 11: 343–353.
Bonander, N., Darby, R.A., Grgic, L., Bora, N., Wen, J., Brogna, S., Poyner, D.R., O'Neill, M.A. and Bill, R.M. 2009. *Microb. Cell Fact.* 8: 10.
Bredell, H., Smith, J.J., Prins, W.A., Görgens, J.F. and van Zyl, W.H. 2016. *FEMS Yeast Res.* 14.
Bro, C., Regenberg, B., Förster, J. and Nielsen, J. 2006. *Metab. Eng.* 8: 102–111.
Brochado, A.R. and Patil, K.R. 2013. *Biotechnol. Bioeng.* 110: 656–659.
Buckholz, R.G. and Gleeson, M.A. 1991. *Biotechnology (N Y)* 9: 1067–1072.
Bundock, P., Mroczek, K., Winkler, A.A., Steensma, H.Y. and Hooykaas, P.J. 1999. *Mol. Gen. Genet.* 261: 115–121.
Bussey, H. and Meaden, P. 1985. *Curr. Genet.* 9: 285–291.
Cai, Z., Zhang, B. and Li, Y. 2012. *Biotechnol. J.* 7: 34–46.
Calık, P., Ata, O., Gunes, H., Massahi, A., Boy, E., Keskin, A., Ozturk, S., Zerze, G.H. and Ozdamar, T.H. 2015. *Biochem. Eng. J.* 95: 20–36.
Calık, P., Bozkurt, B., Zerze, G.H., Inankur, B., Bayraktar, E., Boy, E., Orman, M.A., Acik, E. and Ozdamar, T.H. 2013. *J. Chem. Technol. Biotechnol.* 88: 1631–1640.
Callewaert, N., Laroy, W., Cadirgi, H., Geysens, S., Saelens, X., Jou, W.M. and Contreras, R. 2001. *FEBS Lett.* 503: 173–178.
Camattari, A., Bianchi, M.M., Branduardi, P., Porro, D. and Brambilla, L. 2007. *Appl. Environ. Microbiol.* 73: 922–929.
Celik, E. and Calik, P. 2012. *Biotechnol. Adv.* 30: 1108–1118.
Cereghino, J.L. and Cregg, J.M. 2000. *FEMS Microbiol. Rev.* 24: 45–66.
Chee, M.K. and Haase, S.B. 2012. *G3 (Bethesda).* 2: 515–526.
Chen, Z., Wang, Z., He, X., Guo, X., Li, W. and Zhang, B. 2008. *Appl. Microbiol. Biotechnol.* 79: 545–554.
Cheon, S.A., Kim, H., Oh, D.B., Kwon, O. and Kang, H.A., 2012. *J. Microbiol.* 50: 341–348.
Chiba, Y. and Jigami, Y. 2007. *Curr. Opin. Chem. Biol.* 11: 670–676.
Chiba, Y., Suzuki, M., Yoshida, S., Yoshida, A., Ikenaga, H., Takeuchi, M., Jigami, Y. and Ichishima, E. 1998. *J. Biol. Chem.* 273: 26298–26304.
Chigira, Y., Oka, T., Okajima, T. and Jigami, Y. 2008. *Glycobiology* 18: 303–314.
Choi, B.K., Bobrowicz, P., Davidson, R.C., Hamilton, S.R., Kung, D.H., Li, H., Miele, R.G., Nett, J.H., Wildt, S. and Gerngross, T.U. 2003. *Proc. Natl. Acad. Sci. U S A* 100: 5022–5027.
Colussi, P.A. and Taron, C.H. 2005. *Appl. Environ. Microbiol.* 71: 7092–7098.
Cox, H., Mead, D., Sudbery, P., Eland, R.M., Mannazzu, I. and Evans, L. 2000. *Yeast* 16: 1191–1203.
Crawford, K., Zaror, I., Bishop, R.J. and Innis, M.A. 2003. *US Patent* US6107057.
Cukan, M.C., Hopkins, D., Burnina, I., Button, M. *et al.* 2012. *J. Immunol. Methods* 386: 34–42.
Da Silva, N.A. and Srikrishnan, S. 2012. *FEMS Yeast Res.* 12: 197–214.
Dai, Z., Liu, Y., Huang, L. and Zhang, X. 2012. *Biotechnol. Bioeng.* 109: 2845–2853.
De Pourcq, K., Tiels, P., Van Hecke, A., Geysens, S., Vervecken, W. and Callewaert, N. 2012b. *PLoS One* 7: e39976.
De Pourcq, K., Vervecken, W., Dewerte, I., Valevska, A., Van Hecke, A. and Callewaert, N. 2012a. *Microb. Cell Fact.* 11: 53.
Demain, A.L. and Vaishnav, P. 2009. *Biotechnol. Adv.* 27: 297–306.
Demeke, M.M., Dietz, H., Li, Y., Foulquié-Moreno, M.R., Mutturi, S. *et al.* 2013. *Biotechnol. Biofuels* 6: 89.
Dong, J.X., Xie, X., He, Y.S., Beier, R.C. *et al.* 2013. *PLoS One* 8: e70451.
Donnini, C., Farina, F., Neglia, B., Compagno, M.C. *et al.* 2004. *Appl. Environ. Microbiol.* 70: 2632–2638.
Du, H.X., Xiao, W.H., Wang, Y., Zhou, X., Zhang, Y., Liu, D. and Yuan, Y.J. 2016. *PLoS One* 11: e0146773.

Duquesne, S., Bozonnet, S., Bordes, F., Dumon, C., Nicaud, J.M. and Marty, A. 2014. *PLoS One* 9: e95128.
Edmond, S., Montanier, C., Nicaud, J.M., Marty, A., Monsan, P., André, I. and Remaud-Siméon, M. 2010. *Appl. Environ. Microbiol.* 76: 2684–2687.
Egli, T., Van Dijken, J.P., Veenhuis, M., Harder, W. and Fiechter, A. 1980. *Arch. Microbiol.* 124: 115–121.
Eiden-Plach, A,, Zagorc, T., Heintel, T., Carius, Y., Frank, B. and Manfred, J.S. 2004. *Appl. Envirol. Microbiol.* 70: 961–966.
Eilert, E., Kranz, A., Hollenberg, C.P., Piontek, M. and Suckow, M. 2013. *J. Biotechnol.* 167: 85–93.
Erhart, E. and Hollenberg, C.P. 1983. *J. Bacteriol.* 156: 625–635.
Eskitoros, S and Calik, P. 2014. *J. Chem. Technol. Biotechnol.* 89: 644–651.
Faber, K.N., Haima, P., Harder, W., Veenhuis, M. and Geert, A.B. 1994. *Curr. Genet.* 25: 305–310.
Faber, K.N., Westra, S., Waterham, H.R., Keizer-Gunnink, I., Harder, W., Ab, G. and Veenhuis, M. 1996. *Appl. Microbiol. Biotechnol.* 45: 72–79.
Fang, F., Salmon, K., Shen, M.W., Aeling, K.A., Ito, E., Irwin, B., Tran, U.P.C., Hatfield, G., Da Silva, N.A. and Sandmeyer, S. 2011. *Yeast* 28: 123–136.
Fang, Z., Xu, L., Pan, D., Jiao, L., Liu, Z. and Yan, Y. 2014. *J. Ind. Microbiol. Biotechnol.* 41: 1541–1551.
Fernandes, S. and Murray, P.G. 2010. *Bioeng. Bugs* 1: 424–428.
Ferndahl, C., Bonander, N., Logez, C., Wagner, R., Gustafsson, L., Larsson, C., Hedfalk, K., Darby, R.A. and Bill, R.M., 2010. *Microb. Cell Fact. 9:* 1.
Fickers, P. 2014. *Curr. Res. Microbiol. Biotechnol.* 2: 354–363.
Fickers, P., Benetti, P.H., Wache, Y., Marty, A., Mauersberger, S., Smit, M.S. and Nicaud, J.M. 2005. *FEMS* Yeast *Res.* 5: 527–543.
Fickers, P., Le Dall, M.T., Gaillardin, C., Thonart, P. and Nicaud, J.M. 2003. *J. Microbiol. Methods* 55: 727–737.
Finnis, C.J., Payne, T., Hay, J., Dodsworth, N. *et al.* 2010. *Microb. Cell Fact.* 9: 1.
Fleer, R. 1992. *Curr. Opin. Biotechnol.* 3: 486–496.
Fleer, R., Yeh, P., Amellal, N., Maury, I. *et al.* 1991. *Biotechnology (N Y)* 9: 968–975.
Fogel, S. and Welch, J.W. 1982. *Proc. Natl. Acad. Sci. USA* 79: 5342–5346.
Futcher, A.B. and Cox, B.S. 1983. *J. Bacteriol.* 154: 612–622.
Gasmi, N., Ayed, A., Ammar, B.B.H., Zrigui, R., Nicaud, J.M. and Kallel, H. 2011. *Microb. Cell Fact.* 10: 90.
Gellissen, G. 2000. *App. Microbiol. Biotechnol.* 54: 741–750.
Gellissen, G., Janowicz, Z.A., Weydemann, U., Melber, K., Strasser, A.W. and Hollenberg, C. P. 1992. *Biotechnol. Adv.* 10: 179–189.
Gellissen, G., Kunze, G., Gaillardin, C., Cregg, J.M., Berardi, E., Veenhuis, M. and Klei, I. 2005. *FEMS Yeast Res.* 5: 1079–1096.
Gerngross, T.U. 2004. *Nat. Biotechnol.* 22: 1409–1414.
Gleeson, M.A.G., White, C.E., Meninger, D.P. and Komives, E.A. 1998. *Methods Mol. Biol.* 103: 81–94.
Goffeau, A., Barrell, B.G., Bussey, H., Davis, R.W. *et al.* 1996. *Science* 274: 546–567.
Gomathinayagam, S., Laface, D., Houston-Cummings, N.R., Mangadu, R. *et al.* 2015. *J. Biotechnol.* 208: 13–21.
Gomes, N., Teixeira, J.A. and Belo, I. 2012. *Biotechnol. Lett.* 34: 649–654.
Gong, B., Burnina, I., Stadheim, T.A. and Li, H. 2013. *J. Mass. Spectrom.* 48: 1308–1317.
Gorlani, A., Brouwers, J., McConville, C., van der Bijl, P. *et al.* 2012. *AIDS Res. Hum. Retroviruses* 28: 198–205.
Gritz, L. and Davies, J. 1983. *Gene* 25: 179–188.
Gross, E., Kastner, D.B., Kaiser, C.A. and Fass, D. 2004. *Cell* 117: 601–610.
Guo, Y., Song, H., Wang, Z. and Ding, Y. 2012. *Microbiol. Res.* 167: 246–252.
Haan, G.J., Dijk, R., Kiel, J.A. and Veenhuis, M. 2002. *FEMS Yeast Res.* 2: 17–24.

Haddouche, R., Delessert, S., Sabirova, J., Neuvéglise, C., Poirier, Y. and Nicaud, J.M. 2010. *FEMS* Yeast *Res.* 10: 917–927.
Hadfield, C., Cashmore, A.M. and Meacock, P.A. 1986. *Gene* 45: 149–158.
Hahn-Hagerdal, B., Galbe, M., Gorwa-Grauslund, M.F., Lidén, G. and Zacchi, G. 2006. *Trends Biotechnol.* 24: 549–556.
Hamilton, S.R. and Gerngross, T.U. 2007. *Curr. Opin. Biotechnol.* 18: 387–392.
Hamilton, S.R., Bobrowicz, P., Bobrowicz, B., Davidson, R.C. *et al.* 2003. *Science* 301: 1244–1246.
Hamilton, S.R., Cook, W.J., Gomathinayagam, S., Burnina, I. *et al.* 2013. *Glycobiology* 23: 1192–1203.
Hamilton, S.R., Davidson, R.C., Sethuraman, N., Nett, J.H. *et al.* 2006. *Science* 313: 1441–1443.
Harmsen, M.M., Bruyne, M.I., Raue, H.A. and Maat, J. 1996. *Appl. Microbiol. Biotechnol.* 46: 365–370.
Hartner, F.S. and Glieder, A. 2006. *Microb. Cell Fact.* 5: 1.
Hohenblum, H., Gasser, B., Maurer, M., Borth, N. and Mattanovich, D. 2004. *Biotechnol. Bioeng.* 85: 367–375.
Hollenberg, C.P. and Gellissen, G. 1997. *Curr. Opin. Biotechnol.* 8: 554–560.
Hong, E., Davidson, A.R. and Kaiser, C.A. 1996. *J. Cell Biol.* 135: 623–633.
Horii, K., Adachi, T., Tanino, T., Tanaka, T. *et al.* 2010. *Enzyme Microb. Technol.* 46: 194–199.
Hsieh, H.P. and Da Silva, N.A. 1998. *Appl. Microbiol. Biotechnol.* 49: 147–152.
Huang, E.L. and Demirci, A. 2009. *Food Bioproc. Tech.* 2: 222–228.
Huh, W.K., Falvo, J.V., Gerke, L.C., Carroll, A.S. *et al.* 2003. *Nature* 425: 686–691.
Idiris, A., Tohda, H., Kumagai, H. and Takegawa, K. 2010. *Appl. Microbiol. Biotechnol.* 86: 403–417.
Ilmen, M., Den Haan, R., Brevnova, E., McBride, J. *et al.* 2011. *Biotechnol. Biofuels 4:* 1.
Inan, M., Aryasomayajula, D., Sinha, J. and Meagher, M.M. 2006. *Biotechnol. Bioeng.* 93: 771–778.
Inokuma, K., Yoshida, T., Ishii, J., Hasunuma, T. and Kondo, A. 2015. *Appl. Microbiol. Biotechnol.* 99: 1655–1663.
Ishchuk, O.P., Abbas, C.A. and Sibirny, A.A. 2010. *J. Ind. Microbiol. Biotechnol.* 37: 213–218.
Ishchuk, O.P., Voronovsky, A.Y., Abbas, C.A. and Sibirny, A.A. 2009. *Biotechnol. Bioeng.* 104: 911–919.
Iwata, T., Tanaka, R., Suetsugu, M., Ishibashi, M. *et al.* 2004. *Biotechnol. Lett.* 26: 1803–1808.
Jaafar, L. and Zueco, J. 2004. *Microbiology* 150: 53–60.
Jacobs, P.P., Geysens, S., Vervecken, W., Contreras, R. and Callewaert, N. 2008. *Nat. Protoc.* 4: 58–70.
Jacobs, P.P., Inan, M., Festjens, N., Haustraete, J., Van Hecke, A., Contreras, R., Meagher, M.M. and Callewaert, N. 2010. *Microb. Cell Fact.* 23: 9:93.
Jeong, G.M., Lee, Y.J., Kim, Y.S. and Jeong, K.J. 2014. *J. Ind. Microbiol. Biotechnol.* 41: 989–996.
Jewett, M.C., Hofmann, G. and Nielsen, J. 2006. *Curr. Opin. Biotechnol.* 17: 191–197.
Jimenez, A. and Davies, J. 1980. *Nature* 287: 869–871.
Jo, J.H., Im, E.M., Kim, S.H. and Lee, H.H. 2011. *Biotechnol. Lett.* 33: 1113–1120.
Johnson, E.A. 2013. *Appl. Microbiol. Biotechnol.* 97: 503–517.
Jonson, L., Rehfeld, J.F. and Johnsen, A.H. 2004. *Eur. J. Biochem.* 271: 4788–4797.
Joosten, V., Lokman, C., van Den Hondel, C.A. and Punt, P.J. 2003. *Microb. Cell Fact.* 2: 1.
Jungo, C., Schenk, J., Pasquier, M., Marison, I.W. and von Stockar, U. 2007. *J. Biotechnol.* 131: 57–66.
Juretzek, T., Wang, H.J., Nicaud, J.M., Mauersberger, S. and Barth, G. 2000. *Biotechnol. Bioprocess Eng.* 5: 320–326.
Juretzek, T., Le Dall, M.T., Mauersberger, S., Gaillardin, C., Barth, G. and Nicaud, J.M. 2001. *Yeast* 18: 97–113.
Kamzolova, S.V., Shishkanova, N.V., Morgunov, I.G. and Finogenova, T.V. 2003. *FEMS Yeast Res.* 3: 217–222.

Kang, H.A., Choi, E.S., Hong, W.K., Kim, J.Y. *et al.* 2000. *Appl. Microbiol. Biotechnol.* 53: 575–582.
Katahira, S., Fujita, Y., Mizuike, A., Fukuda, H. and Kondo, A. 2004. *Appl. Environ. Microbiol.* 70: 5407–5414.
Katahira, S., Mizuike, A., Fukuda, H. and Kondo, A. 2006. *Appl. Microbiol. Biotechnol.* 72: 1136–1143.
Kato, M., Fuchimoto, J., Tanino, T., Kondo, A., Fukuda, H. and Ueda, M. 2007. *Appl. Microbiol. Biotechnol.* 75: 549–555.
Katrolia, P., Yan, Q., Jia, H., Li, Y., Jiang, Z. and Song, C. 2011. *J. Mol. Catal. B Enzym.* 69: 112–119.
Khasa, Y.P., Conrad, S., Sengul, M., Plautz, S., Meagher, M.M. and Inan, M. 2011. *Yeast* 28: 213–226.
Khongto, B., Laoteng, K., Tongta, A. 2010. *J. Microbiol. Biotechnol.* 20: 1555–1562.
Kim, M.W., Kim, E.J., Kim, J.Y., Park, J.S. *et al.* 2006. *J. Biol. Chem.* 281: 6261–6272.
Kim, S.R., Skerker, J.M., Kang, W., Lesmana, A. *et al.* 2013. *PLoS One* 8: e57048.
Kim, S.Y., Sohn, J.H., Pyun, Y.R. and Choi, E.S. 2002. *Yeast* 19: 1153–1163.
Kim, S.Y., Sohn, J.H., Pyun, Y.R., Yang, I.S. *et al.* 2007. *J. Microbiol. Biotechnol.* 17: 1308–1315.
Klabunde, J., Kunze, G., Gellissen, G. and Hollenberg, C.P. 2003. *FEMS Yeast Res.* 4: 185–193.
Komeda, T., Sakai, Y., Kato, N. and Kondo, K. 2002. *Biosci. Biotechnol. Biochem.* 66: 628–631.
Kooistra, R., Hooykaas, P.J. and Steensma, H.Y. 2004. *Yeast* 21: 781–792.
Kotrba, P. and Ruml, T. 2010. *Appl. Environ. Microbiol.* 76: 2615–2622.
Kottmeier, K., Müller, C., Huber, R. and Büchs, J. 2010. *Appl. Microbiol. Biotechnol.* 86: 93–101.
Kottmeier, K., Ostermann, K., Bley, T. and Rodel, G. 2011. *Appl. Microbiol. Biotechnol.* 91: 133–141.
Kuroda, K. and Ueda, M., 2006. *Appl. Microbiol. Biotechnol.* 70: 458–463.
Kurylenko, O.O., Ruchala, J., Hryniv, O.B., Abbas, C.A. *et al.* 2014. *Microb. Cell Fact.* 13: 122.
Lashkari, D.A., DeRisi, J.L., McCusker, J.H., Namath, A.F. *et al.* 1997. *Proc. Natl. Acad. Sci. U S A* 94: 13057–13062.
Lazar, Z., Rossignol, T., Verbeke, J., Crutz-Le Coq, A.M. *et al.* 2013. *J. Ind. Microbiol. Biotechnol.* 40: 1273–1283.
Le Dall, M.T., Nicaud, J.M. and Gaillardin, C. 1994. *Curr. Genet.* 26: 38–44.
Lee J.H., Lee, S.G., Do, H., Park, J.C. *et al.* 2013. *Appl. Microbiol. Biotechnol.* 97: 3383–3393.
Lee, J., Choi, S.I., Jang, J.S., Jang, K. *et al.* 1999. *Biotechnol. Prog.* 15: 884–890.
Lee, S.Y., Choi, J.H. and Xu, Z. 2003. *Trends Biotechnol.* 21: 45–52.
Lee, T.I., Rinaldi, N.J., Robert, F., Odom, D.T. *et al.* 2002. *Science* 298: 799–804.
Leite, F.C.B., dos Anjos, R.S.G., Basilio, A.C.M., Leal, G.F.C., Simões, D.A. and de Morais, M. A. 2013. *Plasmid* 69: 114–117.
Li, A., Liu, Z., Li, Q., Yu, L., Wang, D. and Deng, X. 2008. *FEMS Yeast Res.* 8: 6–9.
Li, H., Sethuraman, N., Stadheim, T.A., Zha, D. *et al.* 2006. *Nat. Biotechnol.* 24: 210–215.
Li, H., Wang, Y., Xu, A., Li, S., Jin, S. and Wu, D. 2011a. *FEMS* Yeast *Res.* 11: 160–167.
Li, S., Sing, S. and Wang, Z. 2011b. *Protein Expr. Purif.* 79: 142–148.
Liang, X.X., Wang, B.B., Sun, Y.F., Lin, Y. *et al.* 2013. *Biotechnol. Lett.* 35: 367–374.
Lin, S., Houston-Cummings, N.R., Prinz, B., Moore, R. *et al.* 2012. *J. Immunol. Methods.* 375: 159–165.
Liu, B., Gong, X., Chang, S., Yang, Y. *et al.* 2009a. *J. Biotechnol.* 143: 95–102.
Liu, G., Yue, L., Chi, Z., Yu, W. *et al.* 2009b. *Mar. Biotechnol. (NY)* 11: 619–626.
Liu, Z.L. 2011. *Appl. Microbiol. Biotechnol.* 90: 809–825.
Lodi, T., Neglia, B. and Donnini, C. 2005. *Appl. Environ. Microbiol.* 71: 4359–4363.
Loison, G., Vidal, A., Findeli, A., Roitsch, C. *et al.* 1989. *Yeast* 5: 497–507.
Looser, V., Bruhlmann, B., Bumbak, F., Stenger, C. *et al.* 2015. *Biotechnol. Adv.* 33: 1177–1193.
Lorenz, E., Schmacht, M., Stahl, U. and Senz, M. 2015. *J. Biotechnol.* 216: 131–139.
Macauley-Patrick, S., Fazenda, M,L., McNeil, B. and Harvey, L.M. 2005. *Yeast* 22: 249–270.
Madzak, C. 2015. *Appl. Microbiol. Biotechnol.* 99: 4559–4577.

Madzak, C., Blanchin-Roland, S., Otero, R.R.C. and Gaillardin, C. 1999. *Microbiology* 145: 75–87.
Madzak, C., Gaillardin, C. and Beckerich, J.M. 2004. *J. Biotechnol.* 109: 63–81.
Madzak, C., Otterbein, L., Chamkha, M., Moukha, S., Asther, M., Gaillardin, C. and Beckerich, J. M. 2005. *FEMS* Yeast *Res.* 5: 635–646.
Madzak, C., Tréton, B. and Blanchin-Roland, S. 2000. *J. Mol. Microbiol. Biotechnol.* 2: 207–216.
Maharajh, D., Roth, R., Lalloo, R., Simpson, C. *et al.* 2008. *Appl. Microbiol. Biotechnol.* 79: 235–244.
Matsuoka, M., Matsubara, M., Daidoh, H., Imanaka, T. *et al.* 1993. *Mol. Gen. Genet.* 237: 327–333.
Mattanovich, D., Graf, A., Stadlmann, J., Dragosits, M. *et al.* 2009. *Microb. Cell Fact.* 8: 29.
Matthaus, F., Ketelhot, M., Gatter, M. and Barth, G. 2014. *Appl. Environ. Microbiol.* 80: 1660–1669.
Maury, J., Asadollahi, M.A., Møller, K., Schalk, M. *et al.* 2008. *FEBS Lett.* 582: 4032–4038.
Mayer, A.F., Hellmuth, K., Schlieker, H., Lopez-Ulibarri, R. *et al.* 1999. *Biotechnol. Bioeng.* 63: 373–381.
Mazor, Y., Van Blarcom, T., Carroll, S. and Georgiou, G. 2010. *FEBS J.* 277: 2291–2303.
Medina, V.G., Almering, M.J., van Maris, A.J. and Pronk, J.T. 2010. *Appl. Environ. Microbiol.* 76: 190–195.
Meilhoc, E., Masson, J.M. and Teissié, J. 1990. *Biotechnology (N Y)* 8: 223–227.
Mendoza-Vega, O., Hebert, C. and Brown, S.W. 1994. *J. Biotechnol.* 32: 249–259.
Mergler, M., Wolf, K. and Zimmermann, M. 2004. *Appl. Microbiol. Biotechnol.* 63: 418–421.
Merico, A., Capitanio, D., Vigentini, I., Ranzi, B.M. and Compagno, C. 2004. *J. Biotechnol.* 109: 139–146.
Mitra, N., Sinha, S., Ramya, T.N. and Surolia, A. 2006. *Trends. Biochem. Sci.* 31: 156–163.
Miyajima, A., Miyajima, I., Arai, K. and Arai, N. 1984. *Mol. Cell Biol.* 4: 407–414.
Moon, H.Y., Van, T.L., Cheon, S.A., Choo, J. *et al.* 2013. *J. Microbiol.* 51: 506–514.
Morlino, G.B., Tizzani, L., Fleer, R., Frontali, L. and Bianchi, M.M. 1999. *Appl. Environ. Microbiol.* 65: 4808–4813.
Moussa, M., Ibrahim, M., El Ghazaly, M., Rohde, J. *et al.* 2012. *BMC Biotechnol.* 12: 96.
Mukhopadhyay, A., Redding, A.M., Rutherford, B.J. and Keasling, J.D. 2008. *Curr. Opin. Biotechnol.* 19: 228–234.
Muller, F., Tieke, A., Waschk, D., Muhle, C. *et al.* 2002. *Process Biochem.* 38: 15–25.
Muller, S., Sandal, T., Kamp-Hansen, P. and Dalboge, H. 1998. *Yeast* 14: 1267–1283.
Murray, A.W. and Szostak, J.W. 1983. *Cell* 34: 961–970.
Nagasu, T., Shimma, Y.I., Nakanishi, Y., Kuromitsu, J. *et al.* 1992. *Yeast* 8: 535–547.
Nakamura, N., Yamada, R., Katahira, S., Tanaka, T., Fukuda, H. and Kondo, A. 2008. *Enzyme Microb. Technol.* 43: 233–236.
Nakanishi-Shindo, Y., Nakayama, K.I., Tanaka, A., Toda, Y. and Jigami, Y. 1993. *J. Biol. Chem.* 268: 26338–26345.
Nambou, K., Jian, X., Zhang, X., Wei, L. *et al.* 2015. *Metabolites* 5: 794–813.
Nett, J.H., Cook, W.J., Chen, M.T., Davidson, R.C. *et al.* 2013. *PLoS One* 8: e68325.
Nett. J. and Gerngross, T. 2003. *Yeast* 20: 1279–1290.
Ni, X., Yue, L., Chi, Z., Li, J. *et al.* 2009. *Mar. Biotechnol. (NY).* 11: 81–89.
Nicaud, J.M., Fabre, E. and Gaillardin, C. 1989. *Curr. Genet.* 16: 253–260.
Nicaud, J.M., Madzak, C., Broek, P., Gysler, C. *et al.* 2002. *FEMS* Yeast *Res.* 2: 371–379.
Nishitani, T., Shimada, M., Kuroda, K. and Ueda, M. 2010. *Appl. Microbiol. Biotechnol.* 86: 641–648.
Oh, D.B., Park, J.S., Kim, M.W., Cheon, S.A. *et al.* 2008. *Biotechnol. J.* 3: 659–668.
Okajima, T., Reddy, B.V.V.G., Matsuda, T. and Irvine, K.D. 2008. *BMC Biol.* 6: 1.
Orchard, M.G., Neuss, J.C., Galley, C.M., Carr, A. *et al.* 2004. *Bioorg. Med. Chem. Lett. 14*: 3975–3978.
Orr-Weaver, T.L. and Szostak, J.W. 1983. *Mol. Cell Biol.* 3: 747–749.
Pal, Y., Khushoo, A. and Mukherjee, K.J. 2006. *Appl. Microbiol. Biotechnol.* 69: 650–657.

Parawira, W. and Tekere, M. 2011. *Crit. Rev. Biotechnol.* 31: 20–31.
Park, C.S., Chang, C.C. and Ryu, D.D. 2000a. *Appl. Biochem. Biotechnol.* 87: 1–15.
Park, E.H., Shin, Y.M., Lim, Y.Y., Kwon, T.H. *et al.* 2000b. *J. Biotechnol.* 81: 35–44.
Park, J.N., Song, Y., Cheon, S.A., Kwon, O. *et al.* 2011. *Appl. Environ. Microbiol.* 77: 1187–1195.
Payne, T., Finnis, C., Evans, L.R., Mead, D.J. *et al.* 2008. *Appl. Environ. Microbiol.* 74: 7759–7766.
Phongdara, A., Merckelbach, A., Keup, P., Gellissen, G. and Hollenberg, C.P. 1998. *Appl. Microbiol. Biotechnol.* 50: 77–84.
Pignede, G., Wang, H.J., Fudalej, F., Seman, M., Gaillardin, C. and Nicaud, J.M. 2000. *Appl. Environ. Microbiol.* 66: 3283–3289.
Porro, D., Sauer, M., Branduardi, P. and Mattanovich, D. 2005. *Mol. Biotechnol.* 31: 245–259.
Potgieter, T.I., Cukan, M., Drummond, J.E., Houston-Cummings, N.R. *et al.* 2009. *J. Biotechnol.* 139: 318–325.
Potvin, G., Ahmad, A. and Zhang, Z. 2012. *Biochem. Eng. J.* 64: 91–105.
Prinz, B., Schultchen, J., Rydzewski, R., Holz, C. *et al.* 2004. *J. Struct. Funct. Genomics* 5: 29–44.
Rachubinski, R.A. and Subramani, S. 1995. *Cell* 83: 525–528.
Raimondi, S., Uccelletti, D., Amaretti, A., Leonardi, A., Palleschi, C. and Rossi, M. 2010. *Appl. Microbiol. Biotechnol.* 86: 871–878.
Rakestraw, J.A., Sazinsky, S.L., Piatesi, A., Antipov, E. and Wittrup, K.D. 2009. *Biotechnol. Bioeng.* 103: 1192–1201.
Ramezani-Rad, M., Hollenberg, C.P., Lauber, J., Wedler, H. *et al.* 2003. *FEMS Yeast Res.* 4: 207–215.
Ramon, R., Ferrer, P. and Valero, F. 2007. *J. Biotechnol.* 130: 39–46.
Ravin, N.V., Eldarov, M.A., Kadnikov, V.V., Beletsky, A.V. *et al.* 2013. *BMC Genomics* 14: 837.
Ro, D.K., Ouellet, M., Paradise, E.M., Burd, H. *et al.* 2008. *BMC Biotechnol.* 8: 83.
Ro, D.K., Paradise, E.M., Ouellet, M., Fisher, K.J. *et al.* 2006. *Nature* 440: 940–943.
Robin, S., Petrov, K., Dintinger, T., Kujumdzieva, A. *et al.* 2003. *Mol. Immunol.* 39: 729–738.
Roggenkamp, R., Hansen, H., Eckart, M., Janowicz, Z. and Hollenberg, C.P. 1986. *Mol. Gen. Genet.* 202: 302–308.
Romanos, M.A., Scorer, C.A. and Clare, J.J.1992. *Yeast* 8: 423–488.
Rosa, J.C., Colombo, L.T., Alvim, M.C., Avonce, N., Van Dijck, P. and Passos, F.M. 2013. *Microb. Cell Fact.* 12: 59.
Ryabova, O.B., Chmil, O.M. and Sibirny, A.A. 2003. *FEMS* Yeast *Res.* 4: 157–164.
Rymowicz, W., Fatykhova, A.R., Kamzolova, S.V., Rywińska, A. and Morgunov, I.G. 2010. *Appl. Microbiol. Biotechnol.* 87: 971–979.
Sakamoto, T., Hasunuma, T., Hori, Y., Yamada, R. and Kondo, A. 2012. *J. Biotechnol.* 158: 203–210.
Saliola, M., Mazzoni, C., Solimando, N., Crisà, A., Falcone, C., Jung, G. and Fleer, R. 1999. *Appl. Environ. Microbiol.* 65: 53–60.
Sanda, T., Hasunuma, T., Matsuda, F. and Kondo, A. 2011. *Bioresour Technol. 102*: 7917–7924.
Sauer, U. 2006. *Mol. Syst. Biol.* 2: 62.
Scorer, C.A., Clare, J.J., McCombie, W.R., Romanos, M.A. and Sreekrishna, K. 1994. *Biotechnology (NY)* 12: 181–184.
Shaheen, H.H., Prinz, B., Chen, M.T., Pavoor, T. *et al.* 2013. *PLoS One* 8: e70190.
Shang, F., Wang, Z. and Tan, T. 2008. *Appl. Microbiol. Biotechnol.* 77: 1233–1240.
Shang, F., Wen, S., Wang, X. and Tan, T., 2006. *J. Biosci. Bioeng.* 101: 38–41.
Sherman, D., Durrens, P., Beyne, E., Nikolski, M. *et al.* 2004. *Nucleic Acids Res.* 32: D315-D318.
Shiba, Y., Paradise, E.M., Kirby, J., Ro, D.K. and Keasling, J.D. 2007. *Metab. Eng.* 9: 160–168.
Shusta, E.V., Raines, R.T., Plückthun, A. and Wittrup, K.D. 1998. *Nat. Biotechnol.* 16: 773–777.
Sikorski, R.S. and Hieter, P. 1989. *Genetics* 122: 19–27.
Smith, J.D., Tang, B.C. and Robinson, A.S. 2004. *Biotechnol. Bioeng.* 85: 340–350.
Sonderegger, M. and Sauer, U. 2003. *Appl. Environ. Microbiol.* 69: 1990–1998.
Song, H., Li, Y., Fang, W., Geng, Y. *et al.* 2003. *Biotechnol. Lett.* 25: 1999–2006.

Song, Y., Choi, M.H., Park, J.N., Kim, M.W. *et al.* 2007. *Appl. Environ. Microbiol.*73: 4446–4454.
Spohner, S.C., Müller, H., Quitmann, H. and Czermak, P. 2015. *J. Biotechnol.* 202: 118–134.
St John, T.P. and Davis, R.W. 1981. *J. Mol. Biol.* 152: 285–315.
Stockmann, C., Scheidle, M., Dittrich, B., Merckelbach, A. *et al.* 2009. *Microb. Cell Fact.* 8: 22.
Su, G.D., Huang, D.F., Han, S.Y., Zheng, S.P. and Lin, Y. 2010a. *Appl. Microbiol. Biotechnol. 86*: 1493–1501.
Su, G.D., Zhang, X. and Lin, Y. 2010b. *Biotechnol. Lett.* 32: 1131–1136.
Subramani, S. 1996. *Curr. Opin. Cell Biol.* 8: 513–518.
Swinkels, B.W., van Ooyen, A.J. and Bonekamp, F.J. 1993. *Antonie van Leeuwenhoek* 64: 187–201.
Takors, R., Bathe, B., Rieping, M., Hans, S. *et al.* 2007. *J. Biotechnol.* 129: 181–190.
Tamaru, Y., Ohtsuka, M., Kato, K., Manabe, S. *et al.* 2006. *Biotechnol. Prog.* 22: 949–953.
Tanaka, T., Yamada, R., Ogino, C. and Kondo, A. 2012. *Appl. Microbiol. Biotechnol.* 95: 577–591.
Tang, X., Feng, H. and Chen, W.N. 2013. *Metab. Eng.* 16: 95–102.
Theerachat, M., Emond, S., Cambon, E., Bordes, F. *et al.* 2012. *Bioresour. Technol.* 125: 267–274.
Thodey, K., Galanie, S. and Smolke, C.D. 2014. *Nat. Chem. Biol.* 10: 837–44.
Tomaszewska, L., Rywińska, A. and Gładkowski, W. 2012. *J. Ind. Microbiol. Biotechnol.* 39: 1333–1343.
Tsai, S.L., Goyal, G. and Chen, W. 2010. *Appl. Environ. Microbiol. 76*: 7514–7520.
Ubiyvovk, V.M., Ananin, V.M., Malyshev, A.Y., Kang, H.A. and Sibirny, A.A. 2011. *BMC Biotechnol.* 11:8.
Uccelletti, D., Farina, F., Mancini, P. and Palleschi, C. 2004. *J. Biotechnol.*109: 93–101.
Valkonen, M., Penttilä, M. and Saloheimo, M. 2003. *Appl. Environ. Microbiol.* 69: 2065–2072.
Van den Berg, J.A., van der Laken, K.J., van Ooyen, A.J., Renniers, T.C. *et al.* 1990. *Biotechnology (NY)* 8: 135–139.
van der Klei, I.J., Yurimoto, H., Sakai, Y. and Veenhuis, M. 2006. *Biochim. Biophys. Acta* 1763: 1453–1462.
Van der Vaart, J.M., Te Biesebeke, R., Chapman, J.W., Toschka, H.Y., Klis, F.M. and Verrips, C. T. 1997. *Appl. Environ. Microbiol.* 63: 615–620.
Van Dijk, R., Faber, K.N., Kiel, J.A., Veenhuis, M. and van der Klei, I. 2000. *Enzyme Microb. Technol.* 26: 793–800.
van Ooyen, A.J., Dekker, P., Huang, M., Olsthoorn, M., Jacobs, D.I., Colussi, P.A. and Taron, C. H. 2006. *FEMS* Yeast *Res.* 6: 381–392.
Verbeke, J., Beopoulos, A. and Nicaud, J.M., 2013. *Biotechnol. Lett.* 35: 571–576.
Villas-Boas, S.G., Moxley,J.F., Akesson, M., Stephanopoulos, G. and Nielsen, J. 2005. *Biochem. J.* 388: 669–677.
Vogl, T., Hartner, F.S. and Glieder, A. 2013. *Curr. Opin. Biotechnol.* 24: 1094–1101.
Walsh, D.J. and Bergquist, P.L. 1997. *Appl. Environ. Microbiol.* 63: 3297–300.
Walsh, G. 2014. *Nat. Biotechnol.* 32: 992–1000.
Wang, Q., Li, L., Chen, M., Qi, Q. and Wang, P.G. 2007. *Biotechnol. Lett.* 29: 1561–1566.
Wang, Q., Li, L., Chen, M., Qi, Q. and Wang, P.G. 2008. *Curr. Microbiol.* 56: 352–357.
Wang, Y., Halls, C., Zhang, J., Matsuno, M., Zhang, Y. and Yu, O. 2011. *Metab. Eng.* 13: 455–463.
Wang, Y., Liang, Z.H., Zhang, Y.S., Yao, S.Y., Xu, Y.G. *et al.* 2001. *Biotechnol. Bioeng.* 73: 74–79.
Weirich, J., Goffrini, P., Kuger, P., Ferrero, I. and Breunig, K.D. 1997. *Eur. J. Biochem.* 249: 248–257.
Wen, F., Sun, J. and Zhao, H. 2010. *Appl. Environ. Microbiol.* 76: 1251–1260.
Werten, M.W. and de Wolf, F.A. 2005. *Appl. Environ. Microbiol.* 71: 2310–2317.
Werten, M.W.T., Wisselink, W.H., Jansen-van den Bosch. T.J., de Bruin, E.C. and de Wolf, F.A. 2001. *Protein Eng.* 14: 447–454.

Westfall, P.J., Pitera, D.J., Lenihan, J.R., Eng, D. *et al.* 2012. *Proc. Natl. Acad. Sci. U S A* 109: E111-E118.
Weydemann, U., Keup, P., Piontek, M., Strasser, A.W.M. *et al.* 1995. *Appl. Microbiol Biotechnol. 44:* 377–385.
Xie, W., Ye, L., Lv, X., Xu, H. and Yu, H. 2015. *Metab. Eng.* 28: 8–18.
Xu, P., Raden, D., Doyle, F.J. and Robinson, A.S. 2005. *Metab. Eng.* 7: 269–279.
Xue, Z., Sharpe, P.L., Hong, S.P., Yadav, N.S. *et al.* 2013. *Nat. Biotechnol.* 31: 734–740.
Yang, L.B., Zhan, X.B., Zheng, Z.Y., Wu, J.R. *et al.* 2014. *Bioresour. Technol.* 151: 120–127.
Ye, J., Ly, J., Watts, K., Hsu, A. *et al.* 2011. *Biotechnol. Prog.* 27: 1744–1750.
Yoshida, H., Arai, S., Hara, K.Y., Yamada, R. *et al.* 2011. *Appl. Microbiol. Biotechnol.* 89: 1417–1422.
Youn, J.K., Kim, M.I., Chang, H.N. and Rhee, S.K.2010. *J. Microbiol. Biotechnol.* 20: 1534–1538.
Yu, X.J., Li, H.J., Li, J. and Chi, Z.M. 2010. *Biotechnol. Bioprocess. Eng.* 15: 467–475.
Yu, Z., Du, G., Zhou, J. and Chen, J. 2012. *Bioresour. Technol.* 114: 597–602.
Yue, L., Chi, Z., Wang, L., Liu, J. *et al.* 2008. *J. Microbiol. Methods* 72: 116–123.
Yuzbashev, T.V., Yuzbasheva, E.Y., Vibornaya, T.V., Sobolevskaya, T.I. *et al.* 2012. *Protein Expr. Purif.* 82: 83–89.
Yuzbasheva, E.Y., Yuzbashev, T.V., Laptev, I.A., Konstantinova, T.K. and Sineoky, S.P. 2011. *Appl. Microbiol. Biotechnol.* 91: 645–654.
Zealey, G.R., Goodey, A.R., Piggott, J.R., Watson, M.E. *et al.* 1988. *Mol. Gen. Genet.* 211: 155–159.
Zhang, A.L., Zhang, T.Y., Luo, J.X., Chen, S.C. *et al.* 2007. *J. Ind. Microbiol. Biotechnol.* 34: 117–122.
Zhang, B.Y., Chang, A., Kjeldsen, T.B. and Arvan, P. 2001. *J. Cell Biol.* 153: 1187–1198.
Zhang, W., Zhao, H.L., Xue, C., Xiong, X.H., Yao, X.Q. *et al.* 2006. *Biotechnol. Prog.* 22: 1090–1095.
Zhao, W., Wang, J., Deng, R. and Wang, X. 2008. *J. Ind. Microbiol. Biotechnol.* 35: 189–195.
Zhou, J., Yin, X., Madzak, C., Du, G. and Chen, J. 2012. *J. Biotechnol.* 161: 257–264.
Zhou, X., Yu, Y., Tao, J. and Yu, L. 2014. *J. Biosci. Bioeng.* 118: 420–425.
Zhu, H., Bilgin, M., Bangham, R., Hall, D. *et al.* 2001. *Science* 293: 2101–2105.
Zurek, C., Kubis, E., Keup, P., Horlein, D. *et al.* 1996. *Process Biochem.* 31: 679–689.
Zwart, K.B., Veenhuis, M., Plat, G. and Harder, W. 1983. *Arch. Microbiol.*136: 28–38.

Gene Expression Analysis in *Arxula adeninivorans*: A Nested Quantitative Real Time PCR Approach

Sebastian Worch and Ina Lemke

Abstract During the preparation of RNA traces of genomic DNA are usually co-isolated which might influence downstream applications. We tested several protocols, commercial kits and DNA hydrolysis procedures to remove the DNA contamination and found them to be insufficient. This can raise problems when it comes to gene expression analysis especially when working with intronless genes. Hence, we used a nested quantitative real time PCR approach to avoid amplification from genomic DNA by the use of an artificial anchor sequence introduced at the cDNA synthesis stage. This anchor sequence cannot be found in the genome of *A. adeninivorans* and a first round of amplification using a gene specific oligo in combination with an oligo for the anchor generates fragments which can emerge only from the cDNA. The second PCR step with nested oligos for the gene of interest and the anchor, respectively, significantly increases gene specificity which is crucial particularly when analysing the gene expression status among highly conserved members of a gene family. This second round of amplification represents the actual quantitative real time PCR assay.

Keywords Real-time PCR · RNA isolation · cDNA synthesis · *Arxula adeninivorans*

1 Introduction

Quantitative real-time polymerase chain reaction (qRT-PCR) is an overall accepted tool for detecting and quantifying transcription patterns of desired target genes. This technology is not only used for quantitative genotyping but has been adapted for a large field of applications like diagnosis of disease in human, animals and plants as well as in taxonomy, forensic science and food safety (Deepak et al. 2007).

S. Worch (✉) · I. Lemke
Leibniz Institute of Plant Genetics and Crop Plant Research (IPK), Corrensstr. 3, D-06466 Gatersleben, Germany
e-mail: worch@ipk-gatersleben.de

T. Satyanarayana and G. Kunze (eds.), *Yeast Diversity in Human Welfare*, DOI 10.1007/978-981-10-2621-8_10

These days high throughput techniques for large scale gene expression analyses like microarrays (Maskos and Southern 1992), SAGE studies (Velculescu et al. 1995) or RNA sequencing (Morin et al. 2008) gain increasing significance and qRT-PCR has proven to be a good confirmatory tool to verify the results from such experiments (Puthoff et al. 2003). However, the entire qRT-PCR assay derives from several handling steps beginning with sampling to RNA extraction and purification, cDNA synthesis and analytical methods. All these steps are crucial but not fail-safe and the individual contribution to the overall conclusion seems sometimes neglected. Here we discuss the case when genomic DNA (gDNA) may become a problem as during the preparation of RNA a certain amount of gDNA is usually co-isolated which might result in inaccuracies in quantification. When working with intron-containing genes, residual gDNA is not a big issue because usually the primers for qRT-PCR are located in different exons or even over exon-exon boundaries to avoid amplification from genomic template. Higher eukaryotes tend to have more introns, in human e.g. only 3% of the genes have no intron (Grzybowska 2012). Considering that the majority of *Arxula adeninivorans* genes (88%) are intronless (Kunze et al. 2014) and since enzymatic removal of gDNA is not 100% sufficient we have developed a nested qRT-PCR approach to overcome the problem of contaminating gDNA.

2 Method

It is known that the common RNA isolation procedures yield RNA with significant amounts of gDNA and there are several protocols to remove these contaminations. Besides purification steps like cesium-chloride centrifugation (Glisin et al. 1974) or oligo (dT) chromatography (Aviv and Leder 1972) the enzymatic removal with DNaseI is the most widespread. Even if the different purification attempts may finally succeed none of these treatments will increase the quality of the isolated RNA and if certain transcripts are lost during these procedures the final results will be inaccurate anyway and all the efforts were useless. We tested different approaches and commercial kits to isolate RNA free of gDNA and exemplarily shown in Fig. 2 is a trial with DNaseI and the frequently cutting restriction enzyme CviKI-1 under different conditions (refer to Fig. 2D). According to the supplier (NEB) CviKI-1, derived from CA-1A, a *Chlorella* virus, has 4 expected recognition sites as well as up to eleven relaxed non-cognate sites (star sites) and DNA can be digested to small oligos under "star" conditions. The treatments were performed either "on the column" during the RNA preparation or after the RNA isolation procedure with subsequent purification. Such processed RNA samples were used to synthesize cDNA by reverse transcription reaction with the oligo (dT)-anchor primer (left hand site in Fig. 2A–C). As a control where no cDNA can be generated, the same reactions were set up without the oligo (dT)-anchor primer (right hand site in Fig. 2A–C). The cDNA samples as well as the controls were evaluated by PCR afterwards. Entirely gene specific PCR primers, located within the same exon

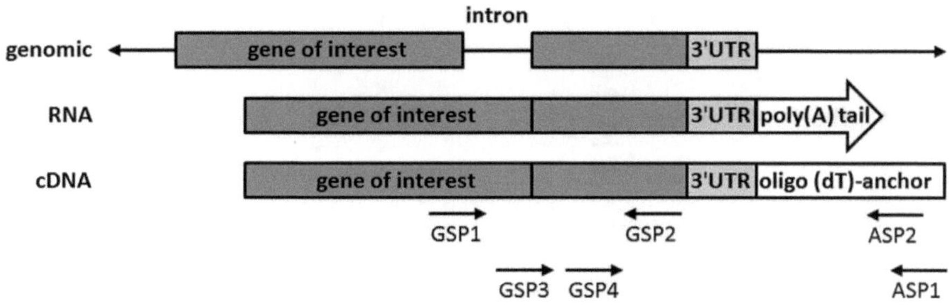

Fig. 1 Principle of the nested quantitative real time PCR approach with positions of gene specific primers (GSP) and anchor specific primers (ASP)

(e.g. GSP2 and GSP4 in Fig. 1), yielded product not only with the cDNA but also with the controls as template indicating that it is virtually impossible to get rid of every last strand of gDNA (Fig. 2A).

Our approach to prevent gDNA emerged products is based on the rapid amplification of cDNA ends (RACE) technique (Frohman 1993). We isolate intact total RNA in as few steps as possible and avoid genomic amplification products during qRT-PCR by the insertion of an artificial anchor sequence at the first strand cDNA synthesis stage. This anchor sequence is part of the oligo (dT)-anchor primer used for cDNA synthesis and cannot be found in the genome of *A. adeninivorans*. Thus, amplification using one gene specific primer (e.g. GSP1 in Fig. 1) in combination with one oligo for the anchor sequence (e.g. ASP1 in Fig. 1) generates fragments which can emerge only from the cDNA. Visible in Fig. 2B is that, in contrast to entirely gene specific primers, PCR products are only gained with cDNA template (left hand site) but no amplification occurs in the controls (right hand site). But since only one of the two primers is designed for the gene of interest unspecific amplification products may occur especially when analysing the gene expression status among highly conserved members of a gene family (Fig. 2B). To increase the specificity of the test a second round of amplification using nested primers (e.g. GSP4 and ASP2 in Fig. 1) is executed subsequently producing a single fragment (Fig. 2C). This second round is performed as the actual qRT-PCR assay while the first round is a standard PCR method. To avoid too many unspecific products during PCR round 1 the number of cycles should be kept as small as possible. As a rule of thumb perform as many as necessary but as few as possible. The optimal number depends on the expression status of the desired target gene and needs to be ascertained for every individual experiment. We usually cycle 5 times in round 1 and use a dilution series of that reaction as template in round 2.

For the design of a nested qRT-PCR assay the contribution of the 3'UTR and the oligo (dT)-anchor primer to the overall PCR product size should be considered. Since the fragments in qRT-PCR assays with intercalating dyes should not be larger than 150 bp we recommend cloning and sequencing the PCR products to gain precise information on the 3' UTR and even on possible alternative polyadenylation events. In case of a large UTR the gene specific primers can be designed to bind to

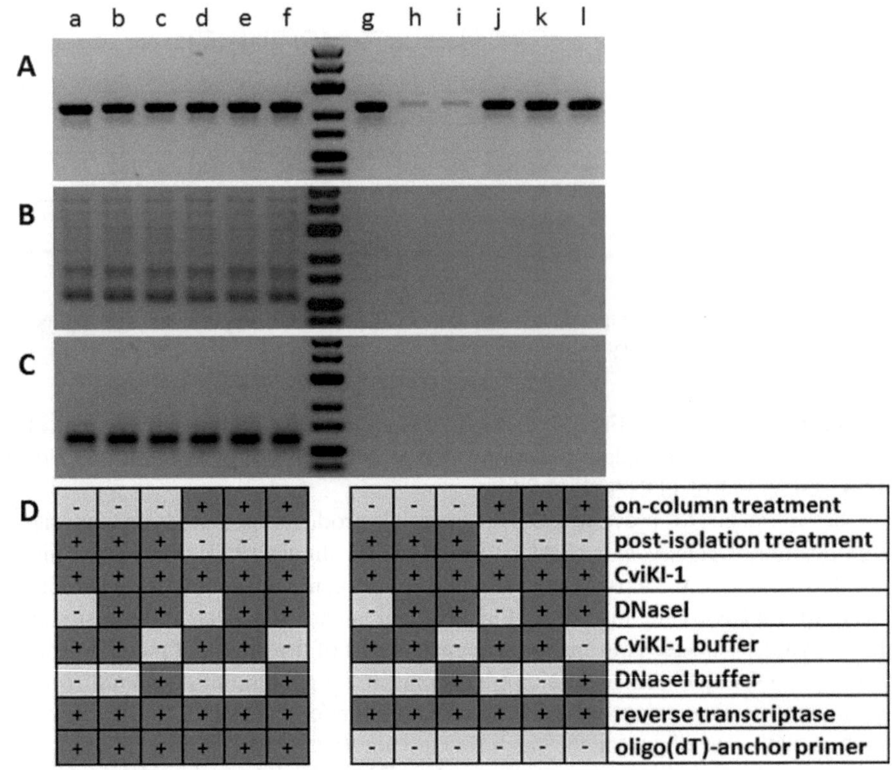

	a	b	c	d	e	f	g	h	i	j	k	l
on-column treatment	-	-	-	+	+	+	-	-	-	+	+	+
post-isolation treatment	+	+	+	-	-	-	+	+	+	-	-	-
CviKI-1	+	+	+	+	+	+	+	+	+	+	+	+
DNaseI	-	+	+	-	+	+	-	+	+	-	+	+
CviKI-1 buffer	+	+	-	+	+	-	+	+	-	+	+	-
DNaseI buffer	-	-	+	-	-	+	-	-	+	-	-	+
reverse transcriptase	+	+	+	+	+	+	+	+	+	+	+	+
oligo(dT)-anchor primer	+	+	+	+	+	+	-	-	-	-	-	-

Fig. 2 Differently treated RNA samples (refer to **D**) used for cDNA synthesis with the oligo (dT)-anchor primer (a–f) and control reactions without this oligo (g–l). Subsequently reverse transcription-PCR was performed with entirely gene specific oligos (**A**) and with one gene specific and one anchor oligo (**B**). Shown in (**C**) is the reamplification of (**B**) with nested oligos. RNA was isolated using a spin column Kit and remaining gDNA was hydrolysed after the isolation procedure or meanwhile on the column. For hydrolysis the frequently cutting restriction enzyme CviKI-1 was used alone or in combination with DNAseI in two buffer systems. The different treatments are summarised in (**D**) and marked with (+) when applied and (−) when not

the untranslated region of the transcript which is, furthermore, usually not as conserved as the open reading frame. In addition, the cloned PCR products may serve as templates for the estimation of the primer efficiency.

To demonstrate the influence of residual gDNA in a conventional qRT-PCR assay, we selected an *A. adeninivorans* gene which contains an intron and designed two oligos which are gene specific with one oligo being located on an exon-exon boundary (e.g. GSP2 and GSP3 in Fig. 1) or both oligos being placed within the same exon (e.g. GSP2 and GSP4 in Fig. 1). As a result the qRT-PCR with the latter pair of primers showed a clearly lower C_t value which is due to additional PCR template provided by gDNA (Fig. 3a). The difference in C_t compared to the assay with one exon-exon boundary located oligo is in the shown case 1.45 suggesting

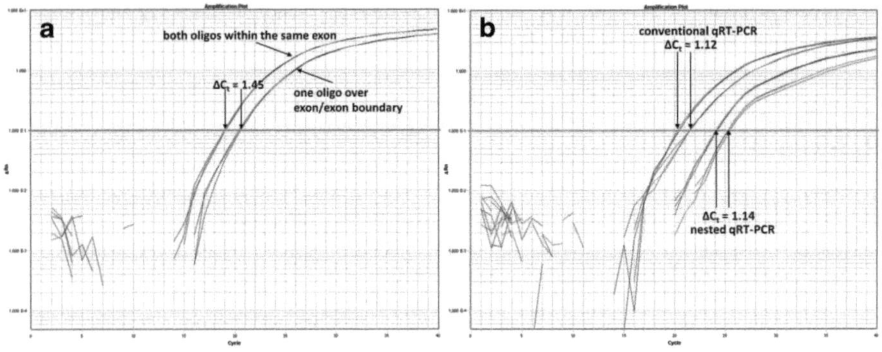

Fig. 3 Amplification plot for a single *A. adeninivorans* gene with oligos which can discriminate cDNA from gDNA (over exon/exon boundary) and those oligos which cannot (within the same exon). In the latter case a significantly lower C_t value indicates the presence of additional template in gDNA form (**a**). Shown in (**b**) is an amplification plot for two *A. adeninivorans* genes with the conventional qRT-PCR compared to the nested qRT-PCR. Similar ΔC_t values indicate that both approaches detect identical gene expression patterns

approximately 3 times more available template for the pair of primers which cannot discriminate cDNA from gDNA.

To validate our experimental design we tested two *A. adeninivorans* genes which do contain introns with a gene specific exon spanning pair of primers (e.g. GSP2 and GSP3 in Fig. 1) and compared the results of this gene expression analysis to a nested qRT-PCR using GSP1 and ASP1 in the first PCR and the nested oligos GSP4 and ASP2 in the qRT-PCR assay. As shown in the amplification plot in Fig. 3b it turned out that the ΔC_t values between two genes was nearly the same with the conventional as with the nested real-time PCR method indicating that both approaches yield similar results in relative gene expression analyses. The difference in C_t values is due to dilution effects after the first PCR and does not affect the relative quantification.

3 Conclusion

The nested qRT-PCR approach with inserted anchor sequence is a suitable method to specifically amplify target cDNAs regardless of contaminating gDNA and yields similar results as the conventional qRT-PCR. Hence, in distinct cases it might supplement other gene expression procedures in terms of plausibility. For example, this method was already successfully applied to verify microarray based data on gene expression during *A. adeninivorans* cultivation with 1-butanol as sole carbon source. Rauter et al. (2016) used nested qRT-PCR to confirm the significant induction of alcohol dehydrogenase 2 (*AADH2*) gene expression upon shift to 1-butanol and revealed its major role in *A. adeninivorans* 1-butanol metabolism.

Acknowledgements This work was supported in part by National Academy of Sciences of Ukraine (Grant Nos 5–15, 6–15 and 35–15).

References

Aviv H. and Leder P. 1972. *Proc. Natl. Acad. Sci. USA* 69: 1408–1412.

Deepak S.A., Kottapalli K.R., Rakwal R., Oros G., Rangappa K.S., Iwahashi H., Masuo Y. and Agrawal G.K. 2007. *Curr. Genomics* 8: 234–251.

Frohman M.A. 1993. *Methods in Enzymology* 218: 340–356.

Glisin V., Crkvenja R. and Byus C. 1974. *Biochemistry-US* 13: 2633–2637.

Grzybowska E.A. 2012. *Biochem. Biophys. Res. Commun.* 424: 1–6.

Kunze G., Gaillardin C., Czernicka M., Durrens P., Martin T., Boeer E., Gabaldon T., Cruz J. A., Talla E., Marck C., Goffeau A., Barbe V., Baret P., Baronian K., Beier S., Bleykasten C., Bode R., Casaregola S., Despons L., Fairhead C., Giersberg M., Gierski P.P., Haehnel U., Hartmann A., Jankowska D., Jubin C., Jung P., Lafontaine I., Leh-Louis V., Lemaire M., Marcet-Houben M., Mascher M., Morel G., Richard G.F., Riechen J., Sacerdot C., Sarkar A., Savel G., Schacherer J., Sherman D.J., Stein N., Straub M.L., Thierry A., Trautwein-Schult A., Vacherie B., Westhof E., Worch S., Dujon B., Souciet J.L., Wincker P., Scholz U. and Neuveglise C. 2014. *Biotechnol. Biofuels* 7.

Maskos U. and Southern E.M. 1992. *Nucl. Acids Res.* 20: 1679–1684.

Morin R.D., Bainbridge M., Fejes A., Hirst M., Krzywinski M., Pugh T.J., McDonald H., Varhol R., Jones S.J.M. and Marra M.A. 2008. *Biotechniques* 45: 81.

Puthoff D.P., Nettleton D., Rodermel S.R. and Baum T.J. 2003. *Plant J.* 33: 911–921.

Rauter M., Kasprzak J., Becker K., Riechen J., Worch S., Hartmann A., Mascher M., Scholz U., Baronian K., Bode R., Schauer F., Vorbrodt H. M. and Kunze G. 2016. *Microb Cell Fact* 15.

Velculescu V.E., Zhang L., Vogelstein B. and Kinzler K.W. 1995. *Science* 270: 484–487.

Development of the Thermotolerant Methylotrophic Yeast *Hansenula polymorpha* as Efficient Ethanol Producer

Kostyantyn Dmytruk, Olena Kurylenko, Justyna Ruchala, Olena Ishchuk and Andriy Sibirny

Abstract Until recently, the methylotrophic yeasts, including *Hansenula polymorpha,* have not been considered as a potential producer of biofuels, particularly, ethanol from lignocellulosics. However it is already known that the thermotolerant methylotrophic yeast *H. polymorpha* is capable to ferment xylose, glucose and cellobiose, the main sugars of lignocellulosic hydrolysates, under elevated temperature. These observations allow considering *H. polymorpha* as a promising organism for high temperature alcoholic fermentation in industrial applications. Although the amount of ethanol produced from xylose by the wild-type strains of *H. polymorpha* is extremely low, the successful approaches of metabolic engineering and classical selection had been developed during last decade, which permitted to increase ethanol accumulation from xylose 30-fold. The available strains accumulate 12.5 g of ethanol per liter from xylose at 45 °C. In this article, we present published and new approaches and main achievements on metabolic engineering and selection of *H. polymorpha* for improved producers of ethanol from xylose, starch, xylan, and glycerol, as well as that of strains with increased tolerance to high temperatures and ethanol.

Keywords Yeasts · Ethanol · *H. polymorpha* · Metabolic engineering · Methylotrophic yeasts

---

K. Dmytruk · O. Kurylenko · A. Sibirny (✉)
Institute of Cell Biology, NAS of Ukraine, Drahomanov Street, 14/16, Lviv 79005, Ukraine
e-mail: sibirny@cellbiol.lviv.ua

J. Ruchala · A. Sibirny
Rzeszow University, Zelwerowicza 4, 35-601 Rzeszow, Poland

O. Ishchuk
Department of Biology, Lund University, Sölvegatan 35, 223 62 Lund, Sweden

T. Satyanarayana and G. Kunze (eds.), *Yeast Diversity in Human Welfare*, DOI 10.1007/978-981-10-2621-8_11

1 Introduction

Our civilization faces increasing energy needs to provide the transport sector, heating, and industrial processes. The intensification of the use of fossil fuels results in an increase in greenhouse gases in the atmosphere, negatively affecting the environment. The instability of oil supply and constant fluctuations in its price lead to an increasing interest in alternative energy sources. These economic and ecological factors determine the interest in the use of renewable energy sources (Schubert 2006).

Ethanol is a promising renewable fuel, which over the next 20 years could become one of the main types of biofuel in the transport sector (Hahn-Hägerdal et al. 2006). Ethanol mixed with gasoline is effectively used in ordinary internal-combustion engines. Pure alcohol is also applied in specialized engines. In contrast to gasoline, ethanol has a higher octane number and an increased level of vaporization heat, which makes it a perfect fuel for hybrid automobiles.

To date, to obtain biofuel in sufficient quantities, dedicated crops, sugarcane and beet have been used as raw materials. The ethanol obtained from starch and sugar (the so-called first-generation ethanol) competes with the food industry and agriculture for feedstocks. In contrast, plant biomass (lignocellulose), in particular, agricultural, industrial wood, and domestic waste is inexpensive alternative source for ‘second-generation’ fuel ethanol production. These raw materials are widespread around the globe and are by-products of human activity. Moreover, such waste does not have alternative uses. Other advantages of the second-generation ethanol obtained from lignocellulose include lower emission of greenhouse gases into the atmosphere and positive impact on the environment, especially by reducing global warming; contribution to increase in the employment of people in rural areas; and an improved energy balance as compared to conventional substrates for fermentation (Hill et al. 2006).

However, lignocellulosic biomass conversion to ethanol is a much more complicated process as compared to the production of ethanol from starch or sugar. At the first stages, lignocellulose requires preliminary thermal, hydrostatic pressure and/or chemical treatment. Then, hydrolysis of raw materials treated with sulfuric acid or hydrolytic enzymes (the optimal enzyme activity is within the temperature range of 50–60 °C) is carried out to obtain monomeric, dimeric (mostly cellobiose) and oligomeric sugars. This process is referred to as saccharification. In the monosaccharide mixture obtained, xylose (five carbonic sugar) is the second most abundant after glucose. The fermentation of monomeric sugars in pilot factories that process lignocellulosic raw materials (IoGen, Abengoa, and others) is performed using the yeast *Saccharomyces cerevisiae*, which can ferment only hexoses. However, researchers focus their efforts on the construction of bacterial (*Escherichia coli, Zymomonas mobilis* and *Klebsiella oxytoca*) and yeast (*S. cerevisiae* and *Pichia stipitis*) strains capable of fermenting all sugars released in hydrolysis, especially xylose. Acid hydrolysis is a sufficiently cheap method, but about 30% of released sugars are transformed into furfural and 5-hydroxymethylfurfural. Accumulation of these components results in: (1) decrease of ethanol yield by 30% due to the drop in sugar amount and (2) significant inhibition of fermentation. These negative effects are not observed during enzymatic hydrolysis, since there is only relatively small accumulation of furfural and 5-hydroxymethylfurfural during the pretreatment process. However, monomeric sugars, which are the end products of this hydrolysis type, inhibit the activities of hydrolytic enzymes (cellulases and hemicellulases). To achieve the complete enzymatic hydrolysis of heteropolymers of lignocellulose, the sugars released should be further converted into ethanol by the microorganisms present in the same reservoir (Olofsson et al. 2008).

Enzymatic hydrolysis (saccharification) and fermentation can be performed simultaneously. Such a process is referred to as Simultaneous Saccharification and Fermentation (SSF) (Hahn-Hägerdal et al. 2006), and during its process hydrolytic enzymes are not inhibited by the sugars released since the latter are immediately metabolized by microorganisms. As most conventional microorganisms used ferment at moderate temperatures of around 25–30 °C the cooling of hydrolysis vessel from the temperature of 60 °C is very expensive at industrial scale. The elimination of this cooling step is of great importance. Additional resources could be saved due to the decrease in the temperature difference during the transition from the fermentation stage to distillation (Abdel-Banat et al. 2010).

To date, the efficiency of the SSF of lignocellulosic biomass has not been studied in detail. One of the unsolved problems concerning the conversion of lignocellulose into ethanol is the absence of eukaryotic microorganisms capable of the conversion of xylose into ethanol at elevated temperatures. One of the most perspective organisms capable of the alcohol fermentation of xylose at a temperature of about 50 °C is the yeast *Hansenula polymorpha* (Ishchuk et al. 2009). However, the efficiency of this process using wild-type strains is too low to apply it in industrial production (0.5 g l^{-1} of alcohol during xylose fermentation).

2 Microorganisms for the Production of Ethanol from Lignocellulosic Biomass

Research in the field of alcoholic fermentation of lignocellulosic biomass is carried out in many developed industrial countries. This interdisciplinary research involves scientists from different fields of study: geneticists, biochemists, chemists, technologists, engineers, mathematicians, economists, and others. It was calculated that commercial ethanol production from plant biomass would make it possible to decrease the import of petroleum products into the United States by one third. Ethanol obtained from agricultural wastes results in the release of greenhouse gas into the atmosphere in significantly smaller quantities than that obtained from cereal crops. According to the data of the United States Department of Energy (http://www.energy.gov/), ethanol from lignocellulose decreases the green-house gas emission by 85% as compared to the use of gasoline.

At the same time, ethanol production from starch (for example, maize) release the greenhouse gas only 18–29% lower as compared to gasoline, since natural gas is used for ethanol production. This is another argument for the development of an effective production technology of fuel ethanol from plant biomass. Despite the efforts of many years, profitable technologies to implement this process have not been developed. For convenience, the development of an effective technology of lignocellulose conversion into ethyl alcohol could be divided into two directions. The first one is the optimization of methods of physical, physicochemical, and enzymatic treatment of lignocellulosic wastes in order to obtain mono- and disaccharides from a very complex mixture of biopolymers constituting lignocellulose (cellulose, hemicellulose, and lignin). The second one is the search and construction of recombinant strains of microorganisms capable of the effective fermentation of all sugars of lignocellulose hydrolysates. It is still challenging to effectively ferment xylose, the second most abundant sugar of lignocellulose hydrolysate (on average, this pentose constitutes 30% of the total sugar content in hydrolysates). Although xylose is widespread in nature, it is almost never found as a free monosaccharide; that is why only a sufficiently limited group of microorganisms can use xylose as a sole carbon source, and even fewer can converse this sugar into ethanol.

Scientists from the United States, Canada, Germany, Sweden, Finland, China, South Africa, the Netherlands and other countries are engaged in the construction and improvement of strains that ferment xylose into ethanol. Some researchers construct corresponding strains based on bacteria (*E. coli* and *Z. mobilis*), while others use yeasts. On the whole, fermentation using yeasts has significant advantages, since yeast cells are larger and could therefore be more easily separated from the fermentation medium. Yeasts are not sensitive to phage lysis. Moreover, an effective technology to obtain high amount of ethyl alcohol from glucose or sucrose with the baker's yeast has been known for centuries. Unfortunately, conventional yeasts are not capable of metabolizing xylose. In many laboratories over the world, metabolic engineering of the baker's yeasts is carried out for the purpose of constructing strains fermenting both glucose and xylose. Alternatively, non-conventional yeast species naturally fermenting xylose and their recombinant strains with further improvements of alcoholic fermentation are developing by others.

Significant progress has been made in the isolation and study of non-conventional yeasts fermenting the main sugars of lignocellulose hydrolysates into ethanol, in particular, *Pachysolen tannophilus*, *Candida shehatae*, *Pichia stipitis* and *Spathaspora passalidarum* (Du Preez and van der Walt 1983; Toivola et al. 1984; Du Preez et al. 1986; Long et al. 2012). The conditions of xylose alcoholic fermentation by strains of these nonconventional yeasts have been defined and initial stages of metabolism for the efficient conversion of this sugar into ethanol has been established (Jeffries and Jin 2004). In particular, it has been found that the effective expression of the *XYL1*, *XYL2*, and *XYL3* genes encoding enzymes of the three initial stages of xylose assimilation, xylose reductase, xylitol dehydrogenase, and xylulokinase, respectively, is of primary importance (Jeffries and Jin 2004). It has also been found out that a significant problem for the effective alcoholic fermentation of xylose in yeasts is the imbalance of nicotinamide coenzymes occurring due to the fact that the enzyme of the first stage of xylose utilization, xylose reductase, prefers NADPH as cofactor, while xylitol dehydrogenase (the second-stage enzyme) is NAD-dependent. Such an imbalance leads to a decrease in the efficiency of ethanol synthesis and accumulation of by product xylitol in the medium (Jeffries and Jin 2004). The formation of xylitol could be reduced by changing the cofactor affinity of xylose reductase (Zeng et al. 2009). Significant progress has been made in the heterologous gene expression of bacteria and fungi encoding xylose isomerase (this enzyme converts xylose directly into xylulose and does not require cofactors) in *S. cerevisiae* cells (Matsushika et al. 2009; Brat et al. 2009).

For the yeast *S. cerevisiae*, it has been found that the enhancement of the gene expression encoding enzymes of non-oxidative part of pentose phosphate pathway plays an important role in the optimization of the alcoholic fermentation of pentoses (Kuyper et al. 2005). A dependence of the efficiency of alcoholic fermentation of xylose on the functioning of components of the respiratory chain has also been observed. Strains of *P. stipitis* with mutated genes that encode cytochrome C or SHAM-sensitive terminal oxidase were characterized by increased productivity of ethanol synthesis from xylose (Jeffries and Shi 2000; Shi et al. 2002). An important area of research is the identification and modification of specific transporters of pentoses for the efficient co-metabolism of sugar mixtures (like glucose and pentoses) (Hahn-Hägerdal et al. 2007). To increase the ability of recombinant strains *S. cerevisiae* to simultaneously metabolize the sugar mixture, a new technique of evolutionary engineering was proposed, which is based on the long-term cultivation of recombinant strains with different sugars (Wisselink et al. 2009). All known xylose transporters are competitively inhibited by glucose. Using a growth-based screening platform, two positions in yeast hexose transporters Hxt7 and Gal2 were identify that can be mutated to yield glucose-insensitive xylose transporters. Among the mutant transporters analyzed, Gal2-N376F had the highest affinity for xylose, along with a moderate transport velocity, and had completely lost the ability to transport hexoses. These transporter versions should prove valuable for glucose/xylose co-fermentation in lignocellulosic hydrolysates by *S. cerevisiae* and other biotechnologically important organisms (Farwick et al. 2014).

Despite significant efforts and certain advances in the field of alcoholic fermentation of the main sugars of lignocellulosic hydrolysates (in particular, xylose), researchers still have not been succeeded in achieving effective economic fermentation of these substrates neither by using natural or recombinant strains. To obtain ethanol from lignocellulose hydrolysates in sufficient quantities, it is necessary to further improve the best available yeast strains. Yeast *H. polymorpha* being considered as one of the most perspective organism for that purpose. Below, we present the main achievements in the field of metabolic engineering of the yeast *H. polymorpha* obtained mainly in the authors' laboratory at the Institute of Cell Biology, National Academy of Sciences of Ukraine and Department of Biotechnology and Microbiology, University of Rzeszow, Poland, and aimed in obtaining improved producers of ethanol from xylose, starch, xylan, and glycerol, as well as strains with increased tolerance to high temperatures and ethanol.

The Institute of Cell Biology, National Academy of Sciences of Ukraine, possesses a number of public collection strains of *H. polymorpha* and other strains of this yeast isolated in the former Soviet Union in the 1970s as a result of the search for methylotrophic yeasts for the source of feed protein. The biochemical analysis of eight *H. polymorpha* strains has shown that all strains ferment glucose, xylose, mannose, maltose, and cellobiose into ethanol, while galactose and L-arabinose are almost useless for the growth of these yeasts. An optimal temperature for glucose and xylose fermentation is 37–40 °C; however, even at 45–48 °C, it is still sufficiently intense, which is an unbroken record for eukaryotic alcoholic fermentation. The fermentation process is the most active under the conditions of moderate aeration and with cells being starved of flavin necessary for cell respiration. *H. polymorpha* appeared to be more resistant to ethanol as compared to *P. stipitis*; however, it was more sensitive to that than *S. cerevisiae* (Ryabova et al. 2003).

3 Development of an Auxonographic Method of Selection of *H. polymorpha* Strains with Improved Parameters of Alcoholic Fermentation of Xylose

To select a mutant with increased (or decreased) ethanol production, it is important to develop a simple and reliable method to detect ethanol formed by yeast colonies during their growth on an agar medium. Auxonography is a group of microbiologic methods based on the test cultures, which are microbial strains with certain growth parameters, to determine the effect of antibiotics, mutagenic activity of drugs, and the ability of microorganisms to break down sugars or other organic substances. To select an *H. polymorpha* mutant capable of increased ethanol synthesis in a medium with xylose, a simple auxonographic method has been developed (Grabek-Lejko et al. 2006). In this method, a test culture of mutant strain of *H. polymorpha xyl1Δ* with damaged xylose reductase does not grow on a medium with xylose; however, it effectively assimilates ethanol as a source of carbon and energy. Thus, during the growth of yeast colonies of interest, which efficiently assimilate xylose on an agar medium with this pentose as the only source of carbon, the ethanol will be produced that will cause the growth zones of *xyl1Δ* strain to be formed around such ethanol producing colonies when *xyl1Δ* is added to the medium. The more ethanol is synthesized by a colony, the more the growth zone of the *xyl1Δ* strain and, vice versa, the growth zone of the *xyl1Δ* strain would be either insignificant or absent around colonies that ineffectively synthesize ethanol or do not produce it at all. Using the method developed, it was possible to select a collection of UV-induced mutants capable of improved alcoholic fermentation of xylose. A biochemical analysis of one of these strains revealed a significant decrease in the specific activity of xylose reductase using NADPH as a cofactor, whereas the specific activity of xylose reductase using NADH remained at the level of the wild-type strain (Grabek-Lejko et al. 2006). Such a peculiarity of mutant xylose reductase was in agreement with our further work on site-directed mutagenesis of the cofactor affinity (Dmytruk et al. 2008a), which resulted in an increase of ethanol production during xylose fermentation.

4 Metabolic Engineering of the Initial Stages of Xylose Catabolism in the Yeast *H. polymorpha* for the Construction of Efficient Producers of Ethanol

During the fermentation of xylose in a yeast cell using NADPH-dependent xylose reductase (XR), xylose is reduced to xylitol, which is further oxidized to xylulose by NAD-dependent xylitol dehydrogenase (XDH) (Fig. 1). These redox reactions under limited aeration during fermentation result in the imbalance of NADP and NADH cofactors, leading to extremely low quantities of ethanol synthesized and simultaneous accumulation of by-product, xylitol, in the medium. To eliminate the imbalance, XR and XDH were replaced with bacterial xylose isomerase (XI), which immediately converts xylose into xylulose and does not require cofactors (Fig. 1).

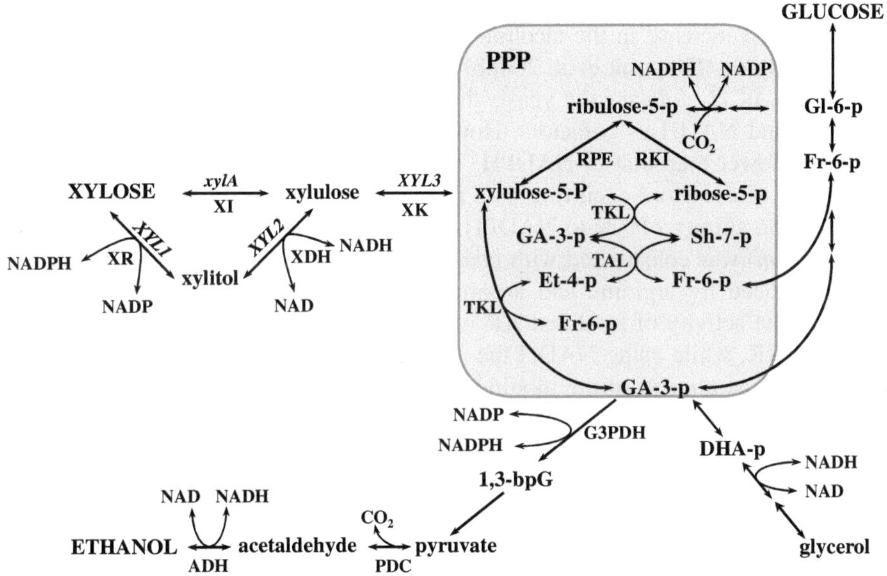

Fig. 1 Pathways of xylose and glucose alcoholic fermentation into ethanol in yeasts. PPP stands for the pentose phosphate pathway; Gl-6-p is glucose 6-phosphate; Fr-6-p is fructose 6-phosphate; Fr-1,6-bp is fructose 1,6-bisphosphate; GA-3-p is glyceraldehyde 3-phosphate; Sh-7-p is sedoheptulose 7-phosphate; Et-4-p is erythrose 4-phosphate; 1,3-bpG is 1,3-bisphosphoglycerate; DHA-p is dihydroxyacetone phosphate; XR is xylose reductase; XDH is xylitol dehydrogenase; XI is xylose isomerase; XK is xylulokinase; TKL is transketolase; TAL is translaldolase; RPE is ribulose 5-phosphate epimerase; RKI is ribose 5-phosphate isomerase; G3PDH is glyceraldehyde 3-phosphate dehydrogenase; PDC is pyruvate decarboxylase; and ADH is alcohol dehydrogenase

For this purpose, *H. polymorpha* strains were constructed with deletions of the *XYL1*, *XYL2-A*, and *XYL2-B* genes, encoding XR and two paralogs of XDH, respectively. The deletion of the *XYL1* gene completely suppressed the activity of XR, while the deletion of the *XYL2-A* gene only reduced the activity of XDH by 30%. An additional deletion of the *XYL2-B* gene reduced the activity of XDH by 65% (Dmytruk et al. 2008b). Further, the *xylA* genes that encode XI of *Escherichia coli* and *Streptomyces coelicolor* were introduced into the genome of the recipient strain (*xyl1Δ xyl2-AΔ xyl2-BΔ*), and expressed from strong constitutive promoter of *H. polymorpha,* which resulted in transformants with functionally active heterologous enzymes. The presence of the heterologous XI protein of *E. coli* in cells of the strains constructed was proven by the visualization of additional bands with a molecular mass of 46 kDa, which corresponds to the theoretically calculated XI mass on electropherograms of cell-free extracts of recombinant strains under denaturing conditions (Voronovsky et al. 2005). The specific activity of XI of recombinant strains reached 80% of the specific activity of the enzyme in bacterial cells of *E. coli.* The yeast strains constructed grew on medium with xylose as the only source of carbon; however, the efficiency of the alcoholic fermentation of xylose remained at a level of the wild-type strain. The increase in the alcoholic fermentation of xylose was promoted by the additional overexpression of homologous gene *XYL3* encoding xylulokinase (XK). The recombinant strain, coexpressing genes *xylA E. coli* and *XYL3 H. polymorpha*, was characterized by a two-fold increase in the specific activity of XK and also by a 3.5- to 4-fold increase in the alcoholic fermentation of xylose as compared to the wild-type strain (Dmytruk et al. 2008b).

The XR of *H. polymorpha* yeasts during an enzymatic reaction could use both NADPH and NADH as cofactors. However, the affinity of XR to NADH is significantly lower than that to NADPH. Thereby, another approach to eliminate the imbalance of cofactors occurring as a result of the actions of XR and XDH is to decrease the affinity of XR to NADPH. Using site-directed mutagenesis, modified XR (*XYL1m*) was constructed with lysine and asparagine at positions 341 and 343 being replaced by arginine and aspartic acid, respectively. Using NADPH as a cofactor, the activity of modified XR was decreased nine-fold as compared to that of native XR, while using NADH the activity of XR remained unchanged in both cases. Thus, as a result of the modification of the primary structure of the protein, the affinity of XR to NADPH was decreased 17-fold as compared to the native enzyme, while the affinity of modified XR to NADH remained unchanged. A recombinant strain of *H. polymorpha* with enhanced expression of modified XR and native XDH and XK was constructed, which was characterized by 7.4-fold increase in the alcoholic fermentation of xylose and a simultaneous five-fold decrease in the production of xylitol as compared to the wild-type strain (Dmytruk et al. 2008a). During xylose fermentation, the recombinant strain of *H. polymorpha* with enhanced expression of modified XR and native XDH and XK significantly improved the strain in converting xylose using bacterial XI and native XK in the quantity of ethanol synthesized. Recombinant strains of *H. polymorpha* with increased production capacity of ethanol during xylose fermentation could be used as microorganisms for a further increase in the efficiency of alcoholic fermentation of basic sugars of lignocellulosic hydrolysates at elevated temperatures using methods of metabolic engineering.

In a more recent work (Kurylenko et al. 2014), the above mentioned approaches for the metabolic engineering were combined with classical selection of ethanol overproducing strain using selection for 3-bromopyruvate resistance. It was demonstrated that there is a positive cumulative effect of the overexpression of engineered XR and native XDH and XK on ethanol production from xylose on the background of $2EthOH^-$ mutant. The *H. polymorpha* mutant $2EthOH^-$ is unable to utilize ethanol as a sole carbon source. It is isolated from strain NCYC495 by UV mutagenesis and characterized by a three-fold increase in ethanol accumulation as compared to parental strain (Ishchuk et al. 2008). The overexpression of the genes *XYL1m, XYL2* and *XYL3* on the background of non-identified mutation in the strain $2EthOH^-$ led to a substantial increase in ethanol accumulation during xylose fermentation (7.44 g l^{-1} at 45 °C relative to 0.6 g l^{-1} in the wild-type strain NCYC495 and 2.05 g l^{-1} in the parental strain $2EthOH^-$—Kurylenko et al. 2014). The further increase in ethanol production from xylose (to 10 g l^{-1} at 45 °C) was obtained in mutants selected on a medium supplemented with toxic concentrations of 3-bromopyruvate, in a similar way to that described for ethanol accumulation form glucose in *S. cerevisiae* (Dmytruk et al. 2016).

The maximal observed level of ethanol produced from xylose by the best isolated strains $2EtOH^-$/XYL1m/XYL2/XYL3/BrPA (near 10 g l^{-1} at 45 °C) represented an ethanol yield of 0.3 g g^{-1} from xylose. These results make *H. polymorpha* close to the other known best xylose-fermenting organisms relative to ethanol production under conditions of well-suitable for the use in SSF process. However, the ethanol yield in *H. polymorpha* is still lower than that reported for other mesophilic xylose fermenting organisms such as *P. stipitis* (0.35–0.44 g g^{-1} xylose) and *S. passalidarum* (0.42 g g^{-1} xylose) (Jeffries et al. 2007; Long et al. 2012), but similar to the best engineered strain of the thermotolerant yeast *Kluyveromyces marxianus* (0.31 g g^{-1} xylose under anaerobic conditions at 45 °C) (Wang et al. 2013). On the other hand, the ethanol productivity in the best isolated strain of *H. polymorpha* is much higher than the best engineered strain of *K. marxianus* (0.179 g l^{-1} h^{-1} versus 0.054 g l^{-1} h^{-1} at 45 °C) (Kurylenko et al. 2014). To be industrially feasible, ethanol yield in *H. polymorpha* has to be further increased to be close to the theoretical maximum yield, which is 0.51 g g^{-1} xylose.

5 Autophagy-Related Gene *ATG13* is Important for Xylose Fermentation

The 3-bromopyruvate appears to be useful antimetabolite for selection of *H. polymorpha* strains with increased performances of xylose alcoholic fermentation (Kurylenko et al. 2014). Insertional mutagenesis was used to shed light on the nature of ethanol overproducing mutants of *H. polymorpha* resistant to 3-bromopyruvate.

Insertion cassette pL2 (Dmytruk et al. 2006) was used for transformation of NCYC495 strain of *H. polymorpha* with subsequent selection of transformants on mineral medium supplemented with 25 mM of 3-bromopyruvate. One of selected transformant revealed reproducible increase in ethanol accumulation during xylose fermentation. Such transformant #63 possessed 40% increase in ethanol production as compared to the parental strain reaching 1.5 g of ethanol l^{-1} (Fig. 2). Southern analysis confirmed the presence of a single copy of insertion cassette in the genome of strain #63 (data not shown). Sequencing of flanking regions revealed that the insertional cassette disrupted the ORF of a gene homologous to the *S. cerevisiae* gene *ATG13* encoding a regulatory subunit of the Atg1p signaling complex, stimulating Atg1p kinase activity, which is required for vesicle formation during autophagy and the cytoplasm-to-vacuole targeting pathway. It was shown that insertion cassette disrupts gene *ATG13* at the position +1272 bp from the initial ATG codon.

Confirmation that the observed increase in xylose alcoholic fermentation performance of strain #63 is a result of insertion cassette integration, rather than the secondary mutation occurring elsewhere in the genome, was an essential part of the study. For construction of deletion cassette, gene *LEU2* of *S. cerevisiae* was flanked with non-coding regions of the *ATG13* gene. The *atg13Δ* strain of *H. polymorpha* was constructed by homologous recombination of the deletion cassette. The *atg13Δ* strain produced elevated amount of ethanol from xylose, similar to that of insertional strain (Fig. 2). Obtained results let us conclude that the autophagy-related gene *ATG13* is somehow involved in regulation of xylose alcoholic fermentation in the yeast *H. polymorpha*. However, mechanisms of such regulations remain to be elucidated. It is important to note that *atg13Δ* mutant showed defects in autophagic degradation of peroxisomal protein alcohol oxidase whereas insertion mutant #63 did not (data not shown). Thus, the role of Atg13p protein in autophagy apparently differs from that in regulation of xylose fermentation.

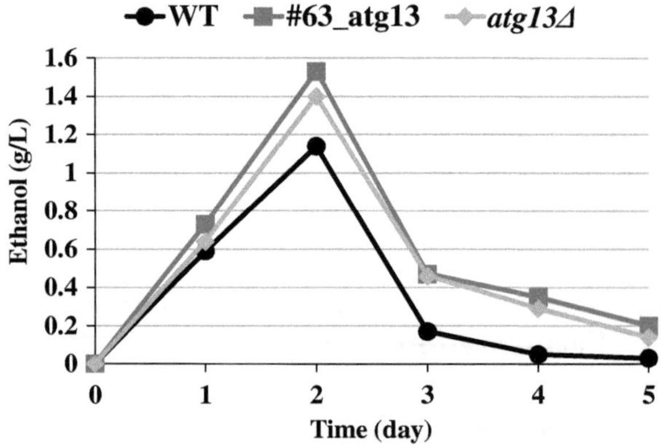

Fig. 2 Ethanol production during fermentation of *H. polymorpha* strains on xylose containing medium. *Circles* represent wild-type strain; *squares and diamonds* represent #63 and *atg13Δ* strains, respectively. The data represent means of typical single cultivation. Alcoholic fermentation was performed in 12% xylose containing medium at 45 °C under semi-anaerobic condition (140 revolutions/min) during 5 days as described elsewhere (Kurylenko et al. 2014)

6 Peroxisomal Transketolase (Dihydroxyacetone Synthase) and Transketolase are Important for Xylose Fermentation in *H. polymorpha*

Further increase in metabolic flux from xylose to ethanol required additional new approaches. All known microorganisms convert xylose to ethanol using the sequential action of the pentose phosphate pathway (PPP) and glycolysis. Xylose is first taken up by cell, and then is converted to xylulose-5-phosphate. In four reactions of the non-oxidative part of the PPP (ribulose-5-phosphate epimerase, ribose-5-phosphate isomerase, transketolase and transaldolase, encoding by *RPE1, RKI1 TKL1* and *TAL1* genes, respectively), xylulose-5-phosphate is converted to different compounds, including the glycolytic intermediates fructose-6-phosphate and glyceraldehyde-3-phosphate (Fig. 1). The last compound can be converted to ethanol. Our interest to PPP is due to the observation of severe decrease in xylose alcoholic fermentation by *H. polymorpha* strains with defects in peroxisome biogenesis (so-called *pex* strains) (Fig. 3). In *pex* strains peroxisomal enzymes are mislocalized to the cytosol which results most frequently in degradation and/or inactivation of these enzymes (Kiel and Veenhuis 2000). Among the proteins of the peroxisomal matrix, our attention was focused on peroxisomal enzymes potentially involved in xylose catabolism. One candidate for such a role is the enzyme peroxisomal transketolase, also known as dihydroxyacetone synthase Das1p that is involved in methanol metabolism but also can catalyze "classical" transketolase reaction (Waites and Quayle 1981). Another candidate is the enzyme peroxisomal transaldolase Tal2p with unknown functions.

To investigate the role of these enzymes in ethanol production during xylose fermentation, the corresponding genes *DAS1* and *TAL2* were overexpressed in *H. polymorpha* NCYC495 strain under control of strong constitutive promoter of *GAP1* gene (encodes glycerol-3-phospate dehydrogenase) that used a plasmid for multicopy integration pGLG61 (Sohn et al. 1999). The recombinant strains overexpressing *DAS1* and *TAL2* revealed 4.6- and 1.5-fold increase in the specific activity of the corresponding enzymes (Table 1).

The performance of xylose alcoholic fermentation by constructed strains was studied. The overexpression of *TAL2* gene resulted in a 1.5-fold increase in ethanol production at fourth day of xylose fermentation as compared to the wild-type strain (Fig. 3). The effect of the overexpression of *DAS1* gene was more pronounced. Strain overexpressing *DAS1* gene synthesized 2.3-fold more ethanol than that of the parental strain after four days of xylose fermentation (Fig. 3). As a consequence, it was shown for the first time that derepression of the peroxisomal enzymes Das1p and Tal2p resulted in the activation of xylose alcoholic fermentation in *H. polymorpha*.

The *H. polymorpha VPS34* gene encoding phosphatidylinositol 3-kinase is involved in endocytosis and vacuolar protein sorting. The *vps34Δ* mutant of *H. polymorpha* is not capable of the selective degradation of peroxisomes (Kiel et al. 1999). The efficiency of alcoholic fermentation of glucose by the *vps34Δ*

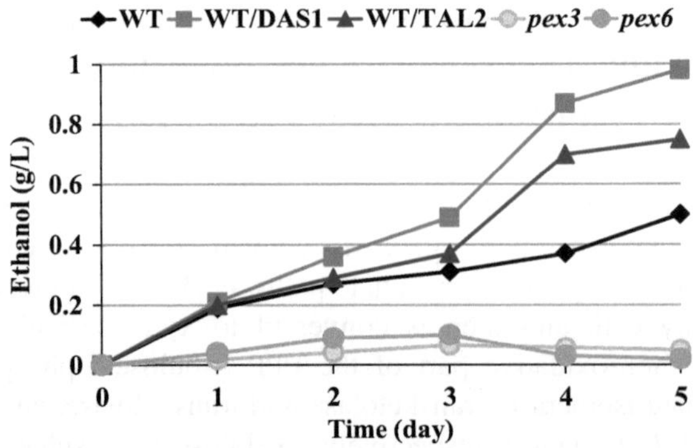

Fig. 3 Ethanol production during fermentation of *H. polymorpha* strains on xylose containing medium. *Diamonds* represent wild-type strain; *squares and triangles* represent strains overexpressing *DAS1* and *TAL2* gene, respectively; *hatched and closed circles* represent *pex3* and *pex6* mutants, respectively. The data represent means of a typical single cultivation. Alcoholic fermentation was performed in 12% xylose containing medium at 37 °C under semi-anaerobic condition (140 revolutions/min) during 5 days as described elsewhere (Kurylenko et al. 2014)

Table 1 The activities of the enzymes dihydroxyacetone synthase (DHAS) and transaldolase (TA) in *H. polymorpha* transformant with overexpressed *DAS1* or *TAL2* gene when compared to the control strain during cultivation in a xylose containing medium

Strain	Activity (U/mg protein)	
	DHAS	TA
WT	0.008 ± 0.001	0.311 ± 0.015
DAS1	0.037 ± 0.009	–
TAL2	–	0.497 ± 0.017

The specific activity of DHAS was determined spectrophotometrically as described before (Waites et al. 1981). The specific activity of transaldolase was measured as previously described (Bergmeyer et al. 1974). The cells used for enzyme assay were taken from the mid-exponential growth phase. All assay experiments were repeated at least twice

– Not determined

H. polymorpha strain was 1.7-fold increased as compared to the wild-type strain (Grabek-Lejko et al. 2011), which also suggests the involvement of peroxisomes in the alcoholic fermentation of carbon substrates.

7 Deletion of Transcription Activator *CAT8* Activates Xylose Alcoholic Fermentation in *H. polymorpha*

The *CAT8* gene encodes a zinc-finger cluster protein that mediates derepression of a number of genes during diauxic shift of the transition between fermentative and nonfermentative metabolism (Hedges et al. 1995). Genomic studies have shown that at least 30 genes, encoding proteins involved in gluconeogenesis, ethanol utilization, and the glyoxylate cycle, are regulated by Cat8p (Haurie et al. 2001; Tachibana et al. 2005).

It was recently shown that the deletion of *CAT8* gene coding for the carbon source-responsive transcriptional regulator in the yeast *Pichia guilliermondii* resulted in the shift from glucose metabolism toward a fermentative metabolism at both metabolic and transcriptional levels. The *cat8Δ* mutant had a 20-fold increase in ethanol production as compared to the wild type under aerobic fermentation of glucose (Qi et al. 2014). Furthermore, deletion of the *CAT8* genes slightly but statistically and significantly improved the fermentation rate of a laboratory yeast strain of *S. cerevisiae* (Watanabe et al. 2013).

A similar strategy was applied to the xylose fermenting yeast *H. polymorpha* to improve its fermentation efficiency and to induce respire-fermentative metabolism. This we believe is due to a defect in the transcription of genes involved in gluconeogenesis that would redirect more xylose toward fermentation. The *cat8Δ* strains of *H. polymorpha* were constructed on the background of the wild-type strain. Gene conferring resistance to the antibiotic nourseothricin was flanked with noncoding regions of *CAT8* gene. A deletion cassette was used for specific site integration via homologous recombination. The deletion mutants had a defect in growth on the gluconeogenic substrates (glycerol, ethanol) whereas growth of this strain on glucose and xylose was not affected. The *cat8Δ* mutants did not show changes in ethanol production in glucose medium whereas accumulated up to three-fold more ethanol in the medium with xylose (Fig. 4). The *cat8Δ* mutants isolated from the most advanced ethanol producer from xylose using hygromycin based deletion cassette also did not show any differences in ethanol production in glucose medium whereas accumulated 25–30% more ethanol in the medium with xylose. Maximal accumulation from xylose reached 12.5 g of ethanol per liter at 45 °C, which exceeds ethanol accumulation in the wild-type strain NCYC495 near

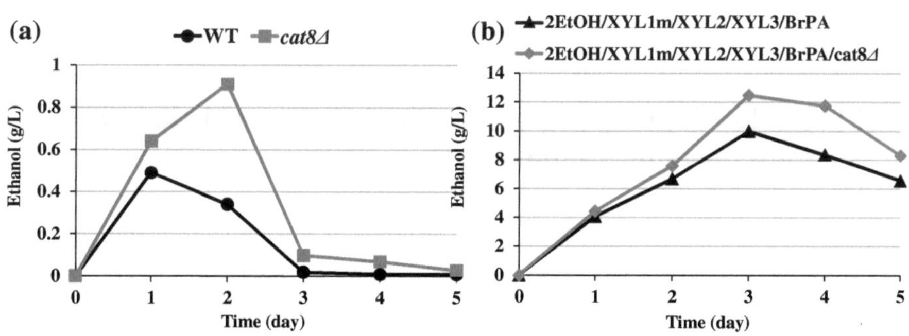

Fig. 4 Ethanol production during fermentation of *H. polymorpha* strains on xylose containing medium. **a** *circles* represent wild-type strain; *squares* represent *cat8Δ* strain, **b** *triangles and diamonds* represent 2EtOH$^-$/XYL1 m/XYL2/XYL3/BrPA and 2EtOH$^-$/XYL1 m/XYL2/XYL3/BrPA/cat8Δ, respectively. The data represent means of typical single cultivation. Conditions of alcoholic fermentation are the same as indicated in Fig. 2

Table 2 Respiration rates of *H. polymorpha* strains. Cells were grown to the late exponential phase in mineral media supplemented with glucose or xylose, collected, washed in cold distilled water and starved in water for 16–18 h

Strain	O_2 consumption (nanomol of O_2 consumed/min/mg of dry mass)	
	Glucose	Xylose
WT	8.49 ± 0.43	13.65 ± 0.71
Δcat8	7.66 ± 0.42	8.61 ± 0.39
2EtOH$^-$/XYL1m/XYL2/XYL3/BrPA	10.51 ± 0.56	5.51 ± 0.28
2EtOH$^-$/XYL1m/XYL2/XYL3/BrPA/Δcat8	10.42 ± 0.57	4.02 ± 0.26

The respiration rate was measured at 30 °C by Yellow Springs Insturment Co. Clark oxygen electrode (model YSI 5300) in a 5 ml reaction vessel. Determinations were performed in distilled air-saturated water with the concentration of cells 0.5 g l^{-1} of dry weight and started by addition of 1% carbon substrate (glucose or xylose). The respiratory rate was expressed as nanomoles of O_2 consumed per minute per mg of cells (dry weight)

25 times (Fig. 4). It was shown that *cat8Δ* mutation resulted in 160 and 140% reduction of respiration rates on xylose as compared to wild-type and 2EtOH$^-$/XYL1m/XYL2/XYL3/BrPA strains, while the respiration capabilities on glucose were not affected (Table 2). It is also important to emphasize that *cat8Δ* mutant did not show any changes in ethanol production during glucose fermentation (not shown), which totally differs from data obtained in *S. cerevisiae* and *P. guilliermondii* (Watanabe et al. 2013; Qi et al. 2014). To summarize, it could be concluded that the transcription regulator Cat8p is apparently involved in repression of xylose alcoholic fermentation and consequently its damage strongly activates this process. Moreover, deletion of *CAT8* strongly reduced cells respiration capacity on xylose directing this substrate from respiration to fermentation processes. Such an approach could be useful for the construction of the industrial xylose fermenting strains with a knock out of the *CAT8* gene to enable further improvement in ethanol production from xylose.

8 Metabolic Engineering of the Final Stages of Ethanol Synthesis During Fermentation

Pyruvate decarboxylase (PDC) is a key enzyme of alcoholic fermentation. This enzyme catalyzed the conversion of pyruvate into acetaldehyde and CO_2 (Fig. 1). Attempts to overexpress the *PDC1* gene encoding PDC in *S. cerevisiae* did not improved the ethanol yield during glucose fermentation (Schaaff et al. 1989; van Hoek et al. 1998).

The *PDC1* gene of *H. polymorpha* was cloned and the effect of overexpression of this gene on the alcoholic fermentation of basic sugars of lignocellulose was studied. At the introduction of multiple copies of expression cassette with the *PDC1* gene under the control of a strong constitutive promoter of the glyceraldehyde

3-phosphate dehydrogenase (*GAPDH*) gene resulted in a 40-fold increase in the specific activity of PDC. At the same time, the recombinant strain constructed was characterized by two-fold increase in the alcoholic fermentation of xylose as compared to the initial strain. The alcoholic fermentation of glucose was also two-fold increased (Ishchuk et al. 2008). Similar results with an increase in the ethanol production from glucose and xylose under the conditions of optimal (37 °C) and elevated (48 °C) temperatures were achieved by overexpression of heterologous gene (*Kluyveromyces lactis PDC1*) in the system of *H. polymorpha* (Ishchuk et al. 2008). Thus, overexpression of the *PDC1* gene could be used as one of the approaches for construction of improved producers of ethanol (*H. polymorpha*) from lignocellulose hydrolysates at an increased temperature.

Yeast alcohol dehydrogenase (ADH) is an enzyme catalyzing the final metabolic reaction during alcoholic fermentation. ADH catalyzes a reversible reducing reaction of acetaldehyde into ethanol with the accompanying oxidation of NADH (Fig. 1). The *ADH* gene sequences are quite conservative in yeasts; however, the regulation, physiological functions, and number of genes differ in different species. Seven *ADH* genes (*ADH1–ADH7*) are identified in *S. cerevisiae*, *P. stipitis*, and *H. polymorpha*, while there is only four *ADH* genes observed in *Kluyveromyces lactis* (*ADH1–ADH4*) (Suwannarangsee et al. 2010). The *ADH1* and *ADH2* genes and their protein products have been studied in detail as they play a key role in alcoholic fermentation, namely, in the synthesis and utilization of ethanol (Denis et al. 1983; Lutstorf and Megnet 1968). In contrast to baker's yeasts, Adh1p *P. stipitis* was activated both at limited aeration (fermentation conditions) and during growth on a respiratory substrate—ethanol (Cho and Jeffries 1998; Passoth et al. 1998). Comparing the catalytic properties of the ADH of different types of yeasts, it has been found that Adh1p *H. polymorpha* exhibits increased catalytic activity during the oxidation of ethanol and a similar level of acetaldehyde reduction (Suwannarangsee et al. 2010). The deletion of the *ADH1* gene in *H. polymorpha* resulted in a decrease in ethanol synthesis during the fermentation of both glucose and glycerol. Overexpression of *ADH1* in *adh1Δ* strain lead to a two-fold increase in the amount of ethanol synthesized during glucose fermentation and to an insignificant improvement of the conversion of glycerol into ethanol (Suwannarangsee et al. 2010). The construction of yeast strains that effectively convert glycerol into ethanol is of growing interest as it is produced in large quantities as a byproduct of biodiesel production. Mitochondrial Adh3p *H. polymorpha* was also identified and characterized (Suwannarangsee et al. 2012). It was shown that deletion of *ADH3* did not affect the cell growth on different carbon sources. However, when *Δadh3* mutant was complemented by an *ADH3* expression cassette fused to a strong constitutive promoter, the resulting strain produced a significantly increased amount of ethanol relative to the wild-type strain in a glucose containing medium. In contrast, in a xylose medium, the ethanol production was substantially reduced in *ADH3* overyexpressing strain as compared to that in the wild-type strain (Suwannarangsee et al. 2012). Taken together, *ADH* genes would be useful engineering target to develop *H. polymorpha* as a substrate specific bioethanol production strain.

9 Study of the Role of the *ATH1* Gene in the Thermotolerance and Mobilization of Trehalose in *H. polymorpha*

Trehalose plays an important role in the thermotolerance of *H. polymorpha*. In this and other yeasts, trehalose synthesis is one of the elements of the response to heat shock. Reinders et al. (1999) have found that at a loss of the activity of trehalose 6-phosphate synthase (the first enzyme of the trehalose synthesis pathway) *H. polymorpha* cells are not able to synthesize trehalose and become sensitive to heat shock. However, for *H. polymorpha*, the effect of an increased intracellular level of trehalose on the thermotolerance has not been studied earlier. In *S. cerevisiae*, the intracellular level of trehalose is maintained due to the balanced action of the synthesis and hydrolysis enzymes of this sugar (Kim et al. 1996). Two enzymes are involved in the trehalose hydrolysis, namely, acid trehalase (encoded by *ATH1*) and neutral trehalase (encoded by the *NTH1* gene) (Londesborough and Varimo 1984). Acid trehalase is required for the growth of *S. cerevisiae* in a medium with trehalose as the only source of carbon and energy (Nwaka et al. 1996). The deletion of the *ATH1* gene in *S. cerevisiae* increases the level of intracellular trehalose more than that of the *NTH1* gene (Kim et al. 1996). It was found that the *ath1Δ* mutant of *S. cerevisiae* is less sensitive to stress factors, such as drying, incubation at low temperatures, ethanol and osmotic stresses (Kim et al. 1996; Parrou et al. 2005). The decrease in the activity of acid trehalase in *S. cerevisiae* increases the tolerance to ethanol and the productivity of alcoholic fermentation (Jung and Park 2005).

To determine the role of the *ATH1* gene in the thermotolerance and mobilization of trehalose in *H. polymorpha*, the *ath1Δ* strain was constructed with the deletion of the corresponding gene. Similarly to the *ath1Δ* mutant of *S. cerevisiae*, the *ath1Δ* strain of *H. polymorpha* does not grow in a medium with trehalose as the only source of carbon and energy. After the loss of the activity of acid trehalase, the *ath1Δ* mutant of *H. polymorpha* accumulates in cells 1.5–2.3 times more trehalose as compared to the wild-type strain at optimal and elevated temperatures. The increase in the intracellular level of trehalose in *H. polymorpha* positively affected the thermotolerance of the cells, since the resistance of the *ath1Δ* mutant of *H. polymorpha* to heat shock had increased two-fold and the ethanol production increased by six-fold during xylose fermentation at 50 °C (Ishchuk et al. 2009).

10 Increase in the Thermotolerance of Recombinant Strains of *H. polymorpha* by Enhanced Expression of the Hsp16p and Hsp104p Heat Shock Proteins

Low molecular mass heat shock proteins belong to the subfamily of proteins with molecular masses of monomers from 12 to 43 kDa (Haslbeck et al. 2005). Heat shock proteins suppress the aggregation of denatured proteins and facilitate the interaction of damaged proteins with other chaperones to restore the native structure (Kitagawa et al. 2002). It was shown that low-molecular mass heat shock protein *S. pombe* Hsp16p is required for the mRNA export under the conditions of heat shock (Yoshida and Tani 2005). The ortholog of this protein in the yeast *S. cerevisiae* is a typical cytosolic heat shock protein functioning as a chaperone under the conditions of normal and increased temperatures (Haslbeck et al. 2004). The closest identified homolog of the corresponding proteins of *S. pombe*—*H. polymorpha* Hsp16p—has a molecular mass of 36.2 kDa. The Hsp104p protein is a representative of the AAA + subfamily of heat shock proteins involved in structural changes in proteins and protein complexes (Cashikar et al. 2005). The Hsp104p of *S. cerevisiae* restores the structure of denaturated proteins using the energy of ATP and a system of co-chaperons—Hsp70p/Hsp40p (Weibezahn et al. 2004). The expression of *S. cerevisiae* Hsp104p is low at normal temperatures and significantly rises at elevated temperature (Lindquist and Kim 1996). The Hsp104p plays a crucial role in the survival of cells during yeast cultivation at extreme temperatures (Parsell and Lindquist 1993). The expression of this gene provides *S. cerevisiae* cells with thermotolerance (Lindquist and Kim 1996). Hsp104p is a highly conserved protein. The amino acid sequence of *H. polymorpha* Hsp104p has 64% homology with *S. cerevisiae* Hsp104p. Guerra et al. (2005) has found that, similar to that of *S. cerevisiae*, the expression of *H. polymorpha* Hsp104p is induced by an increase in the temperature.

To study the effect of hyperexpression of heat shock proteins on the thermotolerance and high temperature fermentation of *H. polymorpha*, strains with enhanced expression of homologous genes *HSP16* and *HSP104* under the control of the promoter of the *GAPDH* gene were constructed. The transformants containing expression cassettes *HSP16* and *HSP104* were characterized by 2- and 10-fold increased survival rate under the conditions of heat shock, respectively. The synergetic effect of the combined overexpression of Hsp16p and Hsp104p was observed at a 12-fold increase in the survival rate under the conditions of heat shock (Ishchuk et al. 2009).

A study of the high temperature fermentation of xylose showed that *H. polymorpha* strains overexpressing genes of heat shock proteins were characterized by improved parameters of alcoholic fermentation under maximal non-lethal temperature; in particular, these strains produced 3–6 times more ethanol at the maximal temperature for *H. polymorpha* (50 °C). At lower temperatures (37 and 48 °C), the production of ethanol in the HSP transformant was at a level of the control strain. Since 50 °C is the maximal temperature that *H. polymorpha* can withstand and the growth at such a temperature is significantly depressed, we suppose that under these conditions a significant quantity of proteins denature. This assumption explains the positive effect of the overexpression of the HSP gene during xylose fermentation at 50 °C (Ishchuk et al. 2009).

11 Identification of *H. polymorpha* Genes Determining Resistance to Ethyl Alcohol

To clone *H. polymorpha* genes providing resistance to ethyl alcohol, experiments using insertional mutagenesis were carried out. After the transformation of the recipient strain of *H. polymorpha* with an insertion cassette, a transformant collection was generated with the cassette being inserted into random chromosomal loci. The minimal toxic concentration of ethanol that suppressed the growth of *H. polymorpha* was determined (it was 7%). After screening transformants on a medium with 7% ethanol, an *H. polymorpha* (7E) insertion mutant was selected that was incapable of growing on such a medium. The 7E strain was 300–500 times more sensitive to exogenous ethanol than the recipient strain. The mutation introduced did not damage the ethanol catabolism but only affected the mechanisms of resistance to increased concentrations of this alcohol.

An analysis of the nucleotide sequences flanking the insertion cassette revealed damage of the gene with a sufficiently high degree of homology (39%) with the *S. cerevisiae MPE1* gene. The *S. cerevisiae MPE1* gene encodes an essential protein—a component of the factor of cleavage and polyadenylation of mRNA at its maturation. In contrast to *S. cerevisiae*, *H. polymorpha MPE1* is not an essential gene and is responsible for resistance to ethanol. Thereby, the *H. polymorpha* gene identified was referred to as *ETT1* (ethanol tolerance); 7E mutant, as *ett1*. The *S. cerevisiae MPE1* does not complement the mutation of *H. polymorpha ett1*; however, partial complementation was provided by the *MPE1* gene of other xylose fermenting yeast, *P. stipitis* fermenting. The data obtained indicated that, similar to *H. polymorpha* Ett1p, *P. stipitis* Mpe1p is responsible for resistance to ethanol, and the *S. cerevisiae* Mpe1p obviously does not have such a function. It has been found that the amino acid sequences of the *S. cerevisiae MPE1*, *H. polymorpha ETT1*, and *P. stipitis MPE1* genes contain common conserved domains: DWNN, an ubiquitin-like domain, and a zinc finger motif. Differences are revealed during a detailed analysis of the DWNN domains of the *S. cerevisiae, P. stipitis* and *H. polymorpha* proteins. According to the amino acid sequences, the DWNN domain has a homology with ubiquitin and could be involved in the covalent modification of proteins by ubiquitin (Pugh et al. 2006). It is known that ubiquitin contains conservative amino acid residues of lysine (K), which are sites of attachment of additional ubiquitin molecules, promoting the formation of polyubiquitin chains (Weissman 2001). After the polyubiquitin chains bound to K48, K11, and K29 have been recognized by the 26S proteasome, the degradation of the protein modified by ubiquitin is observed. The chains bound to K6 and K63 are involved in numerous nonproteolytic processes, e.g. in the response to stress, DNA reparation, and endocytosis (Passmore and Barford 2004). Ubiquitin also contains two conserved glycine residues (GG motif) at the C-end of the molecule. The GG motif is the recognition site of the protease that cleaves the bond between amino acid residues of glycine and triggers the conjugation of ubiquitin (Pugh et al. 2006). A comparison of the ubiquitination sites of the DWNN domains of *S. cerevisiae,*

P. stipitis Mpe1p, and *H. polymorpha* Ett1p revealed differences in the conserved residues of amino acids typical for ubiquitination. The DWNN domain of *H. polymorpha* Ett1p contains the GG motif and K6, *P. stipitis* has only K6, while *S. cerevisiae* totally lacks the corresponding conserved residues for ubiquitination. These data could explain why the *P. stipitis MPE1* gene, but not the corresponding homolog of *S. cerevisiae*, partially complements the *ett1* mutation of *H. polymorpha*.

To study the effect of the *ETT1* expression on the resistance of *H. polymorpha* to ethyl alcohol, a strain with enhanced expression of the corresponding gene was constructed. The overexpression of the *ETT1* gene significantly increased the resistance of *H. polymorpha* to ethanol, resulting in 10- and three-fold improvements in the growth on agar and liquid media with ethanol, respectively.

Beside the resistance to ethanol, the *H. polymorpha ETT1* gene is also responsible for the resistance to different types of stress. The *H. polymorpha* strain with enhanced expression of *ETT1* exhibited increased resistance to the denaturant—azetidine-2-carboxylic acid (AZC) and heat shock and is characterized by improved growth kinetics at increased temperatures of 49 and 50 °C. The *ett1* mutant was not able to grow at an elevated temperature. The quantity of ethanol synthesized by this strain during xylose fermentation was slightly decreased.

The *H. polymorpha* Ett1p is also involved in maintaining the integrity of the cell wall, as the *ett1* mutant does not grow in a medium with sodium dodecyl sulfate (SDS), and the strain with enhanced expression of *ETT1* is characterized by improved growth. Since the overexpression of the *ETT1* gene in *H. polymorpha* is accompanied by increased resistance to a number of stress factors, such as increased temperatures, ethanol, AZC, and SDS, inducing denaturation of proteins, it might be assumed that the *ETT1* expression is especially important under the conditions of denaturation of proteins. This assumption has been proved by data on the growth characteristics of the *H. polymorpha* strains in a medium with rapamycin. The *H. polymorpha* strain with enhanced expression of *ETT1* exhibits increased resistance to rapamycin. It is known that rapamycin inhibits the protein kinase TOR (target of rapamycin), which is involved in the cellular response to starvation. Under the conditions of depletion of nutrients, the signaling pathways of the protein kinase TOR and protein kinase A cooperatively block the cell cycle and activate a cell response to stress (Cardona et al. 2009).

12 Heterologous Expression of the *S. cerevisiae MPR1* Gene in *H. polymorpha*

Beside the denaturation of proteins and the change in the plasma membrane fluidity, ethanol increases the formation of free radicals of oxygen (Costa et al. 1997). The *S. cerevisiae MPR1* gene encoding acetyltransferase is involved in the protection of yeast cells from free radicals of oxygen under the conditions of ethanol stress

(Nomura and Takagi 2004). Since the *H. polymorpha* genome does not contain genes homologous to the acetyltransferase of baker's yeast, the effect of the heterologous *S. cerevisiae MPR1* gene on the tolerance of *H. polymorpha* to increased concentrations of ethanol was studied. The expression cassette contained the *S. cerevisiae MPR1* gene under the control of a strong constitutive promoter of the *GAPDH* gene, which was further introduced into the genome of *H. polymorpha.* The *H. polymorpha* transformants expressing *S. cerevisiae MPR1* had increased resistance to AZC and ethanol as compared to the initial strain (Ishchuk et al. 2010). The resistance level correlated with the number of copies of the expression cassette in the transformant genome. The transformants containing three copies of the *S. cerevisiae MPR1* gene were more resistant to AZC and ethanol than integrants carrying one copy of the gene. Thus, the resistance of *H. polymorpha* to ethyl alcohol could be enhanced by the overexpression of the heterologous *S. cerevisiae MPR1* gene encoding acetyltransferase.

13 Direct Conversion of Starch and Xylan into Ethanol Using Recombinant Strains of *H. polymorpha*

Direct microbial conversion of carbon polymers into ethanol is a perspective technology which could provide profitable production of alcohol from lignocellulosic raw material. One of the key preconditions for the development of this technology is the search, selection, or construction of microorganisms fermenting starch and xylan into ethanol at increased temperatures (Kadam and Schmidt 1997). The optimal temperature for the activity of hydrolases that could be used during direct microbial conversion of polymers into ethanol is about 50 °C. However, most microorganisms used for the construction of producers of ethanol from lignocellulosic and starchy substrates are mesophiles with an optimal temperature of growth and fermentation within 28–40 °C (Fujita and Ito 2004). The remains accumulated after the treatment of grain and maize contain significant amounts of starch. For example, wheat brans contain 20% of starch along with 35% of hemicellulose and 18% of cellulose (Gaspar et al. 2007). Wheat brans are accumulated abundantly as a byproduct of flour production. Thus, the construction of microbial strains capable of the direct conversion of both starch and hemicellulose into ethanol is of great economic importance.

Starch consists of two highly polymeric fractions—amylose and amylopectin. The minor component of starch (20–30%)—amylose—is a linear polysaccharide, which consists of glucose residues linked with α-1,4-glycosidic bonds. Amylopectin represents the major part of starch (70–80%) and contains, beside chains of glucose linked by α-1,4-glycosidic bonds, branched chains linked by α-1,6-glycosidic bonds (Eksteen et al. 2003). Starch is cleaved by the secretory α-amylase and glucoamylase (Piontek et al. 1998). α-Amylase (EC. 3.2.1.1) catalyzes the cleavage of internal α-1,4-glycosidic bonds of starch to maltose, oligosaccharides, and dextrins.

Glucoamylase (EC 3.2.1.3) catalyzes the hydrolysis of glucooligosaccharides and maltose to D-glucose. Recombinant strains of *S. cerevisiae* were constructed that were capable of fermenting starch by the expression of heterologous α-amylase of *Streptococcus bovis* and glucoamylase of *Rhizopus oryzae* (Shigechi et al. 2004). The yeast *Schwanniomyces occidentalis* synthesizes amylolytic enzymes and effectively ferments starch into ethanol (Wang et al. 1989). Secretory α-amylase is encoded by the *SWA2* gene. The *GAM1* gene encodes secretory glucoamylase. This gene was expressed in *H. polymorpha*, leading to the effective secretion of heterologous glucoamylase (Gellissen et al. 1991). Taking into account the ability of the yeast *H. polymorpha* to ferment glucose at increased temperatures (Ishchuk et al. 2009; Voronovsky et al. 2005), the construction of recombinant strains based on this organism that immediately ferment starch into ethanol at elevated temperatures is promising.

The expression cassettes of the *SWA2* (encoding secretory α-amylase) and *GAM1* (encoding secretory glucoamylase) genes under the control of a strong constitutive promoter of *GAPDH* were introduced into the genome of the recipient strain of *H. polymorpha*. Selection of transformants was performed, which were tested with respect to the sizes of the clearing (halo) zones of starch. The presence of a halo is evidence in favor of the effective expression of the secretory forms of α-amylase and glucoamylase. One of the selected strains that formed the largest zone of clearing of starch synthesized more than 3 g l^{-1} of ethanol within less than 48 h during starch fermentation of 3% at pH 5.5 and at a temperature of 48 °C (Voronovsky et al. 2009).

A further increase in the amylolytic activity of the recombinant strains was achieved by the multicopy integration of the plasmid containing the expression cassettes for the *SWA2* and *GAM1* genes. From the collection of the obtained transformants, a strain was selected that contained eight copies of this plasmid. The strain formed the largest clearing zone of starch. The efficiency of alcoholic fermentation of starch was also significantly improved. The quantity of the ethanol synthesized reached 6.5 g l^{-1} which more than twice exceeded the efficiency of the alcoholic fermentation of the strain containing the *SWA2* and *GAM1* genes in one copy. The additional enhancement in the expression of the *PDC1* gene (encodes pyruvate decarboxylase) increased the efficiency of the alcoholic fermentation of starch to 9–10 g l^{-1} (Voronovsky et al. 2009).

β-1,4-Xylan represents a heterogeneous polysaccharide present in the cell wall of plants. Xylose monomers linked by β-1,4-glycosidic bonds are the base of the chain with other sugars binding to it (Torronen et al. 1992). Xylan hydrolysis is catalyzed by endo-β-xylanase (EC 3.2.1.8) and β-D-xylosidase (EC 3.2.1.37). Endo-β-xylanase catalyzes the hydrolytic decomposition of 1,4-β-xylosidic bonds of xylan and xylooligosaccharides. β-D-Xylosidase hydrolyzes xylooligosaccharides to D-xylose (La Grange et al. 1996). Fungi of the genus *Trichoderma* secrete many xylanolytic enzymes. The filamentary mesophilous fungus *Trichoderma reesei* is known for its cellulolytic and xylanolytic activity (Torronen et al. 1992). Two main endoxylanases—Xyn1p and Xyn2p—are isolated from it. During cultivation of *T. reesei* on xylan, Xyn2p constitutes more than 50% of the total amount of xylanolytic enzymes. Representatives of the family *Aspergillus* are also effective producers of cellulolytic and xylanolytic enzymes. The successful expression of *A. niger xlnD* in baker's yeasts provided the synthesis of the secretory forms of β-D-xylosidase (La Grange et al. 1996). Moreover, using coexpression of *T. reesei* endo-β-xylanase and *A. niger* β-D-xylosidase, a recombinant xylan fermenting strains of baker's yeasts are constructed (La Grange et al. 1996).

To create *H. polymorpha* strains fermenting xylan, an integrative plasmid was constructed, which contained the *T. reesei XYN2* (encodes secretory endoxylanase) and *A. niger xlnD* (encodes secretory β-xylosidase) genes under the control of a strong constitutive promoter of the *GAPDH*. After the transformation by this plasmid, screening of *H. polymorpha* was performed analyzing the halo size on the medium supplemented with xylan or p-nitrophenyl-β-xyloside. The specific activity of these enzymes was determined, which correlated with the size of the halo. The strains constructed gained an ability to ferment xylan into ethanol. The amount of the ethanol synthesized was 0.35 g l^{-1} at 37 and 48 °C (Voronovsky et al. 2009).

As a result of the performed work, *H. polymorpha* strains were constructed capable of the direct microbial conversion of starch and xylan into ethanol. The recombinant strains constructed are extremely perspective for the further improvement of the parameters of alcoholic fermentation of carbon polymers and creation of corresponding industrial technology of target microbial conversion that seems to be the most effective technology of alcoholic fermentation of biopolymers.

14 Effect of Glutathione on the Efficiency of Alcoholic Fermentation in *H. polymorpha*

Glutathione (γ-glutamyl-cysteinyl-glycine, GSH) is a biologically active substance of peptide nature playing an important role in a wide spectrum of cellular reactions (Meister 1988). The antioxidant properties of this compound determine its role in maintaining intracellular redox status. Due to the presence of sulfhydric (thiol) groups, GSH in a cell acts as an electron donor and provides the progress of reducing reactions, with it turning into an oxidized form (GSSG). In addition to maintaining the thiol redox status, glutathione is involved in the detoxication of endogenous and exogenous reactive metals and xenobiotics. Thus, GSH plays an important role in the response of a cell to oxidative stress by detoxifying compounds of free radical nature. It is known that during alcoholic fermentation synthesized ethanol also induces oxidative stress of yeast cells (Alexandre et al. 2001), which could in turn limit the efficiency of alcoholic fermentation. To study the effect of GSH as a key factor in the mechanisms of stress response on the efficiency of alcoholic fermentation in *H. polymorpha* yeasts, recombinant strains were constructed with enhanced expression of *GSH2* (encodes γ-glutamyl cysteinyl synthase) (Ubiyvovk et al. 2002) and *MET4* (encodes a transcriptional activator of sulfur metabolism) (Ubiyvovk et al. 2011). The strains overexpressing *GSH2* and

MET4 accumulated about 14 nM of GSH in terms of 1 mg of dry weight of yeast biomass. The glutathione synthesis in the initial strain was 9 nM mg^{-1}. The increase in the intracellular concentration of GSH was in good agreement with the increase in the efficiency of alcoholic fermentation. The ethanol synthesis by the strains with enhanced expression of the *GSH2* and *MET4* genes reached 19 and 17 g l^{-1}, respectively, while the initial strain synthesized only 7 g l^{-1}. Thus, a 2.7- and 2.4-fold increase in the efficiency of alcoholic fermentation of glucose was achieved. The enhancement of the expression of the homologous *GSH1* gene (encodes γ-glutamyl cysteinyl synthase) in the *S. cerevisiae* system did not affect the efficiency of alcoholic fermentation of glucose. The fact that the *H. polymorpha* strains analyzed did not synthesize an increased quantity of ethanol during xylose fermentation is of interest. The recombinant strains of yeasts *H. polymorpha* and *S. cerevisiae* with an increased intracellular pool of GSH appeared to be more sensitive to exogenous ethanol compared to the corresponding parent strains. Thus, glutathione stimulates alcoholic fermentation of glucose in *H. polymorpha* yeasts; however, the mechanisms of this phenomenon require further investigations (Grabek-Lejko et al. 2011).

15 Perspectives

Further work in the field of metabolic engineering of the yeast *H. polymorpha* could make it possible to construct strains with the parameters of alcoholic fermentation of xylose surpassing all strains of microorganisms known to date. Promising directions in the improvement of the parameters of alcoholic fermentation of xylose in *H. polymorpha* could be considered the initial stages of xylose catabolism, transport of this pentose into cells, amplification of the limiting genes of glycolysis and the pentose phosphate pathway, as well as that of the genes determining the resistance to increased temperatures and ethanol. The maximal temperature for the fermentation of *H. polymorpha* (50 °C) should provide the implementation of the SSF of glucose and xylose, as this temperature is close to the optimal one for the activity of cellulases and hemicellulases. The drawback of *H. polymorpha* is its inability to ferment galactose and L-arabinose (it should be noted that, in the collection of microorganisms of the Institute of Cell Biology, National Academy of Sciences of Ukraine, an *H. polymorpha* strain was identified that grew on a medium with L-arabinose, however, did not ferment this sugar). Therefore, the introduction of heterologous genes into this organism would provide the active fermentation of all basic sugars of lignocellulose. At present, the resistance of *H. polymorpha* to toxic products (aldehydes, phenols, and acetic and formic acids) accumulated in lignocellulose hydrolysates under the conditions of acidic hydrolysis has not been studied. Although under the conditions of enzymatic hydrolysis, the amount of such products could be minimized.

Acknowledgements This work was supported in part by National Academy of Sciences of Ukraine (Grant Nos 5–17, 6–17 and 35–17) and Science and Technology Center in Ukraine (STCU) (Grant 6188).

References

Abdel-Banat, B.M., Hoshida, H., Ano, A., Nonklang, S. and Akada, R. 2010. *Appl. Microbiol. Biotechnol.* **85**: 861–867.

Alexandre, H., Ansanay-Galeote, V., Dequin, S. and Blondin, B. 2001. *FEBS Lett.* **498**:98–103.

Bergmeyer, H.U., Gawehn, K. and Grassl, M. (1974) In: Methods of Enzymatic Analysis (ed. H. U. Bergmeyer), Academic Press, Inc., New York, NY, Volume I, pp. 513–514.

Brat, D., Boles, E., and Wiedemann, B. 2009. *Appl. Environ. Microbiol.* **75**: 2304–2311.

Cardona, F., Aranda, A. and del Olmo, M. 2009. *Yeast* **26**: 1–15.

Cashikar, A.G., Duennwald, M. and Lindquist, S.L. 2005. *J. Biol. Chem.* **280**: 23869–23875.

Cho, J.-Y. and Jeffries, T.W. 1998. *Appl. Environ. Microbiol.* **64**: 1350–1358.

Costa, V., Amorim, M.A., Reis, E., Quintanilha, A. and Moradas-Ferreira, P. 1997. *Microbiology* **143**: 1649–1656.

Denis, C.L., Ferguson, J. and Young, E.T. 1983. *J. Biol. Chem.* **258**: 1165–1171.

Dmytruk, K.V., Voronovsky, A.Y. and Sibirny, A.A. 2006 *Curr. Genet.* **50**: 183–191.

Dmytruk, O.V., Dmytruk, K.V., Abbas, C.A., Voronovsky, A.Y. and Sibirny, A.A. 2008a. *Microb. Cell Fact.* **23**: 7–21.

Dmytruk, O.V., Voronovsky, A.Y., Abbas, C.A., Dmytruk, K.V., Ishchuk, O.P. and Sibirny, A.A. 2008b. *FEMS Yeast Res.* **8**: 165–173.

Dmytruk, K., Kshanovska, B., Abbas, C. and Sibirny, A. 2016. *Bioethanol* **2**: 24–31.

Du Preez, J.C. and J.P. van der Walt. 1983. *Biotechnol. Lett.* **5**: 357–362.

Du Preez, J.C., Bosch, M. and Bernard, A. 1986. *Enzyme and Microbial Technol.* **8**: 360–364.

Eksteen, J.M., van Rensburg, P., Cordero, Otero, R.R. and Pretorius, I.S. 2003. *Biotechnol. Bioengineer.* **84**: 639–646.

Farwick, A., Bruder, S., Schadeweg, V., Oreb, M. and Boles, E. 2014. *Proc. Natl. Acad. Sci. USA* **111**: 5159–5164.

Fujita, Y. and Ito, J. 2004. *Appl. Environ. Microbiol.* **70**: 1207–1212.

Gaspar, M., Kalman, G. and Reczey, K. 2007. *Process Biochem.* **42**: 1135–1139.

Gellissen, G., Janowicz, Z.A., Merckelbach, A., Piontek, M., Keup, P., Weydemann, U., Hollenberg, C.P. and Strasser, A.W. 1991. *BioTechnology* **9**: 291–295.

Grabek-Lejko, D., Ryabova, O.B., Oklejewicz, B., Voronovsky, A.Y. and Sibirny, A.A. 2006. *J. Ind. Microbiol. Biotechnol.* **33**: 934–940.

Grabek-Lejko, D., Kurylenko, O.O., Sibirny, V.A., Ubiyvovk, V.M., Penninckx, M. and Sibirny, A.A. 2011. *J. Ind. Microbiol. Biotechnol.* **38**: 1853–9.

Guerra, E., Chye, P.P., Berardi, E. and Piper, P.W. 2005. *Microbiology* **151**: 805–811.

Hahn-Hägerdal, B., Galbe, M., Gorwa-Grauslund, M.F., Lidén, G. and Zacchi, G. 2006. *Trends Biotechnol.* **24**: 549–556.

Hahn-Hägerdal, B., Karhumaa, K., Fonseca, C., Spencer-Martins, I. and Gorwa-Grauslund, M.F. 2007. *Appl. Microbiol. Biotechnol.* **74**: 937–953.

Haslbeck, M., Braun, N., Stromer, T., Richter, B., Model, N., Weinkauf, S. and Buchner, J. 2004. *EMBO J.* **23**: 638–649.

Haslbeck, M., Franzmann, T., Weinfurtner, D. and Buchner, J. 2005. *Nat. Struct. Mol. Biol.* **12**: 842–846.

Haurie, V., Perrot, M., Mini, T., Jenö, P., Sagliocco, F. and Boucherie, H. 2001. *J. Biol. Chem.* **276**: 76–85.

Hedges, D., Proft, M. and Entian, K.D. 1995. *Mol. Cell. Biol.* **15**: 1915–22.

Hill, J., Nelson, E., Tilman, D., Polasky, S. and Tiffany, D. 2006. *Proc. Natl. Acad. Sci. USA* **103**: 11206–11210.
Ishchuk, O.P., Voronovsky, A.Y., Stasyk, O.V., Gayda, G.Z., Gonchar, M.V., Abbas, C.A. and Sibirny, A.A. 2008. *FEMS Yeast Res.* **8**: 1164–1174.
Ishchuk, O.P., Voronovsky, A.Y., Abbas, C.A. and Sibirny, A.A. 2009. *Biotechnol. Bioeng.* **104**: 911–919.
Ishchuk, O.P., Abbas, C.A. and Sibirny, A.A. 2010. *J. Ind. Microbiol. Biotechnol.* **37**: 213–218.
Jeffries, T.W. and Shi, N.Q. 2000. United States Patent; No 6071729.
Jeffries, T.W. and Jin, Y.-S. 2004. *Appl. Microbiol. Biotechnol.* **63**: 495–509.
Jeffries, T.W., Grigoriev, I.V., Grimwood, J., Laplaza, J.M., Aerts, A., Salamov, A., Schmutz, J., Lindquist, E., Dehal, P., Shapiro, H., Jin, Y.-S., Passoth, V. and Richardson, P.M. 2007. *Nat. Biotechnol.* **25**: 319–26.
Jung, Y.J. and Park, H.D. 2005. *Biotechnol. Lett.* **27**: 1855–1859.
Kadam, K.L. and Schmidt, S.L. 1997. *Appl. Microbiol. Biotechnol.* **48**: 709–713.
Kiel, J.A. and Veenhuis, M. 2000. *Cell. Biochem. Biophys.* **32**: 9–19.
Kiel, J.A., Rechinger, K.B., van der Klei, I.J., Salomons, F.A., Titorenko, V.I. and Veenhuis, M. 1999. *Yeast* **15**: 741–754.
Kim, J., Alizadeh, P., Harding, T., Hefner-Gravink, A. and Klionsky, D.J. 1996. *Appl. Environ. Microbiol.* **62**: 1563–1569.
Kitagawa, M., Miyakawa, M., Matsumura, Y. and Tsuchido, T. 2002 *Eur. J. Biochem.* **269**: 2907–2917.
Kurylenko, O.O., Ruchala, J., Hryniv, O.B., Abbas, C.A., Dmytruk, K.V. and Sibirny, A.A. 2014. *Microb. Cell. Fact.* **13**: 122.
Kuyper, M., Hartog, M.M., Toirkens, M.J., Almering, M.J., Winkler, A.A., van Dijken, J.P. and Pronk, J.T. 2005. *FEMS Yeast Res.* **5**: 399–409.
La Grange, D.C., Pretorius, I.S. and van Zyl, W.H. 1996. *Appl. Environ. Microbiol.* **62**: 1036–1044.
Lindquist, S. and Kim, G. 1996. *Proc. Natl. Acad. Sci. USA* **93**: 5301–5306.
Londesborough, J. and Varimo, K. 1984. *Biochem. J.* **219**: 511–518.
Long, T.M., Su, Y.K., Headman, J., Higbee, A., Willis, L.B. and Jeffries, T.W. 2012. *Appl. Environ. Microbiol.* **78**: 5492–500.
Lutstorf, U. and Megnet, R. 1968. *Arch. Biochem. Biophys.* **126**: 933–944.
Matsushika, A., Inoue, H., Kodaki, T. and Sawayama, S. 2009. *Appl. Microbiol. Biotechnol.* **84**: 37–53.
Meister, A. 1988. *J. Biol. Chem.* **263**: 17205–17208.
Nomura, M. and Takagi, H. 2004. *Proc. Natl. Acad. Sci. USA* **101**: 12616–12621.
Nwaka, S., Mechler, B. and Holzer, H. 1996. *FEBS Lett.* **386**: 235–238.
Olofsson, K., Bertilsson, M. and Lidén, G. 2008. *Biotechnol. Biofuels* **1**: 7.
Parrou, J.L., Jules, M., Beltran, G. and François, J. 2005. *FEMS Yeast Res.* **5**: 503–511.
Parsell, D.A. and Lindquist, S. 1993. *Annu. Rev. Genet.* **27**: 437–496.
Passmore, L.A. and Barford, D. 2004. *Biochem. J.* **379**: 513–525.
Passoth, V., Schäfer, B., Liebel, B., Weierstall, T. and Klinner, U. 1998. *Yeast* **14**: 1311–1325.
Piontek, M., Hagedorn, J., Hollenberg, C.P., Gellissen, G. and Strasser, A.W.M. 1998. *Appl. Microbiol. Biotechnol.* **50**: 331–338.
Pugh, D.J., Ab, E., Faro, A., Lutya, P.T., Hoffmann, E. and Rees, D.J. 2006. *BMC Struct. Biol.* **6**: 1.
Qi, K., Zhong, J.J. and Xia, X.X. 2014. *Appl. Environ. Microbiol.* **80**: 3879–3887.
Reinders, A., Romano, I., Wiemken, A. and De Virgilio, C. 1999. *J. Bacteriol.* **181**: 4665–4668.
Ryabova, O.B., Chmil, O.M. and Sibirny, A.A. 2003. *FEMS Yeast Res.* **3**: 157–164.
Schaaff, I., Heinisch. J. and Zimmermann, F.K. 1989. *Yeast* **5**: 285–290.
Schubert, C. 2006. *Nat. Biotechnol.* **24**: 777–784.
Shi, N.Q., Cruz, J., Sherman, F. and Jeffries, T.W. 2002. *Yeast* **19**: 1203–1220.
Shigechi, H., Koh, J., Fujita, Y., Matsumoto, T., Bito, Y., Ueda, M., Satoh, E., Fukuda, H. and Kondo, A. 2004. *Appl. Environ. Microbiol.* **70**: 5037–5040.

Sohn, J., Choi, E., Kang, H., Rhee, J., Agaphonov, M., Ter-Avanesyan, M. and Rhee, S. 1999. *Appl. Microbiol. Biotechnol.* **51**: 800–807.
Suwannarangsee, S., Oh, D.B., Seo, J.W., Kim, C.H., Rhee, S.K., Kang, H.A., Chulalaksananukul, W. and Kwon, O. 2010. *Appl. Microbiol. Biotechnol.* **88**: 497–507.
Suwannarangsee, S., Kim, S., Kim, O.C., Oh, D.B., Seo, J.W., Kim, C.H., Rhee, S.K., Kang, H.A., Chulalaksananukul, W. and Kwon, O. 2012. *Appl. Microbiol. Biotechnol.* **96**: 697–709.
Tachibana, C., Yoo, J.Y., Tagne, J.B., Kacherovsky, N., Lee, T.I. and Young, E.T. 2005. *Mol. Cell. Biol.* **25**: 2138–2146.
Toivola, A., Yarrow, D., van den Bosch, E., van Dijken, J.P. and Scheffers, W.A. 1984. *Appl. Environ. Microbiol.* **47**: 1221–1223.
Torronen, A., Mach, L.R., Massner, R., Gonzales, R., Kalkkinen, N., Harkki, A. and Kubicek, C. P. 1992. *Biothechnology* **10**: 1461–1465.
Ubiyvovk, V.M., Nazarko, T.Y., Stasyk, O.G., Sohn, M.J., Kang, H.A. and Sibirny, A.A. 2002. *FEMS Yeast Res.* **2**: 327–332.
Ubiyvovk, V.M., Ananin, V.M., Malyshev, A.Y., Kang, H.A. and Sibirny, A.A. 2011. *BMC Biotechnol.* **11**: 8.
van Hoek, P., Flikweert, M.T., van der Aart, Q.J., Steensma, H.Y., van Dijken, J.P. and Pronk, J.T. 1998. *Appl. Environ. Microbiol.* **64**: 2133–2140.
Voronovsky, A.Y., Ryabova, O.B., Verba, O.V., Ishchuk, O.P., Dmytruk, K.V. and Sibirny, A.A. 2005. *FEMS Yeast Res.* **5**: 1055–1062.
Voronovsky, A.Y., Rohulya, O.V., Abbas, C.A. and Sibirny, A.A. 2009. *Metab. Eng.* **11**: 234–242.
Waites, M.J. and Quayle, J.R. 1981. *J. Gen. Microbiol.* **124**: 309–316.
Wang, T.T., Lin, L.L. and Hsu, W.H. 1989. *Appl. Environ. Microbiol.* **55**: 3167–3172.
Wang, R., Li, L., Zhang, B., Gao, X., Wang, D. and Hong, J. 2013. *J. Ind. Microbiol. Biotechnol.* **40**: 841–854.
Watanabe, D., Hashimoto, N., Mizuno, M., Zhou, Y., Akao, T. and Shimoi, H. 2013. *Biosci. Biotechnol. Biochem.* **77**: 2255–2262.
Weibezahn, J., Tessarz, P., Schlieker, C., Zahn, R., Maglica, Z., Lee, S., Zentgraf, H., Weber-Ban, E.U., Dougan, D.A., Tsai, F.T., Mogk, A. and Bukau, B. 2004. *Cell* **119**: 653–665.
Weissman, A.M. 2001. *Nat. Rev. Mol. Cell Biol.* **2**: 169–178.
Wisselink HW, Toirkens MJ, Wu Q, Pronk JT, van Maris AJ. 2009. *Appl. Environ. Microbiol.* **75**: 907–914.
Yoshida J, Tani T. 2005. *Cell Struct. Funct.* **29**: 125–138.
Zeng QK, Du HL, Wang JF, Wei DQ, Wang XN, Li YX, Lin Y. 2009. *Biotechnol. Lett.* **31**: 1025–1029.

Ecology, Diversity and Applications of *Saccharomyces* Yeasts in Food and Beverages

Jean-Luc Legras, Virginie Galeote, Carole Camarasa, Bruno Blondin and Sylvie Dequin

Abstract Yeasts from the *Saccharomyces* complex have been used for millennia for the production of fermented food and alcoholic beverages. The availability of large genomic datasets during the past decade has provided new insights into the genetic and phenotypic diversity, population structure and evolutionary history of these yeasts. Studies of these datasets have shown that man-made environments have led to several distinct domesticated variants. Comparative genomics approaches have revealed domestication fingerprints and indicated divergent regions that may explain the adaptation of strains to different ecological niches. In addition, the genetic basis of several technological traits of *S. cerevisiae* has been elucidated through QTL mapping, and strains improved for various industrial traits have been developed through hybridization or evolutionary engineering. The expansion of large-scale genomic and high-throughput phenotypic data on these strains will provide a unique resource for understanding their adaptation to their ecological niches and for elucidating the missing links between genotype and phenotype, paving the way for strain improvement.

Keywords Genetics · Ecology · Evolutionary history · Engineering · *Saccharomyces* · Hybrids

1 Introduction

Fermented foods and beverages have a long and rich history dating back thousands of years, closely linked to the history of agriculture. Fermentation has been an effective way to preserve quality and safety by extending the shelf life of ingredients and improving their tastes and textures. Fermented beverages were used as

J.-L. Legras · V. Galeote · C. Camarasa · B. Blondin · S. Dequin (✉)
SPO, INRA, SupAgro, Université Montpellier, 34060 Montpellier, France
e-mail: sylvie.dequin@inra.fr

T. Satyanarayana and G. Kunze (eds.), *Yeast Diversity in Human Welfare*, DOI 10.1007/978-981-10-2621-8_12

feast drinks but were also used for religious ceremonies and in medicine. Alcoholic beverages appeared during the Neolithic period (8500–4000 BC), when mankind left nomadism for a more settled life and started to cultivate and store cereals. The earliest evidence of an alcoholic beverage made of a combination of rice, honey and fruit, probably grapes, was found in pottery jars from the early Neolithic village (7000–5600 BC) of Jiahu in China (McGovern et al. 2004). Grape wine production dates back to 5400–5000 BC, as shown by the discovery of the presence of calcium tartrate in jars in the Neolithic site of Hajji Firuz in Iran (McGovern et al. 1996). The history of beer and bread is closely linked to that of cereals. The earliest chemically confirmed barley beer to date was discovered at Godin Tepe in the central Zagros Mountains of Iran, ca. 3400–3000 B.C. (Michel et al. 1992).

Although many different yeast species take part in spontaneous fermentations, the genus *Saccharomyces* plays a special role in food and beverage fermentations. In wine-production, the first fermentation stage is dominated by non-*Saccharomyces* yeast activity. However, the growth of these species is limited to the first days of fermentation, and more ethanol-tolerant and strongly fermentative *Saccharomyces* strains, such as *S. cerevisiae*, dominate alcoholic fermentation.

The optimization and diversification of fermentative processes have been performed without understanding the biochemical basis of these transformations. In 1860, the pioneering work of Louis Pasteur identified yeast as the agent responsible for fermentation (Pasteur 1860), opening the way to advanced methods of culturing pure strains. Emile Christian Hansen, from Carlsberg Laboratory (Denmark), developed the first pure yeast culture in 1883, and the first inoculation of a grape must with a pure yeast culture was conducted by Müller-Thurgau in 1890. In 1891, Georges Jacquemin founded the Institute “La Claire”, near Morteau (Jura), where yeast strains were isolated from grapes and produced at small scale before being shipped to France and various European and New World countries to inoculate grape musts. The practice of inoculation with selected pure cultures spread rapidly in the fields of baking and brewing. In oenology, the use of selected yeasts in the form of active dry yeast became widespread in the 1970s. These practices have enabled better control and reliability of the fermentation process, limiting the risk of microbiological changes, and have contributed to improving the quality of fermented food and beverages for approximately 50 years.

In addition to *Saccharomyces cerevisiae*, which is the main species responsible for a large variety of alcoholic fermentation processes, various other *Saccharomyces* species are involved in fermentation processes, and a growing number of interspecies hybrids have recently been identified and characterized. After the biochemical studies of yeast that prevailed until the end of the 20th century, the development of molecular and genetic tools and more recently the fast development of next-generation sequencing (NGS) have provided new insights into the ecology, population structure and diversity of *Saccharomyces* species and hybrids and in the mechanisms involved in the evolution of these strains. This knowledge provides novel directions for future strain improvement. In parallel, much effort has been devoted to the improvement of the strains used in these processes. The improvement of starter cultures has relied on mutagenesis and conventional breeding followed by selection, as well as more targeted approaches based on genetic engineering since the 1980s (Dequin 2001). However, the commercial application of genetically modified (GM) strains has been largely constrained by the lack of public acceptance for this technology. In the past decade, GM-free approaches to exploit natural variation or to generate new diversity have been widely developed, resulting in the design of yeast strains with superior attributes. This chapter focuses on recent advances in our understanding of ecology and biodiversity of *Saccharomyces* yeasts used in traditional fermented food and beverages and on the improvement of yeast starters.

2 Genetic and Phenotypic Diversity of *Saccharomyces* sp.

Despite being isolated from multiple sources, *Saccharomyces* sp. genetic diversity has long been poorly characterized. The advent of molecular tools such as mDNA RFLP (Aigle et al. 1984), pulsed field gel electrophoresis (Vezinhet et al. 1990), PCR-based methods such as RAPD (Paffetti et al. 1995), interdelta PCR (Ness et al. 1993; Legras and Karst 2003), microsatellite tying (Hennequin et al. 2001) or AFLP (Azumi and Goto-Yamamoto 2001) aimed at differentiating yeast strains has revealed the wide diversity of *S. cerevisiae* isolates and paved the way for investigating yeast ecology. However, only the recent use of appropriate genetic markers (such as microsatellite, AFLP, MLST, or genome sequencing) has elucidated the yeast phylogeny, shown the existence of population structures and revealed the historical aspects of some populations, such as sake, wine or beer populations. Interestingly, the correlation between genotypic and phenotypic similarity within *S. cerevisiae* is surprisingly good (correlation coefficient, 0.30; $P = 10^{-26}$) (Liti et al. 2009).

2.1 Genetic Diversity

2.1.1 *Saccharomyces* Complex

From the physiology-based taxonomy of the mid-20th century to the current knowledge of this group, the *Saccharomyces* taxonomy has undergone multiple changes, and even the most recent edition of the reference book for yeast taxonomy, "Yeast, 2011," is outdated. Today's objective classification of this clade relies on data from sexual incompatibilities as proposed by the biological species concept (Naumov 1996; Naumov et al. 2000), molecular phylogeny inferred from sequences of the D1D2 gene regions (Kurtzman et al. 2011), and genomic data (Libkind et al. 2011). This clade contains 6 well defined species: *Saccharomyces cerevisiae, S. paradoxus, S. arboricola, S. mikatae, S. kudriavzevii, S. uvarum* and *S. eubayanus* (Fig. 1). A seventh species, *S. cariocanus*, has been inferred from of sexual

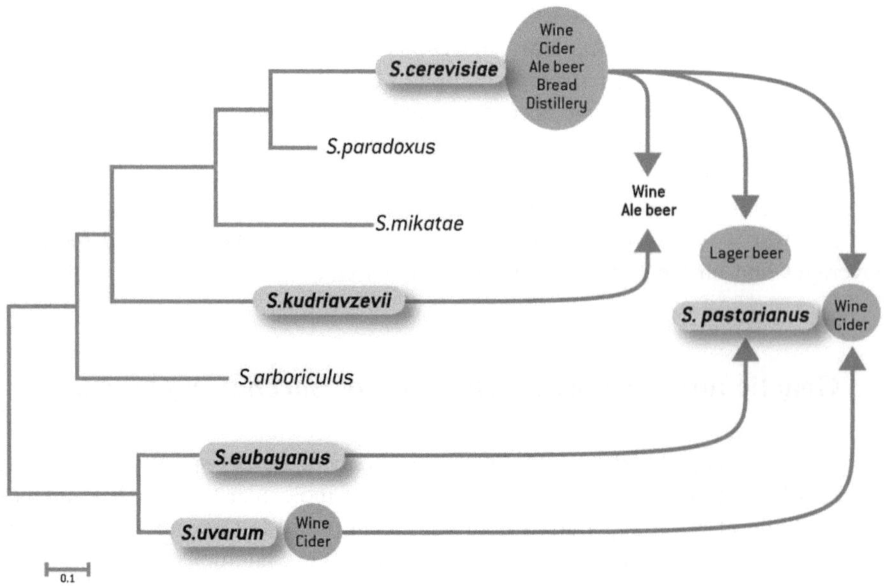

Fig. 1 Relationships between the seven natural species of *Saccharomyces* and their industrial hybrids. Tree topology was obtained using a subset of 25,000 SNPs selected after genome alignment

incompatibility with *S. paradoxus*, but sequencing at several loci revealed identity with this latter species, whereas four reciprocal translocations generated genetic incompatibilities (Liti et al. 2006). Finally, two other species encountered in beer, *S. pastorianus* and *S. bayanus*, are interspecific hybrids of *S. cerevisiae* and *S. eubayanus* and of *S. uvarum* and *S. eubayanus* with some content of *S. cerevisiae*, respectively (Libkind et al. 2011; Nguyen et al. 2011). Indeed, the recent isolation and identification of *S. eubayanus* species, the long-unknown ancestor of lager yeast, and the sequencing of the *S. bayanus* genome (Libkind et al. 2011) have provided major advances in the understanding of *Saccharomyces sp.* phylogeny. These advances also solved the ambiguities and taxonomic debates around *S. uvarum* and *S. bayanus*.

2.1.2 *Saccharomyces* Ecology

When considering *S. cerevisiae* ecology, two opposite concepts have driven ecological studies for long. During the first part of the century, the general idea was that *S. cerevisiae* were ubiquitous on fruits and grapes, whereas at the end of the 20th century, people realized that this idea relied on a biased exploration of diversity caused by enrichment procedures used for yeast isolation (Vaughan-Martini and Martini 1995). Indeed, multiple studies reported the paucity of *S. cerevisiae* isolated from healthy grape berries and fruits, in contrast to cellars. However, *S. cerevisiae* populations could be extremely abundant on damaged berries (Mortimer and Polsinelli 1999; Mortimer 2000). In parallel, many *S. cerevisiae* have been found in natural environments such as oak bark or litter in USA, Europe, or China (Sniegowski et al. 2002; Sampaio and Gonçalves 2008; Wang et al. 2012; Almeida et al. 2015). Interestingly, the other *Saccharomyces* species (*S. paradoxus*, *S. cariocanus*, *S. mikatae*, *S. kudriavzevii*, and *S. uvarum*) have been isolated from the same niche (oak bark), in sympatry with *S. cerevisiae*. However, although *S. paradoxus* and *S. cerevisiae* are isolated at similar temperatures (30 °C), *S. kudriavzevii* and *S. uvarum* are isolated at lower temperature. In addition to this habitat associated with the plant kingdom, *S. cerevisiae* and its closest wild relative *S. paradoxus* have been detected among the most abundant species present in the microflora of insects such as *Drosophila melanogaster* (Chandler et al. 2012), and *S. cerevisiae* has been isolated frequently from *D. melanogaster* captured from vineyards (Buser et al. 2014) and from wasps (Stefanini et al. 2012) or bees (Goddard et al. 2010), indicating that insects play an essential role in the vection of these microorganisms. In contrast to the former species, *S. eubayanus* was first isolated from *Nothofagus* sp. in Patagonia (Libkind et al. 2011) and was subsequently isolated from oaks in the USA and China (Bing et al. 2014; Peris et al. 2014).

2.1.3 Population Structure of *S. cerevisiae*

The past 10 years have seen considerable improvement in the knowledge of *S. cerevisiae* diversity and population structure. A first indication of yeast population structure was given by AFLP data (Azumi and Goto-Yamamoto 2001) for Asian strains, but the MLST studies of Fay and Benavides and the multilocus microsatellite study of Legras et al. (2007) revealed the impact of human activities on yeast diversity. Despite only 12 loci having been used for microsatellite multilocus typing, a clear picture can be seen, and the clustering of yeast strains into different groups corresponds to their isolation source, indicating that the denomination of yeasts as “bread”, “beer”, “wine”, “sake”, or “palm wine” yeasts is highly appropriate (Fig. 2). In this survey of 1000 strains, updated from our former work, several clusters can be detected. The main cluster contains 98% of wine strains, whereas oak isolates can be detected in two clusters, one present mainly in America, Asia and Europe and the second cluster containing strains isolated from countries located around the Mediterranean Sea (Almeida et al. 2015). In addition, a cluster can be observed for flor strains (Legras et al. 2014), or different African beverages (Tapsoba et al. 2015). This global picture of yeast diversity has been confirmed by RADSeq analysis of a set of 262 strains from various origins (Cromie et al. 2013) and, more recently, by whole-genome sequencing (Liti et al. 2009; Schacherer et al. 2009; Strope et al. 2015). In addition, a recent study reported a diversity approximately double that already measured for *S. cerevisiae* for a population isolated from Chinese forests, with the most divergent lineages exhibiting a

Fig. 2 Clustering of 1000 yeast strains isolated from different sources, modified from Legras et al. (2007). This neighbour-joining tree is constructed from the Bruvo's distance between strains based on the polymorphism at 12 loci and is rooted according to the midpoint method. Branches are coloured according to the substrate from which strains have been isolated. Colour code: Wine *dark green*; Flor *light green*; Bread *yellow*; Beer *orange*; Fermented milk *pink*; Sake and miscellaneous Asian origins *dark blue*; sorghum beer or African wines *brown*; Oak tree *blue-green*; distillery from South America and rum from French Indies *purple*; Laboratory strains *red*

nucleotide diversity (π) of 0.0077 (Wang et al. 2012). These population-genomic analyses revealed that *S. cerevisiae* groups could be categorized into 5 main pure lineages (wine/European, sake, USA Oaks, Malaysian, or African), giving admixed groups after crosses such as the bread, rum, and lab strain groups.

A close association between genetic structure and technological origin, implying an initial bottleneck followed by the clonal expansion of a group of strains, can be considered indicative of domestication, especially for wine and sake strains (Fay and Benavides 2005). Also interesting is that the expansion of distant groups suggests expansion from local populations. This conclusion can already be inferred from microsatellite data; as much as 28% of genetic variation can be explained by geographic distance. In agreement with that hypothesis, a population of *S. cerevisiae* has been recently characterized from oaks of the Mediterranean, which is the closest wild genetic resource to the wine population. Inference of demographic parameters from genomic data revealed that the most likely model reflected partial isolation from an ancestral population with asymmetrical migration between the Mediterranean oak population and the wine population and included growth in both populations (Almeida et al. 2015).

The quest to describe the geographic structure and origin of *S. cerevisiae* has been a major focus for wine microbiologists. Interestingly, the first studies comparing yeast populations from different vineyards revealed a wide pattern structuring wine yeast diversity. *S. cerevisiae* strains isolated from wine all over the world belong to one main cluster, but population differentiation measured from pair wise F_{st} distances suggested the importance of migration routes around the Mediterranean sea or along the Danube river and along the Rhone river in France (Legras et al. 2007). In addition, the basal position of a population from Lebanon, as well as other populations from Bulgaria or Romania, suggest a Mesopotamian origin of wine yeast (Legras et al. 2007; unpublished data). In agreement with this human-assisted migration, all estimates of the divergence time obtained from different vineyards or wine populations, or between populations of wine yeasts and wild Mediterranean oak yeasts, indicated that yeast expansion and migration was contemporaneous with the expansion of human agriculture (Fay and Benavides 2005; Legras et al. 2007; Almeida et al. 2015).

Two complementary studies in Portugal and New Zealand provided more information on the scale at which wine population differentiation can be detected (Schuller et al. 2012; Knight and Goddard 2014). In the first Portuguese study, only 3–6.3% of total genetic variation was explained by an interregional effect and the highest effect seemed to occur between the most distant vineyards (180 km). This pattern was clearly confirmed by a wide-scale sampling of the yeast microflora of New Zealand vineyards, in which closely located vineyards presented the lowest differentiation and the highest gene flow between them, whereas the most distant vineyard were the most differentiated. However, the extent of migration between regions did not correlate with geographic location, suggesting that distance was not the sole factor explaining genetic variation (Knight and Goddard 2014).

2.1.4 Population Structure of Other *Saccharomyces* Species

Three very different population structures have been detected for 3 other *Saccharomyces* species: *S. paradoxus*, *S. uvarum* and *S. eubayanus*. In contrast to *S. cerevisiae*, *S. paradoxus* populations are well delineated by geographic origin: Asia, Europe, and America (Liti et al. 2006, 2009; Kuehne et al. 2007). The global divergence observed between the most distant *S. paradoxus* strains is approximately 3.5%, with a much lower nucleotide diversity of 0.1% inside a UK population than the 0.5–0.8% for all the *S. cerevisiae* species and 0.1% inside the wine/European population (Liti et al. 2006, 2009; Bergström et al. 2014). However, besides this highly geographically structured population, a specific population of strains more closely related to European strains has been identified in the northern United States, suggesting secondary contact (Kuehne et al. 2007). In a similar manner to *S. paradoxus*, three distant populations could be identified for *S. uvarum*: a first clade contained Holartic and some South America strains, a second clade contained only South America strains and the third clade strains from an Australasian population. This latter population presented 4.4% divergence from the other lineages, similar to the divergence observed for distant *S. paradoxus* groups. In addition, crosses between this distant lineage and other lineage revealed intermediate levels of spore viability (27–36%), suggesting partial reproductive isolation that may be the result of allopatric divergence (Almeida et al. 2014). Surprisingly, the highest diversity was observed from the Southern Hemisphere population, whereas a much lower diversity was detected among the Holarctic strains (0.14%). This result suggests that the colonization of the Northern Hemisphere by a Patagonian *S. uvarum* population was associated with a shift in habitat from *Nothofagus* to other trees. Interestingly, introgression of European *S. kudriavzevii* as well as of *S. eubayanus* could be detected in the genomes of wine or cider strains from the Northern Hemisphere. The third species, *S. eubayanus*, was isolated very recently in only a few countries in the Americas (Patagonia and USA) and China (Libkind et al. 2011; Bing et al. 2014; Peris et al. 2014). The Patagonian strains could be classified into two lineages by *Structure* analysis, and strains from the USA appeared as mosaics of the two lineages. Interestingly, the *S. eubayanus* haplotype of European lager strains could be associated clearly with one lineage of Patagonian strains, whereas the Tibetan isolates were also very close to the lager yeast. As the Patagonian population is highly diverse, this suggests a possible Southern-Hemisphere origin to the population leading to the Chinese and the lager yeast *S. eubayanus* populations (Peris et al. 2014).

2.1.5 Interspecific Hybrids

Lager beer strains have long been characterized as potential interspecific hybrids (Pedersen 1985), but the number of strains isolated from beer, wines or cider that have been characterized as hybrids has increased drastically during the past 20 years (Masneuf et al. 1998; Bradbury et al. 2006; González et al. 2006, 2008; LeJeune et al. 2007; Lopandic et al. 2008; Erny et al. 2012). These strains are most often encountered in fermentations performed at cool temperature, such as for beer brewing, or wine and cider making, and in this case, they combine the cryophily of *S. uvarum*, *S. kudriavzevii*, or *S. eubayanus* with the fermentation performance of *S. cerevisiae*. Most of these hybrids present variable ploidies and variable content of each of their parental lineages, typically having mosaic chromosomes as detected from comparative genome hybridization (CGH) (Bond et al. 2004; Dunn et al. 2008; González et al. 2008; Bond 2009; Erny et al. 2012; Peris et al. 2012) or genome sequencing (Nakao et al. 2009; Borneman et al. 2011; Baker et al. 2015).

Hybrids Between *S. cerevisiae* and *S. eubayanus*

The identification of the other ancestral species of lager yeast has provided the keys to decipher the hybridization history of several beer hybrids (Libkind et al. 2011), reviewed in Gibson and Liti (2015). A first attempt to compare the genomes within the *S. pastorianus* group by CGH revealed different genomic content for Saaz and Frohberg lager yeasts (Dunn et al. 2008). Although Saaz strains are triploid with 2 copies of the *S. eubayanus* genome and one copy of *S. cerevisiae*, Frohberg lager beer strains are tetraploid with two copies of each parental genome. In addition, CGH data, microsatellite typing, and genome sequencing indicated that each of these two lager yeasts correspond to different hybridization events from two different *S. cerevisiae* (Dunn et al. 2008; Nguyen et al. 2011; Baker et al. 2015). The first genome sequence of the Weihenstephan lager brewing strain revealed a global genome composed of a *S. cerevisiae* and a *S. eubayanus* moiety, with variable number of copies along the genome: one copy of each parent for most mosaic chromosomes with two of chromosomes X and XI (Nakao et al. 2009). In addition, this strain presents some loss of heterozygosity associated with the loss of the *S. eubayanus* genome at the extremities of chromosomes III, VII, XIII and XVI and several translocations (reviewed in Bond 2009). Interestingly, genes coding for efficient α-glucoside transporters are amplified in this strain, whereas genes coding for less efficient transporters are inactivated, indicating adaptation to brewing conditions (Nakao et al. 2009). The genome sequence of the Saaz-type lager brewing strain *S. carlsbergensis* has also been sequenced recently, which confirmed it to be roughly triploid. In contrast to Weihenstephan, *S. carlsbergensis* has 2n content of *S. eubayanus* and n content of *S. cerevisiae*; in this case, loss of heterozygosity associated with the loss of the *S. cerevisiae* moiety was observed for segments of chromosomes II, IV, VIII, XIII and for the whole chromosomes VI, XI, and XII (Walther et al. 2014). Strikingly, resequencing of several lager yeasts recently indicated a complex ploidy dynamic of lager strains with 70–77 chromosomal copies in different cells from the Weihenstephan industrial strain and 49–79 chromosomal copies in the genomes of Frohberg brewing lager strains; less variation was observed among *S. carlsbergensis* isolates (van den Broek et al. 2015).

In addition to lager yeast, *S. bayanus* strains also isolated from beer were characterized as complex hybrids between *S. uvarum* and *S. eubayanus* with introgressions from *S. cerevisiae* (Libkind et al. 2011; Nguyen et al. 2011; Pérez-Través et al. 2014). Interestingly, these strains possess a cluster of genes including the maltotriose transporter *MTY1* in at least 3 copies, suggesting a specific adaptation to the malt environment (Nguyen et al. 2011).

S. cerevisiae × S. uvarum, S. cerevisiae × S. kudriavzevii, *and* S. cerevisiae × S. paradoxus *Hybrids*

In addition to the currently well-characterized beer hybrids, many other strains isolated from wine or cider over time were characterized as hybrids between *S. cerevisiae* and *S. uvarum* or *S. kudriavzevii* and sometimes as triple hybrids between the three species. Among them, the strain CID1 and the hybrid strain S6U were among the first to be identified from cider and wine, respectively (Masneuf et al. 1998). Since then, many several other *S. cerevisiae* × *S. uvarum* hybrids have been isolated from vats fermenting at cool temperature (Antunovics et al. 2005; González et al. 2006; LeJeune et al. 2007), and we could observe many others in our collection isolated from fermentations performed at low temperature in Alsace or in the region of Die (unpublished data). In the case of an Alsatian winery it was even possible to find the two possible *S. cerevisiae* and *S. uvarum* ancestors for the hybrids among the isolates of the winery (LeJeune et al. 2007). *S. cerevisiae* × *S. kudriavzevii* hybrids are often encountered in cool-climate wine-making cellars (Bradbury et al. 2006; González et al. 2006; Lopandic et al. 2007; Peris et al. 2011; Erny et al. 2012), and among Belgian-style beers (González et al. 2008). Interestingly, these strains were mainly triploids though diploids and tetraploids as well as chimeric chromosomes were present (Belloch et al. 2009; Erny et al. 2012; Peris et al. 2012). Unexpectedly, one of these strains had closest relatives that were isolated in vats from vineyards in Hungary, Germany, and Alsace (Erny et al. 2012). Sequencing of the popular industrial strain Vin7 revealed it to have a triploid genome, with even coverage of the *S. kudriavzevii* genome, and interspecific chromosome translocation.

Lastly, *S. cerevisiae* × *S. paradoxus* hybrids have also been isolated from wasp guts (Stefanini et al. 2012) but obviously occur less frequently in wine making.

2.1.6 Introgressions and Horizontal Gene Transfer (HGT)

A number of recent studies have highlighted the unsuspected importance of introgressions and horizontal gene transfer (HGT) in the natural evolution of *Saccharomyces* genomes. Introgressions can be defined as the transfer and incorporation of genetic material between two different species through a process of successful mating and backcrossing, whereas HGT results in gene flow between different species through asexual mechanisms. Several introgressions have been reported between *Saccharomyces* strains. Molecular evidence has demonstrated the introgression of 23 kb from *S. cerevisiae* into the European population of *S. paradoxus* (Liti et al. 2006) encompassing the *KRE1* gene, which confers resistance to killer toxins and might thus confer a selective advantage. A 17 kb region was introgressed from *S. paradoxus* into *S. cerevisiae* that included the tandemly duplicated *ENA1*, *ENA2* and *ENA5* genes, coding for P-type ATPases that transport sodium and lithium out of the cell (Garciadeblas et al. 1993) and may be responsible for differences in lithium sensitivity between strains (Doniger et al. 2008).

In line with these studies, powerful genomic analyses have also revealed numerous introgression events between *Saccharomyces* strains. A multispecies-based microarray that targeted 131 orthologous genes from *S. cerevisiae*, *S. paradoxus*, *S. mikatae*, *S. uvarum*, *S. kudriavzevii*, *Naumovia castellii*, *Lachancea kluyveri* and *Candida glabrata* allowed the detection of multiple introgression events between the closest species *S. paradoxus* and *S. cerevisiae*

(Muller and McCusker 2009). Indeed, analysis of 183 *S. cerevisiae* isolates of various ecological and geographical backgrounds revealed five introgressed *S. paradoxus* DNA fragments of length between 2 and 20 kb in the genomes of three different *S. cerevisiae* isolates. In the same way, Dunn et al. (2012) have created a multispecies microarray platform containing probes covering the genomes of several *Saccharomyces* species (*S. cerevisiae* S288C genome and non-reference sequences identified in other strains, *S. paradoxus*, *S. mikatae*, *S. kudriavzevii*, and *S. uvarum*), along with two more distantly related species, *L. kluyveri* and *N. castellii*. By performing Array CGH (comparative genomic hybridization) on 69 commercial wine strains and 14 strains collected from different environments, they have identified five other *S. paradoxus* introgressions (between 800 bp to 30 kb) on five different chromosomes in 15 strains including 12 wine strains, as well as one introgression event from *S. mikatae*. The largest *S. paradoxus* introgression event, identified in four commercial wine strains and localized in the *S. cerevisiae SUC2* region, contains the *S. paradoxus SUC2* gene, which encodes sucrose-hydrolysing invertase, as well as a gene similar to *S. cerevsisiae HPF1*, a glucan α-1,4-glucosidase that when overexpressed, reduces protein haze formation in white wines (Brown et al. 2007). In addition, the *AWA1* gene, also present in this region, was previously identified in *S. cerevisiae* sake strains and encodes a putative GPI-anchored protein localized to the cell wall, which is involved in foam formation in sake mash by conferring hydrophobicity to the cell surface (Miyashita et al. 2004). These results suggest that some adaptive or industrially desirable qualities might be conferred by *S. paradoxus* genes upon these wine strains (Dunn et al. 2012). More recently, Strope et al. (2015) have sequenced with high quality, assembled and annotated the genomes of 93 *S. cerevisiae* strains from multiple geographic and environmental origins and have shown that introgressed genes are common in the *S. cerevisiae* species, preferentially between the sister species *S. paradoxus* and *S. cerevisiae*.

Other evidence of introgressions in *Saccharomyces* yeasts used in food and beverages has been provided by Almeida et al. (2014). A population genomics approach in the cryotolerant species *S. uvarum*, used in wine and cider fermentation, revealed multiple introgressions from other *Saccharomyces* species. The most prevalent introgressions are from *S. eubayanus* into the genomes of European *S. uvarum* strains; these introgressions are associated with human-driven fermentations (wine and cider). The functional categories overrepresented by the set of genes introgressed into any of the *S. uvarum* strains are nitrogen metabolism, sulfite transport and exopeptidase activities which are relevant functions in the context of wine fermentation. The authors postulated that the anthropic habitats colonized by *S. uvarum* in Europe may have favoured the hybridization of *S. uvarum* with *S. eubayanus*, followed by subsequent introgression by backcrossing to *S. uvarum*. Thereby, the domestication pressures imposed on yeast by anthropic environments could favour the interspecies exchange of genetic material, leading to adaptation of genome content.

Horizontal gene transfer contributes greatly to the evolution of prokaryotes, facilitating the acquisition of new functions and allowing rapid adaptation of these organisms to the environment. In yeasts, gene transfer has long been considered a marginal process in their evolution. However, mounting evidence from next-generation sequencing technologies as well as DNA-seq, RNA-seq and CHIP-seq approaches are revolutionizing our views, demonstrating that this phenomenon has been an important force in generating diversity in yeast. Several cases of horizontal gene transfer from prokaryotes to *Saccharomyces* strains have been demonstrated (Gojkovic et al. 2004; Hall et al. 2005; Hall and Dietrich 2007; Nara et al. 2000). Most of them allow the acquisition of novel adaptive function. The most striking example is the acquisition of the *URA1* gene (dihydroorotate dehydrogenase) from *Lactocococcus lactis*, encoding a cytoplasmic enzyme active even in anaerobic conditions, whereas the ancestral *URA9* gene encoding the strictly aerobic mitochondrial enzyme was secondarily lost in the *Saccharomyces* clade, thus facilitating the anaerobic growth of *S. cerevisiae* (Gojkovic et al. 2004; Hall et al. 2005; Nara et al. 2000). The same scenario has been observed for the biotin biosynthesis pathway in *S. cerevisiae*. This pathway appears to have been lost in an ancestor of *S. cerevisiae* and subsequently rebuilt by a combination of horizontal gene transfer and gene duplication followed by neofunctionalization (Hall and Dietrich 2007). Another example, is the acquisition of the *BDS1* gene (aryl- and alkyl-sulfatase) from *Rhodopseudomonas palustris*, allowing the utilization of sulfate from several organic sources (Hall et al. 2005).

In addition to these HGTs involving bacterial genes, eukaryote-to-eukaryote gene transfers have more recently been identified, and some of them have been shown to be involved in the adaptation of *S. cerevisiae* to environmental conditions.

A striking example is the identification of recurrent gene transfer events from distant yeast species into the genome of wine yeast strains. Genomic analysis of the commercial *S. cerevisiae* wine yeast EC1118 has revealed the presence of three large chromosomal segments, A, B and C (120 kb in total), acquired through independent HGT events from distant yeasts (Novo et al. 2009). The yeasts *Zygosaccharomyces bailii* and *Torulaspora microellipsoides* were identified as the donors of regions B and C, respectively (Novo et al. 2009; Marsit et al. 2015). These two islands exhibit 99.7 and 99.9% shared identity with the homologous regions in *Z. bailii* and *T. microellipsoides*, respectively, suggesting that these HGT events are recent. For the region C, the transfer was dated approximately 2,000 years ago (Marsit et al. 2015). These 3 regions have been found almost exclusively in wine strains and mosaic genomes, suggesting that they might confer an evolutionary advantage in the winemaking context (Borneman et al. 2011; Novo et al. 2009).

Determining the function and the advantage for the recipient strain that could be provided by these horizontal acquisitions remains a major challenge to understanding the role of these events in genomic adaptation. In the example of wine yeasts, the 3 HGT regions comprise 39 genes (including 5 pseudogenes) encoding functions potentially important for winemaking, such as nitrogen and sugar metabolism and transport. Several genes present in the largest of these regions, derived from *T. microellipsoides*, were characterized in detail. *FSY1* encodes a high-affinity fructose/H^+ symporter that might be advantageous at the end of wine fermentation, when fructose is the most abundant sugar (Galeote et al. 2010). Another gene, *XDH1*, encoding a putative xylitol dehydrogenase, was shown to be involved in xylose metabolism (Wenger et al. 2010). Finally, a compelling example of the role of HGT in innovation and adaptation of yeast has been recently provided by the study of two tandem-duplicated *FOT1–2* genes encoding oligopeptide transporters. These transporters confer a strong competitive advantage during grape must fermentation by substantially increasing the number and diversity of oligopeptides transported in yeast. Indeed, oligonucleotides are transported in *S. cerevisiae* via the carrier proteins Ptr2p and Dal5p. As Fot proteins transport different oligopeptides, the presence of Fot transporters allows the utilization by wine yeasts of an extra source of nitrogen, resulting in improved biomass formation, fermentation efficiency and cell viability (Damon et al. 2011; Marsit et al. 2015). Thus, the acquisition of these genes has favoured yeast adaptation to the nitrogen-limited wine fermentation environment. Furthermore, the Fot-mediated peptide uptake substantially affects the central pathways of carbon and nitrogen metabolism, resulting in decreased acetic acid production and increased volatile ester formation, which are considered as positive attributes for the organoleptic balance of wines (Marsit et al. 2016).

Another interesting example of HGT concerns the acquisition by *S. cerevisiae* strains isolated from industrial (brewing, wine, bakery, bioethanol) or laboratory environments of the *ASP3* gene originating from the wine yeast *Wickerhamomyces anomalus* (formerly *Hansenula anomala* and *Pichia anomala*) (League et al. 2012). Asp3 is involved in the degradation pathway of D-asparagine and is induced by nitrogen starvation (Dunlop et al. 1978); thus, its acquisition may have aided yeast adaptation to artificial environments.

Little is known about the mechanisms of HGT in eukaryotes. Natural transformation and conjugation have been proposed in the case of bacterium-to-fungus transfer (Hall et al. 2005), and trans-kingdom DNA transfer between *E. coli* and *S. cerevisiae* by conjugation has been demonstrated (Heinemann and Sprague 1989; Stachel and Zambryski 1989). Similarly, HGT between yeast species could be the result of a transformation of yeast cells by exogenous DNA fragments. Extrachromosomal circular DNAs (eccDNA) molecules containing consensus sequences for autonomous replication origins have been reported in natural and experimental yeast populations (Libuda and Winston 2006; Møller et al. 2015; Thierry et al. 2015) and could be involved in this process. The *S. cerevisiae* region B transferred from *Z. bailli* was found at various chromosomal locations and with different gene organization in different wine yeast strains and was shown to contain an autonomously replicating sequence functional in *S. cerevisiae* (Borneman et al. 2011; Galeote et al. 2011). These observations strongly suggest that region B is prone to circularization and might be transferred to a recipient *S. cerevisiae* by transformation, like a plasmid. Furthermore, the stress conditions in the food environment might facilitate transformation by exogenous DNA by damaging cell membranes. Another possible mechanism for HGT between yeast species could involve DNA transfer from one nucleus to the other during the transient heterokaryotic stage of an abortive hybridization (Morales and Dujon 2012).

Notably, *Z. bailii* and *T. microellipsoides* were isolated from various fruit juices, soft drinks and beverages, including wine (Deak 2007; Kurtzman 2011). Similarly, *W. anomalus* is found in a variety of biotechnological applications, such as wine making, baking, and brewing (Schneider et al. 2012; Walker 2011). Interestingly, recurrent transfers have been recently found in food environments (Cheeseman et al. 2014; League et al. 2012; Ropars et al. 2015). These convergent cases suggest that anthropic environments might offer substantial ecological opportunities for HGT and the sharing of metabolic genes between distantly related yeast species.

Thus, introgressions and HGT events may help yeast evolve to survive fluctuations in the external environment by adapting their metabolic networks to meet the challenges of their ecological niche, hence leading to the evolutionary diversification of strains in foods and beverages.

2.2 Phenotypic Diversity

2.2.1 *S. cerevisiae*

The genetic adaptation of *Saccharomyces* spp. to sugar-rich and oxygen-limited environments following the emergence of fruit-bearing plants gave these yeasts specific phenotypes that made them organisms of choice for fermentation processes. These phenotypes include their ability to produce ethanol, their relatively high tolerance to environmental stresses (ethanol, high temperatures, osmotic conditions) and their facultatively anaerobic growth (Goddard 2008; Piskur et al. 2006). In particular, *S. cerevisiae* combines all these properties that favour its dominance in traditional and industrial fermentations. However, *Saccharomyces* strains exhibit remarkable phenotypic diversity, which is found both among the different *Saccharomyces* species and among *S. cerevisiae* strains from various ecological origins, which could become a huge reservoir for technological innovation in food and beverages industries.

Saccharomyces strains first display an important divergence in their sensitivity to environmental factors that can affect the course of fermentations. *S. cerevisiae* strains are more tolerant to high temperature than *S. bayanus*, *S. kudriavzevii* and *S. paradoxus* isolates (Belloch et al. 2008; Mukherjee et al. 2014). Conversely, *S. bayanus* and *S. kudriavzevii* that efficiently grow at low temperature are classified as cryotolerant yeasts (Belloch et al. 2008; Kishimoto and Goto 1995). These species also differ in their sensitivity to ethanol, *S. cerevisiae* isolates tolerating alcohol better than other yeasts (Arroyo-López et al. 2010; Csoma et al. 2010; Belloch et al. 2008; Mukherjee et al. 2014). A large diversity between strains in their capacity to grow in the presence of high ethanol concentrations has been found within *S. cerevisiae*, that is, interestingly, in line with the ecological niches of the strains (Kvitek et al. 2008; Warringer et al. 2011; Mukherjee et al. 2014). Thus, yeasts from fermentation processes generating ethanol-rich conditions (wine, sake, bioethanol) display a high tolerance for this stressor. Conversely, *S. cerevisiae* strains involved in the production of beverages with lower alcohol content (beer) are more sensitive to ethanol (Steensels and Verstrepen 2014). How resistance to toxic chemical compounds follows the origin of strains is another example of the emergence of phenotypic traits as the result of environmental adaptation. Compared to non-wine strains, wine yeasts have a better resistance to copper and sulfite, which are widely used in vineyards and during winemaking (Fay et al. 2004; Warringer et al. 2011; Pérez-Ortín et al. 2002). In the same way, distillery and ale strains demonstrate an improved resistance to molasses toxin (Ness and Aigle 1995; Borneman et al. 2011).

A large-scale study assessing the phenotypic diversity within *Saccharomyces* also revealed large differences between strains in their ability to utilize some carbon and nitrogen nutrients (Warringer et al. 2011). Differences are particularly evident in the consumption of oligosaccharides that are the major carbon sources of wort. *S. cerevisiae* strains are able to import and catabolize both maltose and maltotriose, whereas strains from *S. eubayanus*, the ancestral species of lager yeast, are unable to grow using maltotriose but display a more efficient use of maltose (Hebly et al. 2015; Krogerus et al. 2015; Duval et al. 2010). In addition, marked diversity exists within *S. cerevisiae* species with regard to the capacity to ferment sugars available in beer wort, with ale strains being more efficient than wild strains or strains from other industrial fermentations (Steensels and Verstrepen 2014), or to catabolize fructose in grape juice, which is enhanced in some wine yeasts thanks to a mutation in the *HXT3* transporter allele (Guillaume et al. 2007) and potentially to the presence of the fructose transporter *FSY1* (Galeote et al. 2010). Also regarding wine-making, a particularly interesting phenotypic trait is the capacity of yeast to assimilate nitrogen sources because nitrogen is the limiting nutrient for yeast growth in most grape juice fermentations. Interestingly, several studies indicate that the adaptation of wine strains to their environment resulted in the emergence of specific features favouring the utilization of the nitrogen resource, such as a more efficient nitrogen metabolism (Crépin et al. 2012, 2014; Jaras et al. 2014) and the ability to utilize different types of di- or tripeptides as nitrogen sources (Homann et al. 2005, Marsit et al. 2015). Extending the analysis to *Saccharomyces* species, Warringer et al. (2011) observed a decreased efficacy of *S. paradoxus* to use threonine and isoleucine as sole nitrogen sources, whereas histidine appears to be a preferred nitrogen source for *S. bayanus* growth. Consistent with the variability within *S. cerevisiae* strains in their ability to efficiently assimilate available nutrients, two opposite strategies of resource utilization have been distinguished: "grasshoppers" refers to strains usually used in industrial processes that reproduce slowly and reach a low carrying capacity but have large cell size in fermentation and a high reproduction rate in respiration, and "ants" refer to strains, isolated from both natural and laboratory environments, which reproduce rapidly and display a large carrying capacity but have a small cell size and a low reproduction rate in respiration (Spor et al. 2008, 2009).

Several studies have focused on the characterization of the diversity between *Saccharomyces* yeasts in the formation of metabolites that are phenotypic traits of considerable interest in the food and beverages industries. First, significant inter-species variability was found both in the formation of intermediates of the central carbon metabolism and in the production of volatile compounds, which are major contributors to fermentative aroma. In general, *S. cerevisiae* strains produce more acetate and less glycerol than *S. uvarum* and *S. kudriavzevii* isolates (Tosi et al. 2009; Csoma et al. 2010; Pérez-Torrado et al. 2015). However, such environmental parameters as temperature may reverse these trends with, for example, an increased formation of acetate by *S. uvarum* observed at high temperature (Pérez-Torrado et al. 2015). In addition, *Saccharomyces* species demonstrate a vast diversity of abilities to produce volatile molecules, just starting to be characterized and exploited in the food and beverages industries. Of particular note are the low formation of diacetyl by *S. eubayanus* compared with *S. cerevisiae*, of interest for beer-making processes (Gibson et al. 2015), and the increased formation of flavour-active higher alcohols and acetate esters by *S. uvarum* and *S. kudriavzevii* (González et al. 2007; Tosi et al. 2009; Csoma et al. 2010; Stribny et al. 2015). Intra-species variability in metabolic features was also revealed, particularly within the *S. cerevisiae* population. For instance, Steensels and Verstrepen (2014) reported that the proportion of ale and sake strains producing phenolic off-flavours, which have a marked negative effect on the quality of beer and sake, during fermentation in ferulic acid-containing wort or malted rice is at least twice lower than that of wine and wild yeasts. Furthermore, the characterization of strains of various origins during wine fermentation showed the emergence of origin-dependent metabolic properties. These include the high formation of acetate by laboratory strains, palm and sake isolates and conversely, the low acetate production of bakers' and wine yeasts, shown by the comparison of metabolic profiles of 72 strains (Camarasa et al. 2011; Fig. 3), and the higher production of fruity aromas by wine strains suggested by a sensorial analysis of wines from 13 *S. cerevisiae* strains (Hyma et al. 2011). The lower diversity within strains used in wine-making than in natural isolates from vineyards regarding metabolic traits of technological interest (acetate production, isoamyl alcohol formation) indicates that wine commercial yeasts are a minimally diverse subset of strains coming from the vineyard environment, likely as a result of human selection (Camarasa et al. 2011). Finally, significant differences between *S. cerevisiae* strains from vineyards are found in their ability to produce higher alcohols and acetate esters (Lopandic et al. 2007; Capece et al. 2010). Interestingly, the variations in metabolic traits allowed their clustering between different groups in line with their isolation area, which support the notion that specific population of native yeast strains can be associated with a terroir (Knight et al. 2015).

2.2.2 Hybrids

One of the main benefits of interspecific hybridization is the huge potential of generating phenotypic diversity and consequently of combining traits of interest from two parents into a single strain and of enabling the emergence of beneficial transgressive phenotypes. Thus, natural and newly generated *Saccharomyces* interspecific hybrids will continue to play a significant role in beer and wine

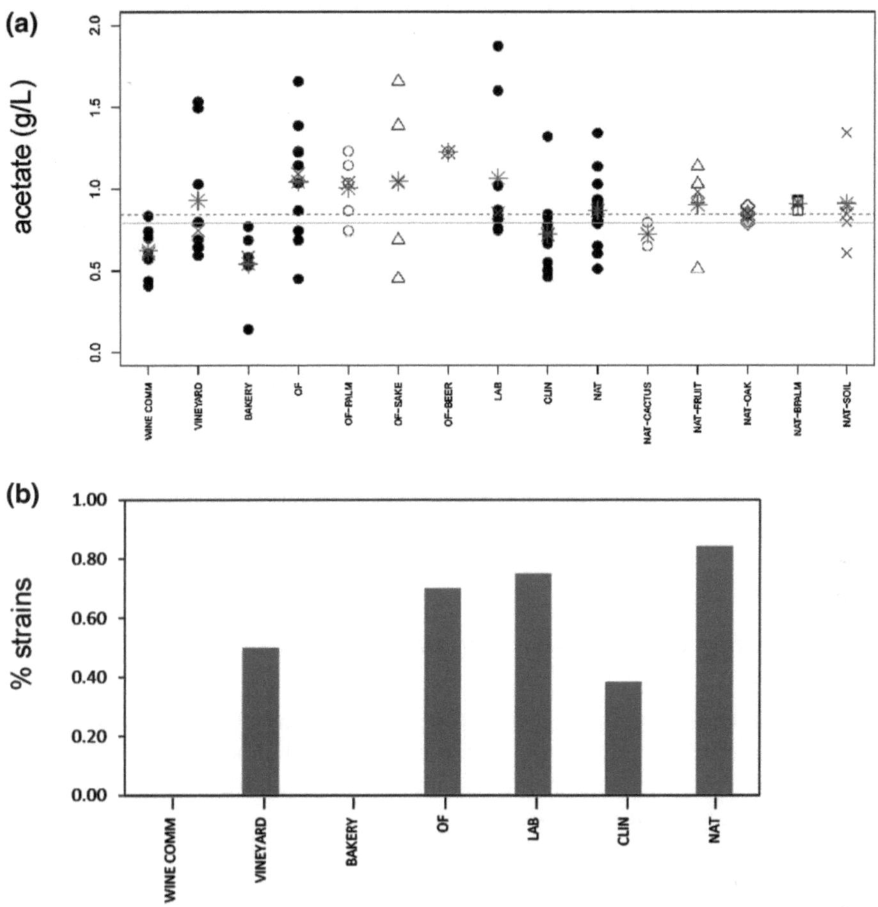

Fig. 3 Acetate production of 72 *S. cerevisiae* strains from different ecological niches. A—Acetate production in the main origin groups: bakery, clinical (CLIN), other fermentation processes (OF), laboratory (LAB), vineyard, natural (NAT) and wine commercial (WINE COMM) (closed symbols). Open symbols correspond to the subdivision of the OF group into beer, palm wine, and sake sub-groups and of the NAT group into oak, Bertam palm, soil, fruit, and cactus sub-groups. The *red* and *blue lines* indicate the mean and median of the entire population, respectively. The mean and the median of groups or sub-groups are indicated by *red* and *blue* stars, respectively. B—Percentage of strains exhibiting acetate production higher than 0.8 g L^{-1} (average value for the overall population) for the 7 main origin groups

making. The enhanced performances (heterosis) in the brewing environment of the natural hybrids *S. pastorianus* compared with those of their parental strains *S. cerevisiae* and *S. eubayanus* has been ascribed to an advantageous combination between the cold tolerance of *S. eubayanus* and the high fermentative capacity of *S. cerevisiae* (Gibson and Liti 2015). Further characterization of the *S. pastorianus* taxon allowed the classification of the strains by *S. cerevisiae*-related or

S. eubayanus-related physiological behaviours, in line with the two genetic groups existing within this species. The Frohberg and Saaz groups are differentiated in particular by their aromatic profile (Gibson et al. 2013). Furthermore, interspecific yeast hybrids isolated from the wine environment (mainly *S. cerevisiae* × *S. bayanus*, *S. cerevisiae* × *S. uvarum* and *S. cerevisiae* × *S. kudriavzevii*) exhibit wider-ranging production profiles of volatile compounds while maintaining the advantageous traits of their parents in terms of fermentation performances and stress resistance (González et al. 2007; Belloch et al. 2008; Pérez-Torrado et al. 2015). In particular, hybrids outperformed *Saccharomyces sensu stricto* species for their low acetate and non-desirable molecule formation, as well as for their increased production of glycerol and volatile metabolites that contribute to wine flavour. Finally, hybridizing *S. cerevisiae* with *S. mikatae*, a species hitherto not associated with industrial fermentations, generated a new breed of *Saccharomyces* strain with an improved aromatic profile (Bellon et al. 2013).

Overall, hybrids appear to have huge potential to deliver increased complexity to beverages' sensory properties and enable the production of alternative styles of wine and beer, without affecting the fermentation course.

2.3 Linking Genetic Diversity and Adaptation

The advent of the genomic era has opened new ways to decipher the genetic variation involved in adaptation. Mutations in several genes conferring improved fitness upon yeast strains in their environment have been identified within the past 20 years. The most striking evidence of yeast adaptation to current wine-making environments is the reactions to chemicals used in the vineyard and winery environments. The amplification of the *CUP1* gene in wine yeast enabling resistance to copper in the media (Fogel et al. 1983) has been associated with the resistance to copper of wine and sake strains (Warringer et al. 2011). The recent re-sequencing of the genome of 100 yeast strains revealed that some strains of the wine/European cluster could possess up to 18 copies of the *CUP1* gene (Strope et al. 2015). Variations have also been gained by wine yeasts as a response to the use of sulfite. Two translocations leading to the overexpression of *SSU1* have been detected and characterized among wine yeasts (Pérez-Ortín et al. 2002; Zimmer et al. 2014).

Another striking evidence of adaptation is the acquisition by wine yeasts by horizontal gene transfer of the gene *FOT1* encoding a dipeptide transporter with a high affinity for glutamate. As glutamate is present in high concentration in grape must, this transporter enables yeast to better exploit the nitrogen resources of the must and provides wine yeast a selective advantage in grape must (Novo et al. 2009; Marsit et al. 2015, 2016). This transfer is also recent, contemporary to the development of wine yeast and can be detected in as many as half of *S. cerevisiae* wine yeast strains. Another hallmark of adaptation has also been detected in the genome of flor yeasts, a family closely related to wine yeasts, with two mutations in the promoter and the coding region of the gene *FLO11* coding for a mucin protein essential for biofilm formation and pseudohyphal growth (Fidalgo et al. 2006; Legras et al. 2014). Even more mutations have been recently detected in genes involved in the regulation of this gene (Coi et al. 2017). Interestingly, flor yeasts also possesses a fructophylic allele of the *HXT3* hexose transporter identified in an industrial strain (Guillaume et al. 2007) and several flor strains have also the high affinity fructose transporter *FSY1*, which may provide an adaptive advantage after alcoholic fermentation when traces of fructose remain in the fermented must whereas glucose is exhausted.

Beer strains also present specific features attesting adaptation to beer malt. The sequencing of the genome of Weihenstephan revealed the amplification of the maltotriose transporter *MTY1*, whereas a less efficient maltotriose transporter *ATG1* of *S. cerevisiae* was suppressed; a similar picture has been found for *S. bayanus* strains NBRC1948 and CBS380 (Nguyen et al. 2011). Last two recent population genomic studies performed on ale beers strains revealed the amplification of the *MAL* locus and loss of function mutations in *PAD1* and *FDC1* genes which are involved in the production of 4-vinyl guaiacol (4-VG), an undesirable off-flavor in beer (Gallone et al. 2016; Gonçalves et al. 2016).

3 Yeast Strain Improvement by Non-GMO Strategies

Although many benefits to wine production have been gained through the selection of strains with suitable properties, such as good fermentation performance and the absence of off-flavours, many efforts remain to be made to improve existing traits or to introduce new desirable properties through systematic strain improvement techniques. The different strategies of strain selection and improvement available have been recently reviewed by Steensels et al. (2014). Here we will focus on the use of rational genetic strategies to exploit existing natural diversity and on non-targeted approaches to generate new diversity, such as evolutionary engineering, which have both proven efficient in the past decade to develop strains with improved traits for fermentative processes.

3.1 QTL Mapping

The identification of the genetic bases underlying yeast phenotypic diversity is a major challenge to better exploit the potential of yeast biodiversity. The high genetic diversity uncovered through the genomic sequencing of yeasts has underscored the unexplored opportunity to improve industrial strains using such a genetic reservoir. The exploitation of yeast diversity for industrial applications requires the identification of genes and alleles with suitable impact on industrial yeast strains' properties. Although in some situations, industrially relevant genes may be inferred directly from sequencing data or functional analyses (Guillaume et al. 2007; Watanabe et al. 2012) such methodologies cannot be generalized to address most industrial phenotypes. Linking phenotype to genotype is not straightforward as most phenotypes are quantitative traits that are controlled by multiple loci with variable effects. The mapping of quantitative trait loci (QTLs) is a method suited for the identification of genes involved in complex traits. Various studies, reviewed by Liti and Louis (2012), have reported the feasibility of QTL mapping in yeast, initially in laboratory strains to map genes involved in model phenotypes such as heat resistance (Steinmetz et al. 2002), sporulation efficiency (Deutschbauer and Davis 2005), gene expression (Brem et al. 2002) or drug resistance (Perlstein et al. 2007). These works have opened a new era of addressing the bases of industrially relevant traits, and these approaches have been applied to map the genes for fermentation traits (Marullo et al. 2007). QTL procedures have been strongly dependent on the availability of genotyping tools to monitor marker distribution in populations. The advent of oligonucleotide arrays that allow the simultaneous detection of thousands of SNPs has been a key advance in QTL mapping (Winzeler et al. 1998). Recently, advances in DNA sequencing techniques have brought new facilities to scan the genomes that can be used in QTL linkage analysis (Parts et al. 2011).

3.1.1 QTL Mapping Strategies

The rationale of QTL mapping is rather simple in its principle because it consists in the establishment of statistical links between the inheritance of a marker in a recombined population and the value of a trait. However, the identification of a QTL can be challenging, not only because some QTLs can have only a weak effect on the phenotype but also because QTLs can interact in a complex manner or have an effect modulated by the environment (Kroymann and Mitchell-Olds 2005; Sinha et al. 2006; Smith and Kruglyak 2008). In many studies, QTLs are found but they do not explain all the variation in the trait and correspond only to major QTLs. Although it has been long considered that most traits were under the control of many QTLs, each with weak effects, various studies have shown that trait variation could be explained by few QTLs, each with a strong effect (Deutschbauer and Davis 2005). In any case, mapping all the QTLs involved in a trait can be challenging.

Until now most QTL mapping studies reported have been performed with laboratory strains, sometimes in crosses with wild or industrial yeasts. Although laboratory strains are useful to address basic questions, they are poorly suited to investigate industrial traits for they carry many deleterious mutations that can impair the mapping of industrially relevant alleles. The use of industrial or wild strains in crosses to build recombined populations for QTL mapping is therefore preferable. However, the complex genomic organization of industrial or natural yeast isolates must also be considered because it can weaken the QTL mapping capacity. Unbalanced set of chromosomes (aneuploidy), polyploidy or chromosomal translocations are quite frequent in these strains and can deteriorate QTL mapping analyses. Those chromosomal alterations may themselves contribute to yeast strains' properties, as suggested by the mapping of a translocation QTL controlling yeast sulfite resistance and a partial disomy governing fermentation rate

(Brion et al. 2013). In addition, chromosomal regions acquired through horizontal transfer that are found in only some industrial yeasts are usually not considered in QTL mapping. As these regions contain genes that may contribute to strains' properties (Marsit and Dequin 2015), monitoring their segregation in populations can be relevant. This concern emphasizes the added value of prior knowledge of a strain's genome before a QTL analysis.

QTL mapping in yeast involves the creation of a recombined haploid population by crossing two haploid strains and letting the resulting hybrid sporulate. A difficulty frequently encountered at this phase when using industrial yeast isolates originates from their ability to switch their mating type after sporulation as this behaviour prevents obtaining a stable haploid clone to perform crosses (Thornton and Eschenbruch 1976). This problem can be solved by deleting the HO gene through gene disruption to obtain a stable haploid line.

A standard way to map QTLs relies on the individual phenotyping and genotyping of a population of haploid segregants (Mackay et al. 2009) (Fig. 4). Such approaches have been used to map QTLs of various phenotypes in laboratory strains as well as in industrial yeasts (Marullo et al. 2007; Katou et al. 2009; Ambroset et al. 2011). The ability to detect QTLs depends strongly both on the population size and on the number of molecular markers available. In addition, the detection power is also influenced by number of chromosomal crossovers in the recombined population, which can be increased by additional rounds of crosses and sporulation (Cubillos et al. 2013). Genotyping has usually been performed using oligonucleotide microarrays, such as Affymetrix DNA chips, which are well suited for QTL mapping as they routinely enable the detection of 2000–4000

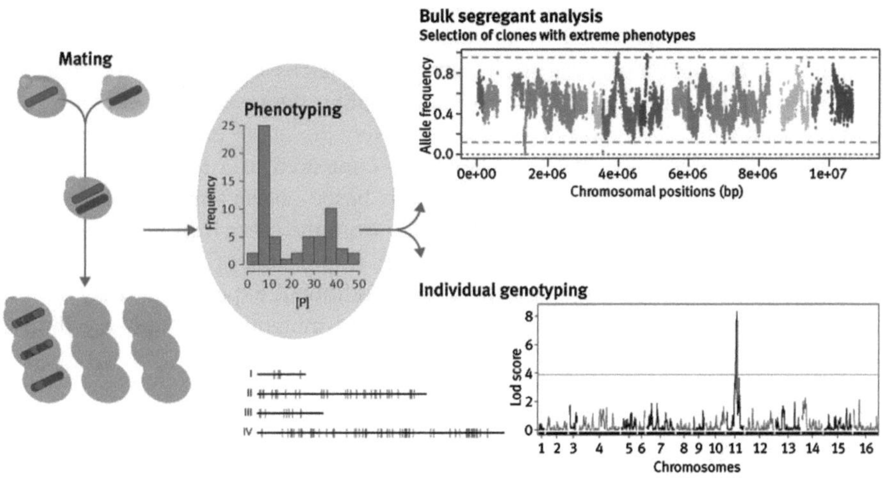

Fig. 4 QTL mapping design. The haploid segregants obtained after sporulation of a hybrid resulting from a cross of the parental strains are phenotyped and are genotyped either individually or as pools of segregants

polymorphisms (Winzeler et al. 1998). The use of DNA microarrays with oligonucleotides optimized in melting temperature is an interesting alternative as the probe design can be customized to specifically detect SNPs of interest (Gresham et al. 2010). More recently, the advent of NGS and the reduction of costs have made the sequencing of the individual clones a relevant genotyping alternative (Rohland and Reich 2012). Despite such progress, genotyping and individual characterization of clones can turn tedious and costly. To circumvent these constraints, an alternative method based on the genotyping of pooled segregants with extreme phenotypes has been proposed (Brauer et al. 2006). This approach, designated "bulk segregants analysis", is founded on the assumption that pools of segregants with extreme phenotypes are enriched in alleles with a given effect on the phenotype. An extension of this methodology (XL-QTL) that enables the assessment of a large number of segregants proved powerful to map genes of fundamental interest (Ehrenreich et al. 2010). However, as many phenotypes of industrial relevance require an individual measure of the trait, this method is not always applicable. In addition, several caveats have been described, such as invasion by diploid strains in enrichment procedures associated with pool analyses (Wilkening et al. 2014).

A limitation of those approaches independent of the mapping device used is that they enable the mapping of only alleles that are present in the two strains used in the cross. To overcome this constraint and expand the genetic diversity that can be addressed, multiparent-based population QTL analyses have been used (Cubillos et al. 2013).

QTL dissection in yeast is facilitated by the possibility to easily perform functional analysis of candidate genes and assess their involvement in a given trait. A popular method to check alleles effects is the reciprocal hemizygosity test (Steinmetz et al. 2002). The test is based on the construction of two hybrids, each harbouring only one active parental allele of the gene investigated, the other parental allele being inactivated by disruption. These hemizygous strains are easily obtained by crossing a parental strain carrying an inactivated copy of the gene with the reciprocal parental strain. This procedure enables the assessment of many genes and facilitates the dissection of QTLs. However, the allele effect is assessed in a diploid context that may alter the phenotype compared to a haploid background. Another method to assess the role of alleles is based on the simple replacement of an allele from one parental strain by the allele from the other haploid parent. Those approaches allow the identification of the genes responsible of variation in traits but not directly to the polymorphism involved. Additional experiments may be necessary including the introduction of the identified mutations through site-directed mutagenesis or related methods.

3.1.2 QTL Approaches for Industrially Relevant Traits

In the past 10 years, a set of publications have reported QTL analyses aiming at the identification of the genetic bases of various industrially relevant traits. They included the production of metabolites that affect fermented products' quality, the yeasts' fermentation capacity or the yeasts' ability to endure industrial stresses.

Genetic Bases of Variations in Metabolites and Aroma Formation

Given that yeast metabolites play a key role in fermented products' quality and are rather easily accessible by analysis, various QTL studies aiming at mapping genes involved in their production have been reported. Acetic acid is an important component of wine quality, and yeast strains can display strong variations in the amount produced. In a pioneering work with wine yeasts, Marullo et al. (2007) described the mapping of QTLs involved in acetic acid production during wine alcoholic fermentation. They showed that variation in the asparaginase gene *ASP1* was responsible for differences in acetic acid production but only when asparagine was the nitrogen source. In a recent study, Salinas et al. (2012) reported the mapping of *ALD6* as a source of variation in acetic acid formation, in agreement with the known role of this gene in its biosynthesis (Saint-Prix et al. 2004). Interestingly, they showed that variations in *ALD6* expression level likely due to mutations in its promoter region were driving the acetic acid production differences. In the same work, Salinas et al. (2012) mapped QTLs for several metabolites and showed that succinic acid production was determined by an epistatic interaction between two genes.

The genetic bases of variations in aroma compounds' biosynthesis have been addressed in sake and wine yeasts. Katou et al. (2009) identified QTLs for higher alcohol and esters production in sake yeasts. Steyer et al. (2012) described a set of QTLs that controlled the formation of 15 compounds including phenyl ethanol, esters, and nerolidol using a cross between a wine yeast and S288C. Remarkably, they showed that variations in the transcription factor *PDR8* were responsible of differences in nerolidol production through a modulation of the expression the gene *QDR2* encoding an exporter protein, later confirmed as a major target of *PDR8* in fermentation by Brion et al. (2013).

Sulfur compounds have a strong impact on wine quality, and variation in the ability of yeasts to release these molecules is well documented. Using a couple of wine yeast strains, Blondin et al. (2013) and Noble et al. (2015) addressed the source of variation in sulfur compounds' production during alcoholic fermentation. They reported the identification of variants of *SKP2* (an ubiquitin protease that targets Met14p for degradation) and *MET2* that modulate the production of SO_2 and H_2S. The two identified alleles were shown especially relevant to reduce sulfur compound formation without accumulation of any intermediate metabolites and to indirectly control the acetaldehyde amount in wines (Noble et al. 2015). Huang et al. (2014) reported the mapping of several genes of the sulfur assimilation pathway, *MET5*, *MET10* and *MET2* that control H_2S production by wine yeasts.

Fermentation Performance

Several studies have addressed the genetic bases of variations in the fermentation performance of industrial yeasts. Zimmer et al. (2014) reported the mapping of a translocation in wine yeasts that reduces the lag phase in sulfited grape musts. The translocation increases the expression of the sulfite export system *SSU1* and confers a selective advantage in the winemaking environment. Jaras et al. (2014) explored the source of differences in nitrogen source consumption between wine yeast and sake isolates. They were able to map a large set of QTLs that affect the use of amino acid and ammonium during fermentation. Interestingly, several of the mapped genes had a pleiotropic effect on nitrogen source utilization that could explain the nitrogen preferences of wine and sake yeasts. Ambroset et al. (2011) identified variation in *ABZ1* (a gene involved *p*-aminobenzoate biosynthesis) between a wine yeast and S288C that modulates the fermentation rate by controlling nitrogen utilization. In addition, it was shown that the laboratory strain S288C was harbouring a defective allele of this gene, obviously due to propagation in rich media.

The ability to ferment in conditions of low nitrogen availability is of major importance to avoid stuck fermentation in wine making. Brice et al. (2014) have investigated the genetic bases of the ability to ferment in conditions of nitrogen starvation. Using a bulk segregant analysis, they reported the mapping of a set of chromosomal regions that control the fermentation rate in starvation. Several genes so identified were involved in nitrogen sensing or signalling, indicating a role for nitrogen signalling in the control of glycolytic flux in starvation. Intriguingly, one of the QTL regions harboured two genes whose alleles had opposite effects on the fermentation capacity (Brice et al. 2014).

The ability of yeast to resist ethanol stress is of key importance for many industrial applications and indeed for bioethanol production. The identification of QTLs controlling ethanol tolerance was reported by Swinnen et al. (2012a, b). Using a pooled-segregant approach they addressed the bases of variation in ethanol resistance in a couple of bioethanol and laboratory yeasts. They showed that ethanol tolerance was modulated by several loci and remarkably revealed that among three genes in a QTL controlling ethanol resistance, one of them was acting in an opposite way to the two others. Yeasts face other stresses during the fermentation of lignocellulosic biomasses such as acetic acid, osmotic pressure, furfural and other inhibitors. Greetham et al. (2014) have addressed the bases of variations in yeast resistance to those stresses and showed that both allelic variation and changes in gene expression were responsible for the phenotypic differences. Discrepancy in resistance to toxic compounds was also noticed by Brion et al. (2013) in wine fermentation who showed that allelic variation in the organic acid export system was involved in octanoic acid resistance.

3.2 Evolutionary Engineering

Evolutionary engineering, also called adaptive, directed, or experimental evolution, is an efficient approach to generate diversity and, as such, is particularly suitable for improving traits showing no or low diversity within the yeast species considered. Furthermore, many phenotypes of technological interest are complex phenotypes, with multi-allelic origin and resulting from multi-level regulations and dynamic interactions in cellular systems. Evolutionary engineering is an interesting alternative to other strain improvement approaches, particularly when the current level of understanding of a trait is insufficient. This approach can result in thorough rewiring of metabolism, with the expression levels of many genes changed simultaneously, mimicking natural evolution. Thus, this non-targeted approach is used as not only a non-GMO approach but also, particularly for non-food applications, in combination with metabolic engineering (Lee et al. 2011). Adaptive evolution is a bottom-up strategy that starts with the generation of a desired phenotype. As no a priori knowledge of the underlying molecular mechanisms is required, this approach can be used to improve a large diversity of yeast strains and species, including hybrids. In the past decade, evolutionary engineering has been increasingly used for improving diverse phenotypes of technological interest. Examples of its application include engineering for the utilization of specific substrates, co-consumption of substrates or tolerance to various stresses or inhibitors.

3.2.1 Evolutionary Engineering Strategies

Experimental evolution is based on the selection of mutations that emerge randomly in a clonal population. These spontaneous mutations are artificially enriched through selection pressure imposed to the cells. Thus, prolonged cultivation under selective conditions is used to select genetic variants having advantageous mutations in the prevailing environment. With respect to their effect on fitness, mutations can be beneficial, neutral, deleterious or lethal. During an evolution experiment, different genotypes can emerge, co-exist and follow one another (Rabbers et al. 2015). Several genotypes are competing, and the final composition will be enriched in the genotype that takes over the whole population by outcompeting others.

Yeast cells have small genomes (e.g., 12 Mb for *S. cerevisiae*), rapid growth rates and can achieve high cell concentrations in a small volume, which makes them ideal organisms for evolutionary engineering. With a mutation rate of 10^{-10}–10^{-8} per base pair (bp) per generation, yeast variants can be obtained after a few hundred generations, in a timeframe varying from a few months to over a year, depending on the growth rate of the strain in the selective conditions used.

Although experimental evolution studies frequently rely on spontaneous mutations, it is also possible to increase genetic diversity using random mutagenesis, mating, protoplast fusion or hybridization. Other approaches using engineering methods, e.g., transposon mutagenesis, deletion mutant libraries, or synthetic biology tools, can also be applied to enhance the creation of phenotypic diversity (David and Siewers 2015), but their utilization generates GMOs, which is a major limitation for food and beverages applications. Generally, although increasing the level of diversity might give a higher chance to acquire complex, multi-allelic traits, it also increases the risk of obtaining deleterious mutations and the difficulty of subsequently identifying beneficial mutation(s).

The choice of selective conditions and experimental design (Fig. 5) is crucial to the outcome of experimental evolutions. Differences in fitness can arise from differences in performance and may be linked to several fitness components, such as growth rate, biomass yield or survival. Various experimental designs can be used in the laboratory: (i) successive batch cultivation (serial transfer); (ii) continuous cultures in chemostat or turbidostat or (iii) single-cell technologies. These modes of cultivation impose different selective pressures and may favour the selection of

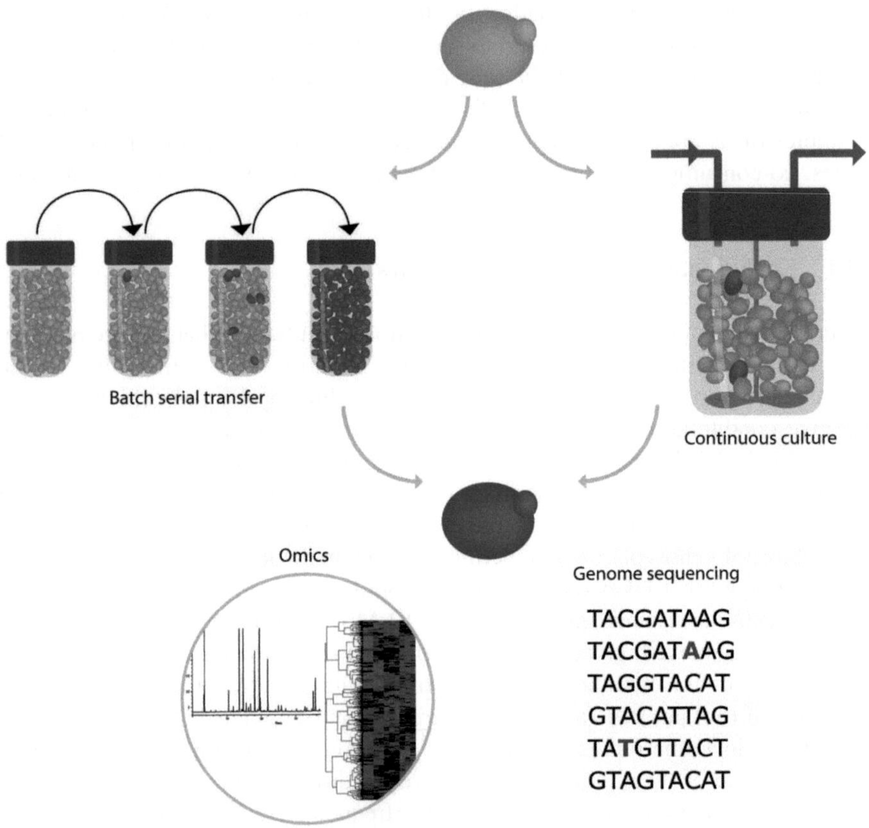

Fig. 5 Evolutionary engineering strategies. Evolved strains are selected during prolonged cultivation in batch or continuous cultures. Evolved strains are analysed by a combination of—omics approaches and whole-genome sequencing to identify the underlying mutations

different phenotypes (Bachmann et al. 2015). For serial batch propagation, after a certain period of growth in liquid culture in flasks or tubes, cells are diluted into fresh medium and several cycles are repeated until evolution is observed. This culturing method is easy to set up and best mimics traditional food or beverage fermentation processes. It usually selects for microorganisms with shorter lag phase or higher growth rate, but certain regimes might also allow the selection of higher biomass formation or a better survival after nutrient depletion (Rabbers et al. 2015; Cadière et al. 2011; Tilloy et al. 2014; Bachmann et al. 2015). To increase the efficiency of the selection, the selection pressure can be progressively increased during the experiment. In contrast to batch cultivation, continuous cultures rely on a constant inflow of nutrients and the constant outflow of random individuals and results in maintaining a constant population size. These conditions favour selection for higher substrate affinity (Bachmann et al. 2015). The use of single-cell technologies, such as emulsion-based approaches has been recently put forward (Bachmann et al. 2013). This approach involves serial propagation of a microbial population in a water-in-oil emulsion and is based on physical separation of individual cells to exclude cell-cell competition. It enables the selection of slow-growing strains with higher biomass yield.

The choice of the cultivation approach is crucial and depends not only on the desired phenotype but also on the fermentation process in which the strain will be used. As the strains to be improved have been selected for their superiority during a typical fermentation process, the cultivation conditions should be as close as possible to those of the target industrial process to generate a new phenotype that could be observable in the industrial conditions as well as to limit the risk of losing desirable traits. Indeed, selection for improved fitness in a specialized environment can often lead to tradeoffs associated with the new phenotype, i.e., improved performance in one condition associated with reduced performance in another condition (Bachmann et al. 2015).

A major challenge in evolutionary engineering is the elucidation of the underlying molecular mechanisms. This is critical to understand the metabolic bases of the evolved phenotype and to better characterize strains, particularly their behaviour and their performance in the various stages of the process, which includes not only the fermentation step but also the biomass production process, storage, etc. The identification of genetic changes and mechanisms that underlie the improved performance of strains generated via non-targeted approaches will also facilitate the rapid transfer of evolved traits among strains, using for instance hybridization approaches and successive back-crosses. High-throughput and genome-wide tools, including methods to profile the genome, transcriptome, proteome, metabolome and fluxome are valuable tools for understanding the engineered phenotypes, which can results from changes at different levels, e.g., gene regulatory networks, enzymes kinetics or localization, transport reactions or metabolic fluxes. Recent advances in whole-genome sequencing technologies have more recently revolutionized the analysis of evolved genotypes, enabling the identification of changes between the genomes of ancestral and evolved strains by direct comparison (Oud et al. 2012). To facilitate the identification of these mutations, it is crucial to run several independent evolutions in parallel as this procedure allows for identifying common evolution patterns and distinguishing beneficial mutations from neutral mutations. In addition, resequencing the ancestral strain is recommended to eliminate variations inherent to sequencing technologies. It is also possible to precisely follow the dynamics of the evolution process using population sequencing.

3.2.2 Evolutionary Engineering of Food-Related Yeasts

Evolutionary engineering, used alone or in combination with metabolic engineering, has proven to be a valuable approach to improving industrially important traits (Sauer 2001; Çakar et al. 2012; Portnoy et al. 2011; Bachmann et al. 2015; Steensels et al. 2014). A growing number of studies have also used evolutionary engineering to improve strains commonly used for food applications (McBryde et al. 2006; Cadiere et al. 2011; López-Malo et al. 2015; Tilloy et al. 2014; Ekberg et al. 2013; Kutyna et al. 2012).

Modifications of Aroma and Ethanol Yield

Various evolutionary engineering approaches have focused on the modification of strain metabolism to increase the sensorial quality of fermented products. An aromatic wine yeast strain was obtained using serial batch transfer on gluconate—a sugar poorly assimilated by *S. cerevisiae*—as the carbon source. Variants exhibiting increased biomass yield on gluconate displayed reduced production of acetic acid, which is undesirable at high levels in wine, and increased production of aroma compounds, particularly acetate ester, contributing to the fruity character of wine (Cadière et al. 2011; Mouret et al. 2015). These improved traits were confirmed in pilot-scale fermentation trials (Cadière et al. 2012). ^{13}C-flux analysis and transcriptome profiling revealed a 1.5-fold higher flux through the pentose phosphate (PP) pathway and increased expression of *GND1* and *TKL1* for the evolved strain compared to the ancestral strain (Cadière 2010; Cadière et al. 2011). The evolved strain also contained more lipids and assimilated phytosterols present in the medium more rapidly (Cadière et al. 2011; Rollero et al. 2016), suggesting that the reduced acetate excretion and increased acetate ester levels of the evolved strain is caused by a greater availability of acetyl-CoA. A higher expression level of genes associated with glycolysis and nitrogen metabolism and repression of genes involved in stress response and respiratory metabolism was also found, revealing a complete rewiring of the metabolic network. Using whole-genome resequencing of several evolved strains, a frameshift mutation in a key regulator of the cAMP-protein kinase A (PKA) pathway, resulting in a constitutively active PKA pathway, was identified (Tilloy V., Cadiere A., Bigey F. and Dequin S., unpublished results).

As alcohol levels have considerably increased during the past 30 years, a major challenge in modern wine industry is to develop wine yeast strains having reduced ethanol yield (Tilloy et al. 2015). Using a laboratory yeast strain and sulfite at alkaline pH as a selection pressure, (Kutyna et al. 2012) isolated a variant with enhanced sulfite tolerance after 150 generations that produces 30% more glycerol than the ancestral strain. However, the production of ethanol in this evolved strain was marginally affected, consistent with the moderate increase in glycerol concentration. (Tilloy et al. 2014) compared various strategies with the aim to reroute yeast metabolism towards glycerol. In addition to its key role in redox balance, glycerol is also the main compatible solute produced by yeast in response to hyperosmotic stress. This response is controlled by the HOG (high osmolarity glycerol pathway) signalling pathway. Serial batch cultivation was carried out using various agents known to stimulate the HOG pathway, such as methylglyoxal, sorbitol or potassium chloride (Tilloy et al. 2014). After prolonged culture under selective salt stress conditions for 200 generations, a variant producing substantially more glycerol and less ethanol than the ancestral strain was obtained. In a second step, this evolutionary strategy was coupled to a genetic approach to amplify the metabolic shift. A population of 150 haploid derivatives from the evolved strain was produced, and the haploid strains with the highest capacity to produce glycerol were selected and mated, generating intra-breed diploid strains (Tilloy et al. 2014). In pilot-scale experiments on different grape musts, this strain decreased the ethanol content of wines by 0.6–1.3% (vol/vol). The wines obtained contained more glycerol and 2,3-butanediol, a polyol having no sensorial impact in wine, as well as more succinic acid. By contrast, the level of acetic acid was markedly reduced and the production of other undesirable by-products such as acetaldehyde and acetoin remained very low. Thus, pilot-scale trials confirmed the reduced alcohol yield and acidification properties of the evolved strain, a combination of traits of particular interest in the context of climate change.

Stress Tolerance and Fermentation Performance

Several studies have used evolutionary strategies to improve stress tolerance and fermentation performances of industrial yeasts. López-Malo et al. (2015) obtained a wine yeast strain better adapted to ferment at low temperature using adaptive evolution. Starting with a pool of 27 commercial wine yeast strains, either EMS-mutagenized or not, serial batch transfer experiments were performed at 12 °C. This strategy enabled the selection of one strain outcompeting the others under both experimental conditions after 200 generations. Transcriptomics profiling showed up-regulation of four mannoprotein-coding genes belonging to the DAN/TIR family. Genome sequencing of the evolved strain identified an allelic variant of the *GAA1* gene that encodes a GPI transamidase complex subunit that adds GPI, which is required for inositol synthesis, to newly synthesized proteins, including mannoproteins. The role of this allele was demonstrated by construction of a site-directed mutant ($GAA1^{Thr108}$) in a derivative haploid of the ancestral strain, which resulted in improved fermentation performance. Unfortunately, the growth and fermentation rate improvement occurred only in the context of the selection medium used and was not observed in natural grape musts. Indeed, the growth advantage was obtained in a selection medium that was limited in inositol and was lost in natural must, which contains excess inositol. Although this work nicely demonstrated the importance of inositol and mannoproteins in the adaptation of yeast at low temperature, it also stresses the importance to perform the experimental evolution under conditions as close as possible to the industrial parameters.

Several studies also tried to improve the fermentation capacity of the lager yeast *S. pastorianus*. To increase the performance of lager yeast in very high-gravity (VHG) wort, Huuskonen et al. (2010) exposed EMS-mutagenized brewing yeast cells to the final stages of VHG, characterized by high concentrations of ethanol and maltose and maltotriose as the sole sources of fermentable sugars. Strains with increased ability to survive in these conditions were isolated and shown to have increased performance at the end of VHG fermentation. In another study, *S. pastorianus* variants exhibiting efficient fermentation rate and vitality in VHG fermentations as well as flavour improvement were obtained using a similar strategy (Yu et al. 2012). By contrast, Ekberg et al. (2013) tried to improve the early steps of VHG fermentation. They used an adaptive evolution strategy based on repetitive culturing of lager yeast in the presence of high concentrations of sorbitol (210 g per litre) to select variants growing rapidly under hyperosmotic stress conditions. Applying this approach to an ethanol-tolerant strain previously isolated from *S. pastorianus* lager yeast, they obtained an osmotolerant variant that fermented faster than the original strains. Interestingly, this variant contains less intracellular trehalose and glycogen than the parent, which led the authors to suggest that an attenuated stress response contributes to the improved fermentation performance.

3.3 Intra- and Inter-specific Breeding

Improvement of industrial yeasts properties can theoretically be attained through simple breeding, which does allow for the transfer of alleles and creation of novel combinations. However, only a few cases of yeast breeding have been reported in literature in the past, given the difficulties associated with breeding. Low sporulation efficiency, low spore viability and homothallism are obstacles frequently encountered in industrial yeast breeding. In addition, because industrially relevant phenotypes are complex traits, the improvement of industrial yeasts through simple breeding has proven difficult. However, breeding has been used in some work to transfer or combine desired traits in industrial yeasts (Thornton 1985; Romano et al. 1985; Prior et al. 2000). Improvements in breeding strategies have been later proposed based on the combination of numerous favourable characters (Marullo et al. 2006). However, without molecular tools to monitor appropriate alleles, breeding remains tedious, given the need to phenotype clones. Recently, the advent of QTL analyses has provided a novel framework to support breeding programmes. By providing information that enables the set-up of molecular tools to monitor relevant alleles in crosses, breeding can be greatly facilitated. Once a QTL has been dissected and the causative gene identified, molecular makers can be defined and crosses guided by a simple PCR-based selection of haploid clones harbouring the preferred alleles. Breeding is usually performed to introduce an additional quality or correct a peculiar trait in an industrial strain. The breeding design must enable both the transfer of the desired alleles and the restoration of the original industrial genome. Such an introgression of alleles can be achieved by a succession of backcrosses with a parental industrial clone. Such a marker-assisted allele transfer enabled Marullo et al. (2007) to introgress several fermentation traits in a wine yeast strain. They showed that after 4 backcrossing cycles, the resulting hybrid conserved 95% of the ancestor strain's genome and was improved in the desired traits.

On the other hand, the construction of interspecific hybrids to generate new combination of genes of potential adaptive value, or strains with new properties, has often been attempted. The first attempts were performed between strains of *S. cerevisiae* and *S. uvarum*, 20 years ago (Zambonelli et al. 1993). The high production of succinic acid or 2-phenyl ethanol and 2-phenyl ethyl acetate is one of most interesting features of these hybrids (Antonelli et al. 1999; Rainieri et al. 1999; Masneuf et al. 2002; Bellon et al. 2015; da Silva et al. 2015). These hybrids inherit their mitochondrial genome from one of the parental genomes, (Pulvirenti et al. 2000; Solieri et al. 2008) and difficulties have been observed in their industrial use that are very likely associated with their mitochondrial genome (Picazo et al. 2015). More recently, other hybrids have been constructed for winemaking purposes with other species: *S. cerevisiae* and *S. kudriavzevii*, *S. cerevisiae and S. mikatae* (Bellon et al. 2013), and *S. cerevisiae* and *S. paradoxus* (Bellon et al. 2011). Despite the low viability of their progenies, interspecific hybrids have also been used as a basis for further breeding programmes (Bizaj et al. 2012), enabling the reduction of H_2S production.

As currently used lager beer strains resulted from two hybridization events that led to a poor diversity of starters, the recent identification of *S. eubayanus* has opened ways for the creation of new interspecific hybrids for beer brewing. In these new hybrids, the combination of the cryophylic properties of *S. eubayanus* with the high production of volatiles (alcohol, esters) of several *S. cerevisiae* strains (Hebly et al. 2015; Mertens et al. 2015; Krogerus et al. 2015) has been achieved.

4 Conclusion/Perspectives

Recent advances in understanding the taxonomy, ecology, genome evolution, and diversity of *Saccharomyces* yeasts have led to a deeper knowledge of their role in the production and quality of foods and beverages. Major progress in yeast strain improvement have also been accomplished, favoured by the development of genetic and genomic approaches, generating superior strains that present new opportunities for the control and exploitation of products and processes.

Further research on the ecology of *Saccharomyces* yeasts in the wild and in various ecosystems, including indigenous fermentations, is worthwhile. Most studies of *Saccharomyces* have been conducted in well-established fermented products such as bread, beer and wine and could be extended for other products. Exploring this yet only partially known biodiversity of natural ecosystems will provide an extended resource for elucidating the mechanisms involved in the adaptation of *Saccharomyces* yeasts to various ecological niches. In addition, this exploration will help identify strains and species with new properties that might be of interest across various food and beverage sectors. The interaction between *Saccharomyces* strains and the ecosystem microflora remains poorly described and understood and will also require further studies. Another issue that has recently attracted interest is the link between diet and health, in both positive and negative contexts. Yeasts have been a component of the human diet for at least 7,000 years. In line with this, it was recently found that members of the human microbiota such as the Gram-negative bacterium *Bacteroides thetaiotaomicron* have evolved a complex machinery to digest and metabolize yeast cell-wall mannans (Cuskin et al. 2015). This finding provides unprecedented insights into the adaptation of the microbiota to yeast domestication in the human diet, reflecting the regular consumption of yeast-leavened bread, as well as yeast-fermented beverages and products such as soy sauce.

Improvement strategies have been so far applied to a small number of species, mainly *S. cerevisiae*, and could be extended to other species and interspecies hybrids. Although the use of QTL mapping strategies has so far been restricted to yeast having a sexual life cycle with viable progeny, alternative approaches based on the Return To Growth (RTG) process have been recently proposed to map complex traits loci (QTLs) in diploid strains without going through sexual reproduction (Laureau et al. 2016). These approaches are based on the property of *S. cerevisiae* to enter the meiotic program, induce double-strand breaks genome-wide, and return to mitotic growth. Moreover, evolutionary strategies, which can be used in any species, should enable many advances, boosted by the development of sequencing approaches that will facilitate the identification of the causative mutations. The growing torrent of yeast sequence data will also provide opportunities to elucidate the missing links between genotypes and phenotypes. Using genome-wide association studies (GWAS) of quantitative traits may offer new perspectives to the mapping of QTLs, despite some difficulties to overcome in yeast species that possess complex population structures. This set of tools and approaches will undoubtedly be an invaluable asset to build a set of strains tailored to meet the challenges of the fermented food and beverage industries of tomorrow.

References

Aigle, M., Erbs, D. and Moll, M. 1984. *J. Am. Soc. Brew. Chem.* **42**: 1–7.

Almeida, P., Gonçalves, C., Teixeira, S., Libkind, D., Bontrager, M., Masneuf-Pomarède, I., Albertin, W., Durrens, P., Sherman, D.J., Marullo, P., Hittinger, C.T., Gonçalves, P. and Sampaio, J.P. 2014. *Nat. Commun.* **5**: 4044.

Almeida, P., Barbosa, R., Zalar, P., Imanishi, Y., Shimizu, K., Turchetti, B., Legras, J.L., Serra, M., Dequin, S., Couloux, A., Guy, J., Bensasson, D., Gonçalves, P. and Sampaio, J.P. 2015. *Mol. Ecol.* Aug **6**.

Ambroset, C., Petit, M., Brion, C., Sanchez, I., Guerin, C., Nicolas, P., Chiapello, H., Bigey, F., Dequin, S. and Blondin, B. 2011. *G3 (Bethesda)* **1**: 263–281.
Antonelli, A., Castellari, L., Zambonelli, C. and Carnacini, A. 1999. *J. Agric. Food Chem.* **47**: 1139–1144.
Antunovics, Z., Irinyi, L. and Sipiczki, M. 2005. *J. Appl. Microbiol.* **98**: 971–979.
Arroyo-López, F.N., Salvadó, Z., Tronchoni, J., Guillamón, J.M., Barrio, E. and Querol, A. 2010. *Yeast* **27**: 1005–15.
Azumi, M. and Goto-Yamamoto, N. 2001. *Yeast* **18**: 1145–1154.
Blondin, B. Noble, J.N. and Sanchez, I. 2013. Patent CA2862399 A1.
Bachmann, H., Fischlechner, M., Rabbers, I., Barfa, N., Branco dos Santos, F., Molenaar, D. and Teusink, B. 2013. *Proc. Natl. Acad. Sci. U S A* **110**:14302–14307.
Bachmann, H., Pronk, J.T., Kleerebezem, M. and Teusink, B. 2015. *Curr. Opin. Biotechnol.* **32**:1–7.
Baker, A.E., Wang, B., Bellora, N., Peris, D., Hulfachor, A.B., Koshalek, J.A., Adams M., Libkind D. and Hittinger, C.T. 2015. *Mol. Biol. Evol.* **32**: 2818–2831.
Belloch, C., Orlic, S., Barrio, E. and Querol, A. 2008. *Int. J. Food Microbiol.* **122**: 188–95.
Belloch, C., Pérez-Torrado, R., González, S.S., Pérez-Ortín, J.E., García-Martínez, J., Querol, A. and Barrio, E. 2009. *Appl. Environ. Microbiol.* **75**: 2534–2544.
Bellon, J.R., Eglinton, J.M., Siebert, T.E., Pollnitz, A.P., Rose; L., de Barros Lopes, M. and Chambers, P.J. 2011. *Appl. Microbiol. Biotechnol.* **91**: 603–612.
Bellon, J.R., Schmid. F., Capone, D.L., Dunn, B.L. and Chambers, P.J. 2013. *PLoS One* **8**: e62053.
Bellon, J.R., Yang, F., Day, M.P., Inglis, D.L. and Chambers, P.J. 2015. *Appl. Microbiol. Biotechnol.* **99**: 8597–8609.
Bergström, A., Simpson, J.T., Salinas, F., Barré, B., Parts, L., Zia, A., Nguyen Ba, A.N., Moses, A.M., Louis, E., Mistonen, V., Warringer, J., Durbin, R. and Liti, G. 2014. *Mol. Biol. Evol.* **31**: 872–888.
Bing, J., Han, P.J., Liu, W.Q., Wang, Q.M. and Bai, F.Y. 2014. *Curr. Biol.* **24**: R380–R381.
Bizaj, E., Cordente, A.G., Bellon, J.R., Raspor, P., Curtin, C.D. and Pretorius, I.S. 2012. *FEMS Yeast Res.* **12**: 456–465.
Bond, U. 2009. In: Advances in applied microbiology (eds L A.I.askin, S. Sariaslani and G.M. Gadd) Vol. **69**, pp. 159–182.
Bond, U., Neal, C., Donnelly, D. and James, T.C. 2004. *Curr. Genet.* **45**: 360–370.
Borneman, A.R., Desany, B.A., Riches, D., Affourtit, J.P., Forgan, A.H., Pretorius, I.S., Egholm, M. and Chambers P.J. 2011. *PLoS Genet.* **7**: e1001287.
Bradbury, J.E., Richards, K.D., Niederer, H.A., Lee, S.A., Rod Dunbar, P. and Gardner, R.C. 2006. *Antonie van Leeuwenhoek* **89**: 27–37.
Brauer, M.J., Christianson, C.M., Pai, D.A. and Dunham, M.J. 2006. *Genetics* **173**: 1813–1816.
Brem, R.B., Yvert, G., Clinton, R. and Kruglyak, L. 2002. *Science* **296**: 752–755.
Brice, C., Sanchez, I., Bigey, F., Legras, J.L. and Blondin, B. 2014. *BMC Genomics* **15**: 495.
Brion, C., Ambroset, C., Sanchez, I., Legras, J.L. and Blondin, B. 2013. *BMC Genomics* **14**: 681.
Brown, S.L., Stockdale, V.J., Pettolino, F., Pocock, K.F., de Barros Lopes, M., Williams, P.J., Bacic, A., Fincher, G.B., Høj, P.B. and Waters, E.J., 2007. *Appl. Microbiol. Biotechnol.* **73**: 1363–1376.
Buser, C.C., Newcomb, R.D., Gaskett, A.C. and Goddard, M.R. 2014. Bonsall M, ed. *Ecol. Lett.* **17**: 1257–1264.
Cadière, A. 2010. Ph.D. thesis, Montpellier SupAgro.
Cadière, A., Ortiz-Julien, A., Camarasa, C. and Dequin, S. 2011. *Metab. Eng.* **13**: 263–271.
Cadière, A., Aguera, E., Caillé, S., Ortiz-Julien, A. and Dequin, S. 2012. *Food Microbiol.* **32**: 332–337.
Çakar ZP, Turanli-Yildiz B, Alkim C, Yilmaz U. 2012. *FEMS Yeast Res.* **12**: 171–182.
Camarasa, C., Sanchez, I., Brial, P., Bigey, F. and Dequin, S. 2011. *PLoS One* **6**: e25147.
Capece, A., Romaniello, R., Siesto, G., Pietrafesa, R., Massari, C., Poeta, C. and Romano, P. 2010. *Int. J. Food Microbiol.* **144**: 187–192.
Chandler, J.A., Eisen, J.A. and Kopp, A. 2012. *Appl. Environ. Microbiol.* **78**: 7327–7336.

Cheeseman, K., Ropars, J., Renault, P., Dupont, J., Gouzy, J., Branca, A., Abraham, A., Ceppi, M., Conseiller, E., Debuchy, R., Malagnac, F., Goarin, A., Silar, P., Lacoste, S., Sallet, E., Bensimon, A., Giraud, T. and Brygoo, Y. 2014. *Nature Communications* **5**: 2876.
Coi, A.L., Bigey, F., Marsit, S., Mallet, S., Zara G., Gladieux, P., Galeote, V., Budroni, M., Dequin, S. and Legras J.L. 2017 *Mol. Ecol.* In press.
Crépin, L., Nidelet, T., Sanchez, I., Dequin, S. and Camarasa, C. 2012. *Appl. Environ. Microbiol.* **78**: 8102–8111.
Crépin, L., Sanchez, I., Nidelet, T., Dequin, S. and Camarasa, C. 2014. *Microb. Cell Fact.* **13**: 109.
Cromie, G.A., Hyma, K.E., Ludlow, C.L., Garmendia-Torres, C., Gilbert, T.L., May, P., Huang, A., Dudley, A.M. and Fay, J.C. 2013. *G3 (Bethesda)* **3**: 2163–2171.
Csoma, H., Zakany, N., Capece, A., Romano, P. and Sipiczki, M. 2010. *Int. J. Food Microbiol.* **140**: 239–248.
Cubillos, F.A., Parts, L., Salinas, F., Bergström, A., Scovacricchi, E., Zia, A., Illingworth, C.J., Mustonen, V., Ibstedt, S., Warringer, J., Louis, E.J., Durbin, R. and Liti, G. 2013. *Genetics* **195**: 1141–1155.
Cuskin, F., Lowe, EC., Temple, M.J., Zhu, Y., Cameron, E.A., Pudlo, N.A., Porter, N.T., Urs, K., Thompson, A.J., Cartmell, A., Rogowski, A., Hamilton, B.S., Chen, R., Tolbert, T.J., Piens, K., Bracke, D., Vervecken; W., Hakki, Z., Speciale, G., Munōz-Munōz, J.L., Day, A., Peña, M. J., McLean, R., Suits, M.D., Boraston, A.B., Atherly, T., Ziemer, C.J., Williams, S.J., Davies, G.J., Abbott, D.W., Martens, E.C. and Gilbert, H.J. 2015. *Nature* **517**: 165–169.
da Silva, T., Albertin, W., Dillmann C., Bely, M., la Guerche, S., Giraud, C., Huet, S., Sicard, D., Masneuf-Pomarede, I., de Vienne, D., and Marullo, P. 2015. *PLoS One* **10**: e0123834.
Damon, C., Vallon, L., Zimmermann, S., Haider, M.Z., Galeote, V., Dequin, S., Luis, P., Fraissinet-Tachet, L. and Marmeisse, R., 2011. *ISME J.* **9**: 67.
David, F. and Siewers, V. 2015. *FEMS Yeast Res.* **15**: 1–14.
Deak, T., 2007. Handbook of food spoilage yeasts. CRC press.
Dequin, S. 2001. *Appl. Microbiol. Biotechnol.* **56**: 577–588.
Deutschbauer, A.M. and Davis, R.W. 2005. *Nat. Genet.* **37**: 1333–1340.
Doniger, S.W., Kim, H.S., Swain, D., Corcuera, D., Williams, M., Yang, S.P. and Fay, J.C. 2008. *PLoS Genet.* **4**: e1000183.
Dunlop, P.C., Meyer, G.M., Ban, D. and Roon, R.J. 1978. *J. Biol. Chem.* **253**: 1297–1304.
Dunn, B., Sherlock, G. and Franklin, B. 2008. *Genome Res.* **18**: 1610–1623.
Dunn, B., Richter, C., Kvitek, D.J., Pugh, T. and Sherlock, G. 2012. *Genome Res.* **22**: 908–924.
Duval, E.H., Alves, S.L., Dunn, B., Sherlock, G. and Stambuk, B.U. 2010. *J. Appl. Microbiol.***109**: 248–259.
Ehrenreich, I.M., Torabi, N, Jia, Y., Kent, J., Martis, S., Shapiro, J.A., Gresham, D., Caudy, A.A., and Kruglyak, L. 2010. *Nature* **464**: 1039–1042.
Ekberg, J., Rautio, J., Mattinen, L., Vidgren, V., Londesborough, J., and Gibson, B. 2013, *FEMS Yeast Res.* **13**: 335–349.
Erny, C., Raoult, P., Alais, A., Butterlin, G., Delobel, P., Matei-Radoi, F., Casaregola, S. and Legras, J.L. 2012. *Appl. Environ. Microbiol.* **78**: 3256–3265.
Fay, J.C. and Benavides, J.A. 2005. *PLoS Genet.* **1**: 66–71.
Fay, J.C., McCullough, H.L., Sniegowski, P.D. and Eisen, M.B.. 2004. *Genome Biol.* **5**: R26.
Fidalgo, M., Barrales, R.R., Ibeas, J.I. and Jimenez, J. 2006 *Proc. Natl. Acad. Sci. U S A.* **103**: 11228–11233.
Fogel, S., Welch J.W., Cathala, G. and Karin, M. 1983. *Curr. Genet.* **7**: 347–355.
Galeote, V., Novo, M., Salema-Oom, M., Brion, C., Valerio, E., Goncalves, P. and Dequin, S. 2010. *Microbiology* **156**: 3754–3761.
Galeote, V., Bigey, F., Beyne, E., Novo, M., Legras, J.L., Casaregola, S. and Dequin, S. 2011. *PLoS One* **6**: e17872.
Gallone, B., Steensels, J., Prahl, T., Soriaga, L., Saels, V., Herrera-Malaver, B., Merlevede, A., Roncoroni, M., Voordeckers, K., Miraglia, L., Teiling, C., Steffy, B., Taylor, M., Schwartz, A., Richardson, T., White, C., Baele, G., Maere, S. and Verstrepen, K J. 2016. *Cell* **166**: 1–14.

Garciadeblas, B., Rubio, F., Quintero, F., Bañuelos, M., Haro, R. and Rodríguez-Navarro, A. 1993. *Mol. Gen. Genet.* **236**: 363–368.
Gibson, B. and Liti, G. 2015. *Yeast* 32: 17–27.
Gibson, B.R., Storgårds, E., Krogerus, K. and Vidgren, V. 2013. *Yeast* **30**: 255–266.
Gibson, B., Krogerus, K., Ekberg, J., Monroux, A., Mattinen, L., Rautio, J. and Vidgren, V. 2015 *Yeast* **32**: 301–316.
Goddard, M.R. 2008. *Ecology* **89**: 2077–282.
Goddard, M.R., Anfang, N., Tang, R., Gardner, R.C. and Jun, C. 2010. *Environ. Microbiol.* **12**: 63–73.
Gojkovic, Z., Knecht, W., Zameitat, E., Warneboldt, J., Coutelis, J.B., Pynyaha, Y., Neuveglise, C., Møller K, Löffler M and Piskur J. 2004. *Mol. Genet. Genomics* **271**: 387–393.
Gonçalves, M., Pontes, A., Almeida, P., Barbosa, R., Serra, M., Libkind, D., Hutzler, M., Gonçalves, P. and Sampaio, J.P. 2016. *Curr. Biol.* **26**: 2750–2761.
González, S.S., Barrio, E., Gafner, J. and Querol, A. 2006. *FEMS Yeast Res.* **6**: 1221–1234.
González, S.S., Gallo, L., Climent, M.A., Barrio, E. and Querol, A. 2007. *Int. J. Food Microbiol.* **116**: 11–18.
González, S.S., Barrio, E. and Querol, A. 2008. *Appl. Environ. Microbiol.* **74**: 2314–2320.
Greetham, D., Wimalasena, T.T., Leung, K., Marvin, M.E., Chandelia, Y., Hart, A.J., Phister, T. G., Tucker, G.A., Louis, E.J. and Smart, K.A. 2014. *PLoS ONE* **9**: e103233.
Gresham, D., Curry, B., Ward, A., Gordon, D.B., Brizuela, L., Kruglyak, L. and Botstein, D. 2010. *Proc. Natl. Acad. Sci. U.S.A.***107**: 1482–1487.
Guillaume, C., Delobel, P, Sablayrolles, J.M. and Blondin, B. 2007. *Appl. Environ. Microbiol.* **73**: 2432–2439.
Hall, C. and Dietrich, F.S. 2007. *Genetics* **177**: 2293–2307.
Hall, C., Brachat, S. and Dietrich, F.S. 2005. *Eukaryot. Cell* **4**: 1102–1115.
Hebly, M., Brickwedde, A., Bolat, I., Driessen, M.R., de Hulster, E.A., van den Broek, M., Pronk, J.T., Geertman, J.M., Daran, J.M. and Daran-Lapujade, P. 2015. *FEMS Yeast Res.* **15**: doi:10.1093/femsyr/fov005.
Heinemann, J.A. and Sprague, G.F. 1989. *Nature* **340**: 205–209.
Hennequin, C., Thierry, A., Richard, G.-F., Lecointre, G., Nguyen, H.-V., Gaillardin, C. and Dujon B. 2001. *J. Clin. Microbiol.* **39**: 551–559.
Homann, O.R., Cai, H., Becker, J.M. and Lindquist, S.L. 2005 *PLoS Genet.* **1**: e80.
Huang, C., Roncorini, M., and Gardner, R. 2014. *Appl. Microbiol. Biotechnol.* **98**: 7125–7135.
Huuskonen, A., Markkula, T., Vidgren, V., Lima, L., Mulder, L. Geurts, W., Walsh, M and Londesborough J. 2010. *Appl. Environ. Microbiol.* **76**: 1563–1573.
Hyma, K.E., Saerens, S.M., Verstrepen, K.J. and Fay, J.C. 2011. *FEMS Yeast Res.* **11**: 540–551.
Jaras, M., Cubillos, F. A., García, V., Salinas, F., Aguilera, O., Liti, G. and Martínez, C. 2014. *PLoS One* **9**: e86533.
Katou, T., Katou, T., Namise, M., Kitagaki, H., Akao, T. and Shimoi, H. 2009. *J. Biosci. Bioeng.* **107**: 383–393.
Kishimoto, M. and Goto, S. 1995. *J. Gen. Microbiol.* **41**: 239–247.
Knight, S. and Goddard, M.R. 2014. *ISME J.* **9**: 361–370.
Knight, S., Klaere, S., Fedrizzi, B. and Goddard, M.R. 2015. *Sci Rep* **5**: 14233.
Krogerus, K., Magalhães, F., Vidgren, V. and Gibson, B. 2015. *J. Ind. Microbiol. Biotechnol.* **42**: 769–778.
Kroymann, J. and Mitchell-Olds, T. 2005. *Nature* **435**: 95–98.
Kuehne, H.A., Murphy, H.A., Francis, C.A. and Sniegowski, P.D. 2007. *Curr. Biol.* **17**: 407–411.
Kurtzman, C.P. 2011. *Antonie Van Leeuwenhoek* **99**: 13–23.
Kurtzman, C.P., Fell, J.W. and Boekhout, T., 2011. The yeasts, a taxonomic study. 5th Edition. Elsevier.
Kutyna, D.R., Varela, C., Stanley, G.A., Borneman, A.R., Henschke, P.A., Chambers, P.J., 2012. *Appl. Microbiol. Biotechnol.* **93**: 1175–1184.
Kvitek, D.J., Will, J.L. and Gasch, A.P. 2008. *PLoS Genet* **4**: e1000223.

Laureau R., Loeillet S., Salinas F., Bergström A., Legoix-Né P., Liti G. and Nicolas A. 2016. *PLoS Genet.* **12**: e1005953.

League, G.P., Slot, J.C. and Rokas, A. 2012. *FEMS Yeast Res.* **12**: 859–863.

Lee, J.W., Kim, T.Y., Jang, Y.-S., Choi, S. and Lee, S.Y. 2011. *Trends Biotechnol.* 29: 370–378.

Legras, J.L. and Karst, F. 2003. *FEMS Microbiol. Lett.* **221**: 249–255.

Legras, J.L., Merdinoglu, D., Cornuet, J.M. and Karst, F. 2007. *Mol. Ecol.* **16**: 2091–2102.

Legras, J.L., Erny, C. and Charpentier, C. 2014. *PLoS One* **9**: e108089.

LeJeune, C., Lollier, M., Demuyter, C., Erny, C., Legras, J.L., Aigle, M. and Masneuf-Pomarède, I. 2007. *FEMS Yeast Res.* **7**: 540–549.

Libkind, D., Hittinger, C.T., Valerio, E., Goncalves, C., Dover, J., Johnston, M., Goncalves, P. and Sampaio, J.P. 2011. *Proc. Natl. Acad. Sci.* U.S.A. **108**: 14539–14544.

Libuda, D.E. and Winston, F. 2006. *Nature* **443**: 1003–1007.

Liti, G., and Louis, E. J. 2012. *PLoS Genet.* **8**: e1002912.

Liti, G., Barton, D.B., Louis, E.J. 2006. *Genetics.* **174**: 839–850.

Liti G, Carter DM, Moses AM, Warringer, J., Parts, L., James, S.A., Davey, R.P., Roberts, I.N., Burt, A., Koufopanou, V., Tsai, I.J., Bergman, C.M., Bensasson, D., O'Kelly, M.J.T., van Oudenaarden, A., Barton, D.B.H., Bailes, E., Nguyen, A.N., Jones, M., Quail, M.A., Goodhead, I., Sims, S., Smith, F., Blomberg, A., Durbin, R. and Louis, E.J. 2009. *Nature* **458**: 337–341.

Lopandic, K., Gangl, H., Wallner, E., Tscheik, G., Leitner, G., Querol, A., Borth, N., Breitenbach, M., Prillinger, H. and Tiefenbrunner, W. 2007. *FEMS Yeast Res.* 7: 953–965.

López-Malo, M., García-Rios, E., Melgar, B., Sanchez, M. R., Dunham, M. J. and Guillamón J. M. 2015. *BMC Genomics* **16**: 537.

Lopandic K, Tiefenbrunner W, Gangl H, Mandl, K., Berger, S., Leitner, G., Abd-Ellah, G.A., Querol, A., Gardener, R., Sterflinger, K., and Prillinger, H. 2008. *FEMS Yeast Res.* **8**: 1063–1075.

Mackay, T. F., Stone, E. A., Ayroles, J. F., 2009. *Nat. Rev. Genet.* **10**: 565–577.

Marsit S. and Dequin, S. 2015. *FEMS Yeast Res.* **15**, fov067.

Marsit, S., Mena, A., Bigey, F., Sauvage, F.-X., Couloux, A., Guy, J., Legras, J.-L., Barrio, E., Dequin, S. and Galeote, V. 2015. *Mol. Biol. Evol.* **32**: 1695–1707.

Marsit, S, Sanchez, I., Galeote, V. and Dequin, S. 2016. *Environ. Microbiol.* **18**: 1148–61.

Marullo, P., Bely, M., Masneuf-Pomarede, I., Pons, M., Aigle, M. and Dubourdieu, D. 2006. *FEMS Yeast Res.* **6**: 268–79.

Marullo, P., Yvert, G., Bely, M., Masneuf-Pomarede, I., Durrens, P., Dubourdieu, D. and Aigle, M. 2007. *FEMS Yeast Res.* **7**: 941–952.

Masneuf, I., Hansen, J., Groth, C., Piskur, J. and Dubourdieu, D. 1998. *Appl. Environ. Microbiol.* **64**: 3887–3892.

Masneuf, I., Murat, M.L., Naumov, G.I., Tominaga, T. and Dubourdieu, D. 2002. *J. Int. Sci. Vigne Vin.* **36**: 205–212.

McBryde C., Gardner J.M., de Barros Lopes M. and Jiranek V. 2006. *Am. J. Enol. Vitic.* **57**: 423–430.

McGovern, P.E., Donald, L., Glusker, L.J.E. 1996. *Nature* **381**: 480–481.

McGovern, P.E., Zhang, J., Tang, J., Zhang, Z., Hall, G.R., Moreau, R.A., Nunez, A., Butrym, E. D., Richards, M.P.,Wang, C.S., Cheng, G., Zhao, Z., Wang, C. 2004. *Proc. Natl. Acad. Sci.* U S A **101**: 17593–17598.

Merten s, S., Steensels, J., Saels, V., De Rouck, G., Aerts, G. and Verstrepen, K.J. 2015. *Appl. Environ. Microbiol.* **81**: 8202–8214.

Michel, R.H., Mc Govern, P.E. and Badler V. R. 1992 Chemical Evidence for Ancient Beer *Nature* **360**: 24.

Miyashita, K., Sakamoto, K., Kitagaki, H., Iwashita, K., Ito, K. and Shimoi, H., 2004. *J. Biosc. Bioeng.* **97**: 14–18.

Møller, H.D., Parsons, L., Jørgensen, T.S., Botstein, D. and Regenberg, B. 2015. *Proc. Natl. Acad. Sci.* U.S.A. **112**: E3114-E3122.

Morales, L. and Dujon, B. 2012. *Microbiol. Mol. Biol. Rev.* **76**: 721–739.

Mortimer, R.K. 2000. *Genome Res.* **10**: 403–409.

Mortimer, R. and Polsinelli, M. 1999. *Res. Microbiol.* **150**: 199–204.

Mouret, J.R., Cadiere, A., Aguera, E., Rollero, S., Ortiz-Julien A, Sablayrolles, J.M., Dequin, S. 2015. *Yeast*. **32**: 257–269.

Mukherjee, V., Steensels, J., Lievens, B., Van de Voorde, I., Verplaetse, A., Aerts, G., Willems, K. A., Thevelein, J.M., Verstrepen, K.J. and Ruyters, S. 2014. *Appl. Microbiol. Biotechnol.* **98**: 9483–9498.

Muller, L.A. and McCusker, J.H., 2009. *FEMS Yeast Res.* **9**: 143–152.

Nakao, Y., Kanamori, T., Itoh, T., Kodama, Y., Rainieri, S., Nakamura, N., Shimonaga, T., Hattori, M. and Ashikari, T. 2009. *DNA Res.* **16**: 115–129.

Nara, T., Hshimoto, T. and Aoki, T. 2000. *Gene* **257**: 209–222.

Naumov, G. 1996. *J. Ind. Microbiol. Biotechnol.* **17**: 295–302.

Naumov, G.I., James, S.A., Naumova, E.S., Louis, E.J. and Roberts, I.N. 2000. *Int. J. Syst. Evol. Microbiol.* **5**: 1931–1942.

Ness, F. and Aigle, M. 1995. *Genetics* **140**: 945–56.

Ness, F., Lavallée, F., Dubourdieu, D., Aigle, M. and Dulau, L. 1993. *J. Sci. Food Agric.* **62**: 89–94.

Nguyen, H.V., Legras, J.L., Neuvéglise, C. and Gaillardin, C. 2011. *PLoS One* **6**: e25821.

Noble, J. Sanchez, I. and Blondin, B. 2015. *Microb. Cell Fact.* **14**: 68.

Novo, M., Bigey, F., Beyne, E., Galeote, V., Gavory, F., Mallet, S., Cambon, B., Legras, J.L., Wincker, P., Casaregola and S., Dequin, S. 2009. *Proc. Natl. Acad. Sci.* U.S.A. **106**: 16333–16338.

Oud, B., van Maris, A.J.A., Daran, J.-M. and Pronk, J.T. 2012. *FEMS Yeast Res.* **12**: 183–196.

Paffetti, D., Barberio, C., Casalone, E., Cavalieri, D., Fani, R., Fia, G., Mori, E. and Polsinelli, M. 1995. *Res. Microbiol.* **146**: 587–594.

Parts, L., Cubillos, F.A., Warringer, J., Jain, K., Salinas, F. Bumpstead, S.J., Molin, M., Zia, A, Simpson, J.T., Quail, M.A., Moses. A., Louis, E.J., Durbin, R. and Liti, G. 2011. *Genome Res.* **21**: 1131–1138.

Pasteur, L. 1860. *Ann. Chim. Phys.* **58**: 323–426.

Pedersen, M. 1985. DNA sequence polymorphisms in the genus Saccharomyces. II. *Carlsb. Res. Commun.* **50**: 262–272.

Pérez-Ortín, J.E., Querol, A., Puig, S., Barrio, E. 2002. *Genome Res.* **12**: 1533–1539.

Pérez-Torrado, R., González, S.S., Combina, M., Barrio, E. and Querol, A. 2015. *Int. J. Food Microbiol.* **204**: 101–110.

Pérez-Través, L., Lopes, C.A., Querol, A. and Barrio, E. 2014. *PLoS One* **9**: e93729.

Peris, D., Belloch, C., Lopandić, K., Álvarez-Pérez, J.M., Querol, A. and Barrio, E. 2011. *Yeast* **29**: 81–91.

Peris, D., Lopes, C.A., Belloch, C., Querol, A. and Barrio, E. 2012. *BMC Genomics* **13**: 407.

Peris, D., Sylvester, K., Libkind, D., Gonçalves, P., Sampaio, J.P., Alexander, W.G. and Hittinger, C.T. 2014. *Mol. Ecol.* **23**: 2031–2045.

Perlstein, E. O., Ruderfer, D. M., Roberts, D. C., Schreiber, S. L. and Kruglyak, L., 2007. *Nat. Genet.* **39**: 496–450.

Picazo, C., Gamero-Sandemetrio, E., Orozco, H., Albertin, W., Marullo, P., Matallana, E. and Aranda, A. 2015. *Lett. Appl. Microbiol.* **60**: 217–222.

Piskur, J., Rozpedowska, E., Polakova, S., Merico, A. and Compagno, C. 2006. *Trends Genet.* **22**: 183–186.

Portnoy, V.A., Bezdan, D. and Zengler, K. 2011. *Curr. Opin. Biotechnol.* **22**: 590–594.

Prior, B.A., Baccari, C., and Mortimer, R.K. 2000. *J. Int. Sc. Vigne Vin.* **33**: 57–65.

Pulvirenti, A.M., Caggia, C., Restuccia, C., Giudici, P., Zambonelli, C. and Levi, V. 2000. *Ann. Microbiol.* **64**: 61–64.

Rabbers, I., van Heerden, J.H., Nordholt, N., Bachmann, H., Teusink, B. and Bruggeman, F. J. 2015. *Metabolites* **5**: 311–343.

Rainieri, S., Zambonelli, C., Tini, V., Castellari, L. and Giudici, P. 1999. *S. Afr. J. Enol. Vitic.* **20**: 47–52.

Rohland, N. and Reich, D. 2012. *Genome Res.* **22**: 939–946.

Rollero **S**, Mouret JR, Sanchez I, Camarasa C, Ortiz-Julien A, Sablayrolles JM, Dequin S. 2016. *Microb Cell Fact*. **9**: 15:32.

Romano, P., Soli, M.G., Suzzi, G., Grazia, L. and Zambonelli, C. 1985. *Appl. Environ. Microbiol.* **50**: 1064–1067.

Ropars, J., Rodríguez de la Vega, R.C., López-Villavicencio, M., Gouzy, J., Sallet, E., Dumas, É., Lacoste, S., Debuchy, R., Dupont, J., Branca, A. and Giraud, T. 2015. *Curr. Biol.* **25**: 2562–2569.

Saint-Prix, F., Bonquist, L. and Dequin, S. 2004. *Microbiology* **150**: 2209–2220.

Salinas, F., Cubillos, F.A., Soto, D., Garcia, V., Bergström, A., Warringer, J., Ganga, M.A., Louis, E.J., Liti, G. and Martinez, C. 2012. *PLoS One* **7**: e49640.

Sampaio, J.P. and Gonçalves, P. 2008. *Appl. Environ. Microbiol.* **74**: 2144–2152.

Sauer, U. 2001. *Adv. Biochem. Eng. Biotechnol.* **73**: 129–169.

Schacherer, J., Shapiro, J.A., Ruderfer, D.M. and Kruglyak, L. 2009. *Nature* **458**: 342–345.

Schneider, J., Rupp, O., Trost, E., Jaenicke, S., Passoth, V., Goesmann, A., Tauch, A. and Brinkrolf, K., 2012. *FEMS Yeast Res.* **12**: 382–386.

Schuller, D., Cardoso, F., Sousa, S., Gomes, P., Gomes, A.C., Santos, M.A.S. and Casal, M. 2012. *PLoS One* **7**: e32507.

Sinha, H., Nicholson, B.P., Steinmetz, L.M., and McCusker, J.H. 2006. *PLoS Genet.* **2**: e13.

Smith, E.N., and Kruglyak, L. 2008. *PLoS Biol.* **6**: e83.

Sniegowski, P.D., Dombrowski, P.G. and Fingerman, E. 2002. *FEMS Yeast Res.* **1**: 299–306.

Solieri, L., Antúnez, O., Pérez-Ortín, J.E., Barrio, E. and Giudici, P. 2008. *Yeast* **25**: 485–500.

Spor, A., Wang, S., Dillmann, C., de Vienne, D. and Sicard, D. 2008. PLoS One. **3**: e1579.

Spor, A., Nidelet, T., Simon, J., Bourgais, A., de Vienne, D. and Sicard, D. 2009. *BMC Evol. Biol.* **9**: 296.

Stachel, S.E. and Zambryski, P.C. 1989. Generic trans-kingdom sex? *Nature* **340**: 190–191.

Steensels, J. and Verstrepen, K.J. 2014. *Annu. Rev. Microbiol.* **68**: 61–80.

Steensels, J., Snoek, T., Meersman, E., Nicolino, M.P., Voordeckers, K. and Verstrepen, K. J. 2014. *FEMS Microbiol. Rev.* **38**: 947–995.

Stefanini, I., Dapporto, L., Legras, J.L., Calabretta, A., Di Paola, M., De Filippo, C., Viola, R., Capretti, P., Polsinelli, M., Turillazzi, S. and Cavalieri, D. 2012. *Proc. Natl. Acad. Sci.* U.S.A. **109**: 13398–13403.

Steinmetz, L. M., Sinha, H., Richards, D. R., Spiegelman, J. I., Oefner, P. J., McCusker, J. H. and Davis, R. W. 2002. *Nature* **416**: 326–330.

Steyer, D., Ambroset, C., Brion, C., Claudel, P., Delobel, P., Sanchez, I., Erny, C., Blondin, B., Karst, F. and Legras J.L. 2012. *BMC Genomics* **13**: 573.

Stribny, J., Gamero, A., Pérez-Torrado, R. and Querol, A. 2015. *Int. J. Food Microbiol.* **205**: 41–46.

Strope, P.K., Skelly, D.A., Kozmin, S.G., Mahadevan, G., Stone, E.A., Magwene, P.M., Dietrich, F.S. and McCusker, J.H., 2015. *Genome Res.* **25**: 762–774.

Swinnen, S., Schaerlaekens, K., Pais, T., Claesen, J., Hubmann, G., Yang, Y., Demeke, M., Foulquié-Moreno, M.R., Goovaerts, A., Souvereyns, K., Clement, L., Dumortier, F. and Thevelein, J.M. 2012. *Genome Res.* **22**: 975–984.

Swinnen, S., Thevelein, J.M. and Nevoigt, E. 2012. *FEMS Yeast Res.* **15**: 215–227.

Tapsoba, F., Legras, J.L., Savadogo, A., Dequin, S. and Traore, A.S. 2015. *Int. J. Food Microbiol.* **211**: 128–133.

Thierry, A., Khanna, V., Créno, S., Lafontaine, I., Ma, L., Bouchier, C. and Dujon, B. 2015. *Nature Communications* **6**: 6154.

Thornton, R.J. 1985. *Am. J. Enol. Vitic.* 36: 47–49.

Thornton, R.J. and Eschenbruch, R. 1976. Antonie van Leeuwenhoek **42**: 503–509.

Tilloy, V., Ortiz-Julien, A. and Dequin, S. 2014. *Appl. Environ. Microbiol.* **80**: 2623–2632.

Tilloy, V., Cadière, A., Ehsani, M. and Dequin, S. 2015. *Int. J. Food Microbiol.* **213**: 49–58.

Tosi, E., Azzolini, M., Guzzo, F. and Zapparoli, G. 2009. *J. Appl. Microbiol.* **107**: 210–218.

van den Broek, M., Bolat, I., Nijkamp, J.F., Ramos, E., Luttik, M. A. H, Koopman, F., Geertman, J. M., de Ridder, D., Pronk, J.T. and Daran, J.M. 2015. *Appl. Environ. Microbiol.* **81**: 6253–6267.

Vaughan-Martini, A. and Martini, A. 1995. *J. Ind. Microbiol.* **14**: 514–522.

Vezinhet, F., Blondin, B. and Hallet, J.N. 1990. *Appl. Microbiol. Biotechnol.* **32**: 568–571.

Walker, G. 2011. *Antonie Van Leeuwenhoek* **99**: 25–34.

Walther, A., Hesselbart, A. and Wendland, J. 2014. *G3* (Bethesda) **4**: 783–793.

Wang, Q.M, Liu, W.Q., Liti, G., Wang, S.A. and Bai, F.Y. 2012. *Mol. Ecol.* **21**: 5404–5417.

Warringer, J., Zörgö, E., Cubillos, F.A., Zia, A., Gjuvsland, A., Simpson, J.T., Forsmark, A., Durbin, R., Omholt, S.W., Louis, E.J., Liti, G., Moses, A. and Blomberg A. 2011. *PLoS Genet.* **7**: e1002111.

Watanabe, D., Araki, Y., Zhou, Y., Maeya, N., Akao, T. and Shimoi, H. 2012. *Appl. Environ. Microbiol.* **78**: 4008–4016.

Wenger, J.W., Schwartz, K. and Sherlock, G., 2010. *PLoS Genetics* **6**: e1000942.

Wilkening, S., Lin, G., Fritsch, E.S., Tekkedil, M.M., Anders, S., Kuehn, R., Nguyen, M., Aiyar, R.S., Proctor, M., Sakhanenko, N.A., Galas, D.J., Gagneur, J., Deutschbauer, A. and Steinmetz, L.M. 2014. *Genetics* **196**: 853–865.

Winzeler, E. A., Richards, D. R., Conway, A. R., Goldstein, A. L., Kalman, S., McCullough, M.J., McCusker; J. L., Stevens, D.A., Wodicka, L., Lockhart, D., J. and Davis, R.W. 1998. *Science.* **281**: 1194–1197.

Yu, Z., Zhao, H., Li, H., Zhang, Q., Lei, H. and Zhao, M. 2012. *Biotechnol. Lett.* **34**: 365–337.

Zambonelli, C., Passarelli, P., Rainieri, S. and Giudici, P. 1993. *Ann. Microbiol. Enzymol.* **43**: 217–223.

Zimmer, A., Durand, C., Loira, N., Durrens, P., Sherman, D.J., and Marullo, P. 2014. *PLoS One* **9**: e86298.

Biotransformation and Detoxification of Environmental Pollutants with Aromatic Structures by Yeasts

Rabea Schlüter and Frieder Schauer

Abstract One of the most important roles of microorganisms in nature is the degradation of naturally occurring or industrially produced organic compounds. Besides bacteria and filamentous fungi, yeasts are abundantly present in many ecosystems and have the ability to degrade numerous organic substances. Naturally occurring substrates for yeasts are polysaccharides, sugars, lipids or proteins, however, in environmental pollutants and xenobiotics structures with aliphatic, alicyclic, aromatic or heterocyclic components dominate. This chapter focuses on the metabolic mechanisms of biotransformation and detoxification of environmental pollutants with aromatic and partially heterocyclic structures by yeasts such as *Candida*, *Debaryomyces*, *Yarrowi*a, and *Trichosporon* species. Aromatic compounds are among the most prevalent and persistent pollutants in the environment. Phenol, chlorinated phenols, methylated and ethylated benzenes, long-chain phenylalkanes and polycyclic aromatic hydrocarbons (some of the latter are contained in mineral oil products), biphenyl and its chlorinated derivatives, dioxins, dibenzofurans, diphenyl ethers and their halogenated derivatives as well as several disinfectants serve as examples of aromatic substrates which undergo degradation or biotransformation by yeasts. Some pollutants can be degraded completely to carbon dioxide and water but many xenobiotics can only be transformed to products many of which have unknown properties and may accumulate in soil and water. The study of these biotransformation mechanisms and the knowledge of the structures and properties of the products formed are of importance for minimising health risks to humans and animals.

Keywords Environmental pollutants · Toxic aromatic compounds · Degradation · Biotransformation

R. Schlüter (✉) · F. Schauer
Institute of Microbiology, Ernst-Moritz-Arndt-University of Greifswald,
Friedrich-Ludwig-Jahn-Str. 15, 17487 Greifswald, Germany
e-mail: rabea.schlueter@uni-greifswald.de

T. Satyanarayana and G. Kunze (eds.), *Yeast Diversity in Human Welfare*, DOI 10.1007/978-981-10-2621-8_13

1 Introduction

Environmental pollutants are substances which are introduced in the environment by human activities and industrialization and occur there in amounts or concentrations sufficient to pose risks for humans and other living beings. Though more than 100,000 chemicals are produced commercially reliable information about their environmental fate or impact on human health exists for only a small fraction of them (Tortella et al. 2005). Environmental pollutants can negatively influence not only the metabolism of humans but also that of animals, plants or microorganisms including bacteria, algae, protists, filamentous fungi or yeasts. On the other hand many heterotrophic microorganisms are part of the so called self-purification potential of soils and water systems. They are able to use a majority of these pollutants as substrates and hence remove these toxic components from the environment. In the best case environmental pollutants can serve as growth substrates for bacteria or fungi and are degraded during the mineralization process to carbon dioxide and water. In many cases, however, the degradation is incomplete and leads to the formation of various metabolic products. These may often be intermediary metabolites, but in some cases no further degradation takes place and such dead end products with unknown properties and toxicity accumulate in the environment.

Yeasts as well as bacteria and filamentous fungi have a considerable potential for the degradation of environmental pollutants. For example, the ability to utilize aliphatic hydrocarbons occurring in oil or oil products is widely distributed among yeasts and approximately 20% of all yeast species are able to use *n*-alkanes as a suitable carbon source for growth. However aromatic components of oil products can also be used as substrates by yeasts and many yeast species have the enzymatic equipment to cleave aromatic ring systems and to degrade a range of environmental pollutants with different aromatic structures or substituents. Aromatic compounds such as phenols, chlorinated phenols, toluene, ethylbenzene or xylenes (TEX), long-chain phenyl alkanes and polycyclic aromatic hydrocarbons as well as biphenyls, diphenyl ethers, dibenzofurans, dibenzo-*p*-dioxins and their halogenated derivatives are among the most prevalent and persistent organic pollutants in the environment (Seo et al. 2009).

2 Biodegradation of Phenols and Their Non-halogenated Derivatives

Phenols and other monoaromatic compounds are common in nature. They can be found in coal tar, in soil as degradation products of lignin, in plants as substances with antimicrobial actions or in the urine of herbivores as degradation products of aromatic nutrients. Furthermore, phenol is an important parent compound for the synthesis of many industrially produced materials, such as phenolic resins, dyes, disinfectants, pharmaceuticals, insecticides, wood preservatives, tanning substances, plasticizers, lubricants, and explosives. For this purpose nearly ten million tons of phenol are chemically produced per year worldwide. Phenol containing waste waters accumulate during the production of ethylene, propylene, acetylene, phenyl glycine, naphthalene, butadiene, detergents, substituted cresols, benzene or toluene. In water phenol shows a relatively high toxicity for fish. In humans chronic phenol intoxication damages blood vessels, liver and kidney. Yeasts are also sensitive to phenols (Ruiz-Ordaz et al. 1998). The bioactivity of phenols is caused by the hydroxyl group attached to a benzene ring. The degree of bioactivity, however, is determined by the substituents present. In several yeasts (e.g. *Candida albicans* or *Candida tropicalis*) the antifungal activity of phenol is relatively low (MIC: >45 mg ml^{-1}) whereas substituted phenols such as *p*-cresol, thymol or eugenol (MIC: 0.44–0.88 mg ml^{-1}) causes markedly higher toxic effects (Gallucci et al. 2014).

The ability of yeasts to degrade phenol has been well-known for more than 50 years (Zimmermann 1958; Harris and Ricketts 1962; Rieche et al. 1962; Wase and Hough 1966; Hashimoto 1970; Rao and Bhat 1971) and the mechanisms of utilization and mineralization have been investigated in detail in selected yeast strains (Neujahr and Varga 1970; Varga and Neujahr 1970b; Gaal and Neujahr 1979, 1981; Hofmann and Kruger 1985; Krug et al. 1985; Mortberg and Neujahr 1985; Krug and Straube 1986). The key enzyme for primary oxidation of phenol to catechol is the inducible phenol hydroxylase, which has been studied in *Trichosporon cutaneum* (Neujahr and Gaal 1973, 1975; Sejlitz and Neujahr 1987; Enroth 2003; Gerginova et al. 2007), *Candida tropicalis* (Hashimoto 1973; Neujahr et al. 1974; Neujahr and Kjellen 1978; Krug and Straube 1986; Paca et al. 2007; Long et al. 2014), and *Candida maltosa* (Hofmann and Kruger 1985; Cejkova et al. 2002). The second step is the intra-diol cleavage of the aromatic ring of catechol by a dioxygenase, the catechol 1,2-oxygenase (Varga and Neujahr 1970a; Shoda and Udaka 1980; Krug and Straube 1986; Ahuatzi-Chacon et al. 2004) to *cis,cis*-muconic acid. The muconic acid is transformed in a third step to (+)-4-carboxymethylenebut-2-en-4-olide (Gaal and Neujahr 1980). This compound is degraded via beta-ketoadipate to succinate and acetyl-CoA and thus enters the intermediary metabolism of the yeast. This *ortho*-cleavage pathway (or beta-ketoadipate pathway) of yeasts characterized by an intra-diol cleavage of the aromatic ring (Powlowski et al. 1985; Gerecova et al. 2015) corresponds in principle to the bacterial *ortho*-cleavage pathway (Ornston and Stanier 1966; Harwood and Parales 1996). However, in bacteria the *meta*-degradation pathway, which involves an extra diol cleavage of the ring system and some other ring cleavage mechanisms, can also occur (Fuchs et al. 2011).

Less than 10% of all yeast species seem to be able to use phenol as a substrate for growth (Table 1). Typical phenol-assimilating yeasts are *Blastobotrys adeninivorans*, *Candida maltosa*, *Candida tropicalis*, *Meyerozyma guilliermondii*, *Cryptococcus humicola*, *Rhodosporidium toruloides* (Fig. 1) or several *Trichosporon* species.

Table 1 Diversity of yeast species able to utilize phenol as carbon source

Ascomycetous yeasts	Basidiomycetous yeasts	References
Blastobotrys adeninivorans[19]	*Cryptococcus humicola*[10,20]	[1]Adav et al. (2007) [2]Aleksieva et al. (2002) [3]Bergauer et al. (2005) [4]Bril'kov et al. (1980) [5]Chrzanowski et al. (2008) [6]Cong et al. (2014) [7]Fialova et al. (2004) [8]Gerecova et al. (2015) [9]Hashimoto (1973) [10]Hofmann and Schauer (1988) [11]Hristov et al. (2012) [12]Iwasaki et al. (2010) [13]Katayamahirayama et al. (1994) [14]Krallish et al. (2006) [15]Krug et al. (1985) [16]Lee et al. (2001) [17]Liu et al. (2011) [18]Margesin et al. (2005) [19]Middelhoven et al. (1992) [20]Middelhoven (1993) [21]Morsen and Rehm (1987) [22]Rigo and Alegre (2004) [23]Rigo et al. (2010) [24]Rocha et al. (2007) [25]Schlueter et al. (2013) [26]Shoda and Udaka (1980) [27]Tsai et al. (2005) [28]Vallini et al. (2001) [29]Neujahr and Varga (1970) [30]Varma and Gaikwad (2008) [31]Wase and Hough (1966)
Candida albicans[8,10,27]	*Cryptococcus laurentii*[20]	
Candida catenulata[10]	*Cryptococcus terreus*[3,10,14,20,21]	
Candida maltosa[7,10]	*Cryptococcus terricola*[3]	
Candida mesenterica[10]	*Leucosporidiella creatinivora*[3,14]	
Candida parapsilosis[10,19,22,23]	*Leucosporidium scottii*[20]	
Candida rugosa[10,24]	*Mastigobasidium intermedium*[3]	
Candida sake[10,30]	*Rhodosporidium lusitaniae*[3]	
Candida tropicalis[1,4,9,10,15,20,30]	*Rhodosporidium toruloides*[10]	
Candida viswanathii[6,28]	*Rhodotorula aurantiaca*[10,20]	
Debaryomyces hansenii[20]	*Rhodotorula glutinis*[10,13,20]	
Debaryomyces subglobosus[31]	*Rhodotorula ingeniosa*[3,20]	
Dipodascus armillariae[20]	*Sporidiobolus ruineniae*[10]	
Geotrichum klebahnii[20]	*Sporobolomyces roseus*[3,20]	
Lodderomyces elongisporus[10]	*Sporobolomyces salmonicolor*[10]	
Magnusiomyces ovetensis[20]	*Trichosporon asahii*[6]	
Magnusiomyces tetrasperma[20]	*Trichosporon cutaneum*[2,20,26,29]	
Meyerozyma guilliermondii[5,10]	*Trichosporon dulcitum*[18,20]	
Pichia membranifaciens[24]	*Trichosporon laibachii*[20]	
Trichomonascus ciferrii[20]	*Trichosporon loubieri*[20]	
Yarrowia lipolytica[5,10,16,30]	*Trichosporon moniliiforme*[12,20]	
	Trichosporon montevideense[11,17]	
	Trichosporon mucoides[25]	

Among ascomycetous yeasts the ability to use phenol seems to be restricted to yeasts with coenzyme CoQ9 and CoQ8—ubiquinones frequently used as phylogenetic markers in yeast taxonomy and in evolutionary studies. In contrast, the typical glucose-fermenting and ethanol-producing yeasts (*Saccharomyces* and related genera) or the methanol-utilizing yeasts express mainly CoQ6 or the CoQ7 and seem not to be able to use phenol as a carbon source (Hofmann and Schauer 1988; Middelhoven et al. 1992). In CoQ9-containing yeasts the ability to use other

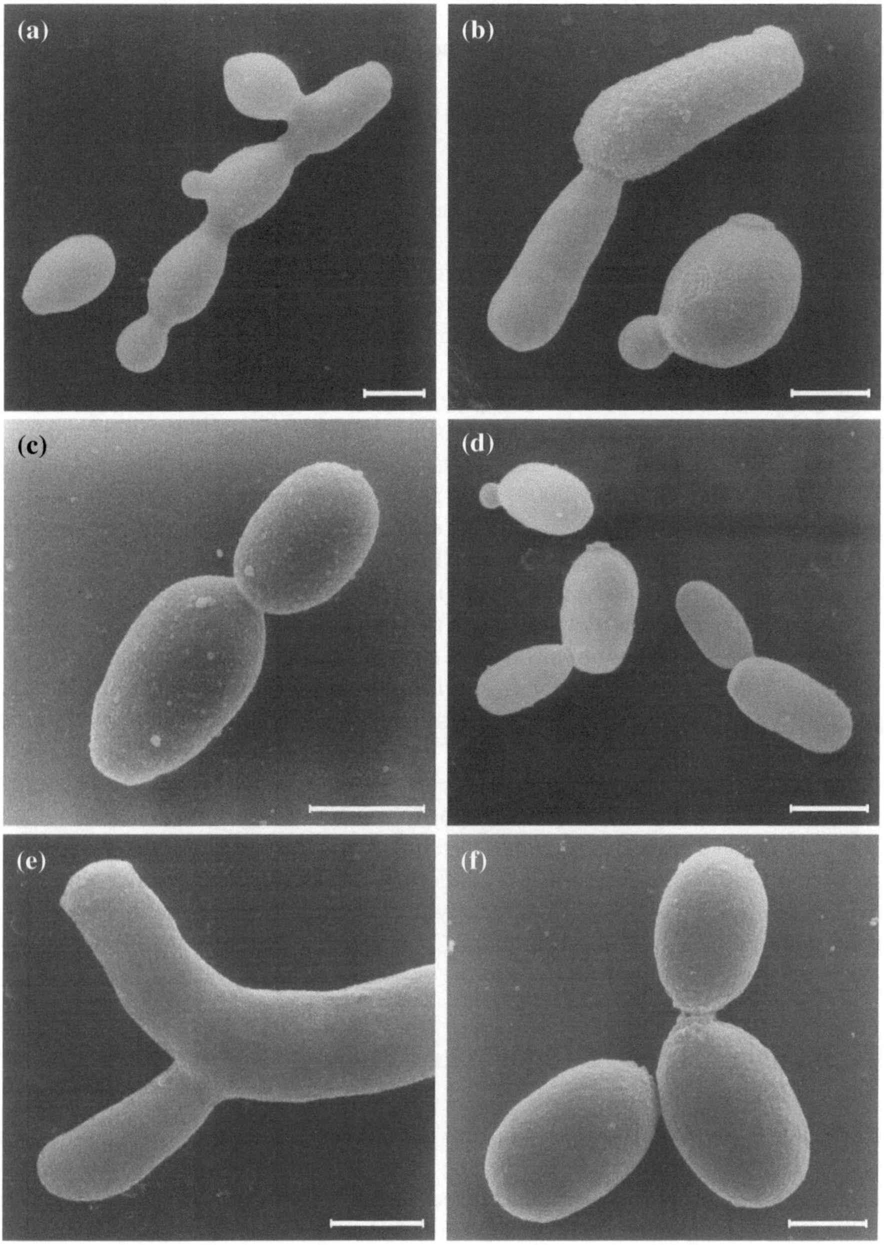

Fig. 1 Phenol-assimilating yeasts visualized by scanning electron microscopy; **a** *Blastobotrys adeninivorans* SBUG 724; **b** *Candida maltosa* SBUG 700; **c** *Candida tropicalis* SBUG 715; **d** *Meyerozyma guilliermondii* SBUG 50; **e** *Cryptococcus humicola* SBUG 517; **f** *Rhodosporidium toruloides* SBUG 137; Scale bar: 2 μm

unconventional carbon sources such as *n*-alkanes, uric acid, putrescine or some other alkylamines is also widely distributed (Bos and Debruyn 1973; Middelhoven et al. 1985; Hofmann and Schauer 1988). Studies by Gerecova et al. (2015) suggest co-evolutionary relationships between genes encoding enzymes from the 3-oxoadipate pathway and genes encoding several alkane hydroxylating P450 forms and certain lipases, monooxygenases or proteins with special mitochondrial functions. In basidiomycetous yeasts the correlation between phenol utilization, coenzyme Q structure and assimilation of further unconventional carbon sources such as *n*-alkanes seems to be not so strong (Hofmann and Schauer 1988).

In addition to phenol catechol, resorcinol, and hydroquinone and some other phenol derivatives can be used as carbon sources by several yeasts or yeast-like fungi, too (Table 2).

Regardless of the inability of many yeasts to grow on toxic phenolic derivatives such as cresols (Hasegawa et al. 1990), some of these substances can be oxidized and partially degraded by several yeast species. *Candida tropicalis* (Hashimoto 1973; Varma and Gaikwad 2008) and *Trichosporon cutaneum* (Aleksieva et al. 2002) have been shown to be able to oxidize *m*-cresol and *Candida tropicalis* (Hashimoto 1973), *Candida maltosa* (Fialova et al. 2004), and *Trichosporon cutaneum* (Powlowski and Dagley 1985) oxidise *p*-cresol.

Of special interest is the degradation of alkylphenols which can be produced during biodegradation of non-ionic surfactants such as alkylphenol polyethoxylates and which are known to be endocrine-disrupting substances that mimic or disrupt estrogenic activity (Terasaka et al. 2006). These compounds appear recalcitrant to further microbial degradation and nonylphenol, for example, increasingly accumulates in sediments, sewage sludge and groundwater (Giger et al. 1984). However, a few yeasts seem to be able to grow on 4-(1-nonyl)phenol as shown for *Candida maltosa* (Corti et al. 1995) and *Candida viswanathii* (Vallini et al. 2001, as *C. aquaetextoris*). The primary attack goes via terminal oxidation of the alkyl chain, followed by *beta*-oxidation. Typical metabolites are 4-acetylphenol and 4-hydroxycinnamic acid. The latter seems to be a growth substrate for *C. viswanathii* (Vallini et al. 2001).

3 Biodegradation of Chlorinated Phenols

Chlorinated phenols are a group of 19 mono-, di-, tri-, tetra-, and pentachlorophenols. With increasing degree of chlorination the solubility in water is decreased, however, the geo- and bioaccumulation is increased. Their toxicity is enhanced in comparison to the unsubstituted phenol. Chlorinated phenols are used as disinfectants, fungicides, herbicides, insecticides, as parent substances for the production of different aromatic chemicals, as intermediate products for the synthesis of dyes or for the production of preservatives for wood, textiles, and leather. Unfortunately, chlorophenol chemicals often contain very toxic polychlorinated dibenzofurans and dibenzo-*p*-dioxins as impurities.

Table 2 Utilization of phenol and some unchlorinated derivatives by selected yeast species

Growth substrate (phenol and phenol derivatives)	*Candida tropicalis*	*Candida maltosa*	*Cryptococcus terreus*	*Trichosporon cutaneum*	*Trichosporon moniliiforme*	*Exophiala jeanselmei**	References
Phenol[a]	+[1,5,6]	+[3]	+[1,7]	+[1,2,8]	+[1]	+[1]	[1]Middelhoven (1993), [2]Hasegawa et al. (1990), [3]Fialova et al. (2004), [5]Hashimoto (1973), [6]Krug et al. (1985), [7]Bergauer et al. (2005), [8]Aleksieva et al. (2002)
Catechol[b]	+[1,5,6]	+[3]	+[1,7]	+[1]	+[1]	+[1]	[1]Middelhoven (1993), [3]Fialova et al. (2004), [5]Hashimoto (1973), [6]Krug et al. (1985), [7]Bergauer et al. (2005), [8]Aleksieva et al. (2002)
Resorcinol[c]	+[1,6]	+[3]	+[1,7]	+[1,2,8]	+[1]	+[1]	[1]Middelhoven (1993), [2]Hasegawa et al. (1990), [3]Fialova et al. (2004), [6]Krug et al. (1985), [7]Bergauer et al. (2005), [8]Aleksieva et al. (2002)
Hydroquinone[d]	+[1,6]		+[1,7]	+[1,2]	+[1]	+[1]	[1] Middelhoven (1993), [2]Hasegawa et al. (1990), [6]Krug et al. (1985), [7]Bergauer et al. (2005)
Pyrogallol[e]	—[1,6]		—[1]	—[1]/+[2]	+[1]	—[1]	[1]Middelhoven (1993),[2]Hasegawa et al. (1990),[6]Krug et al. (1985)
Phloroglucinol[f]	—[1,6]		+[1]	—[1]/+[2]	+[1]	+[1]	[1]Middelhoven (1993), [2]Hasegawa et al. (1990), [6]Krug et al. (1985)
o-Cresol[g]	—[1,5,6]		—[1,7]	—[1,2]	—[1]	(+)[1]	[1]Middelhoven (1993), [2]Hasegawa et al. (1990), [5]Hashimoto (1973), [6]Krug et al. (1985), [7]Bergauer et al. (2005)
m-Cresol[h]	—[1,5,6]		—[1]/(—)[7]	—[1,2]/(+)[8]	—[1]	+[1]	[1]Middelhoven (1993), [2]Hasegawa et al.(1990), [5]Hashimoto (1973), [6]Krug et al. (1985), [7]Bergauer et al. (2005), [8]Aleksieva et al. (2002)
p-Cresol[k]	—[1,5]	—[3]	—[1]/(—)[7]	—[1,2]/+[9]	—[1]	+[1]	[1]Middelhoven (1993), [2]Hasegawa et al. (1990), [3]Fialova et al. (2004), [5]Hashimoto (1973),[7]Bergauer et al. (2005), [9]Powlowski and Dagley (1985)

(continued)

Table 2 (continued)

Growth substrate (phenol and phenol derivatives)	*Candida tropicalis*	*Candida maltosa*	*Cryptococcus terreus*	*Trichosporon cutaneum*	*Trichosporon moniliiforme*	*Exophiala jeanselmei**	References
3-Methylcatechol[l]	—[1]/+[5]		—[1]	(—)[1]	—[1]	—[1]	[1]Middelhoven (1993), [5]Hashimoto (1973)
4-Methylcatechol[m]	—[1]/+[5]		—[1]	(—)[1]	—[1]	+[1]	[1]Middelhoven (1993), [5]Hashimoto (1973)
Orcinol[n]	—[1,6]		+[1]	—[1]	+[1]	+[1]	[1]Middelhoven (1993), [6](Krug et al., 1985)
4-Nitrophenol			—[7]	—[2]/+[8**]			[2]Hasegawa et al. (1990), [7]Bergauer et al. (2005), [8]Aleksieva et al. (2002)
2,6-Dinitrophenol				+[8]			[8]Aleksieva et al. (2002)
4-Ethylphenol				(—)[1]	(—)[1]	+[1]	[1]Middelhoven (1993)
4-(1-Nonyl)phenol		+[4]					[4]Corti et al. (1995)
α-Naphthol		+[4]					[4]Corti et al. (1995)
Salicylic acid[o]	—[1,5,6]	—[3]	+[1,7]	—[1]/+[2]	+[1]	+[1]	[1]Middelhoven (1993), [2]Hasegawa et al. (1990), [3]Fialova et al. (2004), [5]Hashimoto (1973), [6]Krug et al. (1985), [7]Bergauer et al. (2005)

— no growth; (—) very weak growth; (+) weak growth; + good growth; *yeast-like fungus; **3-nitrophenol as substrate

[a](hydroxybenzene)

[b](1,2-dihydroxybenzene)

[c](1,3-dihydroxybenzene)

[d](1,4-dihydroxybenzene)

[e](1,2,3-trihydroxybenzene)

[f](1,3,5-trihydroxybenzene)

[g](2-methylphenol)

[h](3-methylphenol)

[k](4-methylphenol)

[l](1,2-dihydroxy-3-methylbenzene)

[m](1,2-dihydroxy-4-methylbenzene)

[n](1,3-dihydroxy-5-methylbenzene)

[o](2-hydroxybenzoic acid)

Chlorophenols have been classified as priority pollutants and the degradation of mono-, di- and polychlorophenols by bacteria has been frequently reported (Knackmuss and Hellwig 1978; Schmidt et al. 1983; Rojo et al. 1987; Chaudhry and Chapalamadugu 1991; Potrawfke et al. 1998; Hlouchova et al. 2012). Among filamentous fungi mineralization of chlorophenols has so far been shown mainly in certain white-rot fungi and in a few wood-inhabiting species (Fahr et al. 1999; Schlosser et al. 2000). A few yeasts strains seem to be also able to grow on selected monochlorinated hydroxylated benzenes. This was shown for some strains of *Candida tropicalis* with 4-chlorophenol (Jiang et al. 2007; Basak et al. 2013) or 4-chlorocatechol (Krug et al. 1985) as carbon sources. Other strains of *Candida tropicalis* seem not to be able to grow on chlorophenols (Ivoilov and Karasevich 1983; Ahuatzi-Chacon et al. 2004; Wang et al. 2012). However, the addition of phenol improves the growth and enhances the degree of 4-chlorophenol degradation (Wang et al. 2012). The induced phenol hydroxylase from *C. tropicalis* HP15 can oxidize 2-, 3- and 4-chlorophenol to the corresponding chlorinated catechols, however, 2,6-dichlorophenol or 2,3,6-trichlorophenol were not accepted as substrates (Krug and Straube 1986). The catechol 1,2-oxygenase of this yeast species can use the 3-chlorocatechol and 4-chlorocatechol produced as substrates for ring cleavage (Krug and Straube 1986; Basak et al. 2013).

In other yeast species, chlorinated phenols can neither induce the catabolic enzymes needed for their degradation nor can they be oxidized or used as a carbon source. However, in the presence of inducing substrates such as phenol a metabolization of chlorinated phenols can be achieved. In this way, phenol-grown cells of *Rhodotorula glutinis* oxidize 2-, 3- and 4-chlorophenols, 4-bromophenol and 2,4-dibromophenol (Walker 1973; Katayamahirayama et al. 1994) and *Yarrowia lipolytica* cells degrade 4-chlorophenol (Lee et al. 2001). Also the phenol hydroxylase from *Trichosporon cutaneum* is able to transform the substrates 2-, 3- and 4-chlorophenol or 2-, 3- and 4-fluorophenol (Neujahr and Gaal 1973). However, this yeast cannot grow on any of the chlorophenols tested (Hasegawa et al. 1990). Phenol-cultivated cells of *Candida maltosa* are able to break down 3- and 4-chlorophenol very quickly forming 3- and 4-chlorocatechol, which were further degraded via their chlorinated muconic acids to the non-chlorinated ring cleavage product 4-carboxymethylenebut-2-en-4-olide and chloride ions (Polnisch et al. 1992). The dehalogenation step was initiated by the cycloisomerization of the *cis,cis*-chloromuconic acid and achieved during the following dienelactone formation (Fig. 2).

2-Chlorophenol was partially converted by *C. maltosa* to 3-chlorocatechol and 3-chloromuconic acid, without any further dechlorination step. Uninduced cells of *C. maltosa* incubated with mono- or dichlorophenols suffer starvation and under these stress conditions excrete intermediates of aromatic amino acid catabolism (Hammer et al. 1996). Similar observations were recorded for *Saccharomyces cerevisiae* (Yadav et al. 2011). Mechanisms of ring cleavage and dehalogenation of a series of fluorinated phenols by the yeast-like fungus *Exophiala jeanselmei* were described in an excellent study by Boersma et al. (1998).

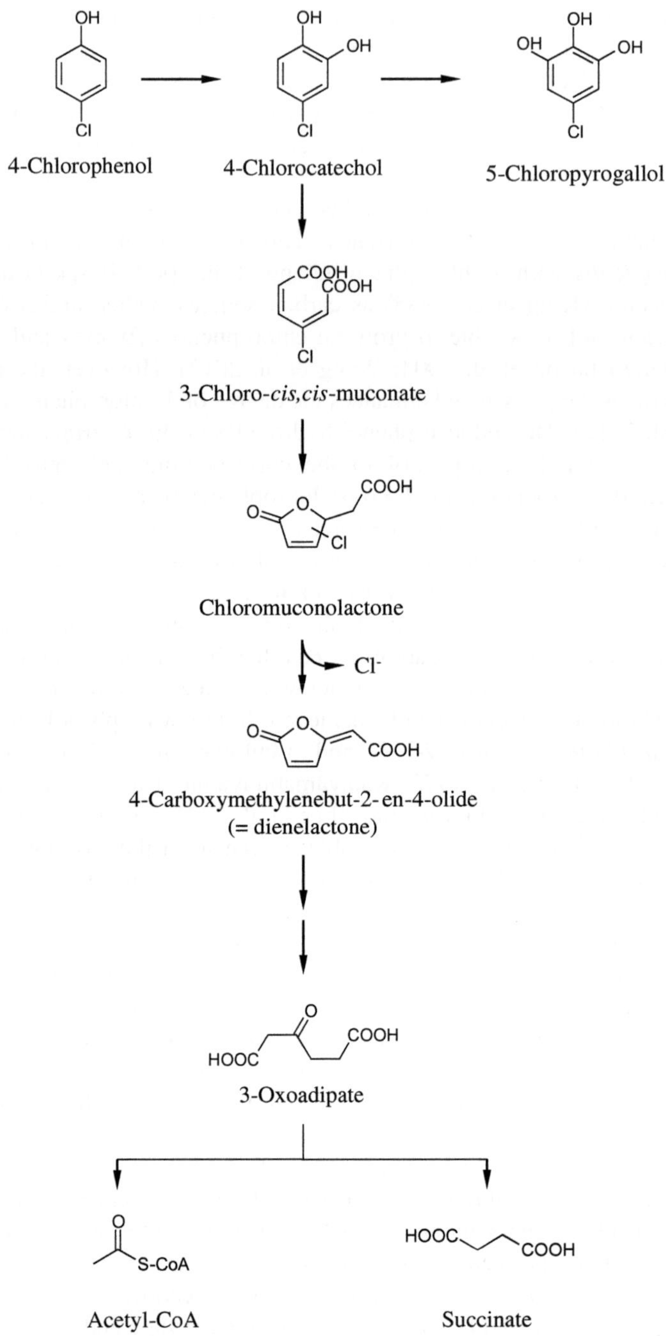

Fig. 2 Degradation of 4-chlorophenol (beta-ketoadipate pathway) by yeasts (according to results of Polnisch et al. 1992; Mazur et al. 1994)

Recent attempts to use yeasts for the elimination of halogenated phenols from waste water resulted in a combination of catalytic hydrodechlorination (HDC) over Pd/C, by which 4-chlorophenol and 2,4-dichlorophenol could be easily and completely dechlorinated under mild conditions, with a biodegradation step by *Candida* yeasts of the phenol formed (Zhou et al. 2012).

Immobilized cells of *Candida tropicalis* in a fluidized bed reactor are suitable for the elimination of higher phenol concentrations and of 4-chlorophenol, (Juarcz-Ramirez et al. 2001; Galindez-Mayer et al. 2008) and cells from *Candida tropicalis* or *Trichosporon cutaneum,* covalently bound to polyacrylamide gel beads or polyamide granules respectively, can be useful for the purification of waste water containing phenols (Godjevargova et al. 1998; Chen et al. 2002).

Further improvements for the elimination of phenol and its chlorinated derivatives were achieved by a combined action of immobilized fungal cells of *Trichosporon cutaneum* and *Aspergillus awamori*. Here it was found not only that the elimination of phenol and 2,4-dichlorophenol was very rapid, but also that 2-, 3- and 4-chlorophenol and 3-methoxyphenol were degraded to an extent of 44–72% respectively within 40 h (Yordanova et al. 2013). Additional efforts have been made by combining *Candida tropicalis* with two bacteria (*Burkholderia* sp. and *Microbacterium phyllosphaerae*). Using phenol as the primary carbon and energy source this microbial consortium co-metabolically degraded mono-, di-, and trichlorophenols, with overall removal efficiencies ranging from 95 to 99.8% and COD removal efficiencies from 85 to 97.8% (Salmeron-Alcocer et al. 2007). A simultaneous decrease of phenolic compounds and heavy metals such as zinc in wastewater was achieved with *Candida tropicalis*, *C. parapsilosis*, *C. maltosa*, and *Yarrowia lipolytica* (Mahgoub et al. 2015).

Furthermore, cells of *Trichosporon beigelii* (*T. cutaneum*) have been used for the development of an amperometric biosensor for the determination of aromatics and their chlorinated derivatives (Riedel et al. 1995). This sensor was more sensitive to chlorophenols, especially to mono- and dichlorinated phenols, than to phenol and was insensitive to benzoate.

4 Biodegradation of Methylated and Ethylated Benzenes (BTEX)

Alkylated benzenes are typical components found in crude oil and its refined products and they continuously enter the environment as pollutants. To this family of aromatic hydrocarbons belong on the one hand alkylbenzenes with short side chains such as toluene (methylbenzene), ethylbenzene, and *o*-, *m*-, and *p*-xylenes (dimethylbenzenes), collectively known (together with benzene) as BTEX compounds. These substances are hazardous volatile organic compounds (VOC) and are among the most abundant components in the water-soluble fraction of oil and refined fuels occurring in polluted water and soil systems or in air or gas streams from petrochemical and chemical industries. On the other hand, hydrophobic alkylated aromatic hydrocarbons with longer alkyl chains (C8–C20), often also designated as phenylalkanes, exhibit quite different chemical properties and degradation mechanisms (see chapter below).

Reports concerning microbial catabolism of short chain alkylated benzenes have mainly focused on bacteria (Sariaslani et al. 1974; Assinder and Williams 1990; Kim et al. 2008; Jiang et al. 2015). In comparison to prokaryotes only little is known about degradation pathways of short chain alkylated benzenes and BTEX in yeasts or filamentous fungi, though these organisms play important roles in ecosystems and in contaminated soil and water systems. It seems that the majority of yeasts is unable to use toluene, ethylbenzene or xylenes as the sole carbon source. None of 32 tested yeast isolates utilized any of the volatile aromatic compounds benzene, toluene, ethylbenzene or *o*-xylene, *m*-xylene, and *p*-xylene (Bergauer et al. 2005). The critical step in degradation seems to be the primary oxidation of these aromatic hydrocarbons. Only a small group of black yeast-like fungi (*Herpotrichiellaceae*, order *Chaetothyriales* and relatives) has been reported as being able to grow effectively on toluene and ethylbenzene and to co-metabolically metabolize xylenes. However, if ring-hydroxylated or alkyl-chain hydroxylated derivatives of alkylated benzenes are used as substrates then improved utilization by several yeasts seems to be possible. Thus, in addition to the yeast-like fungus *Exophiala jeanselmei* the yeasts *Rhodotorula glutinis*, *Trichomonascus ciferrii*, *Trichosporon laibachii*, and *Trichosporon loubieri* are able to grow with 4-ethylphenol (ring-hydroxylated ethylbenzene) and *Geotrichum klebahnii*, *Trichosporon cutaneum*, and *Trichosporon loubieri* can grow with 2-phenylethanol (side-chain oxidized ethylbenzene) as substrate (Middelhoven 1993).

Similar results were obtained with 3- and 4-methylphenol (ring-hydroxylated toluene) and phenylmethanol (side-chain hydroxylated toluene) as carbon sources for yeasts (Table 3).

Growth on unhydroxylated short-chain alkylbenzenes (BTEX) was demonstrated for the yeast-like fungus *Exophiala lecanii-corni* growing on toluene (Woertz et al. 2001; Qi et al. 2002) and ethylbenzene (Qi et al. 2002; Gunsch et al. 2005, 2007). In this case growth on benzene and styrene was also possible (Qi et al. 2002). Furthermore, other related black yeast-like fungi, such as *Exophiala* sp. (Prenafeta-Boldu et al. 2001), *Exophiala oligosperma* (Estevez et al. 2005), *Exophiala sideris* (Seyedmousavi et al. 2011), *Cladophialophora* sp. (Prenafeta-Boldu et al. 2001, 2002, 2004; Nikolova and Nenov 2005), and *Cladophialophora psammophila* (Badali et al. 2011), can grow on toluene and/or ethylbenzene. Xylenes were co-metabolized to different extents. For example *p*-xylene was not degraded by *Cladophialophora* sp. in complex BTEX mixtures but, in combination with toluene, it appeared to be mineralized (Prenafeta-Boldu et al. 2002).

Toluene and ethylbenzene were degraded at the side chain (Uzura et al. 2001) by the same monooxygenase system (Prenafeta-Boldu et al. 2002) to produce phenylmethanol (benzyl alcohol) and 2-phenylethanol, respectively. Styrene was oxidized in *Exophiala jeanselmei* by a cytochrome P450-dependent styrene monooxygenase (Cox et al. 1996). In the fungus *Paecilomyces variotii*, the

Table 3 Utilization of hydroxylated derivatives of toluene and ethylbenzene by selected yeast species

Yeast species	Ring hydroxylated toluene		Side-chain hydroxylated toluene	Ring hydroxylated ethylbenzene	Side-chain hydroxylated ethylbenzene	References
	3-Methylphenol	4-Methylphenol	Phenylmethanol	4-Ethylphenol	2-Phenylethanol	
Candida tropicalis	—[1]	—[1a]	—[1]	—[1]	—[1]	[1]Middelhoven (1993),[a]Hashimoto (1973)
Geotrichum klebahnii	—[1]	—[1]	—[1]	—[1]	+[1]	[1]Middelhoven (1993)
Stephanoascus ciferrii	+[1]	(+)[1]	—[1]	—[1]	—[1]	[1]Middelhoven (1993)
Leucosporidiella creatinivora	(+)[4]	(+)[4]				[4]Bergauer et al. (2005)
Leucosporidium scottii	+[1]	+[1]	+[1]	(—)[1]	(—)[1]	[1]Middelhoven (1993)
Rhodotorula aurantiaca	+[1]	+[1]	+[1]	—[1]	—[1]	[1]Middelhoven (1993)
Rhodotorula glutinis	—[1]	—[1]	(—)[1]	(+)[1]	—[1]	[1]Middelhoven (1993)
Rhodotorula ingeniosa	+[4]	+[4]				[4]Bergauer et al. (2005)
Trichosporon cutaneum	—[1]	—[1,2]	—[1]			
Trichosporon dulcitum	+[1]	+[1]	—[1]	(—)[1]	—[1]	[1]Middelhoven (1993)
Trichosporon laibachii	+[1]	(+)[1]	(+)[1]	+[1]	—[1]	[1]Middelhoven (1993)
Trichosporon loubieri	+[1]	+[1]	(—)[1]	(+)[1]/—[3]	+[1]	[1]Middelhoven (1993), [3]Middelhoven (2003)
Exophiala jeanselmei	+[1]	+[1]	+[1]	+[1]	+[1]	[1]Middelhoven (1993)

— no growth; (—) very weak growth; (+) weak growth; + good growth; [a]but methylcatechol formation and ring cleavage with phenol-grown cells

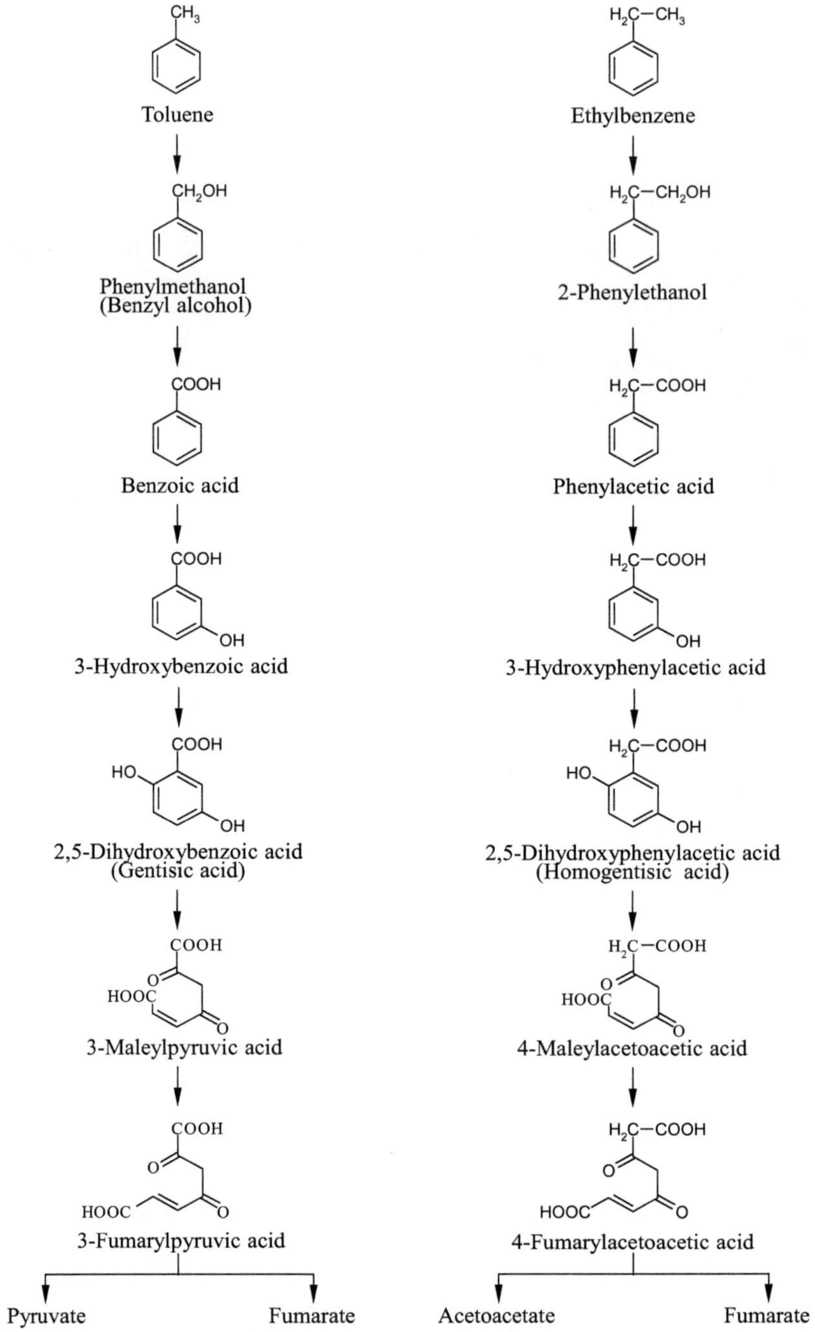

Fig. 3 Degradation of toluene (gentisate pathway) and ethybenzene by yeast-like fungi (*Exophiala* spp.) (according to results of Middelhoven 1993; Gunsch et al. 2005)

oxidation of toluene was initiated by toluene oxygenase (Garcia-Pena et al. 2005). The phenylmethanol (from toluene) and phenylethanol (from ethylbenzene) formed are transformed via an aldehyde intermediate to benzoic acid and phenylacetic acid, respectively. In *Exophiala lecanii-corni* further degradation of phenylacetic acid proceeds via ring hydroxylation to homogentisic acid (2,5-dihydroxyphenylacetic acid) followed by a ring cleavage by homogentisate-1,2-dioxygenase (Gunsch et al. 2005, 2007; Fig. 3). In contrast to ethylbenzene, *o*-xylene does not influence the expression of homogentisate-1,2-dioxygenase (Gunsch et al. 2005) because *o*-xylene and *m*-xylene are converted to phthalates (Prenafeta-Boldu et al. 2002).

Many of the yeast-like fungi capable of growing on alkylated benzenes with short side chains have been used or have been detected in air biofilters for treating BTEX or styrene polluted air. For example *Exophiala jeanselmei* (Cox and Doddema 1996; Cox et al. 1996, 1997), *Exophiala lecanii-corni*, *Exophiala xenobiotica* (Prenafeta-Boldu et al. 2004), *Exophiala oligosperma* (Prenafeta-Boldu et al. 2012), and *Exophiala* sp. (Kennes and Veiga 2004), *Cladophialophora saturnica* (Prenafeta-Boldu et al. 2012), and *Cladophialophora* sp. (Kennes and Veiga 2004) were detected in microbial communities from biofilms of such biofilters. The yeast-like fungi adapt more readily than do many bacteria to adverse environmental conditions of low moisture and low pH (Prenafeta-Boldu et al. 2001; Sun et al. 2002). These environmental factors are relevant for the specific enrichment of such fungi in biofilters (Gallastegui et al. 2013), particularly since the doubling time of these yeast-like fungi with BTEX compounds of at least 2–3 days (Prenafeta-Boldu et al. 2001) are, under non-limiting conditions, clearly longer than for bacteria. Some of the toluene and ethylbenzene supplemented biofilters showed a relatively low biodiversity with similar microbial community structure and dominance of *Exophiala oligosperma*, however, with *p*-xylene the community was more complex and more diverse and encompasses other species of chaethyrialean fungi such as *Cladophialophora saturnica* and *Fonsecaea* species (Prenafeta-Boldu et al. 2012).

5 Biodegradation of Long-Chain Alkylbenzenes (Phenylalkanes)

Phenylalkanes are usually defined as long-chain alkylbenzenes with a dominating hydrophobic alkyl chain (i.e. with a chain length longer than C6). They are typical crude oil components and occur also in refined products such as fuel and diesel oils.

Depending on the geographical origin of the oil monoaromatic compounds make up to 9% of crude oil components, and approximately 20% of these are phenylalkanes with longer side chains (Dutta 2005).

Phenylalkanes enriched in special oil fractions or as pure substances are used as synthetic intermediates, components of detergents, varnishes, and cable oils (Dutta and Harayama 2001). Phenylalkanes are introduced into the environment mainly via oil pollution. Because of their hydrophobicity, they may bind to sludges and sediments and thus enter the food chain by bioaccumulation.

Table 4 Comparison of growth of selected yeasts on glucose, *n*-decane, and different phenylalkanes (according to results of Schauer 1988; Awe et al. 2008)

Yeast species	Glucose	*n*-Decane	Phenylethane	Phenylhexane	Phenyloctane	Phenyldecane
Blastobotrys adeninivorans	+++	+++	–	–	+	++
Candida maltosa	+++	+++	–	–	+	++
Candida tropicalis	+++	+++	–	–	+	++
Candida rugosa	+++	+++	–	–	–	+
Meyeromyces guilliermondii	+++	+++	–	–	+	+
Lindnera jadinii	+++	–	–	–	–	–
Hansenula henricii	+++	–	–	–	–	–
Saccharomyces cerevisiae	+++	–	–	–	–	–

+++ very good growth; ++ good growth; + low growth; — no growth

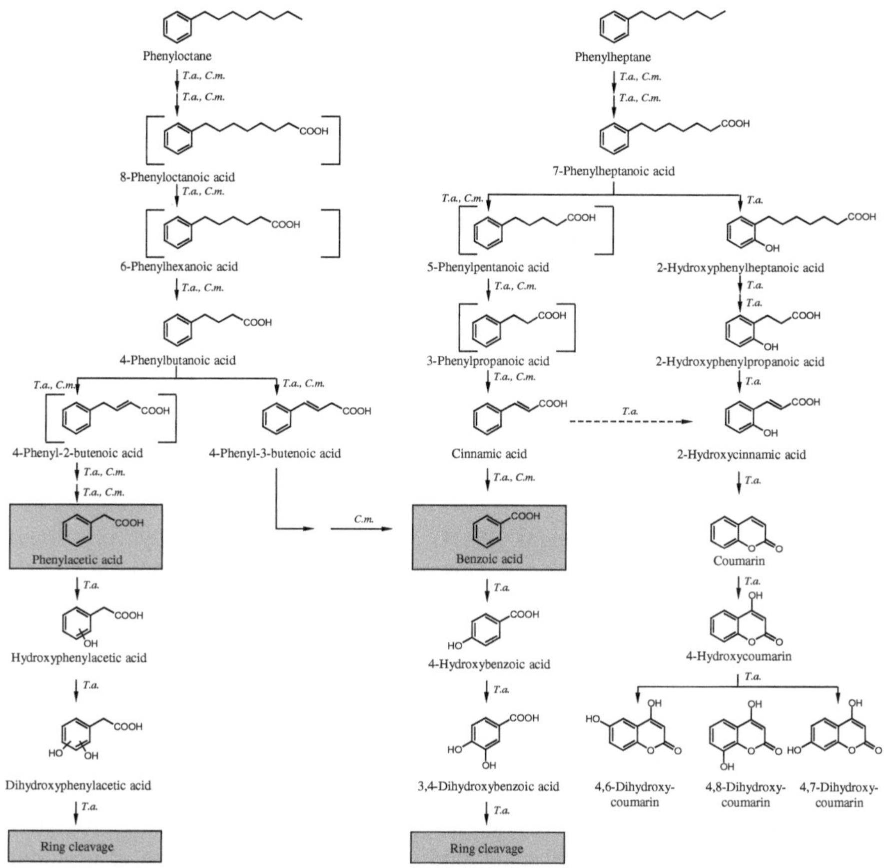

Fig. 4 Degradation of phenyloctane and phenylheptane by the yeasts *Trichosporon asahii* (*T.a.*) and *Candida maltosa* (*C.m.*) (according to results of Awe et al. (2008, 2009); metabolites which have been postulated, but not enriched or clearly analyzed are enclosed in square brackets; pathways marked with dotted lines could not be clearly authenticated)

The degradation mechanisms of phenylalkanes are different from those applicable to short-chain alkylated benzenes (Fedorak and Westlake 1986). Thus, the spectrum of yeast species degrading phenylalkanes differs from that known from short-chain alkylbenzenes. Nearly all hydrocarbon assimilating yeasts are able to oxidize in addition to *n*-alkanes also longer-chained phenylalkanes at the terminal position of the alkyl chain (Table 4).

About 20% of all yeast species are able to use hydrocarbons as substrate (Schauer and Schauer 1986) and the majority of them can also use long-chain phenylalkanes as growth substrates. Normally, degradation is initiated by a primary oxidation of the terminal methyl group of the alkyl chain of phenyl alkanes by the alkane hydroxylating P450-monooxygenase system (Schunck et al. 1987; Scheller et al. 1998). Further degradation of the phenylalkanol products goes via phenylalkanals to the phenylalkanoic acids with the corresponding chain length which are shortened by beta-oxidation. However, only a fraction of the *n*-alkane utilizing yeasts, such as *Trichosporon asahii*, degrades phenylalkanes completely by mineralization. Several yeasts, such as *Candida maltosa*, cannot further transform the accumulated short chain phenylalkanoic acids benzoic acid and phenylacetic acid. Unexpectedly, during degradation of even-chain phenylalkanes small amounts of benzoic acid accumulated in addition to the major product phenylacetic acid (Awe et al. 2008) and during degradation of phenylalkanes by *Trichosporon asahii* toxic coumarines were formed (Awe et al. 2009) as a consequence of ring hydroxylation of the phenylpropionic acid intermediate and its derivatives (Fig. 4).

6 Biotransformation of Polycyclic Aromatic Hydrocarbons

Polycyclic aromatic hydrocarbons (PAH) are a class of structurally very diverse organic compounds. The major concerns regarding PAHs result from the fact that many of these compounds are classified as toxic or carcinogenic for experimental animals (Cerniglia and Crow 1981). PAHs are ubiquitous environmental contaminants introduced into the atmosphere by combustion of fossil fuels, vehicle emissions, forest fires and volcanic activities and into the soil by forest fires and airborne pollution. Fallout of particulate matter in the air, runoff of polluted ground sources as well as pollution of rivers and lakes by municipal and industrial effluents are the causes for the distribution of PAHs in waters (Harvey 1997). Due to the considerable input of PAHs into the environment, ubiquitously distributed yeasts probably developed a range of adaptation strategies over the course of prolonged exposure to the pollutants. As a result, many yeasts able to use PAHs as sole source of carbon and energy or to oxidize polyarenes have been isolated from polluted estuaries, sediments or marine environment. The most commonly isolated genera were *Candida* and *Rhodotorula* followed by *Cryptococcus, Debaryomyces,* and *Trichosporon* (Hagler et al. 1979; Hagler and Mendoncahagler 1981; Kutty and Philp 2008). In particular yeast strains of the genera *Candida* and *Rhodotorula* have a considerable potential for the degradation of PAHs.

The main focus for investigations of microbial biodegradation rested on the regular *cata*-condensed and *peri*-condensed alternant hydrocarbons. The compounds are divided into low-molecular (LMW-PAH) and high-molecular weight PAH (HMW-PAH). In contrast to the chemical definition, PAHs with more than three rings are often referred to as HMW-PAHs in literature concerning microbial biodegradation (Kanaly and Harayama 2000). Figure 5 shows the chemical structures of some commonly studied LMW-PAHs such as naphthalene, fluorene, anthracene and phenanthrene as well as pyrene and benzo(a)pyrene as representatives of the HMW-PAHs.

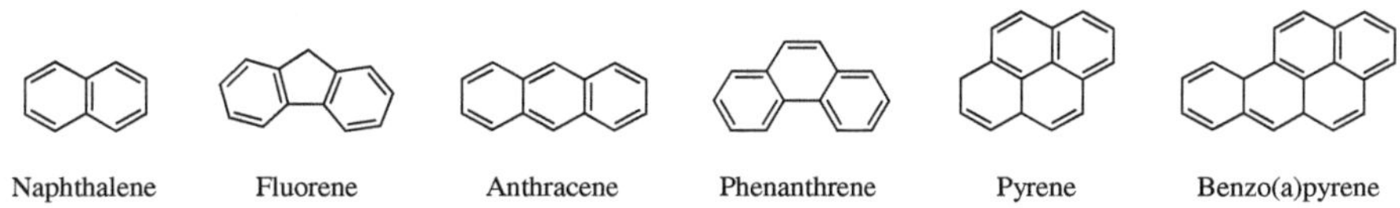

Fig. 5 Structure of selected polycyclic aromatic hydrocarbons

6.1 Biotransformation of LMW-PAHs

Naphthalene is the simplest PAH and consists of two condensed benzene rings. It can be oxidized by yeasts of the genera *Candida, Debaryomyces, Meyerozyma (Pichia), Rhodotorula, Saccharomyces, Sporidiobolus* and *Wickerhamomyces (Pichia)*. Strains isolated from oil-contaminated material and identified as *Wickerhamomyces anomalus (Pichia anomala), Meyerozyma (Pichia) guilliermondii, Rhodotorula dairenensis, Sporidiobolus salmonicolor* and a number of *Candida* species were described as being able to use naphthalene as sole carbon and energy source (Pan et al. 2004; Hesham et al. 2006b; Deng et al. 2010; Ali et al. 2011). When present in a complex growth medium, naphthalene is oxidized to 1-naphthol, 2-naphthol, *trans*-1,2-dihydroxy-1,2-dihydronaphthalene (Fig. 6) and 4-hydroxy-1-tetralone by strains of *Yarrowia (Candida) lipolytica* and to unidentified products by *Candida maltosa, Candida tropicalis* and *Debaryomyces hansenii.* However, none of the organisms tested could use naphthalene alone for growth. The product naphthalene 1,2-oxide, which has so far not been detected, is presumed to be the intermediate in the formation of either 1-naphthol and 2-naphthol or *trans*-1,2-dihydroxy-1,2-dihydronaphthalene (Cerniglia and Crow 1981). Evidence for the involvement of this intermediate is the ratio of 1-naphthol and 2-naphthol of 97:3 (Cerniglia and Crow 1981) as well as the formation of the dihydrodiol by enzymatic hydration of naphthalene 1,2-oxide (Jerina et al. 1970). 1-Naphthol is also formed by *S. cerevisiae* and *Lindnera jadinii (Candida utilis)* (Fig. 6) in the presence of glucose as carbon source suggesting that the oxidation of naphthalene may be independent of the *n*-alkane hydroxylase system (Hofmann 1986).

The degradation of fluorene, a molecule comprising of two benzene rings linked by a cyclopentane ring, has been described for single strains and mixed yeast cultures. Inoculation of a mixed culture of *Candida maltosa*-like yeast, *Wickerhamomyces anomalus* (*Pichia anomala*), *Meyerozyma* (*Pichia*) *guilliermondii, Rhodotorula dairenensis*, and *Sporidiobolus salmonicolor* resulted in the degradation of 91% of the fluorene provided in a mixture of LMW-PAHs and HMW-PAHs (see below) as judged by mass balance analyses (Hesham et al. 2006a).

Garon et al. (2000) tested the ability of 30 fungal strains including two yeasts for their ability to degrade fluorene. While *Rhodotorula mucilaginosa (R. rubra)* did not form any products at all, *Cryptococcus albidus* oxidized fluorene to 9-fluorenol and 9-fluorenone. *Trichosporon mucoides* also oxidized fluorene to the same products but could additionally hydroxylate one aromatic ring of 9-fluorenone at an unknown position as shown by mass-spectrometry analyses (Sietmann 2002).

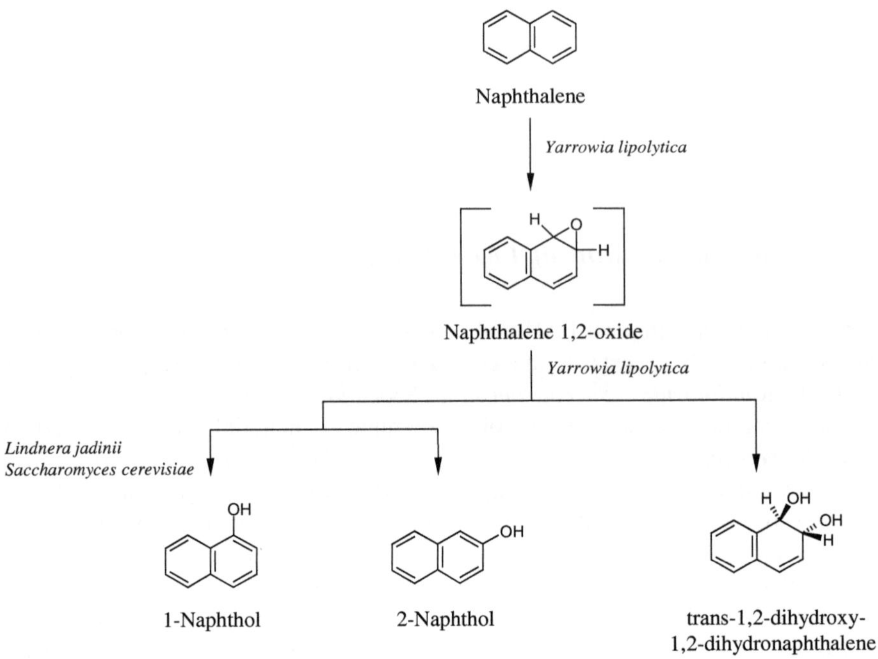

Fig. 6 Proposed pathway for the oxidation of naphthalene by yeasts (according to Cerniglia and Crow 1981)

The next more complex molecule is anthracene which is composed of three condensed benzene rings. All strains of the three yeast genera *Candida, Meyerozyma* (*Pichia*) and *Rhodotorula* which have been described as being able to grow on anthracene were either isolated from contaminated environmental samples or from extreme habitats (Lahav et al. 2002; Margesin et al. 2003; Ali et al. 2011) and exhibited special properties resulting from their adaptation to the environment they were isolated from. The utilization of anthracene as sole carbon source by *Meyerozyma* (*Pichia*) *guilliermondii* and *Rhodotorula mucilaginosa* was demonstrated by visualization of the reduction of the tetrazolium salt 3-(4,5-dimethyl-2-thiazolyl)-2,5-diphenyl-2H-tetrozolium bromide to formazan (Lahav et al. 2002) or, in the case of strains of *Candida* species, was based on the number of colony forming units formed on this substrate (Margesin et al. 2003). However, no information was provided concerning the intermediates formed during growth.

Saccharomyces cerevisiae and *Lindnera jadinii* (*Candida utilis*) did not mineralize anthracene, but were able to oxidize the molecule to unidentified products in the presence of glucose as carbon and energy source (Hofmann 1986).

Phenanthrene also consists of three condensed benzene rings but with a different arrangement to that seen in anthracene. *Rhodotorula glutinis* can grow on up to 200 mg l^{-1} phenanthrene and seemed to be the most efficient yeast strain for phenanthrene degradation (Romero et al. 1998). Other strains of this genus like *R. mucilaginosa (R. rubra)* (MacGillivray and Shiaris 1993), *R. mucilaginosa* (Lahav et al. 2002), *R. dairenensis* (Hesham et al. 2006a) or three unidentified strains isolated from alpine glacier cryoconite (Margesin et al. 2003) are also known to grow with phenanthrene. Various yeast strains of the genera *Candida, Meyerozyma (Pichia), Rhodotorula, Sporidiobolus, Trichosporon, and Wickerhamomyces* (*Pichia*) can also use this PAH as carbon and energy source (MacGillivray and Shiaris 1993; Margesin et al. 2003; Pan et al. 2004; Hesham et al. 2006a, b; Ali et al. 2011).

The degradation pathway is not described but gas chromatographic analyses demonstrated the introduction of one oxygen atom to the molecule by *Candida digboiensis* (Sood and Lal 2009).

Enrichment cultures from sediments of 13 coastal sites in Massachusetts yielded 31 yeasts strains with the most abundant genera isolated being *Candida, Cryptococcus, Rhodotorula,* and *Trichosporon*. On the basis of polar metabolite formation demonstrated by thin-layer chromatography and the evaluation of the release of $^{14}CO_2$, five phenanthrene-degrading yeasts were confirmed and identified as *Candida krusei, Candida zeylanoides, Candida glabrata, Geotrichum klebahnii (Trichosporon penicillatum)* and *Rhodotorula mucilaginosa (R. rubra)*. *Geotrichum klebahnii* rapidly biotransformed phenanthrene to an extent of 8.15 μmol g^{-1} in 120 h (MacGillivray and Shiaris 1993). In sterile sediment slurries inoculated with the phenanthrene-degrading yeast strains *Candida glabrata, Candida rugosa, Yarrowia (Candida) lipolytica*, and *Rhodotorula minuta* and incubated for 120 h, two potential metabolites have been detected by HPLC analyses and identified as 3-phenanthrol and phenanthrene *trans*-3,4-dihydrodiol (MacGillivray and Shiaris 1994). Because less than 3% of the phenanthrene degradation activity was achieved by yeasts the authors assumed that most of the phenanthrene biodegradation in coastal sediment is based on prokaryotic activity.

Another potent habitat for PAH-degrading yeasts is oil-contaminated soil. Rapid growth on agar plates coated with phenanthrene was demonstrated for five yeast strains isolated from oily soil samples in China (Hesham et al. 2006a, b) and these were identified as *Sporidiobolus salmonicolor, Rhodotorula dairenensis, Meyerozyma (Pichia) guilliermondii, Wickerhamomyces anomalus (Pichia anomala),* and a *Candida maltosa*-like strain. Pregrowth with glucose resulted in the removal within 7 days of 50–97% of phenanthrene by the *C. maltosa*-like strain, *M. guilliermondii*, and *W. anomalus* while *S. salmonicolor* and *R. dairenensis* removed only about 16–21% in the same time. However, after cell cultivation with naphthalene there were no differences in phenanthrene degradation rates. The cell surface hydrophobicity of the yeast strains increased after cultivation with naphthalene compared to growth with glucose and reached values comparable with those determined after cultivation with hexadecane. The higher the cell hydrophobicity the higher the PAH degradation rate. Because of the fact that there was no relationship between emulsification activity and PAH degradation rate the authors concluded that increased cell hydrophobicity might be a PAH adaptation strategy of these five yeast strains (Deng et al. 2010).

6.2 *Biotransformation of HMW-PAHs*

Pyrene and benzo(a)pyrene are four- and five-ring HMW-PAHs whose degradation by yeasts has been investigated. Most of the organisms able to degrade LMW-PAHs are also able to oxidize HMW-PAHs as well.

In the case of pyrene most of the degradative yeasts are strains of the genera *Candida* and *Rhodotorula*. *Rhodotorula glutinis* and *R. minuta* isolated from sediments near an oil refinery were shown to be able to mineralize [^{14}C]pyrene in liquid culture. The pyrene consumption rate by *R. glutinis* was improved in the presence of glucose. Both strains degraded 35% of the initial pyrene concentration applied (Romero et al. 2002b) as did *Dekkera bruxellensis* and *Candida* spec. isolated from seawater from Tokyo Bay (Ren et al. 2004). Wang et al. (2007) investigated the influence of immobilization of cells of three *Candida tropicalis* strains on the pyrene degradation potential in a soil slurry bioreactor. Immobilization increased the degradation rate with physical immobilization being more effective than chemical immobilization.

Furthermore, enrichment cultures of acidic oily sludge-contaminated soil resulted in the isolation of *Candida digboiensis* which was able to oxidize pyrene to 1-pyrenol and 2-pyrenol (Sood et al. 2010).

The ability of brewer's yeast to degrade benzo(a)pyrene was investigated already in 1970s. In *Saccharomyces cerevisiae* the cytochrome P450 enzyme system mediated the formation of 3-hydroxybenzo(a)pyrene as the major metabolite. Further metabolites were 9-hydroxybenzo(a)pyrene, 7,8-dihydroxy-7,8-dihydrobenzo(a)pyrene (Fig. 7), a quinone and an unidentified product (Wiseman and Woods 1979). *Yarrowia lipolytica* formed 3-hydroxybenzo(a)pyrene and 9-hydroxybenzo(a)pyrene as well (Cerniglia and Crow 1981). Other yeast strains like *C. tropicalis, C. maltosa, C. krusei* as well as *Debaryomyces hansenii,*

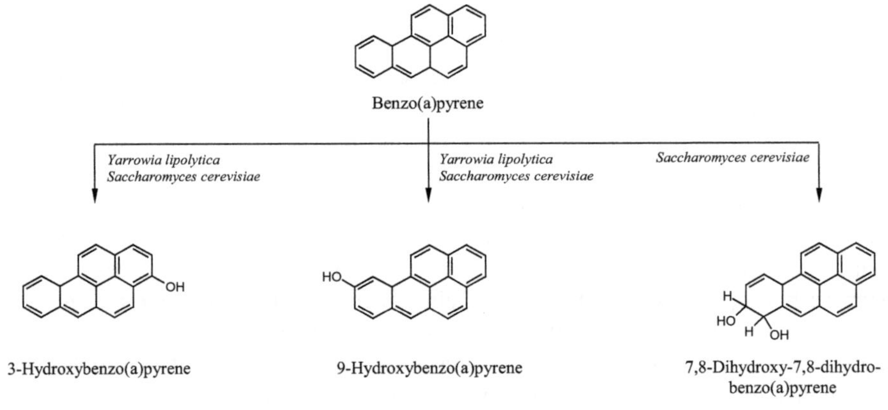

Fig. 7 Proposed pathway for the oxidation of benzo(a)pyrene by *Yarrowia lipolytica* and *Saccharomyces cerevisiae* (according to results of Wiseman and Woods 1979; Cerniglia and Crow 1981)

Table 5 Oxidation and degradation of selected polycyclic aromatic hydrocarbons (PAHs) by different yeast strains

Organism	Utilization of PAHs as carbon and energy source						Oxidation of PAHs						Cometabolic oxidation in the presence of glucose[a] or other PAHs[b,c] in a mixed culture[d]						References
	Naphthalene	Fluorene	Anthracene	Phenanthrene	Pyrene	Benzo (a) pyrene	Naphthalene	Fluorene	Anthracene	Phenanthrene	Pyrene	Benzo (a) pyrene	Naphthalene	Fluorene	Anthracene	Phenanthrene	Pyrene	Benzo (a) pyrene	
Candida digboiensis										+	+								Sood and Lal (2009), Sood et al. (2010)
Candida glabrata										+									MacGillivray and Shiaris (1993, 1994)
Candida krusei										+		+							MacGillivray and Shiaris (1993)
Candida maltosa							+					+							Cerniglia and Crow (1981)
Candida maltosa-like	+			+		+							$+^{d}$	$+^{d}$	$+^{d}$	$+^{d}$	$+^{d}$	$+^{d}$	Hesham et al. (2006a), Deng et al. (2010), Hesham et al. (2012)
Candida rugosa										+									MacGillivray and Shiaris (1994)
Candida spec.	+		+	+	+														Margesin et al. (2003), Ren et al. (2004), Ali et al. (2011)
Candida tropicalis					+		+					+							Cerniglia and Crow (1981), Wang et al. (2007)
Candida viswanathii													$+^{b}$			$+^{b}$	$+^{b}$	$+^{b}$	Hesham et al. (2009)
Candida zeylanoides										+									MacGillivray and Shiaris (1993)
Debaryomyces hansenii							+					+							Cerniglia and Crow (1981)
Dekkera bruxellensis					+														Ren et al. (2004)
Geotrichum klebahnii[1)]										+									MacGillivray and Shiaris (1993)
Lindnera jadinii[2)]													$+^{a}$		$+^{a}$				Hofmann (1986)
Meyerozyma guilliermondii[3)]	+		+	+		+							$+^{d}$	$+^{d}$	$+^{d}$	$+^{d}$	$+^{d}$	$+^{d}$	Lahav et al. (2002), Hesham et al. (2006a), Deng et al. (2010), Hesham et al. (2012)
Saccharomyces cerevisiae												+	$+^{a}$		$+^{a}$				Wiseman and Woods (1979), Hofmann (1986)
Wickerhamomyces anomalus[4)]	+			+		+							$+^{c,d}$	$+^{d}$	$+^{d}$	$+^{c,d}$	$+^{d}$	$+^{c,d}$	Pan et al. (2004), Hesham et al. (2006a), Hesham

(continued)

Table 5 (continued)

Organism	Utilization of PAHs as carbon and energy source						Oxidation of PAHs						Cometabolic oxidation in the presence of glucose[a] or other PAHs[b,c] in a mixed culture[d]						References
	Naphthalene	Fluorene	Anthracene	Phenanthrene	Pyrene	Benzo (a) pyrene	Naphthalene	Fluorene	Anthracene	Phenanthrene	Pyrene	Benzo (a) pyrene	Naphthalene	Fluorene	Anthracene	Phenanthrene	Pyrene	Benzo (a) pyrene	
																			et al. (2006b), Deng et al. (2010), Hesham et al. (2012)
Yarrowia lipolytica							+			+		+							Cerniglia and Crow (1981), MacGillivray and Shiaris (1994)
Rhodotorula dairenensis	+			+		+							+[d]	+[d]	+[d]	+[d]	+[d]	+[d]	Hesham et al. (2006a), Deng et al. (2010), Hesham et al. (2012)
Rhodotorula glutinis				+	+														Romero et al. (1998), Romero et al. (2002b)
Rhodotorula minuta					+					+		+							MacGillivray and Shiaris (1993, 1994), Romero et al. (2002b)
Rhodotorula mucilaginosa			+	+															Lahav et al. (2002)
Rhodotorula mucilaginosa (rubra)										+									MacGillivray and Shiaris (1993)
Rhodotorula spec.			+	+															Margesin et al. (2003)
Sporidiobolus salmonicolor	+			+		+							+[d]	+[d]	+[d]	+[d]	+[d]	+[d]	Hesham et al. (2006a), Deng et al. (2010), Hesham et al. (2012)

[b]Cometabolic oxidation of naphthalene, phenanthrene, pyrene and benzo(a)pyrene according to Hesham et al. (2009);
[c]Cometabolic oxidation of naphthalene, phenanthrene, chrysene and benzo(a)pyrene according to Hesham et al. (2006b);
[d]Cometabolic oxidation of naphthalene, fluorene, phenanthrene, anthracene, pyrene and chrysene by a mixed culture of *Sporidiobolus salmonicolor, Rhodotorula dairenensis, Meyerozyma (Pichia) guilliermondii* [3], *Wickerhamomyces anomalus (Pichia anomala)* [4], *and a Candida maltosa*-like strain according to Hesham et al. (2006a);
[1]*Geotrichum klebahnii (Trichosporon penicillatum)*; [2]*Lindnera jadinii (Candida utilis)*

Rhodotorula minuta and *Yarrowia* (*Candida*) *lipolytica* are also known to be able to oxidize benzo(a)pyrene but the products formed have not so far been identified (Cerniglia and Crow 1981; MacGillivray and Shiaris 1993).

Five yeast strains *Sporidiobolus salmonicolor*, *Rhodotorula dairenensis*, *Wickerhamomyces anomalus* (*Pichia anomala*), *Meyerozyma* (*Pichia*) *guilliermondii*, and a *Candida maltosa*-like strain can grow on agar plates with benzo(a)pyrene as the sole carbon source (Hesham et al. 2012). The results for *Wickerhamomyces anomalus* are contradictory since this strain was described to be able to grow with benzo(a)pyrene on an agar plate whereas in liquid culture only co-metabolic degradation was possible (Hesham et al. 2006b, 2012).

6.3 Biotransformation of a PAH-Mixture

A promising approach to the degradation of PAHs by yeasts is the use of mixed cultures. When LMW-PAHs and HMW-PAHs were given in combination to a mixed culture of *Sporidiobolus salmonicolor*, *Rhodotorula dairenensis*, *Wickerhamomyces anomalus* (*Pichia anomala*), *Meyerozyma* (*Pichia*) *guilliermondii*, and a *Candida maltosa*-like strain, the removal efficiency over a degradation period of 6 weeks was about 90–98% for the LMW-PAHs and about 66–89% for the HMW-PAHs. During this process the five-ring and six-ring PAHs are degraded co-metabolically (Hesham et al. 2006a).

Pure cultures are also known to degrade a mixture of low and high molecular weight PAHs. Thus, *Wickerhamomyces anomalus* (*Pichia anomala*) in liquid culture degraded within 10 days a mixture of naphthalene, phenanthrene, chrysene and benzo(a)pyrene. In comparison to the rate of degradation of single PAHs by this strain, the naphthalene and phenanthrene degradation in the PAH mixture was delayed. From these results the authors concluded that there might be metabolic competition between the LMW-PAHs and the HMW-PAHs (Hesham et al. 2006b). *Candida viswanathii* also efficiently degraded a mixture of PAHs (Hesham et al. 2009).

Table 5 summarizes the mineralization and transformation of selected PAHs by different yeast species.

7 Biotransformation of Biarylic Compounds

The class of biarylic compounds includes biphenyl, dibenzofuran, and diphenyl ether all of which are substances with two aromatic ring systems connected via C–C- and/or C–O-bonds. These compounds and their halogenated derivatives are considered to be toxic and can have a severe impact on human health (Safe 1984; Skene et al. 1989; Connor et al. 1997; Schiestl et al. 1997; Korner et al. 1998; Andersson et al. 1999). All three compounds and their halogenated derivatives are ubiquitous in the environment either because they are extensively used or because of their formation as unwanted side products in industrial processes. Biphenyl is commercially used as a fungistat in packaging of citrus fruits, as a heat transfer fluid, as a dye carrier, and as an intermediate in the synthesis of polychlorinated biphenyls. Diphenyl ether is used as an ingredient of perfume, as a heat transfer medium (Wittich 1992) and as a herbicide component (Yoshimoto et al. 1990). However, though dibenzofuran and halogenated derivatives have never been deliberately produced for industrial usage, they are nonetheless widespread in the environment due to their formation as trace contaminants in many industrial and thermal processes (Fiedler 1996).

Many yeast species with the ability to oxidize biarylic compounds have been isolated from polluted habitats (Hammer et al. 1998; Romero et al. 2001, 2002a), though there are only very few hints, that yeasts might be able to mineralize some biarylic compounds (Romero et al. 2006; Ali et al. 2011). However, biarylic compounds are substrates for biotransformation and many of the products are either no longer toxic or they can serve as substrates for other microorganisms. Thus yeasts can substantially contribute to microbial biodegradation of environmental pollutants.

7.1 *Biotransformation of Diphenyl Ether*

The first report of the oxidation of biarylic compounds by yeasts appeared in 1995 (Schauer et al. 1995). The authors isolated *Trichosporon domesticum* (*T. beigelii*) from soil contaminated with car exhaust and investigated the ability of this strain to oxidize diphenyl ether (DPE). The initial attack of the molecule resulted in a 47:5:48 mixture of 2-hydroxydiphenyl ether, 3-hydroxydiphenyl ether, and 4-hydroxydiphenyl ether. Further oxidation led to the formation of 3,4-dihydroxydiphenyl ether which was a substrate for ring fission resulting in the ring cleavage product 6-carboxy-4-phenoxy-2-pyrone. This was the first report of ring fission of biarylic compounds by yeasts. The potential of other *Trichosporon* yeasts to oxidize DPE was subsequently investigated. The hydroxylation of DPE by 24 strains belonging to 18 species mainly resulted in 2-hydroxydiphenyl ether and 4-hydroxydiphenyl ether. However, five strains did not accumulate 2-hydroxydiphenyl ether (Table 6). These monohydroxylated compounds were oxidized to 2,5-dihydroxydiphenyl ether and 3,4-dihydroxydiphenyl ether. Ring fission of 3,4-dihydroxydiphenyl ether resulted in the formation of 6-carboxy-4-phenoxy-2-pyrone. The ability to carry out ring cleavage was demonstrated for 16 strains belonging to 13 *Trichosporon* species (Sietmann et al. 2002). Mass-spectrometry analyses revealed the formation of additional transformation products formed by *T. mucoides* isolated by Hammer et al. (1998). Besides 6-carboxy-4-phenoxy-2-pyrone, the probable precursor molecule

Table 6 Products formed during the transformation of dibenzofuran and diphenyl ether by different yeast species

Strain	Transformation of dibenzofuran						Transformation of diphenyl ether					References
	Monohydroxylated products				Dihydroxylated product	Ring fission product	Monohydroxylated products		Dihydroxylated products		Ring fission product	
	1-OH[1)]	2-OH[1)]	3-OH[1)]	4-OH[1)]	2,3-diOH[2)]	CMDG	2-OH[3)]	4-OH[3)]	2,5-DiOH[4)]	3,4-DiOH[4)]	CCP	
Candida krusei			+	+								Romero et al. (2002a)
Candida tenuis			+									Romero et al. (2002a)
Candida tropicalis		+	+	+								Romero et al. (2002a)
Cryptococcus curvatus	+	+	+				+	+			+	Sietmann et al. (2002)
Cryptococcus humicola		+	+					+				Sietmann et al. (2002)
Guehomyces pullulans[1)]			+	(+)								Romero et al. (2002a)
Priceomyces haplophilus[2)]			+									Romero et al. (2002a)
Rhodotorula glutinis		(+)	+	+	+	+						Romero et al. (2002a)
Rhodotorula mucilaginosa			+	+								Romero et al. (2002a)
Schwanniomyces vanrijiae[3)]	+	+	+	+	+	+	+	+	+	+	+	Lange (1997)
Trichosporon aquatile	+	+	+			+	+	+			+	Sietmann et al. (2002)
Trichosporon asahii		+	+				+	+				Sietmann et al. (2002)
Trichosporon brassicae							+	+		+	+	Sietmann et al. (2002)
Trichosporon coremiiforme	+	+	+			+	+	+			+	Sietmann et al. (2002)
Trichosporon cutaneum		+	+					+				Sietmann et al. (2002)
Trichosporon domesticum		+	+				+	+	(+)	(+)	+	Sietmann et al. (2002), Schauer et al. (1995)
Trichosporon guehoae		+	+				+	+			+	Sietmann et al. (2002)
Trichosporon inkin	+	+	+				+	+				Sietmann et al. (2002)
Trichosporon japonicum		+			+		+	+	+			Sietmann et al. (2002)
Trichosporon jirovecii		+	+					+				Sietmann et al. (2002)

(continued)

Table 6 (continued)

Trichosporon laibachii		+	+				+	+	+	+	+	Sietmann et al. (2002)
Trichosporon loubieri	+	+	+				+	+			+	Sietmann et al. (2002)
Trichosporon moniliiforme		+	+				+	+	(+)		(+)	Sietmann et al. (2002)
Trichosporon montevideense		+	+			+		+		+	+	Sietmann et al. (2002)
Trichosporon mucoides	+	+	+	+	+	+	+	+		(+)	+	Hammer et al. (1998), Sietmann et al. (2002)
Trichosporon porosum		(+)	(+)				(+)	+			(+)	Sietmann et al. (2002)
Trichosporon sporotrichoides		+	+			+	+	+		+	+	Sietmann et al. (2002)
Trichosporon veenhuisii		+	+				+	+	+		+	Sietmann et al. (2002)
Wickerhamomyces anomalus[4)]			+	+								Romero et al. (2002a)
Yarrowia lipolytica		+	+	+	+	+						Romero et al. (2002a)

+ transformation product detected by different analytical methods according to the references; (+)formation of this product is strain-specific; [1)]OH, hydroxydibenzofuran; [2)]diOH, dihydroxydibenzofuran; CMDG, 2-(1-carboxy methylidene)-2,3-dihydrobenzobfuranylidene glycolic acid; [3)]OH, hydroxydiphenyl ether; [4)]diOH, dihydroxydiphenyl ether; CPP, 6-carboxy-4-phenoxy-2-pyrone; [1)]*Guehomyces pullulans* (*Trichosporon pullulans*); [2)]*Priceomyces haplophilus* (*Pichia haplophila*); [3)]*Schwanniomyces vanrijiae* (*Debaryomyces vanrijiae*); [4)]*Wickerhamomyces anomalus* (*Pichia anomala*)

2-hydroxy-4-phenoxymuconic acid was also detected. This product had earlier been postulated as part of the transformation pathway of DPE by *T. domesticum* (*T. beigelii*) but had not been positively identified (Schauer et al. 1995). Furthermore, 4-phenoxymuconic acid and a triple hydroxylated intermediate with two adjacent hydroxyl groups on one ring and one additional hydroxyl group on the other were shown by mass-spectrometry analyses using 3,4-dihydroxydiphenyl ether as substrate. The oxidation of 2,5-dihydroxydiphenyl ether by *T. mucoides* led to the formation of a trihydroxylated product in which all three hydroxyl groups were assumed to be on one ring (Sietmann 2002; Fig. 8). Furthermore, *Cryptococcus curvatus* and *Schwanniomyces* (*Debaryomyces*) *vanrijiae* (Lange 1997) could also cleave the aromatic ring system of DPE with 6-carboxy-4-phenoxy-2-pyrone as resulting product, while *Cryptococcus humicola* only formed 4-hydroxydiphenyl ether (Sietmann et al. 2002).

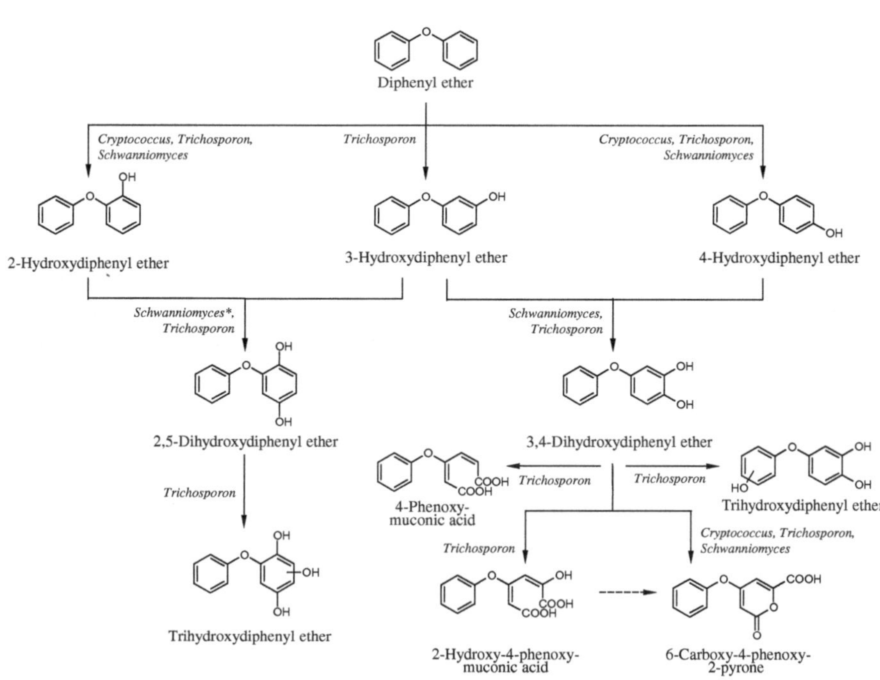

Fig. 8 Main transformation pathway of diphenyl ether by different yeast species of the genera *Cryptococcus, Schwanniomyces (Debaryomyces), and Trichosporon* (according to Schauer et al. 1995; Lange 1997; Sietmann 2002; Sietmann et al. 2002); *2,5-Dihydroxydiphenyl ether is formed by *Schwanniomyces* (*Debaryomyces*) *vanrijiae* only with 2-hydroxydiphenyl ether as substrate (Lange 1997)

7.2 Biotransformation of Dibenzofuran and Chlorinated Derivatives

Dibenzofuran (DBF) consists of a central furan ring combined with two condensed benzene rings and is a much more rigid molecule than is diphenyl ether. No eukaryotic microorganisms have been shown to be able to use DBF for growth. In 1998, a yeast isolated form a dioxin-contaminated soil sample was shown to be able to oxidize DBF and was identified as *Trichosporon mucoides* (Hammer et al. 1998). The initial oxidation step by this strain was the hydroxylation of the molecule at all possible positions. The hydroxylation of 2-hydroxydibenzofuran and 3-hydroxydibenzofuran resulted in the formation of 2,3-dihydroxydibenzofuran. This dihydroxylated compound was then a substrate for ring fission which resulted in 2-(1-carboxymethylidene)-2,3-dihydrobenzo[*b*]furanylidene glycolic acid as product. The formation of this ring fission product required the introduction of a third hydroxyl group into the dihydroxylated intermediate (Hammer et al. 1998). The same ring fission product was formed by various strains of *Rhodotorula glutinis* (Romero et al. 2002a) and *Yarrowia lipolytica* (Romero et al. 2002a; Zinjarde et al. 2014) as well as by *Schwanniomyces* (*Debaryomyces*) *vanrijiae* (Lange 1997). Further investigation with the strain *Trichosporon mucoides* SBUG 801 isolated by Hammer et al. (1998) showed the formation of many more ring fission products with muconic acid as well as lactone structures (Sietmann 2002; Fig. 9).

Different species of the genera *Candida, Cryptococcus, Guehomyces (Trichosporon pullulans), Priceomyces (Pichia haplophila), Rhodotorula, Trichosporon,* and *Wickerhamomyces (Pichia anomala)* are able to introduce one hydroxyl group into the molecule at different positions (Romero et al. 2002a; Sietmann et al. 2002; Table 6; Fig. 9) with 3-hydroxydibenzofuran as the main product. Investigation of 24 strains belonging to 18 species of the genus *Trichosporon* showed that six strains representing the five species *T. sporotrichoides*, *T. montevideense*, *T. mucoides*, *T. aquatile*, and *T. coremiiforme* had the ability to cleave the aromatic ring of DBF (Table 6). However, *T. brassicae* and one strain of *T. porosum* did not transform DBF at all (Sietmann et al. 2002).

There is almost no information about the oxidation of chlorinated DBF by yeasts. *Trichosporon mucoides* formed three monohydroxylated intermediates with 3-chlorodibenzofuran though the position of the hydroxyl group in these products is unknown. Even though no dihydroxylated intermediates were detected, ring fission did take place. The formation of 3-(2-carboxy-vinyl)-6-chloro-benzofuran-2-carboxylic acid and (3-carboxymethylene-6-chloro-3H-benzofuran-2-ylidene)-acetic, respectively, and 2-(2-carboxy-2-chloro-vinyl)-benzofuran-3-carboxylic acid was shown by mass-spectrometry analyses which verified that both the chlorinated and the non-chlorinated ring were starting points for an oxidative attack. Products with lactone structures did not accumulate (Sietmann 2002).

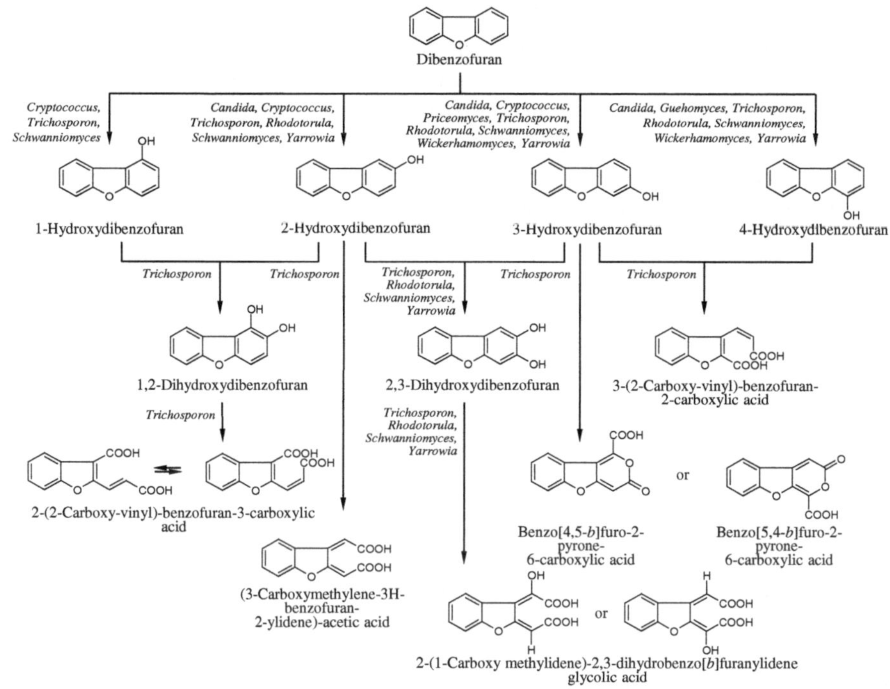

Fig. 9 Main transformation pathway of dibenzofuran by different yeast species of the genera *Candida, Cryptococcus, Guehomyces (Trichosporon pullulans), Priceomyces (Pichia haplophila), Rhodotorula, Schwanniomyces (Debaryomyces vanrijiae), Trichosporon, Wickerhamomyces (Pichia anomala)*, and *Yarrowia* (according to Lange 1997; Hammer et al. 1998; Romero et al. 2002a; Sietmann 2002)

7.3 *Biotransformation of Dibenzo-*p*-Dioxin and Chlorinated Derivatives*

The dibenzo-*p*-dioxins and dioxin-like compounds are tricyclic, essentially planar aromatic and mostly halogenated compounds and include not only the extremely toxic 2,3,7,8-tetrachlorodibenzo-*p*-dioxin (TCDD) but also some higher and less chlorinated congeners of reduced toxicity. The term dioxin-like also include brominated analogues, as well as other halogenated and non-chlorinated dibenzo-*p*-dioxin (DD) compounds. Polychlorinated dibenzo-*p*-dioxins (PCDDs) are ubiquitous environmental contaminants and are emitted into the environment as unwanted byproducts during the production of herbicides, trichlorophenols and pentachlorophenol, during metal smelting, in bleaching processes in the pulp and paper industry, and by incineration of municipal and industrial waste. Since PCDDs are relatively stable and lipophilic, their accumulation in soils and transfer into the food chain and then into humans and animals has been documented.

Aerobic biodegradation of these compounds seems to be restricted to dioxins with a low degree of chlorination. In addition to physicochemical treatment such as high-temperature incineration, elimination of highly chlorinated dioxins can be achieved by anaerobic bacteria which carry out reductive dehalogenation (Mohn and Tiedje 1992; Ballerstedt et al. 1997; Bunge and Lechner 2009; Zanaroli et al. 2015). The products containing fewer halogen substituents are more susceptible to further degradation by anaerobic and aerobic bacteria, filamentous fungi or yeasts (Wittich 1998; Chang 2008; Colquhoun et al. 2012). Degradative pathways for chlorinated dioxins by white-rot fungi demonstrate that oxidative, reductive, and methylation reactions are involved and that P450-oxygenases and ligninolytic enzymes such as lignin peroxidase and manganese peroxidase can play important roles (Valli et al. 1992; Takada et al. 1996; Field and Sierra-Alvarez 2008; Kasai et al. 2010). Reports showing an oxidative or degradative attack on dioxin-like compounds by yeasts are scarce. The yeast strain *Trichosporon domesticum* SBUG 752 (formerly designated as *T. beigelii*) isolated from polluted soil is able to oxidize non-chlorinated dibenzo-*p*-dioxin to hydroxylated derivatives (Henning 1993) as small amounts of hydroxydibenzo-*p*-dioxin were detected by GC-MS analyses. However, because of the trace amounts of this product the exact position of the hydroxyl group on the dioxin molecule could not be determined. After incubation of dibenzo-*p*-dioxin with cells of the yeast *Trichosporon mucoides* SBUG 801 two different monohydroxylated metabolites and one dihydroxylated dibenzo-*p*-dioxin were detected. Again, the exact position of the hydroxyl groups could not be determined because of the small amounts of products formed (Sietmann 2002; Fig. 10).

Fig. 10 Oxidation of dibenzo-*p*-dioxin by *Trichosporon mucoides* (according to results of Sietmann 2002)

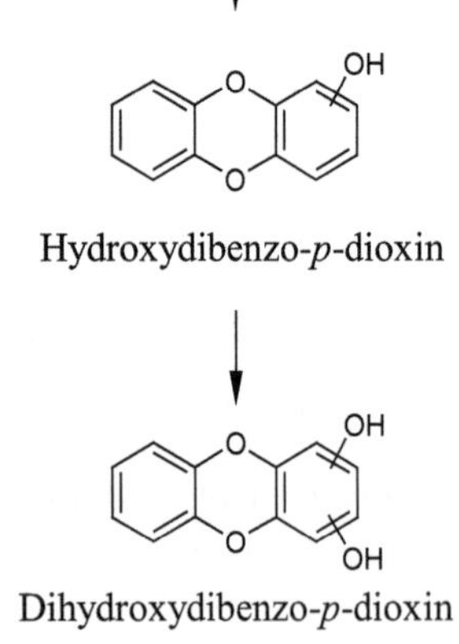

In the yeast *Schwanniomyces (Debaryomyces) vanrijiae* SBUG 770 hydroxylated products from dibenzo-*p*-dioxin were not found, though this yeast was able to oxidize dibenzofuran to 5 different hydroxylated products and one ring cleavage product (Lange 1997). Biodegradation of polychlorinated dibenzo-*p*-dioxins (PCDDs) by recombinant yeast cells expressing cytochrome-P450 proteins from animals was examined by Sakaki et al. (2002). Hydroxylation reactions and cleavage of ether bridges were observed. The authors claim that this study indicates the possibility of the application of yeast cells expressing a foreign cytochrome P450-system to bioremediation of areas contaminated with dioxins.

7.4 Biotransformation of Biphenyl and Chlorinated Derivatives

Until recently, the white rot fungus *Phanerochaete chrysosporium* was the only eukaryotic microorganism known to be able to mineralize biphenyl (Thomas et al. 1992). In 2011, two *Candida* species which may be able to use biphenyl as growth substrate were isolated from sawdust, a waste product of wood processing (Ali et al. 2011). Other yeast species reported in the literature are only able to transform the molecule (Table 7). Yeasts of the genus *Trichosporon* have been best investigated with regard to their oxidation of biarylic compounds (Sietmann et al. 2002). The first step in the oxidation of biphenyl is the introduction of one hydroxyl group into the molecule resulting in either 2-hydroxy-, 3-hydroxy- or 4-hydroxybiphenyl. The oxidation in the *para*-position to 4-hydroxybiphenyl is the favored reaction and is carried out by yeasts of the genera *Candida, Cryptococcus, Guehomyces (Trichosporon pullulans), Priceomyces (Pichia haplophila), Rhodotorula, Saccharomyces, Schwanniomyces (Debaryomyces vanrijiae), Trichosporon*, and *Yarrowia*. In case of rapid further oxidation 4-hydroxybiphenyl is not detected by HPLC analyses as has been shown for some *Trichosporon* species. Further oxidation led to dihydroxylated and trihydroxylated biphenyl derivatives with the second hydroxyl group either on the initially hydroxylated or on the non-hydroxylated ring system (Table 7). The inhibition of primary oxidation steps in the presence of the cytochrome P450 inhibitor 1-aminobenzotriazole indicated that the incorporation of molecular oxygen was catalyzed by a cytochrome P450 enzyme system (Sietmann et al. 2000).

Many yeasts of the genus *Trichosporon* are able to cleave the aromatic ring system. This mechanism has been investigated in detail for *T. mucoides* which utilized 3,4-dihydroxybiphenyl and 2,3-dihydroxybiphenyl with two hydroxyl groups as substrates for ring fission (Sietmann et al. 2001). The *ortho*-ring fission of 3,4-dihydroxybiphenyl resulted in the main ring cleavage products 4-phenyl-2-pyrone-6-carboxylic acid and (5-oxo-3-phenyl-2,5-dihydrofuran-2-yl) acetic acid while 3-phenyl-2-pyrone-6-carboxylic acid is the major product after cleavage of the aromatic ring system of 2,3-dihydroxybiphenyl. The formation of

Table 7 Products formed during the transformation of biphenyl by different yeast species

Strain	Monohydroxylated products			Dihydroxylated products					Ring fission products			References
	2-OH	3-OH	4-OH	2,3-diOH	2,5-diOH	3,4-diOH	2,4′-diOH	4,4′-diOH	3-PPC	4-PPC	OPDA	
Candida krusei		+	+			+						Romero et al. (2001)
Candida tenuis			+									Romero et al. (2001)
Candida tropicalis		+	+			+						Romero et al. (2001)
Cryptococcus curvatus	+		+		+		+				+	Sietmann et al. (2002)
Cryptococcus humicola			+									Sietmann et al. (2002)
Guehomyces pullulans[1)]			+									Romero et al. (2001)
Priceomyces haplophilus[2)]		+	+			+						Romero et al. (2001)
Rhodotorula glutinis		+	+	+		+				+		Romero et al. (2001)
Saccharomyces cerevisiae			+									Wiseman et al. (1975), Layton et al. (2002)
Schwanniomyces vanrijiae[3)]	+		+		+	+				+		Lange et al. (1998)
Trichosporon aquatile	+				+							Sietmann et al. (2002)
Trichosporon asahii	+	+			+	+				+	+	Sietmann et al. (2002)

(continued)

Table 7 (continued)

Strain	Monohydroxylated products			Dihydroxylated products					Ring fission products			References
	2-OH	3-OH	4-OH	2,3-diOH	2,5-diOH	3,4-diOH	2,4′-diOH	4,4′-diOH	3-PPC	4-PPC	OPDA	
Trichosporon brassicae			+									Sietmann et al. (2002)
Trichosporon coremiiforme	+	+			+					+	+	Sietmann et al. (2002)
Trichosporon cutaneum	+		(+)									Sietmann et al. (2002)
Trichosporon domesticum	+		+		+	(+)	+		(+)	+	+	Sietmann et al. (2002)
Trichosporon guehoae	+		+							+		Sietmann et al. (2002)
Trichosporon inkin	+	+	+									Sietmann et al. (2002)
Trichosporon japonicum	+	+			+	+				+		Sietmann et al. (2002)
Trichosporon jirovecii			+									Sietmann et al. (2002)
Trichosporon laibachii			+							+		Sietmann et al. (2002)
Trichosporon loubieri			+		+				+	+	+	Sietmann et al. (2002)
Trichosporon moniliiforme	+		+									Sietmann et al. (2002)
Trichosporon montevideense	+		+		+		+		+	+		Sietmann et al. (2002)

(continued)

Table 7 (continued)

Strain	Monohydroxylated products			Dihydroxylated products					Ring fission products			References
	2-OH	3-OH	4-OH	2,3-diOH	2,5-diOH	3,4-diOH	2,4′-diOH	4,4′-diOH	3-PPC	4-PPC	OPDA	
Trichosporon mucoides	+	+	+	+	+		(+)		(+)	+	+	Sietmann et al. (2000, 2002)
Trichosporon porosum			(+)							(+)		Sietmann et al. (2002)
Trichosporon sporotrichoides			+		+		+		+	+		Sietmann et al. (2002)
Trichosporon veenhuisii			+							+	+	Sietmann et al. (2002)
Yarrowia lipolytica	(+)	(+)	+			(+)		(+)		(+)		Cerniglia and Crow (1981), Romero et al. (2001)

+ transformation product detected by different analytical methods according to the references; (+)formation of this product is strain-specific; *OH* hydroxybiphenyl; *diOH*, dihydroxybiphenyl; *3-PPC* 3-phenyl-2-pyrone-6-carboxylic acid; *4-PPC* 4-phenyl-2-pyrone-6-carboxylic acid; *OPDA* (5-oxo-3-phenyl-2,5-dihydrofuran-2-yl)-acetic acid; 1)*Guehomyces pullulans* (*Trichosporon pullulans*); 2)*Priceomyces haplophilus* (*Pichia haplophila*); 3)*Schwanniomyces vanrijiae* (*Debaryomyces vanrijiae*)

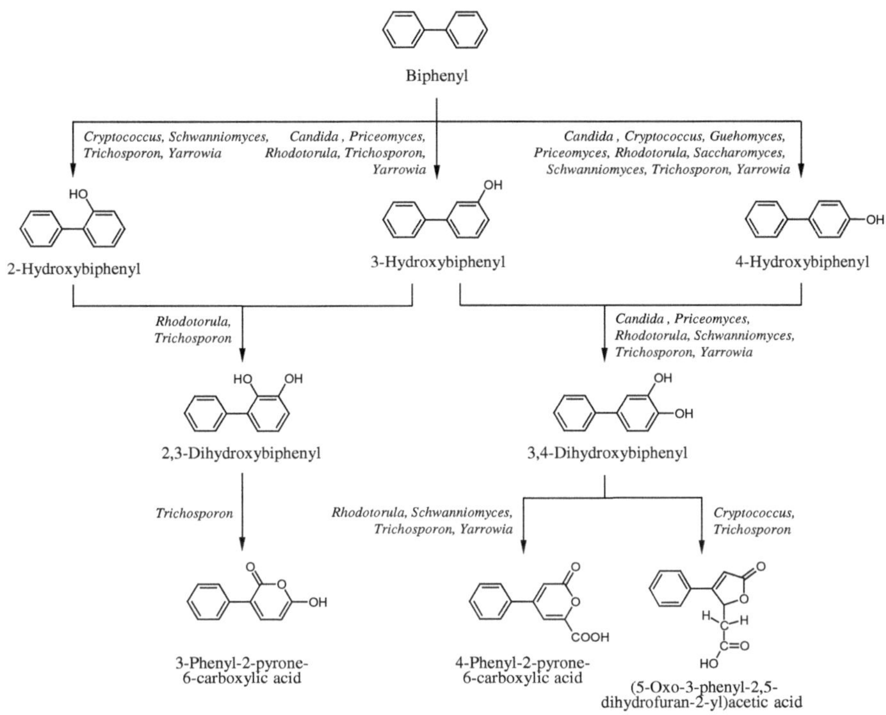

Fig. 11 Main transformation pathway of biphenyl by different yeast species of the genera *Candida, Cryptococcus, Guehomyces (Trichosporon pullulans), Priceomyces (Pichia haplophila), Rhodotorula, Saccharomyces, Schwanniomyces (Debaryomyces vanrijiae), Trichosporon, and Yarrowia* (according to Wiseman et al. 1975; Cerniglia and Crow 1981; Lange et al. 1998; Sietmann et al. 2000; Romero et al. 2001; Sietmann et al. 2001, 2002)

both carboxylic acids required the introduction of a third adjacent hydroxyl group prior to ring fission, while (5-oxo-3-phenyl-2,5-dihydrofuran-2-yl)acetic acid is formed from 3,4-dihydroxybiphenyl as substrate (Schlueter et al. 2013) in a reaction analogous to the cleavage of catechol (Vollmer et al. 1994). There are many other ring cleavage products formed by *T. mucoides* which are unique so far for eukaryotic microorganisms. Ring fission occurred from substrates with hydroxyl groups on one or both aromatic rings resulting in products with hydroxymuconic acid structure. Preliminary stages of the lactones formed were trihydroxylated intermediates with all substituents on one aromatic ring (Sietmann et al. 2001). The main ring cleavage products were also formed by other *Trichosporon* species (Sietmann et al. 2002), *Cryptococcus curvatus* (Sietmann et al. 2002), *Schwanniomyces (Debaryomyces) vanrijiae* (Lange et al. 1998) and *Rhodotorula glutinis* (Romero et al. 2001) (Table 7; Fig. 11).

The mechanism of biphenyl oxidation is comparable to that for the microbial degradation of phenol and phenol-grown cells of *T. mucoides* oxidized biphenyl faster than glucose-grown cells. Because of this, Schlueter et al. (2013) raised the question of whether the enzymes involved in phenol degradation also catalyze biphenyl oxidation. The purification of enzymes with catechol cleavage activity from *T. mucoides* yielded catechol-1,2-dioxygenase and two novel enzymes with high affinity for 3,4-dihydroxybiphenyl. Since only monoaromatic compounds were substrates for the catechol-1,2-dioxygenase, but 3,4-dihydroxybiphenyl was not, the authors concluded that the enzymes of phenol degradation do not catalyze ring cleavage of biphenyl in this yeast.

The yeast strains use ring cleavage of biphenyl as a mechanism to detoxify the substrate. Biphenyl passes the cell envelope because of its hydrophobicity and must then be transformed into water-soluble intermediates which can be actively excreted. The hydroxylated intermediates formed by *T. mucoides* were even more toxic than biphenyl, but the ring fission products did not have toxic effects. In this way the yeast cells effectively detoxify the biarylic compound (Sietmann et al. 2001). Conjugate formation, which has been described as a detoxification mechanism of biarylic compounds (Dodge et al. 1979; Golbeck et al. 1983) and PAHs (Pothuluri et al. 1990; Sutherland et al. 1991) by filamentous fungi, has not been described for yeasts.

Not only biphenyl but also chlorinated derivatives might be growth substrates for yeasts. Enrichment cultures with 2,4-dichlorobiphenyl, 2,3′,4- and 2,4′,5-trichlorobiphenyl as sole carbon source, inoculated with surface sediments, yielded three different yeast strains identified as *Schwanniomyces capriottii* (*Debaryomyces castellii*), *Debaryomyces marasmus*, and *Dipodascus aggregatus* (Romero et al. 2006). However, these results are difficult to assess because media supplements as well as phenol in the biphasic system might also allow microbial growth.

Other studies demonstrated the transformation of chlorinated biphenyls by *Saccharomyces cerevisiae* and *Trichosporon mucoides*. *S. cerevisiae* introduced a hydroxyl group in the *para*-position of the non-chlorinated ring in 4-chlorobiphenyl and di- and trichlorobiphenyl derivatives (Layton et al. 2002). In contrast, *T. mucoides* additionally formed *ortho*- and *meta*-hydroxylated derivatives and can hydroxylate the chlorinated as well as the unchlorinated ring (Sietmann et al. 2006). Starting from a compound chlorinated at the C4 position, *T. mucoides* hydroxylated the non-chlorinated ring to form mono- and dihydroxylated intermediates which are substrates for ring fission of the non-halogenated ring resulting in hydroxymuconic acids and the corresponding lactones (Fig. 12). This transformation pathway occurred analogous to the oxidation of biphenyl by this yeast (Sietmann et al. 2000, 2001).

2-Chloro-4-hydroxybiphenyl is also a substrate for ring fission by *T. mucoides* (Fig. 12) but lactones are not formed (Sietmann 2002) probably due to (i) steric hindrance and (ii) the requirement of a third hydroxyl group at this position to form one of the phenyl-pyrone-carboxylic acids (see above).

While *S. cerevisiae* was able to oxidize di- and trichlorobiphenyls as well (Layton et al. 2002), such compounds, with the exception of 2,2′-dichlorobiphenyl, could not be oxidized by *T. mucoides* (Sietmann 2002, 2006).

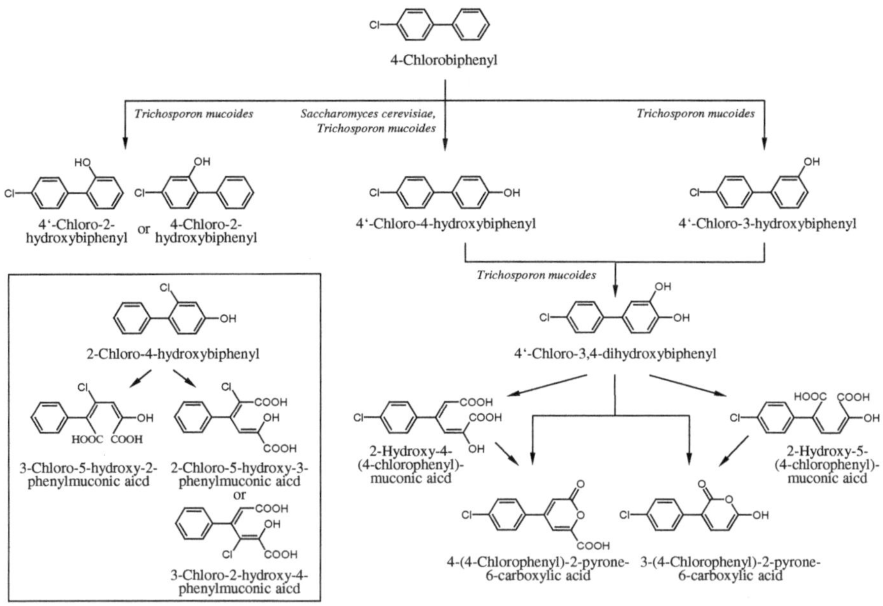

Fig. 12 Main transformation pathway of 4-chlorobiphenyl by *Saccharomyces cerevisiae* (according to Layton et al. 2002) and *Trichosporon mucoides* (according to Sietmann et al. 2006) and transformation of 2-chloro-4-hydroxybiphenyl by *Trichosporon mucoides* (according to Sietmann 2002)

8 Biotransformation of disinfectants

Hospital disinfection is crucial to prevent nosocomial infections and thus the use of disinfectants is indispensable. However, disinfectants are used not only in hospitals but also in households, in the food industry and in livestock farming. The varied use results in an accumulation of disinfectants in municipal waste water. Studies in 8 hospitals by Gartiser et al. (2000) have shown that the average total use of disinfectants—excluding alcohol-based ones—is around 4.4 g per bed and day, which leads to a final concentration of about 9 mg per liter in the waste water.

These disinfectants reach the sewage treatment plant with the wastewater. If they are not degraded they can either directly access the ground water and drinking water or they may be introduced into the soil via sewage sludge and reach the ground and drinking water via this route as well.

With regard of the selection of disinfectants, there is still a great imbalance between considerations of effectiveness and environmental impact. The antimicrobial effect is obviously much more important than questions of environmental sustainability.

The biodegradability of the active ingredients of different disinfectants is very varied. While alcohols, aldehydes and peroxide compounds can be easily degraded, aromatic compounds and halogenated aromatics have a medium biodegradability while certain other compounds are scarcely biodegradable at all.

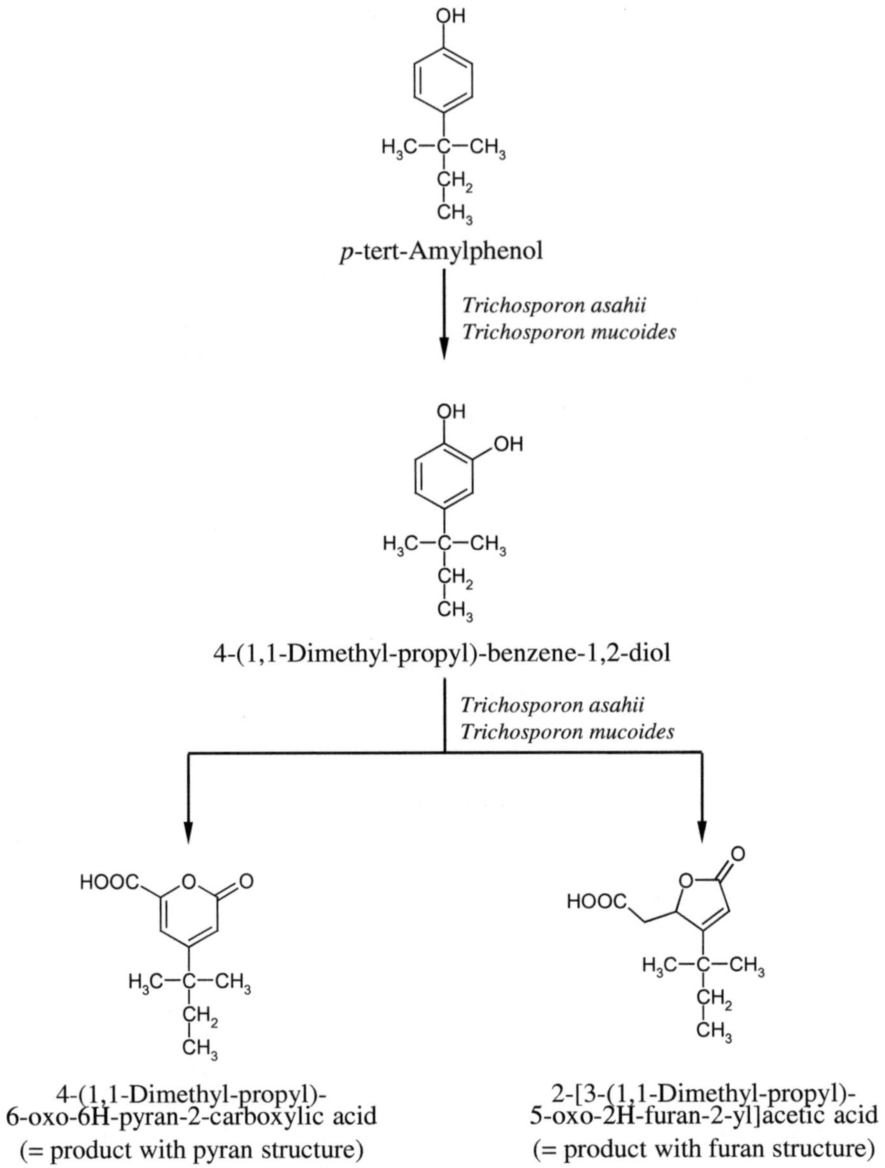

Fig. 13 Main transformation pathway of *p*-tert-amylphenol by *Trichosporon asahii* and *Trichosporon mucoides* (according to Schlueter et al. 2014)

Classical aromatic disinfectants such as phenol, chlorinated phenols, cresols or xylenols are possible aromatic substrates for several yeasts. The degradation mechanisms of phenol and its derivatives (cresols or chlorinated phenols) have been described in the previous sections (see above). Schlueter et al. (2014) focused attention on the biotransformation of *p*-tert-amylphenol as a disinfectant with phenolic structure by different organisms. The substance *p*-tert-amylphenol is used as the active ingredient of surface disinfectants. While *Candida maltosa* could not oxidize this compound at all, *Trichosporon asahii* and *Trichosporon mucoides* hydroxylated *p*-tert-amylphenol at the *ortho*-position of the hydroxyl group to form 4-(1,1-dimethyl-propyl)-benzene-1,2-diol. The dihydroxylated intermediate is a substrate for ring fission and two lactones with different structures were formed: one with a pyran structure and one with a furan structure (Fig. 13). Both ring fission products are novel structures, unknown so far as transformation products of disinfectants. This transformation mechanism serves as a detoxification strategy of *Trichosporon* species, since the disinfectant was toxic to these organisms, but one of the ring fission products was not.

In addition to disinfectants, numerous environmentally important pharmaceuticals with aromatic structures have been produced as medical products and these may well be transformed by yeast strains capable of oxidizing environmental pollutants with aromatic structures.

9 Conclusions and Future Prospectives

The examples of the degradation of aromatic pollutants presented in this review demonstrate that yeasts are able to degrade and mineralize many significant pollutants with high efficiency. For recalcitrant molecules, such as those with condensed ring systems or with many halogen substituents, yeasts can at least contribute to a primary oxidation or biotransformation. In this process, various intermediates get accumulated and excreted which serve as substrates for other microorganisms such as bacteria and filamentous fungi. Thus yeasts, which are widespread, highly abundant in particular ecosystems and equipped with enhanced metabolic activity, represent a very important group of microorganisms. They contribute importantly to bioremediation by eliminating environmental pollutants in air, in soil or in various waters systems. Thus yeasts offer advantages compared to bacteria in terms of environmental remediation particularly in slightly acidic milieus or under increased cellular stress conditions (shear force, solvents, etc.). Importantly, yeasts in general also have shorter generation times than do filamentous fungi. Thus, in the future yeasts will find increasing applications in the elimination and detoxification of environmental pollutants.

References

Adav, S. S., Chen, M.-Y., Lee, D.-J. and Ren, N.-Q. 2007. *Biotechnol. Bioeng.* **96**: 844–852.

Ahuatzi-Chacon, D., Ordorica-Morales, G., Ruiz-Ordaz, N., Cristiani-Urbina, E., Juarez-Ramirez, C. and Galindez-Mayer, J. 2004. *World J. Microbiol. Biotechnol.* **20**: 695–702.

Aleksieva, Z., Ivanova, D., Godjevargova, T. and Atanasov, B. 2002. *Process Biochem.* **37**: 1215–1219.
Ali, N., Eliyas, M., Al-Sarawi, H. and Radwan, S. S. 2011. *Chemosphere.* **83**: 1268–1272.
Andersson, P. L., Blom, A., Johannisson, A., Pesonen, M., Tysklind, M., Berg, A. H., Olsson, P. E. and Norrgren, L. 1999. *Arch. Environ. Con. Tox.* **37**: 145–150.
Assinder, S. J. and Williams, P. A. 1990. *Adv. Microb. Physiol.* **31**: 1–69.
Awe, S., Mikolasch, A., Hammer, E. and Schauer, F. 2008. *Int. Biodeterior. Biodegrad.* **62**: 408–414.
Awe, S., Mikolasch, A. and Schauer, F. 2009. *Appl. Microbiol. Biotechnol.* **84**: 965–976.
Badali, H., Prenafeta-Boldu, F. X., Guarro, J., Klaassen, C. H., Meis, J. F. and De Hoog, G. S. 2011. *Fungal Biol.* **115**: 1019–1029.
Ballerstedt, H., Kraus, A. and Lechner, U. 1997. *Environ. Sci. Technol.* **31**: 1749–1753.
Basak, B., Bhunia, B., Dutta, S. and Dey, A. 2013. *Int. Biodeterior. Biodegrad.* **78**: 17–23.
Bergauer, P., Fonteyne, P. A., Nolard, N., Schinner, F. and Margesin, R. 2005. *Chemosphere.* **59**: 909–918.
Boersma, M. G., Dinarieva, T. Y., Middelhoven, W. J., van Berkel, W. J. H., Doran, J., Vervoort, J. and Rietjens, I. 1998. *Appl. Environ. Microbiol.* **64**: 1256–1263.
Bos, P. and Debruyn, J. C. 1973. *A. Van Leeuw. J. Microb.* **39**: 99–107.
Bril'kov, A. V., Pechurkin, N. S. and Litvinov, V. V. 1980. *Mikrobiologiia* **49**: 466–72.
Bunge, M. and Lechner, U. 2009. *Appl. Microbiol. Biotechnol.* **84**: 429–444.
Cejkova, A., Masak, J., Jirku, V., Fialova, A. and Moenandar, D. 2002. Waste Management and the Environment, WIT Press, Ashurst, England, pp. 99–106.
Cerniglia, C. E. and Crow, S. A. 1981. *Arch. Microbiol.* **129**: 9–13.
Chang, Y.-S. 2008. *J. Mol. Microb. Biotech.* **15**: 152–171.
Chaudhry, G. R. and Chapalamadugu, S. 1991. *Microbiol. Rev.* **55**: 59–79.
Chen, K. C., Lin, Y. H., Chen, W. H. and Liu, Y. C. 2002. *Enzyme Microb. Tech.* **31**: 490–497.
Chrzanowski, L., Bielicka-Daszkiewicz, K., Owsianiak, M., Aurich, A., Kaczorek, E. and Olszanowski, A. 2008. *World J. Microbiol. Biotechnol.* **24**: 1943–1949.
Colquhoun, D. R., Hartmann, E. M. and Halden, R. U. 2012. *J. Biomed. Biotechnol.* Article number: 408690.
Cong, L. T. N., Mai, C. T. N., Thanh, V. T., Nga, L. P. and Minh, N. N. 2014. *Water Sci. Technol.* **70**: 329–336.
Connor, K., Ramamoorthy, K., Moore, M., Mustain, M., Chen, I., Safe, S., Zacharewski, T., Gillesby, B., Joyeux, A. and Balaguer, P. 1997. *Toxicol. Appl. Pharm.* **145**: 111–123.
Corti, A., Frassinetti, S., Vallini, G., Dantone, S., Fichi, C. and Solaro, R. 1995. *Environ. Pollut.* **90**: 83–87.
Cox, H. H. J. and Doddema, H. J. 1996. European Patent EP-0710147
Cox, H. H. J., Faber, B. W., VanHeiningen, W. N. M., Radhoe, H., Doddema, H. J. and Harder, W. 1996. *Appl. Environ. Microbiol.* **62**: 1471–1474.
Cox, H. H. J., Moerman, R. E., vanBaalen, S., vanHeiningen, W. N. M., Doddema, H. J. and Harder, W. 1997. *Biotechnol. Bioeng.* **53**: 259–266.
Deng, Y., Zhang, Y., Hesham, A. E.-L., Liu, R. and Yang, M. 2010. *Appl. Microbiol. Biotechnol.* **86**: 1933–1939.
Dodge, R. H., Cerniglia, C. E. and Gibson, D. T. 1979. *Biochem. J.* **178**: 223–230.
Dutta, T. 2005. *Environ. Geochem. Hlth.* **27**: 271–284.
Dutta, T. K. and Harayama, S. 2001. *Environ. Sci. Technol.* **35**: 102–107.
Enroth, C. 2003. *Acta Crystallogr. D.* **59**: 1597–1602.
Estevez, E., Veiga, M. C. and Kennes, C. 2005. *J. Ind. Microbiol. Biot.* **32**: 33–37.
Fahr, K., Wetzstein, H. G., Grey, R. and Schlosser, D. 1999. *FEMS Microbiol. Lett.* **175**: 127–132.
Fedorak, P. M. and Westlake, D. W. S. 1986. *Appl. Environ. Microbiol.* **51**: 435–437.
Fialova, A., Boschke, E. and Bley, T. 2004. *Int. Biodeterior. Biodegrad.* **54**: 69–76.
Fiedler, H. 1996. *Chemosphere.* **32**: 55–64.
Field, J. A. and Sierra-Alvarez, R. 2008. *Chemosphere.* **71**: 1005–1018.
Fuchs, G., Boll, M. and Heider, J. 2011. *Nat. Rev. Microbiol.* **9**: 803–816.

Gaal, A. and Neujahr, H. Y. 1979. *J. Bacteriol.* **137**: 13–21.
Gaal, A. and Neujahr, H. Y. 1980. *Biochem. J.* **191**: 37–43.
Gaal, A. and Neujahr, H. Y. 1981. *Arch. Microbiol.* **130**: 54–58.
Galindez-Mayer, J., Ramon-Gallegos, J., Ruiz-Ordaz, N., Juarez-Ramirez, C., Salmeron-Alcocer, A. and Poggi-Varaldo, H. M. 2008. *Biochem. Eng. J.* **38**: 147–157.
Gallastegui, G., Barona, A., Rojo, N., Gurtubay, L. and Elias, A. 2013. *Process Saf. Environ.* **91**: 112–122.
Gallucci, M. N., Carezzano, M. E., Oliva, M. M., Demo, M. S., Pizzolitto, R. P., Zunino, M. P., Zygadlo, J. A. and Dambolena, J. S. 2014. *J. Appl. Microbiol.* **116**: 795–804.
Garcia-Pena, I., Hernandez, S., Auria, R. and Revah, S. 2005. *Appl. Environ. Microbiol.* **71**: 4280–4285.
Garon, D., Krivobok, S. and Seigle-Murandi, F. 2000. *Chemosphere.* **40**: 91–97.
Gartiser, S., Stiene, G., Hartmann, A. and Zipperle, J. 2000. *UBA-Texte 1–00.* Umweltbundesamt, Berlin.
Gerecova, G., Nebohacova, M., Zeman, I., Pryszcz, L. P., Tomaska, L., Gabaldon, T. and Nosek, J. 2015. *Fems Yeast Research* **15**: fov006.
Gerginova, M., Manasiev, J., Shivarova, N. and Alexieva, Z. 2007. *Z. Naturforsch. C* **62**: 83–86.
Giger, W., Brunner, P. H. and Schaffner, C. 1984. *Science* **225**: 623–625.
Godjevargova, T., Aleksieva, Z., Ivanova, D. and Shivarova, N. 1998. *Process Biochem.* **33**: 831–835.
Golbeck, J. H., Albaugh, S. A. and Radmer, R. 1983. *J. Bacteriol.* **156**: 49–57.
Gunsch, C. K., Cheng, Q., Kinney, K. A., Szaniszlo, P. J. and Whitman, C. P. 2005. *Appl. Microbiol. Biotechnol.* **68**: 405–411.
Gunsch, C. K., Kinney, K. A., Szaniszlo, P. J. and Whitman, C. P. 2007. *Biotechnol. Bioeng.* **98**: 101–111.
Hagler, A. N. and Mendoncahagler, L. C. 1981. *Appl. Environ. Microbiol.* **41**: 173–178.
Hagler, A. N., Santos, S. S. and Mendoncahagler, L. C. 1979. *Rev. Microbiol.* **10**: 36–41.
Hammer, E., Kneifel, H., Hofmann, K. and Schauer, F. 1996. *J. Basic Microb.* **36**: 239–243.
Hammer, E., Krowas, D., Schafer, A., Specht, M., Francke, W. and Schauer, F. 1998. *Appl. Environ. Microbiol.* **64**: 2215–2219.
Harris, G. and Ricketts, R. W. 1962. *Nature* **195**: 473–474.
Harvey, R. G. 1997. Polycylic aromatic hydrocarbons. Wiley-VCH, New York.
Harwood, C. S. and Parales, R. E. 1996. *Annual Review of Microbiology* **50**: 553–590.
Hasegawa, Y., Okamoto, T., Obata, H. and Tokuyama, T. 1990. *J. Ferment. Bioeng.* **69**: 122–124.
Hashimoto, K. 1970. *J. Gen. Appl. Microbiol.* **16**: 1–13.
Hashimoto, K. 1973. *J. Gen. Appl. Microbiol.* **19**: 171–187.
Henning, K. 1993. PhD thesis. Ernst-Moritz-Arndt-University Greifswald, Greifswald, Germany.
Hesham, A. E.-L., Khan, S., Liu, X., Zhang, Y., Wang, Z. and Yang, M. 2006a. *Yeast* **23**: 879–887.
Hesham, A. E.-L., Wang, Z., Zhang, Y., Zhang, J., Lv, W. and Yang, M. 2006b. *Ann. Microbiol.* **56**: 109–112.
Hesham, A. E.-L., Alamri, S. A., Khan, S., Mahmoud, M. E. and Mahmoud, H. M. 2009. *Afr. J. Biotechnol.* **8**: 2218–2223.
Hesham, A. E.-L., Khan, S., Tao, Y., Li, D., Zhang, Y. and Yang, M. 2012. *Environ. Sci. Pollut. R.* **19**: 3568–3578.
Hlouchova, K., Rudolph, J., Pietari, J. M. H., Behlen, L. S. and Copley, S. D. 2012. *Biochemistry* **51**: 3848–3860.
Hofmann, K. H. 1986. *J. Basic Microb.* **26**: 109–111.
Hofmann, K. H. and Kruger, A. K. 1985. *J. Basic Microb.* **25**: 373–379.
Hofmann, K. H. and Schauer, F. 1988. *A. Van Leeuw. J. Microb.* **54**: 179–188.
Hristov, A., Gouliamova, D., Nacheva, L. and Tsekova, K. 2012. *Cr. Acad. Bulg. Sci.* **65**: 335–340.
Ivoilov, V. S. and Karasevich, I. N. 1983. *Mikrobiologiia* **52**: 956–961.

Iwasaki, Y., Gunji, H., Kino, K., Hattori, T., Ishii, Y. and Kirimura, K. 2010. *Biodegradation* **21**: 557–564.
Jerina, D. M., Daly, J. W., Witkop, B., Zaltzman, P. and Udenfrie, S. 1970. *Biochemistry* **9**: 147–156.
Jiang, B., Zhou, Z., Dong, Y., Tao, W., Wang, B., Jiang, J. and Guan, X. 2015. *Appl. Biochem. Biotech.* **176**: 1700–1708.
Jiang, Y., Wen, J., Lan, L. and Hu, Z. 2007. *Biodegradation* **18**: 719–729.
Juarez-Ramirez, C., Ruiz-Ordaz, N., Cristiani-Urbina, E. and Galindez-Mayer, J. 2001. *World J. Microbiol. Biotechnol.* **17**: 697–705.
Kanaly, R. A. and Harayama, S. 2000. *J. Bacteriol.* **182**: 2059–2067.
Kasai, N., Ikushiro, S.-i., Shinkyo, R., Yasuda, K., Hirosue, S., Arisawa, A., Ichinose, H., Wariishi, H. and Sakaki, T. 2010. *Appl. Microbiol. Biotechnol.* **86**: 773–780.
Katayamahirayama, K., Tobita, S. and Hirayama, K. 1994. *Water Sci. Technol.* **30**: 59–66.
Kennes, C. and Veiga, M. C. 2004. *J. Biotechnol.* **113**: 305–319.
Kim, J. M., Le, N. T., Chung, B. S., Park, J. H., Bae, J.-W., Madsen, E. L. and Jeon, C. O. 2008. *Appl. Environ. Microbiol.* **74**: 7313–7320.
Knackmuss, H. J. and Hellwig, M. 1978. *Arch. Microbiol.* **117**: 1–7.
Korner, W., Hanf, V., Schuller, W., Bartsch, H., Zwirner, M. and Hagenmaier, H. 1998. *Chemosphere.* **37**: 2395–2407.
Krallish, I., Gonta, S., Savenkova, L., Bergauer, P. and Margesin, R. 2006. *Extremophiles* **10**: 441–449.
Krug, M., Ziegler, H. and Straube, G. 1985. *J. Basic Microb.* **25**: 103–110.
Krug, M. and Straube, G. 1986. *J. Basic Microb.* **26**: 271–281.
Kutty, S. N. and Philp, R. 2008. *Yeast* **25**: 465–483.
Lahav, R., Fareleira, P., Nejidat, A. and Abeliovich, A. 2002. *Microbial. Ecol.* **43**: 388–396.
Lange, J. 1997. PhD thesis. Ernst-Moritz-Arndt-University Greifswald, Greifswald, Germany.
Lange, J., Hammer, E., Specht, M., Francke, W. and Schauer, F. 1998. *Appl. Microbiol. Biotechnol.* **50**: 364–368.
Layton, A. C., Sanseverino, J., Gregory, B. W., Easter, J. P., Sayler, G. S. and Schultz, T. W. 2002. *Toxicol. Appl. Pharm.* **180**: 157–163.
Lee, J. S., Kang, E. J., Kim, M. O., Lee, D. H., Bae, K. S. and Kim, C. K. 2001. *J. Microbiol. Biotechn.* **11**: 112–117.
Liu, H., Yu, Q. J., Wang, G., Ye, F. and Cong, Y. 2011. *Process Biochem.* **46**: 1678–1681.
Long, Y., Yang, S., Xie, Z. and Cheng, L. 2014. *Can. J. Microbiol.* **60**: 585–591.
MacGillivray, A. R. and Shiaris, M. P. 1993. *Appl. Environ. Microbiol.* **59**: 1613–1618.
MacGillivray, A. R. and Shiaris, M. P. 1994. *Appl. Environ. Microbiol.* **60**: 1154–1159.
Mahgoub, S., Abdelbasit, H. and Abdelfattah, H. 2015. *Desalin. Water Treat.* **53**: 3381–3387.
Margesin, R., Gander, S., Zacke, G., Gounot, A. M. and Schinner, F. 2003. *Extremophiles* **7**: 451–458.
Margesin, R., Fonteyne, P. A. and Redl, B. 2005. *Res. Microbiol.* **156**: 68–75.
Mazur, P., Pieken, W. A., Budihas, S. R., Williams, S. E., Wong, S. and Kozarich, J. W. 1994. *Biochemistry* **33**: 1961–1970.
Middelhoven, W. J. 1993. *Anton. Leeuw. Int. J. G.* **63**: 125–144.
Middelhoven, W. J. 2003. *Mycoses* **46**: 7–11.
Middelhoven, W. J., Dekievit, H. and Biesbroek, A. L. 1985. *A. Van Leeuw. J. Microb.* **51**: 289–301.
Middelhoven, W. J., Coenen, A., Kraakman, B. and Gelpke, M. D. S. 1992. *Anton. Leeuw. Int. J. G.* **62**: 181–187.
Mohn, W. W. and Tiedje, J. M. 1992. *Microbiol. Rev.* **56**: 482–507.
Morsen, A. and Rehm, H. J. 1987. *Appl. Microbiol. Biotechnol.* **26**: 283–288.
Mortberg, M. and Neujahr, H. Y. 1985. *J. Bacteriol.* **161**: 615–619.
Neujahr, H. Y. and Gaal, A. 1973. *Eur. J. Biochem.* **35**: 386–400.
Neujahr, H. Y. and Gaal, A. 1975. *Eur. J. Biochem.* **58**: 351–357.
Neujahr, H. Y. and Kjellen, K. G. 1978. *J. Biol. Chem.* **253**: 8835–8841.

Neujahr, H. Y. and Varga, J. M. 1970. *Eur. J. Biochem.* **13**: 37–44.
Neujahr, H. Y., Lindsjo, S. and Varga, J. M. 1974. *A. Van Leeuw. J. Microb.* **40**: 209–216.
Nikolova, N. and Nenov, V. 2005. *Water Sci. Technol.* **51**: 87–93.
Ornston, L. N. and Stanier, R. Y. 1966. *J. Biol. Chem.* **241**: 3776–3786.
Paca, J., Jr., Kremlackova, V., Turek, M., Sucha, V., Vilimkova, L., Paca, J., Halecky, M. and Stiborova, M. 2007. *Enzyme Microb. Tech.* **40**: 919–926.
Pan, F., Yang, Q. X., Zhang, Y., Zhang, S. J. and Yang, M. 2004. *Biotechnol. Lett.* **26**: 803–806.
Polnisch, E., Kneifel, H., Franzke, H. and Hofmann, K. H. 1992. *Biodegradation* **2**: 193–199.
Pothuluri, J. V., Freeman, J. P., Evans, F. E. and Cerniglia, C. E. 1990. *Appl. Environ. Microbiol.* **56**: 2974–2983.
Potrawfke, T., Timmis, K. N. and Wittich, R. M. 1998. *Appl. Environ. Microbiol.* **64**: 3798–3806.
Powlowski, J. B. and Dagley, S. 1985. *J. Bacteriol.* **163**: 1126–1135.
Powlowski, J. B., Ingebrand, J. and Dagley, S. 1985. *J. Bacteriol.* **163**: 1136–1141.
Prenafeta-Boldu, F. X., Kuhn, A., Luykx, D., Anke, H., van Groenestijn, J. W. and de Bont, J. A. M. 2001. *Mycol. Res.* **105**: 477–484.
Prenafeta-Boldu, F. X., Vervoort, J., Grotenhuis, J. T. C. and van Groenestijn, J. W. 2002. *Appl. Environ. Microbiol.* **68**: 2660–2665.
Prenafeta-Boldu, F. X., Ballerstedt, H., Gerritse, J. and Grotenhuis, J. T. C. 2004. *Biodegradation* **15**: 59–65.
Prenafeta-Boldu, F. X., Guivernau, M., Gallastegui, G., Vinas, M., de Hoog, G. S. and Elias, A. 2012. *FEMS Microbiol. Ecol.* **80**: 722–734.
Qi, B., Moe, W. M. and Kinney, K. A. 2002. *Appl. Microbiol. Biotechnol.* **58**: 684–689.
Rao, B. V. and Bhat, J. V. 1971. *Anton. Van Lee. J. M. S.* **37**: 303–312.
Ren, H. F., Zanma, S., Urano, N., Endo, H., Mineki, S. and Hayashi, T. 2004. *Nippon Suisan Gakk.* **70**: 687–692.
Rieche, A., Hilgetag, G., Lorenz, M. and Martini, A. 1962. In: Continuous Cultivation of Microorganisms. Proceedings of the Second Symposium (ed. I. Malek), Czechoslovak Academy of Science, Prague, pp. 293–299.
Riedel, K., Beyersdorfradeck, B., Neumann, B. and Schaller, F. 1995. *Appl. Microbiol. Biotechnol.* **43**: 7–9.
Rigo, M. and Alegre, R. M. 2004. *Folia Microbiol.* **49**: 41–45.
Rigo, M., Alegre, R. M., Mazile Vidal Bezerra, J. R., Coelho, N. and Bastos, R. G. 2010. *Braz. Arch. Biol. Techn.* **53**: 481–486.
Rocha, L. L., Cordeiro, R. d. A., Cavalcante, R. M., do Nascimento, R. F., Silveira Martins, S. C., Santaella, S. T. and Maciel Melo, V. M. 2007. *Mycopathologia* **164**: 183–188.
Rojo, F., Pieper, D. H., Engesser, K. H., Knackmuss, H. J. and Timmis, K. N. 1987. *Science* **238**: 1395–1398.
Romero, M. C., Cazau, M. C., Giorgieri, S. and Arambarri, A. M. 1998. *Environ. Pollut.* **101**: 355–359.
Romero, M. C., Hammer, E., Cazau, M. C. and Arambarri, A. M. 2001. *World J. Microbiol. Biotechnol.* **17**: 591–594.
Romero, M. C., Hammer, E., Cazau, M. C. and Arambarri, A. M. 2002a. *Environ. Pollut.* **118**: 379–382.
Romero, M. C., Salvioli, M. L., Cazau, M. C. and Arambarri, A. M. 2002b. *Environ. Pollut.* **117**: 159–163.
Romero, M. C., Reinoso, E. H., Moreno Kiernan, A., Urrutia, M. I. and Kiernan, A. M. 2006. *Electron. J. Biotechn.* **9**: 221–226.
Ruiz-Ordaz, N., Hernandez-Manzano, E., Ruiz-Lagunez, J. C., Cristiani-Urbina, E. and Galindez-Mayer, J. 1998. *Biotechnol. Progr.* **14**: 966–969.
Safe, S. 1984. *CRC Crit. Rev. Toxicol.* **13**: 319–395.
Sakaki, T., Shinkyo, R., Takita, T., Ohta, M. and Inouye, K. (2002) Arch. Biochem. Biophys. **401**: 91–98.
Salmeron-Alcocer, A., Ruiz-Ordaz, N., Juarez-Ramirez, C. and Galindez-Mayer, J. 2007. *Biochem. Eng. J.* **37**: 201–211.

Sariaslani, F. S., Harper, D. B. and Higgins, I. J. 1974. *Biochem. J.* **140**: 31–45.
Schauer, F. 1988. Thesis of Habilitation. Ernst-Moritz-Arndt-University Greifswald, Greifswald, Germany.
Schauer, F. and Schauer, M. 1986. *Wissenschaftliche Zeitschrift der Ernst-Moritz-Arndt-Universität Greifswald, Mathematisch-Naturwisenschaftliche Reihe* **35**: 13–23.
Schauer, F., Henning, K., Pscheidl, H., Wittich, R. M., Fortnagel, P., Wilkes, H., Sinnwell, V. and Francke, W. 1995. *Biodegradation* **6**: 173–180.
Scheller, U., Zimmer, T., Becher, D., Schauer, F. and Schunck, W. H. 1998. *J. Biol. Chem.* **273**: 32528–32534.
Schiestl, R. H., Aubrecht, J., Yap, W. Y., Kandikonda, S. and Sidhom, S. 1997. *Cancer Res.* **57**: 4378–4383.
Schlosser, D., Grey, R., Höfer, C. and Fahr, K. 2000. Bioremediation of contaminated soils. Marcel Dekker, New York.
Schlueter, R., Lippmann, R., Hammer, E., Salazar, M. G. and Schauer, F. 2013. *Appl. Microbiol. Biotechnol.* **97**: 5043–5053.
Schlueter, R., Roeder, A., Czekalski, N., Gliesche, D., Mikolasch, A. and Schauer, F. 2014. *Appl. Microbiol. Biotechnol.* **98**: 373–384.
Schmidt, E., Hellwig, M. and Knackmuss, H. J. 1983. *Appl. Environ. Microbiol.* **46**: 1038–1044.
Schunck, W. H., Mauersberger, S., Huth, J., Riege, P. and Muller, H. G. 1987. *Arch. Microbiol.* **147**: 240–244.
Sejlitz, T. and Neujahr, H. Y. 1987. *Eur. J. Biochem.* **170**: 343–349.
Seo, J.-S., Keum, Y.-S. and Li, Q. X. 2009. *Int. J. Environ. Res. Public Health* **6**: 278–309.
Seyedmousavi, S., Badali, H., Chlebicki, A., Zhao, J., Xavier Prenafeta-Boldu, F. and De Hoog, G. S. 2011. *Fungal Biol.* **115**: 1030–1037.
Shoda, M. and Udaka, S. 1980. *Appl. Environ. Microbiol.* **39**: 1129–1133.
Sietmann, R., Hammer, E., Moody, J., Cerniglia, C. E. and Schauer, F. 2000. *Arch. Microbiol.* **174**: 353–361.
Sietmann, R., Hammer, E., Specht, M., Cerniglia, C. E. and Schauer, F. 2001. *Appl. Environ. Microbiol.* **67**: 4158–4165.
Sietmann, R. 2002. PhD thesis. Ernst-Moritz-Arndt-University Greifswald, Greifswald, Germany.
Sietmann, R., Hammer, E. and Schauer, F. 2002. *Syst. Appl. Microbiol.* **25**: 332–339.
Sietmann, R., Gesell, M., Hammer, E. and Schauer, F. 2006. *Chemosphere.* **64**: 672–685.
Skene, S. A., Dewhurst, I. C. and Greenberg, M. 1989. *Hum. Toxicol.* **8**: 173–203.
Sood, N. and Lal, B. 2009. *J. Environ. Manage.* **90**: 1728–1736.
Sood, N., Patle, S. and Lal, B. 2010. *Environ. Sci. Pollut. R.* **17**: 603–610.
Sun, Y. M., Quan, X., Chen, J. W., Yang, F. L., Xue, D. M., Liu, Y. H. and Yang, Z. H. 2002. *Process Biochem.* **38**: 109–113.
Sutherland, J. B., Selby, A. L., Freeman, J. P., Evans, F. E. and Cerniglia, C. E. 1991. *Appl. Environ. Microbiol.* **57**: 3310–3316.
Takada, S., Nakamura, M., Matsueda, T., Kondo, R. and Sakai, K. 1996. *Appl. Environ. Microbiol.* **62**: 4323–4328.
Terasaka, S., Inoue, A., Tanji, M. and Kiyama, R. 2006. *Toxicol. Lett.* **163**: 130–141.
Thomas, D. R., Carswell, K. S. and Georgiou, G. 1992. *Biotechnol. Bioeng.* **40**: 1395–1402.
Tortella, G. R., Diez, M. C. and Duran, N. 2005. *CRC Cr. Rev. Microbiol.* **31**: 197–212.
Tsai, S. C., Tsai, L. D. and Li, Y. K. 2005. *Biosci. Biotech. Bioch.* **69**: 2358–2367.
Uzura, A., Katsuragi, T. and Tani, Y. 2001. *J. Biosci. Bioeng.* **91**: 217–221.
Valli, K., Wariishi, H. and Gold, M. H. 1992. *J. Bacteriol.* **174**: 2131–2137.
Vallini, G., Frassinetti, S., D'Andrea, F., Catelani, G. and Agnolucci, M. 2001. *Int. Biodeterior. Biodegrad.* **47**: 133–140.
Varga, J. M. and Neujahr, H. 1970a. *Plant and Soil* **33**: 565–571.
Varga, J. M. and Neujahr, H. Y. 1970b. *Eur. J. Biochem.* **12**: 427–434.
Varma, R. J. and Gaikwad, B. G. 2008. *Enzyme Microb. Tech.* **43**: 431–435.
Vollmer, M. D., Fischer, P., Knackmuss, H. J. and Schlomann, M. 1994. *J. Bacteriol.* **176**: 4366–4375.

Walker, N. 1973. *Soil Biol. Biochem.* **5**: 525–530.
Wang, J., Ma, X., Liu, S., Sun, P., Fan, P. and Xia, C. 2012. In: Seventh International Conference on Waste Management and Technology (eds. L. Jinhui and H. Hualongeds), pp. 299–303.
Wang, X., Gong, Z., Li, P. and Zhang, L. 2007. *B. Environ. Contam. Tox.* **78**: 522–526.
Wase, D. A. J. and Hough, J. S. 1966. *J. Gen. Microbiol.* **42**: 13–23.
Wiseman, A. and Woods, L. F. J. 1979. *J. Chem. Technol. Biot.* **29**: 320–324.
Wiseman, A., Gondal, J. A. and Sims, P. 1975. *Biochem. Soc. T.* **3**: 278–281.
Wittich, R. M. 1992. *BioEngineering* **10**: 33–40.
Wittich, R. M. 1998. *Appl. Microbiol. Biotechnol.* **49**: 489–499.
Woertz, J. R., Kinney, K. A., McIntosh, N. D. P. and Szaniszlo, P. J. 2001. *Biotechnol. Bioeng.* **75**: 550–558.
Yadav, V., Shitiz, K., Pandey, R. and Yadav, J. 2011. *Yeast* **28**: 81–91.
Yordanova, G., Godjevargova, T., Nenkova, R. and Ivanova, D. 2013. *Biotechnol. Biotec. Eq.* **27**: 3681–3688.
Yoshimoto, T., Garashi, K., Fujita, T. and Harayama, T. 1990. *J. Pestic. Sci.* **15**: 341–352.
Zanaroli, G., Negroni, A., Haeggblom, M. M. and Fava, F. 2015. *Curr. Opin. Biotech.* **33**: 287–295.
Zhou, S., Jin, X., Sun, F., Zhou, H., Yang, C. and Xia, C. 2012. *Water Sci. Technol.* **65**: 780–786.
Zimmermann, R. 1958. *Naturwissenschaften* **45**: 165–165.
Zinjarde, S., Apte, M., Mohite, P. and Kumar, A. R. 2014. *Biotechnol. Adv.* **32**: 920–933.

Phytase of the Unconventional Yeast *Pichia anomala*: Production and Applications

Swati Joshi and Tulasi Satyanarayana

Abstract Most of cereal and legume seeds and their by-products contain 1–2% phytic acid (*myo*-inositol-1,2,3,4,5,6-hexakis dihydrogen phosphate) that represents >60% of total phosphorus content in plant products. A large portion of phytic acid in various plant seeds is in the form of salts called phytates. The phytate phosphorus is scarcely available to monogastrics (non-ruminants) because of insufficient levels of phytic acid hydrolysing gastric enzymes in them. Inorganic phosphorus (Pi) is supplemented in diets designed for such animals, which include swine, poultry, and fish to meet their Pi requirement. The unutilized phytate P is excreted that becomes an environmental pollutant in areas of intensive animal rearing. Excessive phytate P from soils flows to lakes and the sea that causes eutrophication, algal blooms and death of aquatic organisms. Furthermore, the negatively charged phosphate groups on phytic acid chelates positively charged metal ions of nutritional importance including Fe^{2+}, Ca^{2+}, Zn^{2+}, Cu^{2+}, Mg^{2+}, Mn^{2+}, causing poor absorption of the bound metals in gastrointestinal (GI) tract. This is one of the significant factors for causing wide-spread nutritional deficiencies of calcium, iron, and zinc in countries where plant based diets are predominantly consumed. Challenges in the areas of environmental sustainability, animal and human nutrition and health have prompted research on phytases. This review focuses on the production, properties and applications of native and recombinant phytases of *Pichia anomala* and scope of improving its potential through protein engineering.

Keywords *Pichia anomala* · Phytic acid · Native phytase · Recombinant phytase · Dephytinization

---

S. Joshi (✉)
National Institute of Occupational Health (NIOH), Meghani Nagar (Opp. Old IGP office), Ahmedabad, Gujarat, India
e-mail: swati.joshi.aj@gmail.com

T. Satyanarayana
Division of Biological Sciences and Engineering, Netaji Subhas Institute of Technology, Azad Hind Fauz Marg, Sector 3, Dwarka, Delhi 110078, India
e-mail: tsnarayana@gmail.com

T. Satyanarayana and G. Kunze (eds.), *Yeast Diversity in Human Welfare*, DOI 10.1007/978-981-10-2621-8_14

1 Introduction

Phytases constitute a subgroup of phosphatases which liberate at least one phosphate from *myo*-inositol hexakisphosphate. Most of the time, degradation ends in generation of *myo*-inositol phosphates with lower degree of phosphorylation (Hara et al. 1985; Kerovuo et al. 2000; Sajidan et al. 2004; Casey and Walsh 2004). International Union of Pure and Applied Chemistry and the International Union of Biochemistry (IUPAC–IUB) distinguish two categories of phytate-hydrolyzing enzymes, 3-phytases and 6-phytases, which start the hydrolysis of phosphoester bond at 3rd and 6th positions of phytates, respectively. These enzymes are widely distributed in nature, and these are reported from animals, plants and microbes. For instance, phytate-hydrolyzing enzymes were found in the calf serum (McCollum and Hart 1908), fishes, reptiles, and birds (Rapoport et al. 1941), and in plants including maize (Huebel and Beck 1996), rice (Maugenest et al. 1999), and wheat (Nakano et al. 1999). However, predominantly research has been focused on phytases of microbial origin, mainly on those from filamentous fungi such as *Aspergillus fumigatus* (Pasamontes et al. 1997) or *Rhizopus oligosporus* (Casey and Walsh 2004), *Mucor piriformis* (Howson and Davis 1983), *Cladosporium* sp. (Quan et al. 2004) and thermophilic mould *Sporotrichum thermophile* (Singh and Satyanarayana 2008; Maurya et al. 2017). In the past few decades, phytate-hydrolyzing enzymes of bacteria such as *Escherichia coli* (Greiner et al. 1993), *Klebsiella* spp. (Tambe et al. 1994; Sajidan et al. 2004), *Pseudomonas* spp. (Kim et al. 2003; Cho et al. 2003), various species of *Bacillus* (Kerovuo et al. 1998; Kim et al. 1998; Wang et al. 2001) and different yeasts such as *Arxula adeninivorans* (Sano et al. 1999), *Schwanniomyces occidentalis* (Segueilha et al. 1992), *Pichia anomala* (Vohra and Satyanarayana, 2001; 2002a) have been investigated. Oh et al. (2004) and Konietzny and Greiner (2002) reviewed occurrence and biochemical features of phytases. Physicochemical characteristics and prospective applications of phytases have been reviewed by Rao et al. (2009). *P. anomala* is among the most important unconventional yeasts as it has been found associated with food and beverages, and is useful in different biotechnological fields such as biopreservation (Ingvar et al. 2011) and generation of metabolites of low molecular mass (Van Eck et al. 1993; Fredlund et al. 2004). This chapter reviews developments in the production of phytase by this yeast and its potential applications.

2 Isolation of Phytase Producing Strain of *P. anomala*

Many yeast isolates were collected by isolation from dried buds of *Madhuca latifolia* and *Woodfordia fruticosa* flowers, collected from diverse geographical locations of India (Vohra and Satyanarayana 2002b). Yeasts species are capable of surviving at high sugar concentrations present in sugary nectar of flowers due to their osmophilic nature (Spencer and Spencer 1997). Vohra and Satyanarayana (2002b) screened these isolates for extracellular, intracellular and cell-bound phytases. As cell bound fractions showed higher titers of phytase, further studies on phytase producers were focused on the isolates showing higher phytase production. The maximum titres of phytase were recorded in a yeast isolate, which was identified as *Pichia anomala*. Various physiological, morphological and biochemical features were taken into account to ascertain the identity of this yeast (Vohra and Satyanarayana 2001). The yeast isolate has been deposited at the Microbial Type Culture Collection, Chandigarh (India) [MTCC-4133].

3 Native Phytase of *P. anomala*

3.1 Formulation of the Phytase Production Medium

Reduction in the production cost and time are critical factors in order to use any microbial product. Efforts have been made to achieve higher phytase production in reduced time span. Both physical and nutritional parameters were optimized by employing various statistical approaches. One-variable-at-a-time (OVAT) approach increased phytase titres from the initial 2.5–68 Ug^{-1} Dry Yeast Biomass corresponding to a 27-fold improvement, when yeast cultivation was carried out for 24 h at 20 °C. The medium formulated upon optimization contained Fe^{2+} (0.15 mM), beef extract (1.0%) and glucose (4.0%) (Vohra and Satyanarayana, 2001). Phytase production was further increased by response surface methodology (RSM), which minimized the overall production cost. Density of the inoculum proved to be an insignificant variable as it did not show any significant effect (Vohra and Satyanarayana, 2002a). Production in laboratory fermentor is advantageous over production in shake flasks as the former allows mechanical control of parameters of production process such as pH and aeration, mixing of nutrients, oxygen and heat transfer. When *P. anomala* was cultivated in a 22 L laboratory fermentor, the biomass increased from 4.2 to 8 g L^{-1}. Complete abolition of lag phase was seen during cultivation in fermentor, and highest production was attained in 16 h in contrast to 24 h in shake flasks (Vohra et al. 2006). By fed batch fermentation, further increase in phytase production was achieved. For reduction of the production cost, cane molasses was used as one of the media components, which is one of the low-priced carbohydrate sources. Besides sugar (~50%), molasses contain nitrogenous substances, trace elements and vitamins (Huang and Tang 2006). When statistical approaches were used for optimizing fermentation variables using cane molasses, higher enzyme yields were attained in the medium that contained glucose and beef extract (Vohra and Satyanarayana 2004). The medium formulated for achieving high phytase titre contained: 8.0% cane molasses, 0.4% urea and 2.0% inoculum density. About 5-fold increase in phytase production was attained due to optimization (Kaur and Satyanarayana 2005). Further improvement in production in phytase production was achieved by cyclic fed-batch strategy of fermentation in comparison with batch and fixed volume fermentations (Verma and Satyanarayana 2012). In cyclic fed batch fermentation, phytase production was sustainable over seven days in cane molasses medium (Verma and Satyanarayana 2012). Among detergents and chemicals used for permeabilization of *P. anomala* cells, Triton X-100 was identified as the most efficient. A 15.0% increase in PPHY activity was attained upon treatment with 5.0% Triton X-100 for 30 min. Permeabilization led to shrinkage of cell protoplasm as revealed by electron microscopy. In storage at 4 °C, the permeabilized cells retained phytase activity for 2 months and at 60 °C, for a month. Immobilization of permeabilized *P. anomala* cells in alginate allowed reuse of biomass with sustained phytase activity (Kaur and Satyanarayana 2010).

3.2 Phytase Purification and Determination of N-Terminal Sequence

The phytase of *P. anomala* naturally occurs as a cell-bound enzyme. Mechanical disruption of yeast cells by French press and subsequent analysis of enzyme activity revealed that 78.0% of cell-bound enzyme is cell wall associated. Purification of this enzyme was carried out by acetone precipitation and column chromatography using anion exchanger DEAE-Sephadex matrix. The purified protein was of 64 kDa molecular mass on SDS-PAGE that formed a homo-hexameric quaternary structure in native conditions with a pI of 4.5. The kinetic parameters K_m and V_{max} values for substrate phytic acid were recorded as 0.20 mM and 6.34 μmol mg^{-1} protein min^{-1}, respectively (Vohra and Satyanarayana 2002c).

The N-terminal amino acid (aa) sequence of the PPHY is'VAIQKALVPG', which shows similarity with N-terminus of phytases from *Schwanniomyces occidentalis* and *Debaryomyces castellii.* Based on sequence similarity of N-terminal aa and consensus sequences, complete *PPHY* of *P. anomala* with both its promoter and terminator sequence was PCR amplified (Kaur et al. 2010). PPHY sequence showed similarity with other yeast phytases from *S. occidentalis* and *D. castellii.* The occurrence of the conserved active site heptapeptide motif i.e. RHGERYP and the catalytically significant dipeptide (HD) towards the C-terminal confirmed PPHY to be a phytase of Histidine Acid Phosphatase (HAP) group. Only one copy of *PPHY* gene was detected in *P. anomala* genome (Kaur et al. 2010).

3.3 Biochemical Characteristics of Native Phytase

The endogenous phytase of *P. anomala* shows broad substrate specificity, as it can hydrolyse glucose-6-phosphate, p-nitrophenyl phosphate, ADP and ATP besides phytic acid. The enzyme efficiently hydrolyzes insoluble phytates of calcium and magnesium, but not iron phytate. The cell-free purified phytase exhibits optimum activity at 60 °C and pH 4.0, while the membrane bound phytase at 75 °C and pH 4.0. The half-life values ($T_{1/2}$) of purified phytase at 60, 70 and 80 °C were 7 days, 48 h and 5 min, respectively. Higher thermostability was exhibited by the cell-bound form of the phytase than the cell free purified enzyme. As a prerequisite for feed pelleting, enzymes used as a feed supplement must withstand temperature as high as 60–90 °C for a few seconds (Wyss et al. 1998). In the presence of the substrate phytic acid, $T_{1/2}$ increased from 5 to 45 min at 80 °C (Vohra and Satyanarayana 2002c). Phytase activity was reduced to half in the presence of various metal ions such as Cu^{2+}, Zn^{2+}, Hg^{2+} (1 mM), while Fe^{3+} drastically reduced (~90%) phytase activity. None of the cations displayed a stimulatory effect on phytase.

4 Recombinant Phytase of *P. anomala*

4.1 Heterologous Expression of PPHY

For heterologous expression of proteins of eukaryotic origin, eukaryotic expression systems are preferred over prokaryotic systems, as the former have similar post translational machinery. *P. anomala* phytase is an ideal candidate to be used as a feed and food additive. For certain applications such as selective hydrolysis of phosphoester bond and concomitant removal of P from IP6 in immobilized phytase bio-reactors, dephytinization of soy milk and as a bread additive, soluble enzyme is preferable. In order to express PPHY extracellularly, Kaur et al. (2010) cloned and expressed *P. anomala* phytase in yeasts such as *A. adeninivorans* G1212, *Saccharomyces cerevisiae* Cl3ABYS86 and *Hansenula polymorpha* RB11. A major proportion of recombinant phytase, however, remained in cell-bound form. Among three yeast strains used, a high titre of recombinant phytase was produced only by *H. polymorpha*. Phytase thus produced by recombinant *H. polymorpha* RB11/Xplor1-URA3-SwARS-FMD-PPHY and wild-type *P. anomala* were purified and characterized. *PPHY* gene was placed downstream of an inducible *H. polymorpha* derived formate dehydrogenase promoter and induced using glycerol. Induction in *S. cerevisiae* Cl3ABY86/pYES2-PPHY was done in YMM-galactose medium and in *A. adeninivorans* G1212/YRC102-PPHY in YMM-glucose medium. Both the wild type and recombinant phytases exhibited a molecular mass of ~380 kDa which corresponds to a hexamer. The temperature and pH optima for the activity of the phytase were 60 °C and 4.0, respectively. The recombinant phytase exhibited a broad substrate spectrum and exhibited activity on phytic acid, 1-naphthylphosphate, glucose-6-phosphate, *p*-nitrophenylphosphate, ADP, sodium pyrophosphate, AMP and ATP (Kaur et al. 2010). After unsuccessful attempts of extracellular expression of PPHY in *S.cerevisiae*, *A. adeninivorans* and *H. polymorpha*, further attempts have been made for extracellular expression of PPHY in methylotrophic host *Pichia pastoris*. In two protein expression studies, different promoters (i.e. *AOX* and *GAP*) were used to control the expression of *PPHY* gene. Most exploited alcohol oxidase gene promoter (*AOX*) offers the tight regulation of gene expression upon the addition of methanol, while *GAP* promoter is constitutive and is known to allow expression of foreign genes at almost equal levels to that attained under *AOX* promoter (Latiffi et al. 2013, Waterham et al. 1997). *PPHY* gene was cloned in *E. coli-P. pastoris* shuttle vectors pPICZαA and pGAPZαA and recombinant *P. pastoris* harbouring *PPHY* gene were generated in electrocompetent host cells. Once the phytase producing recombinant *P. pastoris* clones were generated, the clones secreting high enzyme titres were selected for maximizing enzyme production.

Optimization of fermentation variables is known to improve recombinant protein production in *P. pastoris* (Wang et al. 2009). Employing Plackett–Burman (PB) design and RSM, the expression under *AOX* promoter has been increased ~22-fold as compared to the endogenous yeast (Joshi and Satyanarayana 2014). In case of *PPHY* expression under *GAP* promoter, OVAT and two factorial design were employed for optimizing the enzyme secretion that resulted in ~44-fold improvement in PPHY production in comparison over the native host (Joshi and Satyanarayana 2015a).

4.2 Bioinformatic Analysis of PPHY

In silico analysis revealed several molecular features of phytase dictated by its amino acid sequence. First feature that could be observed with the aid of PROSITE software was the presence of a catalytically important conserved heptapeptide (RHGERYP) and a dipeptide (HD), which are a characteristic feature of Histidine Acid Phosphatases (HAPs). Moreover, upon comparison and phylogenetic analysis of the sequence with other known yeast and fungal phytases, PPHY has been shown to belong to 3-phytase (Fig. 1). Molecular docking of PPHY with its substrate and vanadate (inhibitor) displayed that vanadate interacts with the same aa residues in tertiary structure of the enzyme, paving the way for further experiments related to bifunctioanlity of the phytase as a vanadium dependent virtual haloperoxidase (Fig. 1) (Joshi and Satyanarayana 2015b).

4.3 Biochemical Characteristics of Recombinant Phytase

The recombinant phytase is a hexameric glycoprotein of ~420 kDa (monomeric protein is of ~70 kDa), where glycan represents 24.3% of the phytase. The temperature and pH optima of rPPHY are 60 °C and 4.0, similar to the endogenous enzyme.

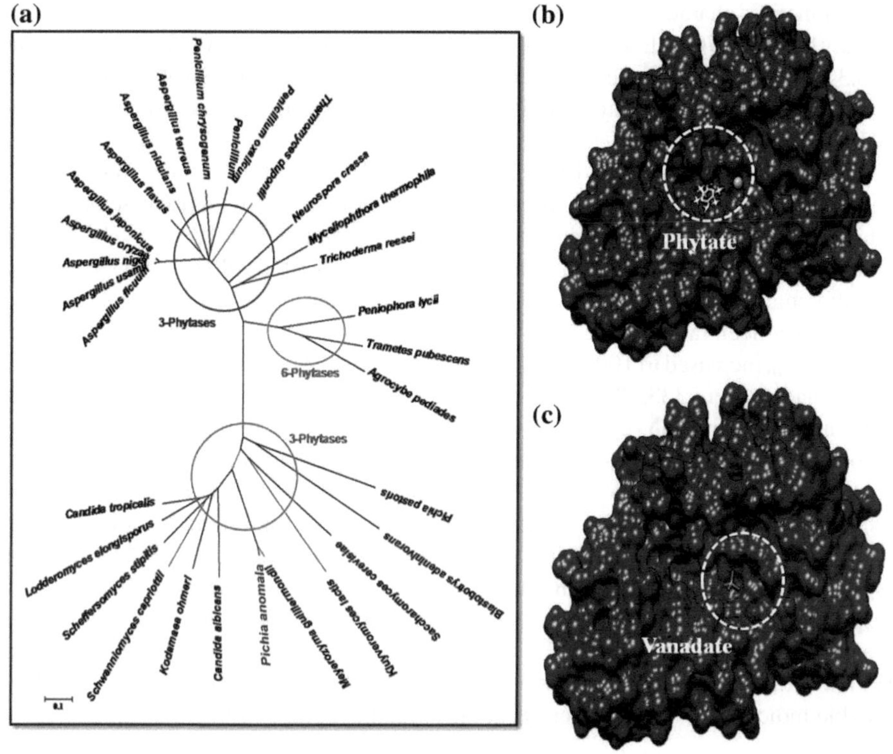

Fig. 1 Bioinformatic analysis of PPHY: Phylogenetic position of *P. anomala* phytase in comparison with the other yeast and fungal phytases (**a**) [Accession numbers of various phytases used in construction of the neighbour joining phylogenetic tree are: *Aspergillus terreus* (GenBank ID: AAB52507), *A. nidulans* (GenBandk ID: AAB96871), *A. ficuum* (GenBank ID: AAG40885), *A. niger* (GenBank ID: BAA74433), *A. usamii* (GenBank ID: ABA42097.1), *A. oryzae* (GenBank ID: AAT12504), *A. japonicas* (GenBank ID: ACE79228), *A. flavus* (NCBI ref. no.: XP002376973), *Penicillium* sp. Q7 (GenBank ID: ABM92788), *P. oxalicum* (GenBank ID: AAL55406), *P. chrysogenum* (NCBI ref. no.: XP002561094), *Myceliophthora thermophila* (GenBank ID: AAB52508), *Neurospora crassa* (GenBank ID: AAS94253), *Trichoderma reesei* (GenBank ID: EGR47873), *Trametes pubescens* (GenBank ID: CAC48234), *Agrocybe pediades* (GenBank ID: CAC48160), *Peniophora lycii* (GenBank ID: CAC48195), *P. anomala* (GenBank ID: FN641803), *P. pastoris* (Swiss-Prot: P52291), *Lodderomyces elongisporus* (NCBI ref no.: XP001527604), *Candida tropicalis* (NCBI ref. no.:XP002546108), *C. albicans* (NCBI ref. no.: XP713452), *Kodamaea ohmeri* (GenBank ID: ABU53001), *Saccharomyces cerevisiae* (GenBank ID: EDN64708), *Kluyveromyces lactis* (GenBank ID: CAA83964), *Talaromyces dupontii* (GenBank ID: AAB96873.1), *Scheffersomyces stipitis* CBS 6054 (NCBI ref. no.: XP001385108), *Meyerozyma guilliermondii* (GenBank ID: CAL69849.1), *Blastobotrys adeninivorans* (GenBank ID: CAJ77470.1), *Schwanniomyces capriottii* (GenBank ID: ABN04184.1)]; Docking of sodium phytate (**b**) and vanadate (**c**) with rPPHY structure. Position occupied by sodium phytate in the substrate binding cleft is indicated by *dotted circles*

The kinetic characteristics K_m, V_{max}, k_{cat}, and k_{cat}/K_m of rPPHY are 0.2 ± 0.03 mM,78.2 ± 1.43 nmol mg^{-1} s–, $65{,}655 \pm 10.92$ s^{-1}, and 328.3 ± 3.12 μM^{-1} s^{-1}, respectively (Joshi and Satyanarayana 2014). The half-life, activation energy, temperature quotient, activation energy of thermal denaturation, and ΔH^0_d i.e. enthalpy of the phytase are 4.0 min (at 80 °C), 27.72 kJ mol^{-1}, 2.1, 410.62 kJ mol^{-1}, and 407.8 kJ mol^{-1} (65–80 °C), respectively. The ΔG^o_d i.e. free energy of the process (increases from 49.56 to 71.58 kJ mol^{-1} with rise in temperature, while ΔS^0_d (entropy of inactivation) remains unchanged at 1.36 kJ mol^{-1} K^{-1} (Joshi and Satyanarayana 2015b). Vanadate competitively inhibited rPPHY activity, as V_{max} at different concentrations of vanadate remained constant (78.13 η kat mg^{-1} s^{-1}). When the concentration of the inhibitor was increased gradually from 0 to 500 μM, the apparent K_m increased to 1000 μM. K_i values at 50, 100, and 500 μM concentrations of metavanadate were 333, 200, and 1.2 μM, respectively. In the presence of metavanadate and H_2O_2, rPPHY exhibited haloperoxidase activity as it changed red-orange colour of phenol red to blue-violet (Joshi and Satyanarayana 2015b). It has been reported that Ag^{2+}, Al^{2+}, Pb^{3+}, and Sn^{2+} completely destroyed rPPHY activity at 5 mM, while Ba^{2+}, Cu^{2+}, Fe^{2+} and Zn^{2+} exerted a lower degree of inhibition, Na^{+} did not affect rPPHY. The cations Ca^{2+} and Ni^{2+} slightly stimulated the activity of rPPHY. Melting temperature (T_m) values of both native and recombinant phytase were 73 and 70 °C respectively, suggesting that the equilibrium between folded and unfolded forms reaches quickly in case of rPPHY than the native. High T_m of both the native and rPPHY indicates toward their higher temperature stability as T_m is known as one of the reliable indicators of protein thermostability (Kumar et al. 2000).

5 Applications of *P. anomala* Phytase

Phytase is commonly used in phytate reduction in the industries engaged primarily in feed and food preparation, and in generating myoinositol phosphates. Besides this, day by day more stress is being laid on its application in the field of aquaculture, soil amendment and in semisynthesis of various peroxidases. About 66.7% of phosphorus of vegetable (plant) origin feedstuffs is present as phytates (Harland and Morris 1995). Ruminants harbor gut microflora which catalytically liberate inorganic phosphorus (Pi) from phytic acid. Humans and monogastric farm animals including chickens and pigs produce negligible amounts of phytase in their GI tracts (Vohra et al. 2006). Their nutritional requirement of P is fulfilled by supplementing plant based meals with phosphate. The phytin phosphorus gets disposed in the form of animal excreta (Mullaney et al. 2000). Soil and water-borne microorganisms and enzymatic activity cleave the phytic acid found in the manure of these animals. Eutrophication problem occurs upon transport of the liberated phosphorus into the aquatic systems (Vohra and Satyanarayana 2003). This culminates in O_2 depletion due to excessive growth of algal biomass. According to Nelson et al. (1971), the availability of P can be improved by adding microbial phytase to the animal feed or by utilizing phytase-rich cereal based diets. Improvement in the growth and phosphorus retention in the body of broiler chicks was observed upon supplementation of their feed with yeast biomass with phytase activity. Additionally, there was considerable decrease in the phosphorus excretion in the faecal matter of chicks fed with phytase-fortified diets. Furthermore, the feed conversion (FC) ratio declined from 2.272 in the control to 1.949 (Vohra and Satyanarayana 2003; Vohra et al. 2006). As an alternative to fish meal, *Glycine max* (soybeans) derived products are some of the promising feed ingredients (Swick 2002) owing to their price, aa composition and easy availability. In soybean based meal, 60–80% of P occurs as phytate that remains unavailable to monogastric fishes (Pallauf and Rimbach 1997; Raboy 1997). During feeding trials of milk fish with soybean based feed, the P excretion was higher in the control fishes as compared to the same feed supplemented with thousand units per kilogram of cell-bound phytase, clearly suggesting that soybean based fish diet with phytase supplementation can be used for the mass cultivation of marine milkfish (Hassan et al. 2009). The fishes *Labeo rohita* and *Clarias batrachus* fed with phytase improved feed registered better rate of survival and growth due to ameliorated P and better protein assimilation, besides mitigating excretion of NH_3 and Pi. The NH_3 and P excretion were also minimized in the phytase fed fishes. The membrane-bound PPHY could effectively dephytinize wheat flour, rice flour and soybean flour and wheat bran to a varied extent, with improved dephytinization at 60 than at 37 °C. The permeabilized *P. anomala* cells exhibited higher efficacy over the non-permeabilized yeast cells. Since permeabilization allowed immobilization and reuse of the *P. anomala* cells for sustained dephytinization of soymilk, in near future it can be used in the development of a continuous system (Kaur and Satyanarayana 2010). Phytases are renowned for their significant role in environmental protection by reducing the P levels in animal excreta and reducing the need to supplement diets of monogastric farm animals with additional phosphorus. The use of phytase in aquaculture is growing, since it permits the use of cost-effective plant based meals.

Applicability of recombinant phytase was also tested in the treatment of soy protein. The treatment with rPPHY for 3 h led to the separation of glycinin and liberation of phytate-bound phosphorous (Joshi and Satyanarayana 2014). Upon treatment with rPPHY, glycinin was precipitated due to the hydrolysis and removal of phytate. The removal of phytate reduced the solubility of glycinin that led to its precipitation, while β-conglycinin fraction remained soluble in the supernatant (Fig. 2) due to its glycoprotein nature (Saito *et al.* 2001). This finding may pave the way for the use of PPHY in producing soy products with improved quality as well as special foods for people who are allergic to the specific components of soy proteins. The treatment of poultry feeds with rPPHY for 2 h led to the liberation of inorganic phosphate (Joshi and Satyanarayana 2015a). Among the feeds, broiler starter feed was rapidly hydrolysed, followed by pre-starter and finisher feeds (Fig. 2). This suggests that ingredients of pre-starter and finisher feeds are not resistant to dephytinization by rPPHY, but may need a slight modification or pretreatment for efficient phytate hydrolysis. Wheat flour contains up to 4 mg g^{-1} phytic acid (Garcia-Estepa et al. 1999) that lowers bioavailability of minerals. The supplementation of bread dough with rPPHY increased Pi, reducing sugars and

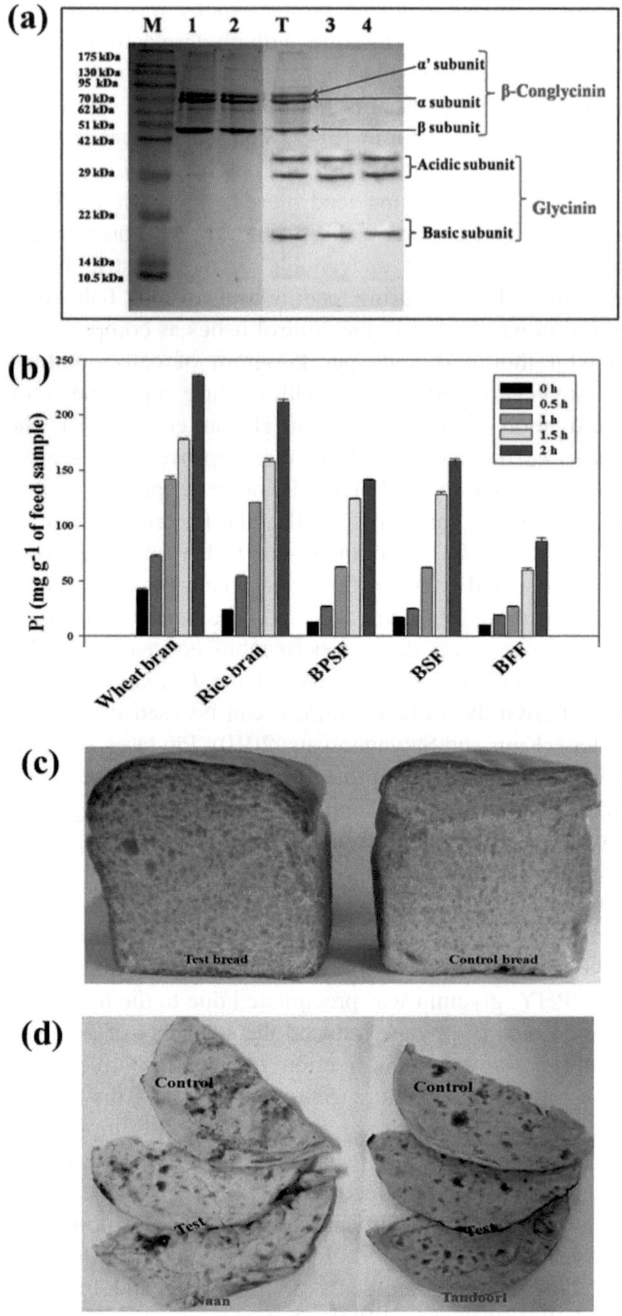
(a)
M 1 2 T 3 4
175 kDa
130 kDa
95 kDa
70 kDa
62 kDa
51 kDa
42 kDa
29 kDa
22 kDa
14 kDa
10.5 kDa
α' subunit
α subunit
β subunit
β-Conglycinin
Acidic subunit
Basic subunit
Glycinin
(b)
Pi (mg g-1 of feed sample)
0 h
0.5 h
1 h
1.5 h
2 h
Wheat bran
Rice bran
BPSF
BSF
BFF
(c)
Test bread
Control bread
(d)
Control
Test
Control
Test
Naan
Tandoori

◂**Fig. 2** Applications of recombinant phytase: **a** fractionation of allergenic soy proteins viz. β-conglycinin and glycinin. SDS-PAGE profile of separated glycinin and β-conglycinin by the action of rPPHY; M: molecular weight markers, L1-L2: β-conglycinin fraction, LT: total soy protein, L3-4: glycinin fraction; **b** The liberation of phytate phosphorous by the hydrolysis of different poultry feeds by rPPHY under simulated gut conditions; **c** Texture of whole wheat breads prepared with (test bread) and without the addition (control bread) of rPPHY; **d** Texture of unleavened flat Indian breads prepared with (test bread) and without the addition (control bread) of rPPHY

content of soluble protein in bread as compared to bread made with commercial enzymes without any change in the texture of bread (Fig. 2). The supplementation of dough rPPHY resulted in considerable (72.5%) reduction in phytic acid content of bread (Joshi and Satyanarayana 2015a). Apart from the applications of rPPHY as food and feed additive, the recombinant enzyme exhibits haloperoxidase activity upon inhibition with vanadate ions, therefore, its potential as vanadium-dependent haloperoxidase can be explored further.

Both endogenous phytase and its recombinant counterpart are of immense utility due to applicability as a supplement in chick and fish feeds, in fractionation of allergenic soy protein, in reducing phytate content of wheat flour and soymilk, and in the preparation of both fermented whole wheat and unleavened flat Indian breads.

6 Conclusions

Pichia anomala produces a cell-bound phytase that exhibits desirable features such as acidstability, thermostability, resistance to digestive proteases and broad substrate spectrum. Optimization of production parameters and concomitant use of low cost media components led to cost effective production of high phytase titres. Heterologous expression of PPHY in methylotrophic yeast *P. pastoris* followed by process optimization led to high extracellular expression of recombinant phytase, which simplifies downstream processing of the enzyme. Both native and recombinant phytases could be used as an additive to different feeds and foods. Novel applications of recombinant phytase in the fractionation of allergenic soy proteins and in generating vanadate-dependent haloperoxidase are expected to lead to novel applications.

References

Casey, A. 2004. *J. Biotechnol.* 110:313–322.

Cho, J.S., Lee, C.W., Kang, S.H., Lee, J.C., Bok, J.D., Moon, Y.S., Lee, H.G., Kim, S.C. and Choi, Y.J. 2003. *Curr. Microbiol.* 47:290–294.

Fredlund, E., Broberg, A., Boysen, M.E., Kenne, L. andSchn¨urer, J. 2004b. *Appl. Microbiol. Biotechnol.* 64:403–409.

Garcia-Estepa, R.M., Guerra-Hernandez, E. and Garcia-Villanova, B. 1999. *Food Res. Int.* 32:217–21.

Greiner, R., Konietzny, U. and Jany, K.D. 1993. *Arch. Biochem. Biophys.* 303:107–113.

Hara, A., Ebina, S., Kondo, A. and Funagua, T. 1985. *Agric. Biol. Chem.* 49:3539–3544.

Harland, B.F. and Morris, E.R. 1995. *Nutr. Res.* 15:733–754.

Hassan, S., Altaff, K. and Satyanarayana, T. 2009. *Pak. J. Nutr.* 8:341–344.

Howson, S.J. and Davis, R.P. 1983. *Enzyme Microb. Technol.* 5:377–389.

Huang, W.C. and Tang, I.C. 2006. In: Bioprocessing for Value-Added Products from Renewable Resources: New Technologies and Applications (ed.Yang, S.T), Elsevier, The Netherlands, pp. 185–224.

Huebel, F. and Beck, E. 1996. *Plant Physiol.* 112:1429–1436.

Ingvar, S. and Petter, M. 2011. *Antonie van Leeuwenhoek* 99:113–119.

Pallauf, J. And Rimbach, G. 1997. *Arch. Anim. Nutr.* 50:301–31.

Joshi, S. and Satyanarayana, T. 2014. *J. Ind. Microbiol. Biotechnol.* 41:977–87.

Joshi, S. and Satyanarayana, T. 2015a. *J. Ind. Microbiol. Biotechnol.*42:1389–1400.

Joshi, S. and Satyanarayana, T. 2015b. *Appl. Biochem. Biotechnol.* 176:1351–69.

Kaur, P. and Satyanarayana, T. 2005. *Process Biochem.* 40:3095–3102.

Kaur, P. and Satyanarayana, T. 2010. *J. Appl. Microbiol.* 108:2041–2049.

Kaur, P., Singh, B., Böer, E., Straube, N., Piontek, M., Satyanarayana, T. and Kunze, G. 2010. *J. Biotechnol.* 149:8–15.

Kerovuo, J., Lauraeus, M., Nurminen, P., Kalkkinen, N. and Apajalahti, J. 1998. *Appl. Environ. Microbiol.* 64:2079–2085.

Kerovuo, J., Rouvinen, J. and Hatzack, F. 2000. *Biochem. J.* 352:623–628.

Kim, Y.O., Kim, H.K., Bae, K.S., Yu, J.H. and Oh, T.K. 1998. *Enzyme Microb. Technol.* 22:2–7.

Kim, H.W., Kim, Y.O., Lee, J.H., Kim, K.K. and Kim, Y.J. 2003.c. *Biotechnol. Lett.* 25:1231–1234.

Konietzny, U. and Greiner, R. 2002. *Int. J. Food. Sci. Technol.* 37:91–812.

Kumar, S., Tsai, C.J. and Nussinov, R. 2000. *Protein Eng.* 13:179–191.

Latiffi, A.A., Salleh, A.B., Rahman, R.N., Oslan, S.N. and Basri, M. 2013. *Genes Genet. Syst.* 88 :85–91.

Maugenest, S., Martinez, I., Godin, B., Perez, P. and Lescure, A.M. 1999. *Plant Mol. Biol.* 39:503–514.

Maurya A.K., Parashar D. and Satyanarayana T. 2017. *Int. J. Biol. Macromol.* 94:36–44.

McCollum, E.V. and Hart, E.B. 1908. *J. Biol. Chem.* 4:497–500.

Mullaney, E.J., Daly, C.B. and Ullah, A.H. 2000. *Adv. Appl. Microbiol.* 47:157–99.

Nakano, T., Joh, T., Tokumoto, E. and Hayakawa, T. 1999. *Food Sci. Technol. Res.* 5:18–23.

Nelson, T.S., Sheih, T.R., Wodzinski, R.J. and Ware, J.H. 1971. *J. Nutr.* 101:1289–1294.

Oh, B.C., Choi, W.C., Park, S., Kim, Y.O. and Oh, T.K. 2004. *Appl. Microbiol. Biotechnol.* 63:362–372.

Pasamontes, L., Haiker, M., Wyss, M., Tessier, M. and Van Loon, A.P.G.M. 1997. *Appl. Environ. Microbiol.* 63:1696–1700.

Quan, C.S., Tian, W.J., Fan, S.D. and Kikuchi, Y.I. 2004. *J. Biosci. Bioeng.* 97:260–266.

Raboy,V. 1997. In: Cellular and Molecular Biology of Plant Seed Development(eds. B.A. Larkins and I.K. Vasil), Kluwer Academic Publishers, Dordrecht, pp. 441–447.

Rao, D.E.C.S., Rao, K.V., Reddy, T.P. and Reddy, V.D. 2009. *Crit. Rev. Biotechnol.* 29:182–198.

Rapoport, S., Leva, E. and Guest, G.M. 1941. *J. Biol. Chem.* 139:621–632.

Sajidan, A., Farouk, A., Greiner, R., Jungblut, P., Mueller, E.C. and Borriss, R. 2004. *Appl. Microbiol. Biotechnol.* 65:110–118.

Sano, K., Fukuhara, H. and Nakamura, Y. 1999. *Biotechnol. Lett.* 21:33–38.

Satio, T., Kohno, M., Tsumura, K., Kugimiya, W. and Kito, M. 2001. *Biosci. Biotechnol. Biochem.* 65:884–887.

Segueilha, L., Lambrechts, C., Boze, H., Moulin, G. and Galzy, P. 1992. *J. Ferment. Bioeng.* 74:7–11.

Singh, B. and Satyanarayana, T. 2008. *Bioresour. Technol.* 99:2824–2830.

Spencer, J.F.T. and Spencer, D.M. 1997.In: Yeasts in Natural and Artificial Habitats (eds. J.F.T. Spencer and D.M. Spencer), Springer, Berlin, pp. 33–58.

Swick, R.A. 2002. *Glob. Aquacult. Advocate.* 5:46–49.

Tambe, S.M., Kaklij, G.S., Kelkar, S.M. and Parekh, L.J.1994. *J. Ferment. Bioeng.* 77:23–27.

Van Eck, J.H., Prior, B.A. and Brandt, E.V. 1993. *J. Gen. Microbiol.* 139:1047–1054.

Verma, D. and Satyanarayana, T. 2012. *Afr. J. Biotechnol.* 11:13705–13709.

Vohra, A. 2002b. Production, purification, characterization and application of phytase from *Pichia anomala* (Hansen) Kurtzman. PhD thesis.

Vohra, A. and Satyanarayana, T. 2001. *Biotechnol. Lett.* 23:551–554.

Vohra, A. and Satyanarayana, T. 2002a. *Process Biochem.* 37:999–1004.

Vohra, A. and Satyanarayana, T. 2002c. *World J. Microbiol. Biotechnol.*18:687–691.

Vohra, A. and Satyanarayana, T. 2003 *Crit. Rev. Biotechnol.* 23:29–60.

Vohra, A. and Satyanarayana, T. 2004. *J. Appl. Microbiol.* 97:471–476.

Vohra, A., Rastogi, S.K. and Satyanarayana, T. 2006. *World J. Microbiol. Biotechnol.* 22:553–558.

Wang, Y., Yao, B., Zeng, H., Shi, X., Cao, S., Yuan, T. and Fan, Y. 2001. *Wei sheng wu xue bao* 41:198–203.

Wang, Y., Wang, Z., Du, G., Hu, Z., Liu, L., Li, J. and Chen, J. 2009. *Bioresour. Techol.* 100:1343–1349.

Waterham, H.R., Digan, M.E., Koutz, P.J., Lair, S.V. and Cregg, J. 1997. *Gene* 186:37–44.

Wyss, M., Pasamontes, L., Remy, R., Kohler, J., Kusznir, E., Gadient, M., Muller, F. and Van loon, A.P.G.M. 1998. *Appl. Environ. Microbiol.* 64:4446–4451.

Conventional and Non-conventional Yeasts for the Production of Biofuels

Volkmar Passoth

Abstract Liquid biofuels have great potential to replace fossil transportation fuels. Bioethanol, the current major biofuel, is mainly produced using the yeast *Saccharomyces cerevisiae*. However, present biofuels are largely generated from first generation, i.e. food grade, raw materials. Second generation, lignocellulosic biomass represents an abundant and cheap feedstock for bioethanol production; however, sugars present in cellulose and hemicellulose are difficult to access and not all of these sugars can be assimilated by *S. cerevisiae*. Other yeast species have been explored to ferment hemicellulose sugars such as xylose or to be specifically competitive in ethanol production. There are still attempts to introduce those species in industrial processes. The major approach has been to express heterologous genes enabling xylose assimilation in *S. cerevisiae*. Recently, by metabolic and evolutionary engineering in industrial isolates, *S. cerevisiae* strains applicable for industrial ethanol production from lignocellulose have been constructed. Apart from bioethanol, yeasts have been manipulated to produce butanol. Production of biodiesel and other advanced biofuels by oleaginous yeasts or engineered *S. cerevisiae* from lignocellulose has been proposed. Yeasts can also be used in side processes of biofuel production, such as preservation and pretreatment of biomass, or co-conversion of biomass to high value products.

Keywords Biofuels · Bioethanol · Biodiesel · Butanol · Terpenes · Alkanes · Lignocellulose

---

V. Passoth (✉)
Department of Molecular Sciences, Swedish University of Agricultural Sciences, Box 7015 SE-75007, Uppsala, Sweden
e-mail: volkmar.passoth@slu.se

T. Satyanarayana and G. Kunze (eds.), *Yeast Diversity in Human Welfare*, DOI 10.1007/978-981-10-2621-8_15

1 Introduction

Liquid fuels are currently mainly derived from non-renewable mineral oil. Ongoing depletion of mineral oil resources, environmental problems generated by the massive use of fossil raw materials, and concerns about political and economic dependencies of the global community on a few mineral oil exporting countries have been the main drivers to find alternative resources to generate fuels and chemicals. Biofuels can represent such alternative resources, as they are produced from renewable biomass. Bioethanol, biodiesel and biogas are presently the major biofuels that are used in transportation. Ethanol is usually produced by fermenting sugars to ethanol with the help of yeasts (mainly *Saccharomyces cerevisiae*), and biogas is produced by a complex consortium of bacteria and archaea under anaerobic conditions, whereas biodiesel is generated from oil that is extracted from certain oil plants. Currently, biofuels account for less than 5% of the total energy spent in transportation (Cheng and Timilsina 2011; Gnansounou 2010; Solomon et al. 2007). Moreover, the great majority of current biofuels is produced from "first generation" feedstock, i.e. biomass that can also be used to produce human food. In that way, since the production of first generation feedstock often requires a high input of fossil energy, effects of biofuels on fossil fuel replacement are rather low or they may even increase fossil fuel consumption in some specific cases (Gnansounou 2010; Hill et al. 2006). Therefore, biofuels generated from non-edible plant parts, lignocellulose, may represent a more sustainable replacement for fossil fuels. Lignocellulose is the most abundant biomass on earth, and biofuels and chemicals produced from this material have the potential to replace a substantial number of mineral oil based products (Vanholme et al. 2013). However, lignocellulose is the key structural element in plants and has evolved to be recalcitrant to degradation. Its utilisation for any biotechnological application requires a variety of thermochemical and enzymatic pretreatments, which are costly and generate inhibitors that can affect the subsequent fermentation processes. This is one of the reasons why second generation biofuels currently cannot compete with first generation biofuels or fossil fuels (Stephen et al. 2012).

Biofuel production, both from first and second generation substrates, includes a variety of steps, including production, harvest and storage of the plant biomass, pretreatment, fermentation, concentration of the fuel, and handling and valorisation of the side products; in some of these steps, yeasts can be utilised (Passoth 2014; Petrovic 2015; Vanholme et al. 2013). Ethanol, which represents the major biofuel on the global market, has for millennia been produced using the conventional yeast *Saccharomyces cerevisiae*. *S. cerevisiae* has also become a eukaryotic model organism and a major tool for molecular biology. Consequently, there are now methods to engineer this organism to, for instance, utilise unusual substrates which are present in lignocellulose, be resistant against fermentation inhibitors, or to generate other biofuels or high value products other than ethanol (Hong and Nielsen 2012; Passoth 2014; Petrovic 2015). Nevertheless, there is also a huge and largely unexplored potential among yeasts other than *S. cerevisiae*, the so called non-conventional yeasts. Yeasts span a higher phylogenetic distance than the whole tribe of chordates (Fischer et al. 2006), and we have just started to utilise their amazing metabolic diversity to convert lignocellulose or waste products to biofuels and valuable chemicals.

2 The Food Versus Fuel Debate

Plant biomass is the basis of both food and biofuel production. Currently, almost all commercially produced liquid biofuels are derived from first generation substrates, i.e. substrates that can also be converted to human food. Is it ethically acceptable to convert food to biofuels, while a considerable number of people still suffer from hunger and malnutrition? These types of questions are not easy to handle, and a growing number of scientific publications dealing with the ethics of biofuels illustrates the ongoing debate (Buyx and Tait 2011; Mohr and Raman 2013; Pols and Spahn 2014).

Utilisation of biofuels has increased in recent years; in parallel, an increase in food prices has been observed. An increased demand for agricultural products due to biofuel production is indeed one driver of increased food prices; however, other factors and especially the price for mineral oil greatly impact food prices because modern agriculture is largely reliant on inputs from fossil resources. Another factor is the global increase in meat consumption. Meat production requires up to tenfold more input energy per produced nutritive calorie compared to plant based food, which is also pushing up the general prices for food (Pelletier et al. 2011). One may even argue that high food prices do not necessarily cause hunger among the poorest. According to the United Nations, about 80% of people with an income of less than one Euro per day are involved in agricultural food production. Thus, an increase in food price may improve the income of such people (Thompson 2012). On the other hand, examples abound where poor farmers have been evicted from their land when there was an increasing demand for biofuels and food production. In many cases, land entitlements of poor farmers are not sufficiently protected by national laws, or those farmers lack the ability to enforce their rights (Mohr and Raman 2013).

There is a certain degree of agreement that second generation biofuels will avoid the food versus fuel trade-off. However, utilisation of lignocellulose as raw material will most probably not solve all problems correlated to biofuel production. Ethanol is produced from sugars present in the plants. The major reason that second generation ethanol has not reached the global market (with a few exceptions) is the high costs of extracting the sugar from cellulose and hemicellulose. If there is a technology for cheaply obtaining the sugar-monomers from these macro-molecules, these sugars could also be used in food production (Thompson 2012). Non-food plants can grow on non-arable, marginal land that is not suitable for food production. However, biofuel crops usually grow better on arable land, and changes in land use have been observed when the price for those energy plants can compete with that of food plants (Mohr and Raman 2013). Moreover, not all marginal lands are suited for energy crop production, for instance erosion can increase when steep slopes are cultivated or saturated soils can emit high levels of greenhouse gases when drained and utilized for energy crop production (Gollany et al. 2015).

Arising from the different debates and views on biofuels, the Nuffield Council on Bioethics has formulated six principles for evaluating biofuel production. These principles state that biofuels (i) shall not be produced at the expense of essential human rights, including access to sufficient food and water, health rights, work rights and land entitlements, (ii) shall be produced in a sustainable way, including careful use of water, land and other resources, and keeping biodiversity, and (iii) shall decrease greenhouse gas emission. People involved in biofuel production shall (iv) obtain just reward, (v) costs and benefits of biofuels shall be fairly shared. If all principles (i–v) are fulfilled, (vi) an ethical duty is formulated, to develop this biofuel (Buyx and Tait 2011). At the current stage, many of these principles are not put into practice by major consumers and producers of biofuels, such as European Union, USA, or China. This includes issues of human rights, sustainability, and just rewards and fair trade, which are in practice difficult to control when biofuels are imported (Mortimer 2013). Moreover, the impact of biofuel production on greenhouse gas emission is difficult to determine. For instance, removal of non-edible, lignocellulosic parts of biomass from agricultural or forest soils over a certain limit can result in erosion, degradation of soil organic matter, and in a long term, an increase in greenhouse gas emissions (Gollany et al. 2015). Nevertheless, the growing awareness of problems related to biofuel production is positive, and some policies have been adjusted towards obtaining ethically acceptable circumstances for biofuel production (Mortimer 2013). Indeed, problems of mineral oil consumption still persist and the replacement of mineral oil-derived fuels by biofuels may reduce our dependency on fossil carriers of energy. However, biofuels alone will not solve the problem of dependency in mineral oil, and lifestyle shifts towards consuming less energy will likely be necessary (Pols and Spahn 2014).

3 Ethanol Production

3.1 Ethanol as a Fuel

Bioethanol is currently the major biofuel that is produced worldwide, in 2014, 88.2 million m^3 were produced worldwide (US Energy Information Administration 2016). Ethanol, the product of the fermentative metabolism of yeasts, has been used by humankind for millennia in beverages, as drug or disinfectant. Although it is in modern times rather consumed for its inebriating effect, it was earlier an almost essential agent for conservation of beverages, due to limited access to clean water resources (Vallee 1998). Ethanol also has a history as biofuel. The engine constructed by Nikolaus Otto in 1860, on which most modern gasoline-based engines are based, ran on ethanol. Henry Ford's automobiles built at the turn of the

19th–20th centuries ran on ethanol (Solomon et al. 2007). However, in the 1920s, ethanol as a fuel was replaced by the cheaper gasoline. The oil crisis of the 1970s prompted a temporary increase in interest in ethanol as a fuel, but when oil prices decreased, this interest more or less disappeared, except in Brazil, where there is a substantial production of sugar and thus the opportunity for large-scale ethanol production. Towards the end of the 20th century, ethanol was utilised to replace lead compounds as octane booster, but later it was also seen as a gasoline replacement. Currently, ethanol is blended with gasoline and plays an important role in decreasing the dependency on fossil resources in the European Union and the USA (Gnansounou 2010; Solomon et al. 2007).

3.2 First- and Second Generation Feedstocks for Ethanol Production

Ethanol is produced by the fermentative degradation of sugar by the yeast. Thus, plants accumulating sugar or polysaccharides that can be degraded to simple sugars are common feedstocks for ethanol production. This includes sugar plants such as sugar cane in Brazil and sugar beets in Europe, and starch containing plants such as corn in the USA and wheat in Europe. However, the greatest proportion of biomass is comprised of lignocellulose. A major component of lignocellulose consists of cellulose and hemicellulose, which are both polysaccharides and, thus, potential sugar resources for ethanol production. However, releasing sugars from lignocellulose is generally difficult, as it is the key structure component of the plants and is recalcitrant to degradation. Cellulose is a poly-glucose molecule, with 1,4-βglycosidic bonds. Due to its characteristics, cellulose forms a crystalline structure that is poorly accessible for degrading enzymes (Carpita and Gibeaut 1993; Mosier et al. 2005). Hemicellulose is a branched heteropolysaccharide and its composition differs among plant species (Girio et al. 2010). In most plants, especially crops and hardwood, the major component of hemicellulose is xylan, which is a polymer of the pentose xylose. Other abundant sugar monomers in hemicellulose are the pentose arabinose and the hexoses mannose and galactose. The hemicellulose is wrapped around the cellulose fibre, providing an additional element of stability. Apart from this, the cellulose is bound to the hetero-poly-aromatic compound lignin, which forms a scaffold for the polysaccharides. Lignin is hydrophobic and due to this provides protection against enzymatic hydrolysis. The structure of lignocellulose can only be broken by a massive thermochemical pre-treatment. However, this pre-treatment, apart from opening the structure of the polysaccharides for subsequent enzymatic hydrolysis, also produces compounds that are inhibitory for subsequent fermentation processes (Jönsson et al. 2013; Ko et al. 2015). These inhibitors include dehydrogenation products of pentoses and hexoses, furfural and hydroxy-methyl-furfural (HMF), respectively, acetic acid, which arises from acetyl-groups in hemicellulose and lignin, other weak acids, which are degradation products of sugars, and phenolic compounds from lignin degradation. After thermochemical pre-treatment, the polysaccharides are degraded by the corresponding cellulases and hemicellulases. A variety of enzymes is now commercially available for the degradation of lignocellulose biomass (Chandel et al. 2012); however, enzyme costs are still a crucial factor of costs of bioethanol, and obtaining profit from lignocellulose ethanol production may require the establishment of innovative approaches such as on-site enzyme production (Liu et al. 2016).

3.3 Manipulating S. cerevisiae *to Produce Ethanol from Lignocellulose*

The conventional yeast *S. cerevisiae* can efficiently ferment hexoses, however, pentoses cannot be utilised. A variety of non-conventional yeast species can naturally ferment xylose to ethanol, including *Scheffersomyces stipitis*, *Scheffersomyces shehatae*, *Spathaspora passalidarum*, *Pachysolen tannophilus*, or *Hansenula* (*Ogataea*) *polymorpha* (Kurtzman et al. 2011; Ryabova et al. 2003). However, these species have a rather low tolerance to fermentation inhibitors released during thermochemical pre-treatment and to high ethanol concentration. Moreover, these species are respiratory yeasts that cannot grow under anaerobic conditions, only ferment under oxygen limitation, and re-assimilate the formed ethanol even under limited aeration (e.g. Görgens et al. 2005; Passoth et al. 2003). Thus, it is very difficult to establish an industrial ethanol process using these yeasts. Therefore, considerable research has focused on genetically engineering *S. cerevisiae* to ferment xylose and other sugars released from hemicellulose. Fungi usually assimilate xylose by a two-step mechanism, firstly reducing the sugar to xylitol and secondly re-oxidising it to xylulose, which is further metabolised via the non-oxidative pentose-phosphate pathway. In most fungi, xylose reduction is performed by a NADPH-dependent xylose reductase (XR), while xylitol oxidation to xylulose by xylitol dehydrogenase (XDH) requires NAD^+. This co-factor dependency generates a redox imbalance, which is the reason that, although a variety of yeasts can assimilate xylose, only a few can ferment it to ethanol. Xylose-fermenting yeasts have a dual cofactor utilisation of NADH and NADPH by XR. Expression of the *S. stipitis* XR and XDH genes were the basis of the first successful attempts to manipulate *S. cerevisiae* to ferment xylose (Kötter and Ciriacy 1993). Since then, a number of engineered strains have been generated, mainly based on the laboratory strain CEN.PK (see Passoth 2014 for a recent review). More recently, genetic engineering was also performed in industrial isolates, which can perform fermentations of lignocellulose hydrolysates under industrial production conditions. Garcia Sanchez et al. (2010a, b) generated a strain, TMB 3130, based on the diploid wine isolate USM21 from South Africa (Westhuizen and Pretorius 1992). In this strain, *XYL1*, *XYL2* (encoding XR and XDH, respectively from *S. stipitis*), and *XKS1* (encoding *S. cerevisiae* xylulo-kinase) were over expressed (Fig. 1). In addition, the metabolic pathway for arabinose assimilation, encoded by araA from *Bacillus subtilis*, araB and araD from *Escherichia coli* was expressed in the yeast strain. The initially obtained strain was evolutionarily engineered by continuous cultivation over about 65 generations on xylose and arabinose as carbon sources. The resulting strain produced a considerable amount of ethanol from xylose; however, the undesirable side product xylitol was also produced to a substantial extent. Arabinose was consumed, but almost completely converted to arabitol (Garcia Sanchez et al. 2010b). Several attempts were made to express a cofactor-independent xylose isomerase (XI) in *S. cerevisiae* and by this to overcome the problem with the redox imbalance. However, in most cases, enzyme activities were too low for appropriate xylose conversion (Bergdahl et al. 2012). However, using a codon-optimised *XylA* gene from *Clostridium phytofermentans*, growth on xylose could be obtained. The XI encoded by the modified *XylA* was not inhibited by xylitol, in contrast to eukaryotic xylose isomerases. Xylitol is always formed to some extend in genetically engineered *S. cerevisiae*, due to the activity of non-specific aldose reductases (Brat et al. 2009). This codon optimised *XylA* gene was then used to engineer the industrial Ethanol Red strain to ferment xylose. Apart from *XylA*, a variety of other genes was expressed in the strain, including a *S. cerevisiae* sugar transporter (*HXT7*), which was modified to transport both glucose and xylose, genes of the pentose phosphate pathway (Fig. 1), and of the arabinose assimilation pathway. The resulting transformants were still not adequate for xylose fermentation. They were therefore further manipulated by chemical mutagenesis, and genome shuffling, i.e. mass meeting of isolated spores with each other and with those of the parental strain. Several further crossing segregation experiments were performed with strains displaying superior inhibitor tolerance, and finally, strains with good tolerance to spruce hydrolysate, better glucose consumption rates, and improved final ethanol concentrations compared to the original strain were obtained (Demeke et al. 2013a, b). Genome comparisons revealed that in the evolved strain the heterologous *XylA*-gene had become part of an extrachromosomal circular DNA element (eccDNA), which finally resulted in a multiple (nine-fold) integration of the gene into the genome of the manipulated strain (Demeke et al. 2015).

Attempts have been made to generate *S. cerevisiae* strains which can directly convert cellulose and other polysaccharides to ethanol, using an approach called surface engineering. In surface engineered strains, polymer-degrading enzymes such as cellulases, hemicellulases or amylases are fused to a glycosylphosphatidylinositol anchoring system, resulting in the display of these enzymes on the surface of the engineered cells. Direct conversion of lignocellulosic biomass to ethanol and other compounds by those strains represents a first step towards consolidated bioprocessing of lignocellulose to ethanol. However, development of those strains also requires the engineering of the secretary pathways of *S. cerevisiae* and in general the generation of strains able to cope with the stress related to the fermentation of lignocellulose substrates (Hasunuma et al. 2015).

Fig. 1 Genes overexpressed in industrial isolates to obtain a xylose-fermenting *S. cerevisiae* suitable for commercial ethanol production from lignocellulose hydrolysate. Manipulated steps include xylose transport, xylose-reductase (*XR*), xylitol dehydrogenase (*XDH*), and the pentose phosphate pathway (*PPP*). *HXT7* encodes a *S. cerevisiae* sugar transporter, but was mutated to transport both glucose and xylose), *XYL1* and *XYL2* are from *S. stipitis*, *XKS1*, *TKL1* and -*2*, and *TAL1* and -*2* are *S. cerevisiae*-genes, and *XylA* is a a codon optimised gene from *Clostridium phytofermans* (Garcia Sanchez et al. 2010a, b; Demeke et al. 2013a, b)

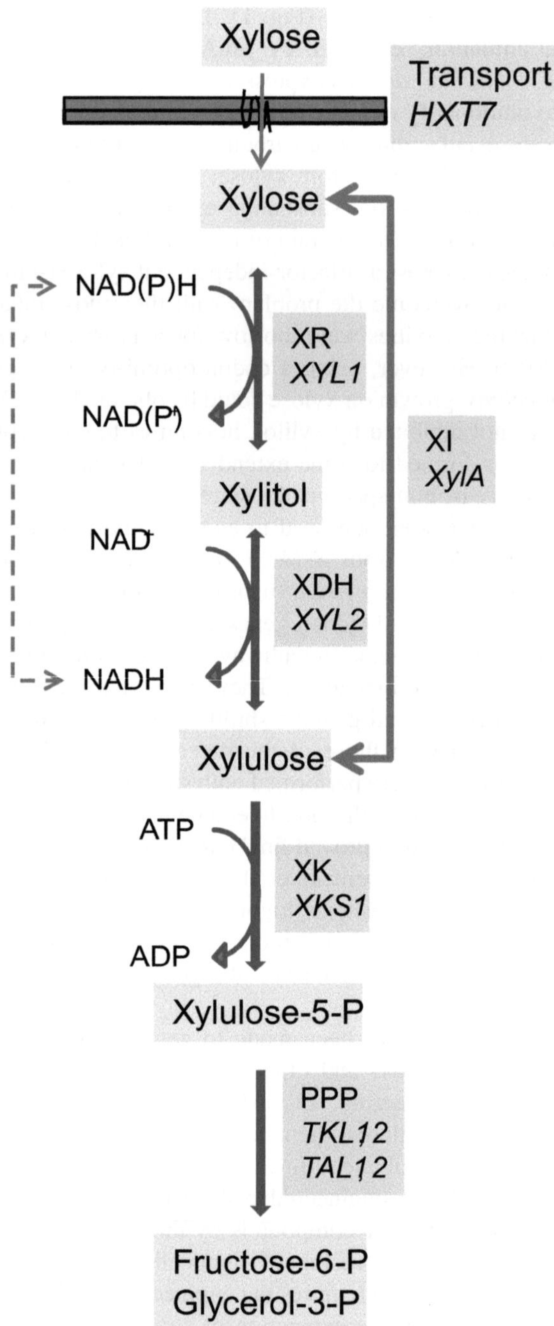

3.4 *Genetic Engineering of* S. cerevisiae *for Increased Inhibitor Tolerance*

Apart from manipulating *S. cerevisiae* to ferment xylose to ethanol, a variety of attempts have been made to obtain strains with increased resistance to the fermentation inhibitors generated during thermochemical pre-treatment of lignocellulose. Furfural and HMF are detoxified by conversion to the less toxic alcohols furfuryl alcohol and 2, 5-bis hydroxymethylfuran, which occur mainly by NADH and NADPH dependent pathways, respectively. Strains resistant to the inhibitors in lignocellulose hydrolysate showed enhanced NADH and NADH-dependent furfural reduction activities (Sanchez et al. 2012). Metabolic engineering towards an enhanced activity of the pentose—phosphate pathway resulted in strains with higher tolerance to furfural and HMF, most probably due to increased production of NADPH and, due to this, increased ability to reduce these compounds. NADPH can also be used to detoxify reactive oxygen species (ROS), which are formed by the cell in response to an exposure to furfural and HMF (Allen et al. 2010). Other attempts to overproduce NADPH included the overexpression of an $NADP^+$-dependent cytosolic aldehyde dehydrogenase (Zheng et al. 2012). However, this enzyme converts acetaldehyde to acetic acid, which is itself an inhibitor of fermentation.

Acetic acid is a weak organic acid, which is frequently used as antimicrobial agent for food preservation. At a pH below its pK_a (4.76), more than 50% of it is protonated and can freely diffuse through the cell membrane; it can also be transported into the cell by the facilitator Fps1. Inside the cell it dissociates, which impacts the intracellular pH, disrupting various important metabolic pathways, including glycolysis and amino acid transport (Sousa et al. 2012). Cells exposed to acetic acid undergo programmed cell death, which is probably due to the formation of ROS (Ludovico et al. 2001). Resistant cells of a genetically modified, xylose-fermenting *S. cerevisiae* have been isolated by long term cultivation at a pH below the pK_a and at increasing concentrations of acetic acid. The resulting isolates could tolerate up to 6 g/l acetate. During selection, high xylose consumption was observed, probably because of the high energy demand of detoxification (Wright et al. 2011). Genetic engineering of *S. cerevisiae* to overproduce ascorbic acid resulted in increased tolerance against acetic acid, probably because of detoxification of ROS (Martani et al. 2013). Overexpression of the transcription factor *HAA1* also increased acetate tolerance. This transcription factor down-regulates expression of the facilitator Fps1, among other effects (Tanaka et al. 2012). Expressing an NADH-dependent aldehyde dehydrogenase in *S. cerevisiae* enabled the resulting strain to use acetate as electron acceptor, converting it to ethanol. Thus, in this case detoxification was combined with product generation (Guadalupe-Medina et al. 2013). When nitrophenylphosphatase was inactivated in a xylose-fermenting, engineered *S. cerevisiae* strain by deleting *PHO13*, xylose fermentation was improved (Van Vleet et al. 2008). Apart from this, the deletion strain also produced more ethanol compared to the wild—type in the presence of common inhibitors such as acetate, formic acid, furfural and HMF, and in lignocellulose (rice straw) hydrolysate. As response to *pho13* deletion, increased expression of genes involved in NADPH-production, the pentose phosphate cycle, glycolysis and alcoholic fermentation has been observed. The increase in xylose fermentation ability may thus be due to increased activity of the pentose phosphate pathway, which is crucial for xylose assimilation and the increased tolerance against acetic acid may be achieved by increased NADPH-production (Fujitomi et al. 2012; Kim et al. 2015). Further manipulations included overexpression of a formate dehydrogenase gene (*FDH1*), to obtain increased resistance against formic acid and expression of a laccase gene from *Trametes versicolor* (lcc2), to obtain a strain resistant to degradation products of lignin (Larsson et al. 2001).

3.5 Manipulations for Improved Fermentation of First Generation Substrates

Commercially, ethanol is currently almost exclusively produced from first generation substrates. Accordingly, efforts have been made to improve the conversion of first generation substrates to ethanol. Sucrose is the major substrate of sugar cane and sugar beet-based ethanol production, and thus, improving the ability of yeast to ferment sucrose would have a substantial effect on sustainability of global ethanol production. *S. cerevisiae* hydrolyses most of the sucrose extracellularly by secreting invertase. There is also some capacity to directly transport sucrose into the cell via a proton symport system. This system consumes energy, hence, less biomass would be formed and thus more ethanol. Basso and colleagues constructed a strain where the signal sequence for export out of the cell was removed from the invertase-gene. This strain hydrolysed most of the sucrose intracellularly; however, growth rate was diminished and the residual sucrose concentration in the medium was high. Selection in anaerobic, sucrose-limited continuous fermentation finally resulted in a strain with higher affinity and higher growth and ethanol production rates and increased ethanol yield compared to the wild type (Basso et al. 2011). This indicates that even on a conventional substrate such as sucrose there is potential for process improvement.

Yeast cells form glycerol during alcoholic fermentation to re-oxidise NADH formed, for instance, during amino acid synthesis (van Dijken and Scheffers 1986). Glycerol is also produced in response to extracellular stress; it serves as a compatible solute to balance osmotic stress (Ansell et al. 1997). However, glycerol production removes carbon from ethanol formation, decreasing the yield of ethanol production, and the cell loses energy, since ATP used for phosporylation of the glucose will not be regenerated. Various attempts have been made to engineer strains for decreased glycerol production (summarised in Passoth 2014). However, glycerol is also involved in stress response; strains without glycerol production were sensitive to osmotic and other stresses and lacked the ability to grow anaerobically (Ansell et al. 1997; Valadi et al. 1998), and would thus not be suitable for industrial use. Recently, experiments of genetic manipulation have been combined with genome shuffling by multiparental protoplast fusion of strains with desired phenotypes, and a variety of stress-tolerant, low glycerol producing strains could be isolated (Guo et al. 2011; Tao et al. 2012; Wang et al. 2012).

3.6 Ethanol Production with Non-conventional Yeasts

Usually, ethanol is produced using the conventional yeast *S. cerevisiae*; it is almost a dogma that this yeast dominates high ethanol environments because of its exceptional ethanol tolerance (Woolfit and Wolfe 2005). However, a few years ago it was observed that *S. cerevisiae* had been outcompeted by *D. bruxellensis* in a Swedish industrial ethanol production plant. In this plant, the production process obviously went on for several years, driven by this non-conventional yeast (Passoth et al. 2007). In contrast, *D. bruxellensis* frequently occurs in Brazilian ethanol plants, but there its occurrence is correlated with a decrease in ethanol productivity and—yield (de Souza Liberal et al. 2007). Further investigation showed that this yeast produces less glycerol than *S. cerevisiae*, has a higher biomass yield and an ethanol yield comparable or even higher than industrial *S. cerevisiae* strains. On the other hand, its specific ethanol production- and growth rates were only about 20% of those found for *S. cerevisiae* (Blomqvist et al. 2010). To date it is not completely clear how this yeast can outcompete the faster growing *S. cerevisiae*. However, there are some common principles for outcompetition. *D. bruxellensis* has been shown to dominate fermentations where the sugar concentration was quite low, such as in ethanol plants running continuous fermentations with cell recirculation (de Souza Liberal et al. 2007; Passoth et al. 2007), towards the end of wine production, where it acts as a spoilage yeast, or in the last stages of lambic beer production (Blomqvist and Passoth 2015; Schifferdecker et al. 2014; Steensels et al. 2015). In a glucose-limited continuous co-cultivation, *S. cerevisiae* was outcompeted within a few days. This can be due to a higher affinity of *D. bruxellensis* glucose transporters for the substrate compared to those of *S. cerevisiae*, a higher energy efficiency of metabolism, or a combination of both (Blomqvist et al. 2012). When cultivated under glucose-limitation, *D. bruxellensis* expressed several potential high-affinity sugar transporters (Tiukova et al. 2013), which may be responsible for higher affinity towards the limited substrate. High affinity sugar transporters of *D. bruxellensis* have been shown to have a 30–60-fold higher affinity towards glucose compared to those of *S. cerevisiae* (van Urk et al. 1989). There was also no detectable expression of glycerol-phosphate-phosphatase, which may explain the low production of glycerol and thus improve energy efficiency for the cell. Moreover, genes of the respiratory chain, including that of complex I were expressed, which may enable the cell to utilize even low amounts of oxygen in the medium for oxydative phosphorylation (Tiukova et al. 2013).

Apart from high affinity and/or high energy efficiency of metabolism, other factors may also contribute to *D. bruxellensis*' competitive ability. Strains isolated from Brazilian ethanol plants were able to assimilate nitrate, and there was a co-incidence of increasing nitrate concentrations in the sugar cane juice and the occurrence of *D. bruxellensis* in the industrial processes (de Barros Pita et al. 2011). However, assimilation of nitrate cannot explain all episodes of outcompetition by *D. bruxellensis*, as for instance, the Swedish isolates were not able to assimilate nitrate (Schifferdecker et al. 2014), and no nitrate was found in the Swedish ethanol process (Blomqvist 2011). Competitiveness may also be conferred by the interaction with other microorganisms. In most ethanol plants in which *D. bruxellensis* was found in substantial amounts, the lactic acid bacterium *Lactobacillus vini* was co-isolated (Blomqvist and Passoth 2015). The basis of this interaction is not really clear. In batch co-fermentations, ethanol production of both *S. cerevisiae* and *D. bruxellensis* were negatively influenced (Tiukova et al. 2014a). On the other hand, in subsequent batch cultures with recycled cells and sugar cane juice as substrate, no negative effect of *L. vini* was observed, rather a slightly stimulatory effect on *D. bruxellensis* (de Souza et al. 2012). Nevertheless, the bacterium obviously induced co-flocculation with the yeast cells, and flocs formed by *D. bruxellensis* and *L. vini* showed another shape and lower stability than those formed by *S. cerevisiae* and *L. vini*. Maybe these characteristics are advantageous in the cell recirculation systems (Tiukova et al. 2014a). However, outcompetition of *S. cerevisiae* by *D. bruxellensis* has also been shown in co-cultures without any bacteria (Blomqvist et al. 2012), thus, interaction with bacteria does not appear to be essential.

D. bruxellensis can ferment cellobiose, a di-saccharide released upon cellulose degradation, to ethanol (Blomqvist et al. 2010; Spindler et al. 1992). Moreover, some strains have even been shown to be able to assimilate xylose, the main monosaccharide of hemicelluloses (Schifferdecker et al. 2014). This makes this yeast an interesting candidate for converting lignocellulose to ethanol, although it is less tolerant than *S. cerevisiae* to inhibitors released during thermochemical pre-treatment of lignocellulose. By pertinent pre-cultivations, the yeast could be adapted to the inhibitors, resulting in a similar inhibitor tolerance to that of *S. cerevisiae* (Blomqvist et al. 2011; Tiukova et al. 2014b). Thus, *D. bruxellensis* may represent an interesting alternative for developing first- and second generation ethanol production processes based on continuous fermentations with cell recirculation. In any case, the yeast can serve as a model to understand microbial interactions in ethanol fermentations. This model character is supported by the availability of the genome sequences of a variety of strains from wine-, beer-, and ethanol production (Borneman et al. 2014; Crauwels et al. 2014, 2015; Olsen et al. 2015), and the development of tools for genetic manipulation (Miklenic et al. 2015).

Several other non-conventional yeasts have been considered for ethanol production. Yeasts that ferment sugars released from hemicellulose, such as xylose, to ethanol are of special interest for establishing sustainable lignocellulose-based ethanol production. However, as mentioned above, those yeasts are not able to grow anaerobically, re-assimilate the formed ethanol as soon the sugar is consumed, and are very sensitive to fermentation inhibitors and high ethanol concentrations. Nevertheless, targeted evolution by repeated culturing in concentrated hydrolysates and ethanol-challenged continuous xylose-based culturing has recently been applied to a *S. stipitis* strain. The resulting isolate could grow in lignocellulose hydrolysates with up to 20% dry matter loads and produced 40 g/l ethanol in less than 167 h (Slininger et al. 2015). Co-culturing *S. cerevisiae* and *S. stipitis* for converting both hexoses and pentoses in lignocellulose has also been investigated, and a final ethanol concentration of 46 g/l and an ethanol productivity of 0.49 g/lh from lignocellulose hydrolysate could be achieved by optimising the cell ratio of both species (Unrean and Khajeeram 2015).

Thermotolerant yeasts can be very interesting for industrial ethanol production, as they can ferment at temperatures that are close to the temperature optima of most cellulose-degrading enzymes, and thus can be used to establish a process of simultaneous saccharification and fermentation (SSF). Apart from this, fermentations at high temperatures also minimise the risk of contamination by undesirable microbes, and cooling costs can be reduced (Abdel-Banat et al. 2010). In 2003, it was discovered that the thermotolerant yeast *Ogataea polymorpha* (syn. *Hansenula polymorpha*) can ferment xylose to ethanol at high temperatures (Ryabova et al. 2003). However, the amount of ethanol formed was quite low, and far below formation rates, yields and final concentrations described for other xylose-fermenting yeasts or recombinant *S. cerevisiae* strains. Therefore, a variety of manipulations has been applied. Expression of a bacterial XI-gene (*xylA*of *Escherichia coli*) resulted in an active protein in this yeast, in contrast to earlier experiments in *S. cerevisiae*. When this gene was expressed in a strain where the native XR-XDH-dependent xylose assimilation pathway was inactivated, higher ethanol production compared to the wild type was obtained (Dmytruk et al. 2008). Overexpressing the *PDC1*- or *ATH1*-genes also increased ethanol production from xylose. *ATH1* encodes an acid trehalase, and its overexpression resulted in an even more thermotolerant strain (Ishchuk et al. 2008; Ishchuk et al. 2009). Fermentation of starch and xylane by *O. polymorpha* could be achieved by expressing genes encoding for α-amylase, glucoamylase, xylanase and β-xylosidase (Voronovsky et al. 2009), and even glycerol could be converted to ethanol in strains overexpressing *PDC1* and *ADH1* (Kata et al. 2016).

A thermotolerant strain of *Kluyveromyces marxianus* has been engineered to display *Trichoderma resii* endoglucanase and *Aspergillus aculeatus* β-glucosidase at its cell surface. The resulting strain converted glucan to ethanol at 48 °C with a yield close to the theoretical maximum (0.47 g per g cellulose) (Yanase et al. 2010).

Thus, non-conventional yeasts may have some potential for industrial ethanol production. However, ethanol productivities and tolerance of most of these yeasts are still far below that of *S. cerevisiae*, so the performance of those yeasts has to be greatly improved to be relevant for future industrial processes.

4 Production of Advanced Biofuels

4.1 Butanol

Butanol is a four-carbon alcohol, which has some advantages over ethanol, such as higher energy density, better capacity to be blended with gasoline, or lower corrosiveness. Both n- and iso-butanol and 2-butanol can be used as fuel (Buijs et al. 2013; Generoso et al. 2015). Currently, butanol is mainly produced by chemical synthesis, although as early as 1912, a fermentative process based on solventogenic clostridia was commercialised. However, this process is currently not competitive with chemical synthesis from fossil resources (Green 2011). A yeast-based process may be advantageous, since yeasts are in general more robust in industrial conditions and are less sensitive to oxygen than clostridia. Production of n-butanol starts with the catalysis of two acetyl-CoA by thiolase into aceto-acetyl-CoA. Then, n-butanol is produced by a series of reductive reactions and dehydration (Steen et al. 2008, summarised in Passoth 2014). The highest obtained concentration of butanol was 2.5 mg/l, which was below the concentrations obtained by engineered *E. coli* or the *Clostridium*-based fermentation (Steen et al. 2008). Recently, n-butanol production was further enhanced by improving CoA-supply by overexpressing the *E. coli coaA* gene encoding panthothenate kinase. Addition of pantothenate, optimising enzymes involved in n-butanol synthesis and disruption of alcohol dehydrogenases resulted in a final butantol concentration of 130 mg/l (Schadeweg and Boles 2016).

Production of iso-butanol has been investigated more intensively than that of n-butanol. Yeasts naturally produce iso-butanol during degradation of valine via the Ehrlich pathway, where valine is deaminated to α-ketoisovaleriate, which is decarboxylated and reduced to isobutanol (Hazelwood et al. 2008). Biosynthesis of valine starts from pyruvate, and thus from glycolysis. By several steps, α-ketoisovalerate is synthesised, which is then converted to valine by transamination (summarised in Passoth 2014). Thus, valine synthesis and—degradation are connected by α-ketoisovalerate, so it is possible to generate iso-butanol from glucose. However, valine synthesis and—degradation take place in different cellular compartments, synthesis in the mitochondrial matrix, and degradation in the cytoplasm. Overexpression of three genes of the mitochondrial biosynthesis pathway (*ILV2*, *ILV5*, and *ILV3*) and one gene of the cytoplasmatic valine degradation pathway (*BAT2*) yielded for the first time an overproduction of isobutanol (increase from 0.28 mg/g glucose to 4.12 mg/g in complex medium) (Chen et al. 2011). To overcome limitations associated with compartmentalisation of the pathways, the corresponding enzymes have been expressed in the cytoplasm (Brat et al. 2012; Buijs et al. 2013; Matsuda et al. 2012). Further improvements have been achieved by enhancing the pyruvate levels by inactivating *PDC*-genes, expressing α-ketoacid decarboxylases and alternative alcohol dehydrogenases, and blocking the mitochondrial valine synthesis pathway (Brat et al. 2012; Kondo et al. 2012; Matsuda et al. 2012). Expressing codon optimised *ILV2*, *ILV5* and *ILV3* with truncated mitochondrial import signal sequences in an appropriate strain background increased the production of iso-butanol to a final concentration of 0.63 g/l, with a yield of 15 mg/g glucose (Brat et al. 2012). Overexpression of xylose isomerase, transaldolase, and xylulokinase in this strain resulted in the conversion of xylose to isobutanol, with a final concentration of 1.36 mg/l and a yield of 0.16 mg isobutanol/g xylose (Brat and Boles 2013). An alternative to expressing the isobutanol pathway in the cytoplasm, is to express it in the mitochondrial matrix. When *S. cerevisiae ARO10* (encoding α-ketoacid decarboxylase) and the *Lactococcus lactis* AdhA (encoding *L. lactis* ADH7) were fused with mitochondrial targeting sequences and overexpressed together with *ILV2*, *ILV5* and *ILV3*, a maximum isobutanol concentration of 630 mg/l was obtained in complete medium (Avalos et al. 2013). In several patent descriptions, final iso-butanol concentrations of up to 18.6 g/l and yields of up to 0.33 g/g glucose (i.e. about 80% of the theoretical maximum) have been reported (Buijs et al. 2013).

S. cerevisiae is a fermentative yeast, which means that its metabolism is mainly directed towards ethanol production. This implies that it is difficult to obtain sufficient amounts of acetyl-CoA for the synthesis of butanol and other advanced biofuels (Chen et al. 2013). Other, respiratory yeasts might thus be more appropriate for butanol production. The yeast, *Arxula* (*Blastobotrys*) *adeninivorans* has been manipulated to produce n-butanol by introducing the relevant genes, mainly from *Clostridium acetobutylicum*. The resulting strain was able to produce 1 g/l n-butanol from starch (Kunze et al. 2014), the highest amount produced reported for a yeast to date.

4.2 *Biodiesel Production by Yeasts*

4.2.1 Importance of Biodiesel Production

Biodiesel can be regarded as an advanced biofuel because of its higher energy density compared to ethanol. It has been estimated that there is a growing potential for biodiesel especially in Europe, due to the expanding use of diesel engines for transportation fleets (Gnansounou 2010). Currently, biodiesel is commercially produced from triglycerides extracted from oil plants, such as soy, oil palms or oilseed rape. In a transesterification reaction, the glycerol is replaced by short chain alcohols, usually methanol, to form fatty acid methyl ester (FAME). Arable land is used for cultivating the oil plants, and since the energy yield per hectare of oil plants is relatively low compared to sugar plants, replacing a substantial proportion of mineral diesel with biodiesel may considerably impact food production. Moreover, oil plants can be cultivated on cleared rain forest areas; indeed, deforestation in such areas for oil plant production has been reported (Azócar et al. 2010; Graham-Rowe 2011).

Oleaginous yeasts may represent a good alternative to oil plants. They can grow on a variety of substrates, for instance on sugar-monomers of hemicellulose, to a

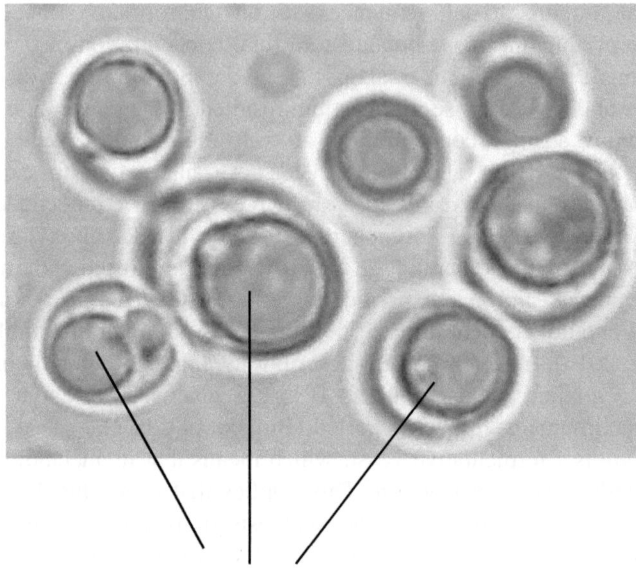

Fig. 2 Formation of intracellular lipid bodies by *Lipomyces starkeyi* grown under nitrogen limitation (1000-fold magnification, photograph Nils Mikkelsen)

high cell density, and with a cellular lipid content of at least 20% and up to more than 70% (Fig. 2). Oleaginous yeasts have been grown on residues from the paper industry (spent sulphite liquor) to produce fat for food purposes (fat for baking or as butter replacement) in Germany during both world wars. About 70 oleaginous yeast species have been discovered, both within ascomycetes (e.g. *Lipomyces starkeyi* or *Yarrowia lipolytica*) and basidiomycetes (e.g. *Rhodotorula glutinis* or *Rhodosporidium toruloides*). Novel oleaginous species could be identified by targeted screening, for instance by cultivating on carbon source-free medium, on medium with glycerol as sole carbon source, or on nitrogen poor medium (Sitepu et al. 2014b, and references therein).

4.2.2 Physiology of Lipid Accumulation in Oleaginous Yeasts

Lipid accumulation in oleaginous yeasts usually occurs in conditions of concomitant carbon surplus and limitation in other compounds, such as nitrogen, phosphorus or sulphur. Lipid accumulation process under nitrogen limitation is the most studied, where the intracellular concentration of AMP drops to 5% of the cellular concentration under C-limitation due to its deamination to IMP. Since AMP is an essential cofactor of the tricarbonic acid cycle (TCC)-enzyme isocitrate-dehydrogenase (IDH), isocitrate is not further metabolised and equilibrates with citrate. Citrate is transported out of the mitochondria. In the cytoplasm, citrate is degraded to acetyl-CoA and oxaloacetate, by the reaction of the ATP citrate lyase (ACL). Oxaloacetate is transported back to the mitochondria, while acetyl-CoA is the initial precursor of fatty acid synthesis (Ratledge and Wynn 2002). From acetyl-CoA, malonyl-CoA is formed by the action of acetyl-CoA carboxylase; malonyl-CoA is the substrate of the fatty acid synthase (FAS)-enzyme complex, which elongates the acyl-CoA chain by two carbons in each step (Tehlivets et al. 2007). Apart from acetyl-CoA, fatty acid production requires NADPH, and it is not completely clear from which reaction in the cell this is formed. Malic enzyme, which converts malic acid to pyruvate, CO_2 and NADPH, has long been regarded as the source of NADPH in the cytoplasm which is used for fatty acid synthesis. Indeed, an increased level of malic enzyme has been observed during lipid accumulation in *R. toruloides* (Zhu et al. 2012). However, in many oleaginous yeasts, no cytoplasmic malic enzyme could be identified. In *L. starkeyi*, cytoplasmatic malic enzyme has a preference for NADH instead of NADPH (Tang et al. 2010), and *Y. lipolytica* possesses only a mitochondrial malic enzyme, of which overexpression did not result in enhanced lipid accumulation (Beopoulos et al. 2011). Instead, enhanced production of 6-P-gluconate dehydrogenase has been observed (Liu et al. 2009, 2011). There are still many factors of lipid accumulation that are not understood, including species and strain specificities, influence of pH, or C:N ratio of the medium (see recent reviews of the physiology of lipid accumulation by yeasts, e.g. Passoth 2014; Ploier et al. 2014; Sitepu et al. 2014b).

4.2.3 Lipid Production from Lignocellulose Hydrolysates

Lignocellulose has a high C/N ratio (Hyvönen et al. 2000; Reinertsen et al. 1984), which is advantageous for biolipid production. However, the inhibitors present in lignocellulose hydrolysate are also toxic for oleaginous yeasts. High-cell density cultures of *R. toruloides* on glucose reached a lipid content 72.7 g/l (Li et al. 2007), whereas cultures on hydrolysates of lignocelluloses such as wheat and rice straw, corn stover or sewage sludge only yielded about 10 g/l (Angerbauer et al. 2008; Galafassi et al. 2012; Huang and Wu 2009; Yu et al. 2011). Various oleaginous yeasts have been tested for their response to fermentation inhibitors. HMF obviously influenced lipid accumulation only at higher concentrations and only for some strains. Yet many strains showed some lag-phase upon exposure to HMF. Furfural and vanillin were highly inhibitory (Chen et al. 2009; Hu et al. 2009; Sitepu et al. 2014a). Acetic acid had some ambiguous impact on lipid production. At a concentration of 5 g/l, acetic acid completely inhibited growth of strains belonging to *L. starkeyi*, *R. glutinis* and *R. toruloides. Trichosporon curaneum* showed some growth at this concentration (Chen et al. 2009). However, inhibition by acetic acid and weak organic acids in general depends on the pH of the medium. Twelve out of 48 tested oleaginous strains, belonging to species such as *Cryptococcus humicola*, *Trichosporon coremiiforme*, *Cyberlindnera saturnus* or *Schwanniomyces occidentalis* were able to grow in medium containing 2.5 g/l acetate at an initial pH of 3.5, however, the impact on lipid accumulation has not been tested (Sitepu et al. 2014a).

Several lignocellulose substrates have been tested for their potential for lipid production, including spent sulphite liquor, straw, corn stover hydrolysate, sewage sludge or birch hemicellulose hydrolysate (Angerbauer et al. 2008; Brandenburg et al. 2016; Calvey et al. 2016; Galafassi et al. 2012; Huang and Wu 2009; Lundin 1950; Yu et al. 2011). In some of these fermentations, lipid concentrations above 10 g/l were reached, e.g. 15.7 g/l from sugar cane bagasse using *Trichosporon fermentans* (Huang and Wu 2012), 12.3 g/l from corn cob hydrolysate by *Trichosporon cutaneum* (Gao et al. 2014), or 11.5 g/l by *T. fermentans* from rice straw hydrolysate (Huang and Wu 2009). Acid hydrolysis of lignocellulose separates cellulose and lignin into the solid phase, while the hemicellulose is present in the liquid phase. This enables the conversion of the hemicellulose sugars to lipids, as most oleaginous yeasts can assimilate pentoses. On the other hand, hemicellulose hydrolysate is rich in soluble inhibitors such as furfural and acetic acid, requiring the establishment of a cultivation method that enables adaptation of the production yeasts to the inhibitors. Recently, using a pH-regulated fed-batch culture, a lipid concentration of 8 g/l could be obtained from birch-hemicellulose hydrolysate, which is one of the highest lipid concentrations that have been obtained from hemicellulose hydrolysate (Brandenburg et al. 2016).

4.2.4 Manipulation of Yeasts to Obtain Enhanced Lipid Amounts

For most of the above-mentioned oleaginous yeast species, except *Y. lipolytica*, methods for molecular manipulations are unavailable or only poorly developed (Beopoulos et al. 2009). *Y. lipolytica* can convert glucose, glycerol and hydrophobic substances such as alkanes to lipids, but not xylose and other lignocellulose-derived sugars. Its intracellular lipid content rarely exceeds 30%, making it less attractive for the production of biodiesel from lignocellulose (Sitepu et al. 2014b). However, it has great potential for producing chemicals such as organic acids, and it has been manipulated for production of long-chain polyunsaturated fatty acid such as the Ω-3-fatty acids docosahexaenoic acid (DHA) and eicosapentoic acid (EPA); and it is also a model organism to understand the molecular physiology of lipid accumulation (Beopoulos et al. 2009). Inactivating *GUT2*, which encodes for a glycerol-3-P dehydrogenase, and *POX1-6*, genes involved in β-oxidation of fatty acids, resulted in significant increase in lipid production by *Y. lipolytica* (Beopoulos et al. 2008). Evolutionary engineering-based approaches have been tested in *R. glutinis* and *L. starkeyi*. Cells were randomly mutagenised using UV or radioactive radiation and selected on medium containing cerulenin, an inhibitor of lipid synthesis. Colonies with increased diameter were formed by mutants with increased amounts of intracellular lipids (Tapia et al. 2012; Wang et al. 2009).

As methods for manipulating oleaginous yeasts are underdeveloped, much focus has been placed on obtaining a lipid-producing *S. cerevisiae*. *S. cerevisiae* is not an oleaginous yeast; in fact, it can grow even when production of storage lipids is blocked (Sandager et al. 2002). However, it is an established industrial organism that is very robust to the conditions of industrial fermentations. Moreover, the toolbox for its molecular manipulation is very well developed, so that it can at least serve as a model organism for metabolic engineering of increased lipid production in yeasts. In contrast to oleaginous yeasts, *S. cerevisiae* cannot export surplus acetyl-CoA, the precursor of lipid synthesis, from the mitochondria to the cytoplasm via the citrate-oxaloacetate shuttle (Beopoulos et al. 2011). Cytoplasmatic acetyl-CoA is produced by the reactions of pyruvate decarboxylase (PDC), aldehyde dehydrogenase (AlDH) and acetyl-CoA synthase (Holzer and Goedde 1957). When PDC is inactivated, C2-compounds such as ethanol or acetate have to be provided to enable the yeast cell to grow (Pronk et al. 1996). Similarly to butanol production, re-directing the metabolic flux from ethanol production towards acetyl-CoA and thus providing sufficient amounts of precursors for lipid production is one of the major challenges when manipulating *S. cerevisiae* for high production of lipids (Buijs et al. 2013). Overexpressing the AlDH gene *ALD6* and a modified acetyl-CoA synthetase gene from *Salmonella enterica* increased the acetyl-CoA level in *S. cerevisiae* (Shiba et al. 2007). In another approach, apart from expressing these two genes, *ADH2*, encoding the assimilatory alcohol dehydrogenase of *S. cerevisiae*, was overexpressed. This engineering of intracellular acetyl-CoA will enable the production of lipids and other, even more advanced biofuels (Chen et al. 2013). Overexpression of fatty acid synthase (FAS), acetyl-CoA carboxylase, and diacyl-glycerol acyltransferase encoding genes (*FAS1* and *FAS2*, *ACC1*, and *DGA1*, respectively), resulted in a strain accumulating up to 17% lipid content of dry matter (Runguphan and Keasling 2014). Introducing *ACL* and disrupting *IDH1* and *IDH2* resulted in a total fatty acid content of 21% (Tang et al. 2013). Further manipulations included disruption of the transcription factor Snf2 (Kamisaka et al. 2006), the fatty acyl-CoA synthetase encoded by *FAA3* and overexpression of a modified variant of *DGA1*, which resulted in an increase in lipid content up to 45% of cell dry matter (Kamisaka et al. 2013), converting *S. cerevisiae* to an oleaginous yeast. Interestingly, disruption of *SNF2* had a higher effect than overexpression of the fatty acid synthesis genes (summarised in Zhou et al. 2014). Snf2 is a transcriptional factor involved in the activation of glucose repressed genes at glucose depletion (Abrams et al. 1986), and the finding that lipid production increases upon its inactivation indicates that this factor also plays an important role in the control of the expression of genes involved in lipid formation.

4.2.5 Lipid Extraction from Yeast Cells

Lipids are stored in lipid bodies inside the cells, and thus, need to be extracted to convert them to biodiesel. Lipid extraction from the cells represents one major challenge among many in establishing yeast-based biodiesel production, because the yeasts' cell walls are a barrier for extraction, and because lipophilic compounds other than triacylglycerides are present in the lipid bodies. When lipids for human consumption were produced, lipids were released by autolysis at 50 °C (Hesse 1949). Various extraction methods have been tested at pilot scale, in most cases, based on ethanol-hexane mixtures. Further optimisation is required for lipid extraction from yeasts to obtain a commercially viable process (Ageitos et al. 2011; Jacob 1992). Yeast lipids also contain a comparatively large amount of free fatty acids, which can lead to undesirable saponification reactions during transesterification, which is achieved under alkaline conditions (Azócar et al. 2010; Robles-Medina et al. 2009).

One approach to the extraction problem is to engineer strains which excrete the formed lipids. In bacteria, ABC transporters have been expressed to export hydrophobic molecules out of the cell (Dunlop et al. 2011). ABC-transporters are ubiquitous among all kingdoms of life, and thus those transporters may also be active in yeasts. When expressing the transporters under the control of an inducible promoter, lipid-accumulation and secretion can be separated. Cells can be transferred to a biphasic system with an aquatic and an organic phase, where secreted lipids will accumulate in the organic phase, while the cells stay in the liquid phase. This would even allow re-use of cells in subsequent fermentations (Doshi et al. 2013). Moreover, it has been shown for *Candida tropicalis* that the cells secrete lipids under oxygen limited conditions, and thus transfering lipid accumulating yeast cells to oxygen limitation may represent a simple method of getting lipids out of the cells (Phadnavis and Jensen 2013).

Fatty acid ethyl esters (FAEE) are naturally excreted from yeast cells, and thus their production would solve both issues with transesterification and secretion of the biofuel into the growth medium. *S. cerevisiae* was manipulated to produce FAEE by expressing heterologous wax ester synthases. Overexpressing a mutated *ACC1*, encoding an acetyl CoA carboxylase, which could not be inactivated by Snf1 mediated phosphorylation, also increased FAEE production, and finally a FAEE concentration of 34 mg/l could be obtained (Shi et al. 2014). Additional improvements were obtained by disrupting fatty acid consuming pathways, such as β-oxidation and synthesis of sterylesters and TAGs, increasing fatty acid biosynthesis, and activating the ethanol consumption pathway (de Jong et al. 2014; Runguphan and Keasling 2014; Valle-Rodríguez et al. 2014; Zhou et al. 2014). First attempts were performed to obtain a stable FAEE-producing strain by integrating all constructs into the genome, but the productivity was lower compared to other strains (de Jong et al. 2015). The highest FAEE-concentrations to date were generated from glycerol as carbon source. Overexpressing the glycerol assimilation pathway and a glycerol transport protein, combined with disrupting the genes of di-hydroxy acetone phosphate degradation and glycerol export, forced ethanol production from glycerol in *S. cerevisiae*. When an acyltransferase from *Acinetobacter baylyi* was also overexpressed, the engineered strain was able to condense the formed ethanol with externally added oleic acid to 0.52 g/l ethyl oleate (Yu et al. 2012).

4.3 Terpenes, Alkanes and Other Highly Reduced Biofuels

Increasing the intracellular pool of acetyl-CoA can, apart from producing fatty acid for biodiesel production, lead to the generation of even more advanced biofuels, including terpenes and alkanes. Those biofuels have a lower oxygen content even compared to biodiesel and have, therefore a very high energy density (Zhang et al. 2011).

Terpenes, also called isoprenoids are built up from five carbon subunits of isopren ($CH_2 = C(CH_3)CH = CH_2$) and its derivatives. The isopren-derivatives isopentenyl pyrophosphate (IPP) or dimethyl-allyl-pyrophosphate (DMAP) are condensed to the prenyl-pyrophosphates geranyl pyrophosphate (GPP, C10), farnesyl pyrophosphate (FPP, C15) and geranyl-geranyl pyrophosphate (GGPP, C20), from which monoterpenes (C10), sesquiterpenes (C15), or diterpenes (C20) are formed (Buijs et al. 2013; Zhang et al. 2011). Yeasts produce various compounds from terpenes, including the steroid ergosterol (Parks and Casey 1995) and carotenoids (Mata-Gomez et al. 2014). *S. cerevisiae* has been engineered to overproduce terpene-derived substances, such as the anti-malaria drug artemisin (Ro et al. 2006). IPP, GPP and FPP can be hydrolysed to alcohols that can be used as fuel additives (Zhang et al. 2011). Those alcohols, especially isopentanol, have been proposed as gasoline additives, since they show all the advantageous characteristics of ethanol regarding increasing the octane number of the gasoline and reducing the amount of harmful pollutants in the exhaust emissions, but lack the negative features, as they do not increase the vapour pressure as do gasoline-ethanol blends, have a better tolerance to water, and do not increase the fuel consumption (Hull et al. 2006). The sesquiterpene alcohol farnesol and the terpene farnesene have been suggested as diesel fuels (Zhang et al. 2011). An FPP-overproducing *S. cerevisiae* strain, with overexpressed FPP production pathway and reduced activity of a competing synthesis pathway (squalene synthesis) has been engineered to generate bisabolene, by overexpressing bisabolene synthase. The resulting strain produced more than 900 mg bisabolene per litre. Bisabolene can be chemically hydrogenated to bisabolane, a biosynthetic alternative to D2 diesel (the diesel fuel usually used in cars). Bisabolane was shown to have similar or partially even better characteristics compared to D2 diesel (Peralta-Yahya et al. 2011).

Alkanes and alkenes are the major components of fossil oil, and they are usually regarded as non-renewable, or at least only renewable within very long time frames. However, findings that several microorganisms can naturally produce hydrocarbons, has to some extent challenged this view (Fu et al. 2015; Ladygina et al. 2006). Production of long-chain (C_{10}–C_{31}) and volatile (C_2–C_5) n-alkanes by yeasts was discovered already during the 1960s. These yeasts include several *Saccharomyces* species, including *S. oviformis* and *S. ludwigii*, which accumulated long chain n-alkanes up to 10.2% of their cell dry weight, *Candida tropicalis* (up to 0.031% n-alkanes related to cell dry weight), as well as *R. glutinis*, *D. bruxellensis*, *Saccharomyces octosporus* and *Schizosaccharomyces pombe*, which produced ethane in the range from 21.5 to 11.4 ml/l and h, and *Rhodotorula minuta* var.

texenis, which produced 16.4 ml isobutene per l and h (Ladygina et al. 2006, and references therein). Probably due to the comparatively low final concentrations and production rates, the natural abilities of yeasts to produce n-alkanes have not been received much attention in the past. However, a strain belonging to *Aureobasidium pullulans* var. *melanogenum*, isolated from a mangrove ecosystem, was recently shown to produce 32.5 g/l heavy oil from glucose/corn steep liquor. This heavy oil comprised fatty acids and about 60% alkanes—mainly tetradecane, tetracosane, hexacosane, heptacosane, octacosane, and tetratetracontane (Liu et al. 2014). Metabolic pathways for n-alkane production in yeasts are largely unknown (Fu et al. 2015). Since very long chain alkanes, such as heptacosane (C27) or tetratetracontane (C44) were formed by *A. pullulans* var. *melanogeum*, head-to-head condensation, where two fatty acids condense at their carboxyl groups (Fig. 3a) has been assumed to be responsible for alkane formation in this yeast (Liu et al. 2014). In *Arabidopsis thalliana*, very long chain alkanes are produced from very long chain acyl-CoAs (VLC-acyl-CoA), which are reduced to fatty aldehydes by VLC-acyl-CoA reductase (encoded by CER3) and then further converted to alkanes by VLC-aldehyde decarbonylase (encoded by CER1) (Bernard et al. 2012; Fu et al.

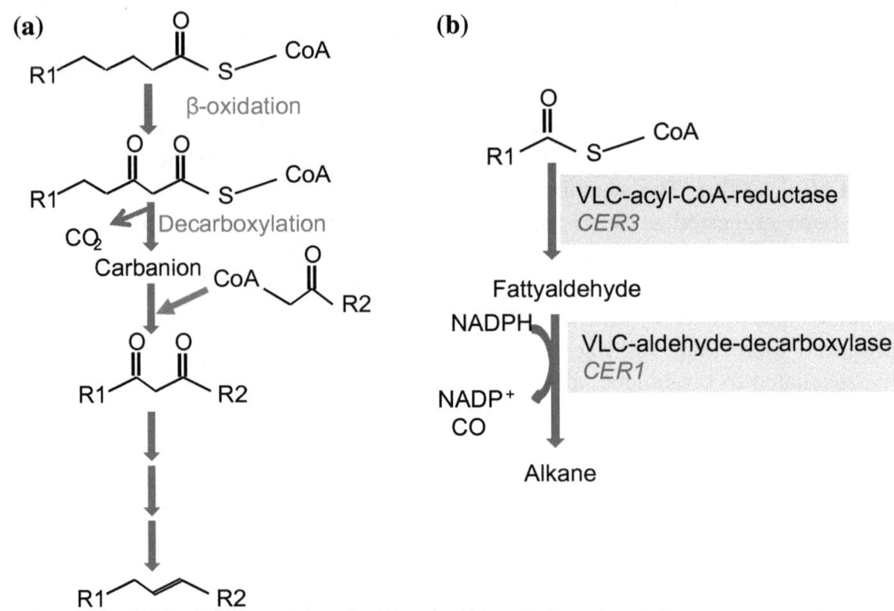

Fig. 3 Potential pathways of alkane synthesis in yeasts (Fu et al. 2015; Ladygina et al. 2006). **a** Head-to head condensation of fatty acids, which has been suggested for *Aureobasidium pullulans* var. *melanogenum*, but has currently only been found in bacteria. **b** Very long chain (VLC) fatty acid decarboxylation. VLC-acyl-CoA-reductase and VLC-aldehyde-decarboxylase have a specifity for *R1* of C24 and longer. The pathway generates n-alkanes in plants, and expression of *CER3* and *CER1* in *S. cerevisiae* resulted in production of C29 alkanes. *CER3* and *CER1*-like genes have been found in *A. pullulans* var. *melanogenum*, indicating the presence of the pathway also in this yeast

2015) (Fig. 3b). Interestingly, CER like genes containing motifs similar to the His-clusters that are essential for CER1 function in plants have been found in the genome of *A. pullulans* var. *melanogenum* (Gostincar et al. 2014), which indicates, in contrast to the above-mentioned assumption, the presence of a plant-like alkane synthesis pathway in this yeast (Fu et al. 2015).

Several efforts have been made to genetically engineer yeasts to produce hydrocarbons. The *A. thalliana* VLC alkane synthesis pathway (Fig. 3b) was established in *S. cerevisiae* by expressing CER1 and CER3 in a strain with mutated fatty acid elongase (*SUR4#*), which produces elongated (C28 and C30) fatty acids. Co-expression of long-chain acyl-CoA synthetase I (LACS1) and CYTB5-B, a heme-protein that serves as electron transporting factor, additionally enhanced alkane production (mainly nonacosane, C29) by the engineered yeast strain, which finally reached 86 μg per g yeast biomass (dry weight) (Bernard et al. 2012). The above-mentioned very long n-alkanes are may be less attractive for the fuel industry, since their chain length is too big. An alkane biosynthetic pathway was introduced into *S. cerevisiae* by expressing a fatty acyl-ACP-reductase (FAR) gene and a fatty aldehyde deformylating oxygenase (FADO) gene from the cyanobacterium *Synechoccocus elongatus*. FAR catalyses the reduction of fatty acyl-CoA to fatty aldehyde, which is further reduced to alkanes or alkenes by FADO. In addition, *E. coli* ferredoxine and ferredoxine reductase were expressed, to provide the required redox cofactors for FADO in the cytoplasm, and *HFD1*, encoding a hexadecanal dehydrogenase that obviously competed with FADO for the fatty aldehydes, was disrupted. The resulting strain produced up to 22 μg heptadecane per g dry weight (Buijs et al. 2015). In *Y. lipolytica*, a pathway for pentane production has been introduced. This was achieved by expressing soybean lipoxygenase I (encoded by a codon-optimised Gmlox1 gene). Lipoxygenase I converts linoleic acid into 13-hydroperoxylinoleic acid, which is converted to pentane and 13-oxo-cis-9, trans 11-tridecadienoic acid. The latter conversion can occur spontaneously, or by the action of lipoxygenase I and/or a hydroperoxide lyase. Pentane production (3.28 mg/l) was observed after expression of Gmlox1. Production could be increased to 4.98 mg/ml by optimising the growth medium and knocking out genes involved in β-oxidation (Blazeck et al. 2013).

Industrial advanced biofuel production by yeasts would have a great impact on the replacement of fossil fuels by renewable fuels, since advanced biofuels such as butanol, biodiesel, terpenes and alkanes have significant advantages compared to ethanol. However, production of those advanced biofuels is far away from levels that are interesting for industry. Native producers of those biofuels, such as oleaginous yeasts are still poorly investigated and may have lower robustness in the harsh conditions of industrial processes. Moreover, in most cases there are no or only poorly developed tools for manipulating those yeasts in order to improve their efficiency. On the other hand, established fermentation organisms such as *S. cerevisiae* have a naturally low level of acetyl-CoA, which is the precursor of all of these biofuels. In spite of very sophisticated approaches of metabolic engineering, the attainable final concentrations and productivities do not yet allow commercial production of those advanced biofuels by yeasts (Buijs et al. 2013; Fu et al. 2015;

Zhang et al. 2011). Further research, both on manipulating established systems and investigating non-conventional yeasts, is required to establish yeasts as production organisms for advanced biofuels.

5 Integration of Biofuel Processes

Biofuel production offers great opportunities to at least partially reduce the demands of the global society for non-renewable fossil fuels. However, in many cases biofuels are more expensive to produce, especially when it comes to production from second generation feedstocks (Cheng and Timilsina 2011; Stephen et al. 2012). Biorefineries can be made more efficient and thus more competitive with fossil based processes, when the partial processes of biofuel production are integrated.

Biofuel production includes not only the fermentation processes, but also growth of biomass, harvest, storage, pretreatment, and removal of production residues. Valorisation of product residues would improve the environmentally and economically sustainable of biofuel processes (Liguori et al. 2013; Vanholme et al. 2013). The amazingly diverse metabolic capacities of yeasts can be used to improve a variety of different steps during biofuel production (Passoth 2014; Petrovic 2015).

Efficient preservation is an important step in biofuel production. Plant biomass is seasonally generated, and has to be preserved to be used in the continuously running biofuel refinery. Drying is a common method for preservation of first generation substrate; however, this can require a substantial amount of energy (Olstorpe and Passoth 2011). Lignocellulose such as straw is usually passively dried on the field, and high moisture content can result in substantial losses of produced biomass (Nilsson 2000). Energy saving ensilation techniques or airtight storage of moist biomass with addition of biocontrol yeasts has been developed for material that is used as animal feed (Olstorpe and Passoth 2011; Zheng et al. 2011). Moist stored biomass has also been tried for biofuel production. *Wickerhamomyces anomalus* and *S. stipitis* have been tested as biocontrol organisms. It has been shown that moist storage improved bioethanol production from the material, probably because the sugar polymers were more accessible to the enzymes during pretreatment. Moreover, addition of *S. stipitis* to stored wheat straw had an additional positive effect on degradation of cellulose, probably because of a partial degradation of the hemicellulose during storage by the xylanase activity of the yeast (Passoth et al. 2009, 2013). Thus, energy saving biomass preservation can at the same time also act as pretreatment of the biomass. Biomass pretreatment accounts for a large proportion of the total energy input (Sassner et al. 2008), and sophisticated storage technology which decreases energy inputs may have a great impact on developing a sustainable biofuel production process.

A strategy to save enzyme costs would be to develop a process of combined biomass degradation and fermentation. This kind of consolidated bioprocessing can be achieved by using a surface-engineered fermentation yeast strain, expressing polysaccharide-degrading enzymes (Hasunuma et al. 2015).

Utilisation of all fractions of biomass is essential to obtain a sustainable process. As mentioned above, substantial efforts have been made to engineer *S. cerevisiae* to convert the pentoses of the hemicellulose fraction to ethanol. Converting the pentoses to biodiesel, using pentose-assimilating oleaginous yeasts may be another way of obtaining a sustainable biofuel refinery (Brandenburg et al. 2016). It is also possible to engineer yeasts to produce high-value chemicals out of the plant biomass (Buijs et al. 2013; Fillet et al. 2015; Zhang et al. 2011). Co-production of high value chemicals and biofuels can substantially improve the economy of the whole process and thus contribute to make it more attractive compared to fossil fuel production. Conversion of the lignin fraction to biofuel or high value compounds is difficult because of its hydrophobicity and to some extent unpredictable structure. Typically it is burned to obtain process energy. It is also possible to convert it to lignosulfonate during biomass pretreatment. Lignosulfonate represents a valuable side product, as it can be used in a variety of applications, e.g. as plasticiser in concrete (Zhu and Pan 2010). Lignin can also be used as fertilizer, and in this function it can play an important role for the carbon balance of soil (Jarecki and Lal 2003).

Fermentation residues from ethanol production can represent an important cost factor, since they are rich in nutrients and their disposal is expensive (Wilkie et al. 2000). When these residues can instead be utilised to generate value, this would in general improve the sustainability of biofuel processes. Residues from first-generation ethanol production (distillers' grain) are a high quality additive to animal feed. In the EU, the contribution of distillers' grain to animal feed is roughly equivalent to that of maize cultivated on 0.7 Mio ha (Özdemir et al. 2009). Conversion of the distillers' grain to animal feed was essential to obtain a positive energy balance for the whole process of corn-based ethanol production, as it contributed to about one sixth of the total energy output (Hill et al. 2006). Residues from second-generation ethanol production have less potential as animal feed, due to inhibitors and lignin-derived compounds that may influence palatability. However, extraction of proteins for animal or human nutrition has been suggested (Chiesa and Gnansounou 2011). An easier way to generate value from fermentation residues is the production of biogas. Some studies indicate that the total energy output from a combined ethanol/biogas production was higher than biogas production alone, and the biogas production rate was also enhanced, although the positive impact of ethanol production on a subsequent biogas process varied (Dererie et al. 2011; Kreuger et al. 2011; Theuretzbacher et al. 2015). In any case, a combined ethanol/biogas production can diversify the products out of a biofuel refinery, which can provide an economically more competitive process. When advanced biofuels such as biolipids are produced, subsequent biogas production is even more crucial to obtain a sustainable process, as those fermentations are aerobic processes, which generate a higher yeast biomass with corresponding energy content (Karlsson et al. 2016). Residues of the biogas process are very good fertilisers, and can thus help to increase the production of raw materials for biofuel and food production (Odlare et al. 2011).

6 Outlook

Currently, the major biofuel- ethanol—is produced by the conventional yeast *S. cerevisiae*. This yeast has been established as fermentation organism for millennia and is efficient and robust under the harsh conditions of industrial fermentation. However, ethanol production will probably soon reach a limit. Ethanol is for several reasons (see above) not the ideal fuel, and conversion of some components of second generation biomass to ethanol is difficult. Yeasts have a great potential to contribute to future sustainable biorefineries. *S. cerevisiae* is a platform organism that can be utilised as a multipurpose cell factory, using the highly developed tool box for engineering its metabolism to produce high value chemicals and advanced biofuels. On the other hand, non-conventional yeasts represent an amazing resource of diversity to metabolise unusual substrates and to generate high value products. They may serve as gene resources to construct a *S. cerevisiae*-based cell factory, or they may be used as cell factories themselves when techniques for their engineering become available. The global society needs to decrease its dependency on non-renewable fossil raw materials for the generation of fuels and chemicals—this represents a great challenge and chance for yeast research both in conventional and non-conventional systems.

Acknowledgements Biofuel related research in my group was funded by the Swedish Energy Authority (STEM), the MicroDrive-program at the Swedish University of Agricultural Sciences, the Swedish Research Council for Environment, Agricultural Sciences and Spatial Planning (Formas) and the Visby-program of the Swedish Institute. I thank Dr. Su-Lin Leong for linguistic advice.

References

Abdel-Banat, B. A., Hoshida, H., Ano, A., Nonklang, S. and Akada, R. 2010. *Appl Microbiol Biotechnol* 85: 861–867.

Abrams, E., Neigeborn, L. and Carlson, M. 1986. *Mol Cell Biol* 6: 3643–51.

Ageitos, J., Vallejo, J., Veiga-Crespo, P. and Villa, T. 2011. *Appl Microbiol Biotechnol* 90: 1219–1227.

Allen, S. A., Clark, W., McCaffery, J. M., Cai, Z., Lanctot, A., Slininger, P. J., Liu, Z. L. and Gorsich, S. W. 2010. *Biotechnol Biofuels* 3: 2.

Angerbauer, C., Siebenhofer, M., Mittelbach, M. and Guebitz, G. M. 2008. *Bioresource Technology* 99: 3051–3056.

Ansell, R., Granath, K., Hohmann, S., Thevelein, J. M. and Adler, L. 1997. *EMBO J* 16: 2179–2187.

Avalos, J. L., Fink, G. R. and Stephanopoulos, G. 2013. *Nat Biotech* 31: 335–341.

Azócar, L., Ciudad, G., Heipieper, H. J. and Navia, R. 2010. *Appl Microbiol Biotechnol* 88: 621–636.

Basso, T. O., de Kok, S., Dario, M., do Espirito-Santo, J. C., Müller, G., Schlölg, P. S., Silva, C. P., Tonso, A., Daran, J. M., Gombert, A. K., van Maris, A. J., Pronk, J. T. and Stambuk, B. U. 2011. *Metab Eng* 13: 694–703.

Beopoulos, A., Mrozova, Z., Thevenieau, F., Le Dall, M. T., Hapala, I., Papanikolaou, S., Chardot, T. and Nicaud, J. M. 2008. *Appl Environ Microbiol* 74: 7779–89.
Beopoulos, A., Cescut, J., Haddouche, R., Uribelarrea, J. L., Molina-Jouve, C. and Nicaud, J. M. 2009. *Prog Lipid Res* 48: 375–387.
Beopoulos, A., Nicaud, J. M. and Gaillardin, C. 2011. *Appl Microbiol Biotechnol* 90: 1193–1206.
Bergdahl, B., Heer, D., Sauer, U., Hahn-Hägerdal, B. and van Niel, E. W. 2012. *Biotechnol Biofuels* 5: 34.
Bernard, A., Domergue, F., Pascal, S., Jetter, R., Renne, C., Faure, J.-D., Haslam, R. P., Napier, J. A., Lessire, R. and Joubès, J. 2012. *The Plant Cell* 24: 3106–3118.
Blazeck, J., Liu, L., Knight, R. and Alper, H. S. 2013. *J Biotechnol* 165: 184–194.
Blomqvist, J., Eberhard, T., Schnürer, J. and Passoth, V. 2010. *Appl Microbiol Biotechnol* 87: 1487–1497.
Blomqvist, J. 2011. Ph.D. thesis. Department of Microbiology, Swedish University of Agricultural Sciences, Uppsala
Blomqvist, J., South, E., Tiukova, I., Momeni, M. H., Hansson, H., Ståhlberg, J., Horn, S. J., Schnürer, J. and Passoth, V. 2011. *Lett Appl Microbiol* 53: 73–78.
Blomqvist, J., Nogue, V. S., Gorwa-Grauslund, M. and Passoth, V. 2012. *Yeast* 29: 265–274.
Blomqvist, J. and Passoth, V. 2015. *FEMS Yeast Res* 15: fov021.
Borneman, A. R., Zeppel, R., Chambers, P. J. and Curtin, C. D. 2014. *PLoS Genet* 10: e1004161.
Brandenburg, J., Blomqvist, J., Pickova, J., Bonturi, N., Sandgren, M. and Passoth, V. 2016. *Yeast* 33: 451–462.
Brat, D., Boles, E. and Wiedemann, B. 2009. *Appl Environ Microbiol* 75: 2304–2311.
Brat, D., Weber, C., Lorenzen, W., Bode, H. B. and Boles, E. 2012. *Biotechnol Biofuels* 5: 65.
Brat, D. and Boles, E. 2013. *FEMS Yeast Res* 13: 241–244.
Buijs, N. A., Siewers, V. and Nielsen, J. 2013. *Curr Opin Chem Biol* 17: 480–488.
Buijs, N. A., Zhou, Y. J., Siewers, V. and Nielsen, J. 2015. *Biotechnol Bioeng* 112: 1275–1279.
Buyx, A. and Tait, J. 2011. *Science* 332: 540–541.
Calvey, C.H., Su, Y.-K., Willis, L.B., McGee, M. and Jeffries, T.W. 2016. *Bioresource Technol* 200: 780–788.
Carpita, N. C. and Gibeaut, D. M. 1993. *The Plant Journal* 3: 1–30.
Chandel, A. K., Chandrasekhar, G., Silva, M. B. and Silverio da Silva, S. 2012. *Crit Rev Biotechnol* 32: 187–202.
Chen, X., Li, Z., Zhang, X., Hu, F., Ryu, D. Y. and Bao, J. 2009. *Appl Biochem Biotechnol* 159: 591–604.
Chen, X., Nielsen, K. F., Borodina, I., Kielland-Brandt, M. C. and Karhumaa, K. 2011. *Biotechnol Biofuels* 4: 21.
Chen, Y., Daviet, L., Schalk, M., Siewers, V. and Nielsen, J. 2013. *Metab Eng* 15: 48–54.
Cheng, J. J. and Timilsina, G. R. 2011. *Renew Energ* 36: 3541–3549.
Chiesa, S. and Gnansounou, E. 2011. *Bioresour Technol* 102: 427–436.
Crauwels, S., Zhu, B., Steensels, J., Busschaert, P., De Samblanx, G., Marchal, K., Willems, K. A., Verstrepen, K. J. and Lievens, B. 2014. *Appl Environ Microbiol* 80: 4398–4413.
Crauwels, S., Van Assche, A., de Jonge, R., Borneman, A. R., Verreth, C., Troels, P., De Samblanx, G., Marchal, K., Van de Peer, Y., Willems, K. A., Verstrepen, K. J., Curtin, C. D. and Lievens, B. 2015. *Appl Microbiol Biotechnol* 99: 9123–9134.
de Barros Pita, W., Leite, F. C., de Souza Liberal, A. T., Simoes, D. A. and de Morais, M. A., Jr. 2011. *Antonie van Leeuwenhoek* 100: 99–107.
de Jong, B. W., Shi, S., Siewers, V. and Nielsen, J. 2014. *Microb Cell Fact* 13: 39.
de Jong, B. W., Shi, S., Valle-Rodriguez, J. O., Siewers, V. and Nielsen, J. 2015. *J Ind Microbiol Biotechnol* 42: 477–486.
de Souza Liberal, A. T., Basilio, A. C., do Monte Resende, A., Brasileiro, B. T., da Silva-Filho, E. A., de Morais, J. O., Simoes, D. A. and de Morais, M. A., Jr. 2007. *J Appl Microbiol* 102: 538–547.
de Souza, R. B., dos Santos, B. M., de Fatima Rodrigues de Souza, R., da Silva, P. K., Lucena, B. T. and de Morais, M. A., Jr. 2012. *J Ind Microbiol Biotechnol* 39: 1645–1650.

Demeke, M., Dumortier, F., Li, Y., Broeckx, T., Foulquie-Moreno, M. and Thevelein, J. 2013a. *Biotechnol Biofuels* 6: 120.
Demeke, M. M., Dietz, H., Li, Y., Foulquie-Moreno, M. R., Mutturi, S., Deprez, S., Den Abt, T., Bonini, B. M., Lidén, G., Dumortier, F., Verplaetse, A., Boles, E. and Thevelein, J. M. 2013b. *Biotechnol Biofuels* 6: 89.
Demeke, M.M., Foulquié-Moreno, M.R., Dumortier, F. and Thevelein, J.M. 2015. *PLoS Genetics* 11: e1005010
Dererie, D. Y., Trobro, S., Momeni, M. H., Hansson, H., Blomqvist, J., Passoth, V., Schnürer, A., Sandgren, M. and Ståhlberg, J. 2011. *Bioresour Technol* 102: 4449–4455.
Dmytruk, O. V., Voronovsky, A. Y., Abbas, C. A., Dmytruk, K. V., Ishchuk, O. P. and Sibirny, A. A. 2008. *FEMS Yeast Res* 8: 165–173.
Doshi, R., Nguyen, T. and Chang, G. 2013. *Proc Natl Acad Sci U S A* 110: 7642–7647.
Dunlop, M. J., Dossani, Z. Y., Szmidt, H. L., Chu, H. C., Lee, T. S., Keasling, J. D., Hadi, M. Z. and Mukhopadhyay, A. 2011. *Mol Syst Biol* 7: 487.
Fillet, S., Gibert, J., Suarez, B., Lara, A., Ronchel, C. and Adrio, J. L. 2015. *J Ind Microbiol Biotechnol* 42: 1463–1472.
Fischer, G., Rocha, E. P. C., Brunet, F., Vergassola, M. and Dujon, B. 2006. *PLoS Genet* 2: e32.
Fu, W. J., Chi, Z., Ma, Z. C., Zhou, H. X., Liu, G. L., Lee, C. F. and Chi, Z. M. 2015. *Appl Microbiol Biotechnol* 99: 7481–7494.
Fujitomi, K., Sanda, T., Hasunuma, T. and Kondo, A. 2012. *Bioresour Technol* 111: 161–166.
Galafassi, S., Cucchetti, D., Pizza, F., Franzosi, G., Bianchi, D. and Compagno, C. 2012. *Bioresour Technol* 111: 398–403.
Gao, Q., Cui, Z., Zhang, J. and Bao, J. 2014. *Bioresour Technol* 152: 552–556.
Garcia Sanchez, R., Hahn-Hägerdal, B. and Gorwa-Grauslund, M. 2010a. *Biotechnol Biofuels* 3: 19.
Garcia Sanchez, R., Karhumaa, K., Fonseca, C., Sànchez Nogué, V., Almeida, J. R., Larsson, C. U., Bengtsson, O., Bettiga, M., Hahn-Hägerdal, B. and Gorwa-Grauslund, M. F. 2010b. *Biotechnol Biofuels* 3: 13.
Generoso, W.C., Schadeweg, V., Oreb, M. and Boles, E. 2015. *Curr Opin Biotechnol* 33: 1–7.
Girio, F. M., Fonseca, C., Carvalheiro, F., Duarte, L. C., Marques, S. and Bogel-Lukasik, R. 2010. *Bioresour Technol* 101: 4775–4800.
Gnansounou, E. 2010. *Bioresour Technol* 101: 4842–4850.
Gollany, H. T., Titus, B. D., Scott, D. A., Asbjornsen, H., Resh, S. C., Chimner, R. A., Kaczmarek, D. J., Leite, L. F., Ferreira, A. C., Rod, K. A., Hilbert, J., Galdos, M. V. and Cisz, M. E. 2015. *Environ Manage* 56: 1330–1355.
Gostincar, C., Ohm, R. A., Kogej, T., Sonjak, S., Turk, M., Zajc, J., Zalar, P., Grube, M., Sun, H., Han, J., Sharma, A., Chiniquy, J., Ngan, C. Y., Lipzen, A., Barry, K., Grigoriev, I. V. and Gunde-Cimerman, N. 2014. *BMC Genomics* 15: 549.
Graham-Rowe, D. 2011. *Nature* 474: S6-S8.
Green, E. M. 2011. *Curr Opin Biotechnol* 22: 337–343.
Guadalupe-Medina, V., Metz, B., Oud, B., van Der Graaf, C. M., Mans, R., Pronk, J. T. and van Maris, A. J. 2013. *Microb Biotechnol*
Guo, Z. P., Zhang, L., Ding, Z. Y. and Shi, G. Y. 2011. *Metab Eng* 13: 49–59.
Görgens, J. F., Passoth, V., van Zyl, W. H., Knoetze, J. H. and Hahn-Hägerdal, B. 2005. *FEMS Yeast Res* 5: 677–683.
Hasunuma, T., Ishii, J. and Kondo, A. 2015. *Curr Opin Chem Biol*29: 1–9.
Hazelwood, L. A., Daran, J. M., van Maris, A. J., Pronk, J. T. and Dickinson, J. R. 2008. *Appl Environ Microbiol* 74: 2259–2266.
Hesse, A. 1949. In: Advances in Enzymology and Related Areas of Molecular Biology (eds. John Wiley & Sons, Inc., pp. 653–704
Hill, J., Nelson, E., Tilman, D., Polasky, S. and Tiffany, D. 2006. *Proc Natl Acad Sci U S A* 103: 11206–11210.
Holzer, H. and Goedde, H. W. 1957. *Biochem Z* 329: 175–191.
Hong, K.-K. and Nielsen, J. 2012. *Cell Mol Life Sci* 69: 2671–2690.

Hu, C., Zhao, X., Zhao, J., Wu, S. and Zhao, Z. K. 2009. *Bioresour Technol* 100: 4843–4847.
Huang, C., Zong, M.-h., Wu, H. and Liu, Q.-p. 2009. *Bioresour Technol* 100: 4535–4538.
Huang, C., Wu, H., Li, R.-f. and Zong, M.-h. 2012. *New Biotechnology* 29: 372–378.
Hull, A., Golubkov, I., Kronberg, B., Marandzheva, T. and van Stam, J. 2006. *International Journal of Engine Research* 7: 203–214.
Hyvönen, R., Olsson, B. A., Lundkvist, H. and Staaf, H. 2000. *Forest Ecol Manag* 126: 97–112.
Ishchuk, O. P., Voronovsky, A. Y., Stasyk, O. V., Gayda, G. Z., Gonchar, M. V., Abbas, C. A. and Sibirny, A. A. 2008. *FEMS Yeast Res* 8: 1164–1174.
Ishchuk, O. P., Voronovsky, A. Y., Abbas, C. A. and Sibirny, A. A. 2009. *Biotechnol Bioeng* 104: 911–919.
Jacob, Z. 1992. *Crit Rev Biotechnol* 12: 463–491.
Jarecki, M. K. and Lal, R. 2003. *Crit Rev Plant Sciences* 22: 471–502.
Jönsson, L. J., Alriksson, B. and Nilvebrant, N.-O. 2013. *Biotechnol Biofuels* 6: 16.
Kamisaka, Y., Noda, N., Tomita, N., Kimura, K., Kodaki, T. and Hosaka, K. 2006. *Biosci Biotechnol Biochem* 70: 646–653.
Kamisaka, Y., Kimura, K., Uemura, H. and Yamaoka, M. 2013. *Appl Microbiol Biotechnol* 97: 7345–7355.
Kata, I., Semkiv, M.V., Ruchala, J., Dmytruk, K.V. and Sibirny A.A. 2016. *Yeast* 33: 471–478.
Karlsson, H., Ahlgren, S., Sandgren, M., Passoth, V., Wallberg, O., Hansson, P.-A. 2016. *Biotechnol Biofuels* 9: 229.
Kim, S.R., Xu, H., Lesmana, A., Kuzmanovic, U., Au, M., Florencia, C., Oh, E.J., Zhang, G., Kim, K.H. and Jin, Y.-S. 2015. *Appl Environ Microbiol* 81: 1601–1609
Kondo, T., Tezuka, H., Ishii, J., Matsuda, F., Ogino, C. and Kondo, A. 2012. *J Biotechnol* 159: 32–37.
Kreuger, E., Sipos, B., Zacchi, G., Svensson, S. E. and Björnsson, L. 2011. *Bioresour Technol* 102: 3457–3465.
Kunze, G., Hähnel, U. and Riechen, J. H. 2014. Patent European Union EP2694646 A1
Kurtzman, C. P., Fell, J. W. and Boekhout, T. 2011. The Yeasts, a taxonomic study, fifth edition, Elsevier, Amsterdam, The Netherlands.
Ko, J.K., Um, Y., Park, Y.-C., Seo, J.-H. and Kim, K.H. 2015. *Appl Microbiol Biotechnol* 99: 4201–4212.
Kötter, P. and Ciriacy, M. 1993. *Appl Microbiol Biotechnol* 38: 776–783.
Ladygina, N., Dedyukhina, E. G. and Vainshtein, M. B. 2006. *Process Biochemistry* 41: 1001–1014.
Larsson, S., Cassland, P. and Jönsson, L. J. 2001. *Appl Environ Microbiol* 67: 1163–1170.
Li, Y., Zhao, Z. and Bai, F. 2007. *Enzyme Microb Technol* 41: 312–317.
Liguori, R., Amore, A. and Faraco, V. 2013. *Appl Microbiol Biotechnol* 97: 6129–6147.
Liu, G., Zhang, J. and Bao, J. 2016. *Bioprocess Biosyst Engin* 39: 133–140
Liu, H., Zhao, X., Wang, F., Li, Y., Jiang, X., Ye, M., Zhao, Z. K. and Zou, H. 2009. *Yeast* 26: 553–566.
Liu, H., Zhao, X., Wang, F., Jiang, X., Zhang, S., Ye, M., Zhao, Z. K. and Zou, H. 2011. *FEMS Yeast Res* 11: 42–51.
Liu, Y. Y., Chi, Z., Wang, Z. P., Liu, G. L. and Chi, Z. M. 2014. *J Ind Microbiol Biotechnol* 41: 1329–1337.
Ludovico, P., Sousa, M. J., Silva, M. T., Leão, C. and Côrte-Real, M. 2001. *Microbiology* 147: 2409–2415.
Lundin, H. 1950. *J I Brewing* 56: 17–28.
Martani, F., Fossati, T., Posteri, R., Signori, L., Porro, D. and Branduardi, P. 2013. *Yeast* 30: 365–378.
Mata-Gomez, L. C., Montanez, J. C., Mendez-Zavala, A. and Aguilar, C. N. 2014. *Microb Cell Fact* 13: 12.
Matsuda, F., Kondo, T., Ida, K., Tezuka, H., Ishii, J. and Kondo, A. 2012. *Biosci Biotechnol Biochem* 76: 2139–2141.
Miklenic, M., Zunar, B., Stafa, A. and Svetec, I. K. 2015. *FEMS Yeast Res* 15:

Mohr, A. and Raman, S. 2013. *Energy Policy* 63: 114–122.
Mortimer, N. D. 2013. *Biofuels* 4: 501–509.
Mosier, N., Wyman, C., Dale, B., Elander, R., Lee, Y. Y., Holtzapple, M. and Ladisch, M. 2005. *Bioresource Technology* 96: 673–686.
Nilsson, D. 2000. *J Agric Eng Res* 76: 27–36.
Odlare, M., Arthurson, V., Pell, M., Svensson, K., Nehrenheim, E. and Abubaker, J. 2011. *Appl Energ* 88: 2210–2218.
Olsen, R. A., Bunikis, I., Tiukova, I., Holmberg, K., Lotstedt, B., Pettersson, O. V., Passoth, V., Käller, M. and Vezzi, F. 2015. *Gigascience* 4: 56.
Olstorpe, M. and Passoth, V. 2011. *Antonie van Leeuwenhoek* 99: 57–62.
Özdemir, E. D., Härdtlein, M. and Eltrop, L. 2009. *Energy Policy* 37: 2986–2996.
Parks, L. W. and Casey, W. M. 1995. *Annu Rev Microbiol* 49: 95–116.
Passoth, V., Cohn, M., Schäfer, B., Hahn-Hägerdal, B. and Klinner, U. 2003. *Yeast* 20: 39–51.
Passoth, V., Blomqvist, J. and Schnürer, J. 2007. *Appl Environ Microbiol* 73: 4354–4356.
Passoth, V., Eriksson, A., Sandgren, M., Ståhlberg, J., Piens, K. and Schnürer, J. 2009. *Biotechnol Biofuels* 2: 16.
Passoth, V., Tabassum, M. R., Nair, H. A., Olstorpe, M., Tiukova, I. and Ståhlberg, J. 2013. *Enzyme Microb Technol* 52: 105–110.
Passoth, V. 2014. In: Molecular Mechanisms in Yeast Carbon Metabolism (eds. J. Piškur and C. Compagno), Springer Berlin Heidelberg, pp. 217–259
Pelletier, N., Audsley, E., Brodt, S., Garnett, T., Henriksson, P., Kendall, A., Kramer, K. J., Murphy, D., Nemecek, T. and Troell, M. 2011. In: Annual Review of Environment and Resources (eds. A. Gadgil and D. M. Liverman), ANNUAL REVIEWS, 4139 EL CAMINO WAY, PO BOX 10139, PALO ALTO, CA 94303–0897 USA, pp. 223-246
Peralta-Yahya, P. P., Ouellet, M., Chan, R., Mukhopadhyay, A., Keasling, J. D. and Lee, T. S. 2011. *Nat Commun* 2: 483.
Petrovic, U. 2015. *Yeast* 32: 583–93.
Phadnavis, A. G. and Jensen, P. R. 2013. Patent USA
Ploier, B., Daum, G. and Petrovič, U. 2014. In: Molecular Mechanisms in Yeast Carbon Metabolism (eds. J. Piškur and C. Compagno), Springer Berlin Heidelberg, pp. 169–215
Pols, A. J. K. and Spahn, A. 2014. In: Enzyclopedia of food and agricultural ethics (eds. P. B. Thompson and D. M. Kaplan), Springer, Dordrecht, pp. 211–217
Pronk, J. T., Steensma, H. Y. and van Dijken, J. P. 1996. *Yeast* 12: 1607–1633.
Ratledge, C. and Wynn, J. P. 2002. *Adv Appl Microbiol* 51: 1–51.
Reinertsen, S. A., Elliott, L. F., Cochran, V. L. and Campbell, G. S. 1984. *Soil Biol Biochem* 16: 459–464.
Ro, D. K., Paradise, E. M., Ouellet, M., Fisher, K. J., Newman, K. L., Ndungu, J. M., Ho, K. A., Eachus, R. A., Ham, T. S., Kirby, J., Chang, M. C., Withers, S. T., Shiba, Y., Sarpong, R. and Keasling, J. D. 2006. *Nature* 440: 940–3.
Robles-Medina, A., González-Moreno, P. A., Esteban-Cerdán, L. and Molina-Grima, E. 2009. *Biotechnology Advances* 27: 398–408.
Runguphan, W. and Keasling, J. D. 2014. *Metabol Engin* 21: 103–113.
Ryabova, O. B., Chmil, O. M. and Sibirny, A. A. 2003. *FEMS Yeast Res* 4: 157–164.
Sanchez, I. N. V., Bettiga, M. and Gorwa-Grauslund, M. F. 2012. *AMB Express* 2: 68.
Sandager, L., Gustavsson, M. H., Ståhl, U., Dahlqvist, A., Wiberg, E., Banas, A., Lenman, M., Ronne, H. and Stymne, S. 2002. *J Biol Chem* 277: 6478–6482.
Sassner, P., Galbe, M. and Zacchi, G. 2008. *Biomass Bioen* 32: 422–430.
Schadeweg, V. and Boles, E. 2016. *Biotechnol Biofuels* 9: 44.
Schifferdecker, A. J., Dashko, S., Ishchuk, O. P. and Piskur, J. 2014. *Yeast* 31: 323–332.
Shi, S., Valle-Rodriguez, J. O., Siewers, V. and Nielsen, J. 2014. *Biotechnol Bioeng* 111: 1740–1747.
Shiba, Y., Paradise, E. M., Kirby, J., Ro, D.-K. and Keasling, J. D. 2007. *Metabol Engin* 9: 160–168.

Sitepu, I., Selby, T., Lin, T., Zhu, S. and Boundy-Mills, K. 2014a. *J Ind Microbiol Biotechnol* 41: 1061–70.
Sitepu, I. R., Garay, L. A., Sestric, R., Levin, D., Block, D. E., German, J. B. and Boundy-Mills, K. L. 2014b. *Biotechnology Advances* 32: 1336–1360.
Slininger, P. J., Shea-Andersh, M. A., Thompson, S. R., Dien, B. S., Kurtzman, C. P., Balan, V., da Costa Sousa, L., Uppugundla, N., Dale, B. E. and Cotta, M. A. 2015. *Biotechnol Biofuels* 8: 60.
Solomon, B. D., Barnes, J. R. and Halvorsen, K. E. 2007. *Biomass Bioenerg* 31: 416–425.
Sousa, M. J., Ludovico, P., Rodrigues, F., Leão, C. and Côrte-Real, M. 2012. In: Cell homeostasis and stress response (eds. P. Bubulya), InTech, Rijeka, Croatia, pp. 73–100
Spindler, D. D., Wyman, C. E., Grohmann, K. and Philippidis, G. P. 1992. *Biotechnol Lett* 14: 403–407.
Steen, E. J., Chan, R., Prasad, N., Myers, S., Petzold, C. J., Redding, A., Ouellet, M. and Keasling, J. D. 2008. *Microb Cell Fact* 7: 36.
Steensels, J., Daenen, L., Malcorps, P., Derdelinckx, G., Verachtert, H. and Verstrepen, K. J. 2015. *Int J Food Microbiol* 206: 24–38.
Stephen, J. D., Mabee, W. E. and Saddler, J. N. 2012. *Biofuel Bioprod Bior* 6: 159–176.
Tanaka, K., Ishii, Y., Ogawa, J. and Shima, J. 2012. *Appl Environ Microbiol* 78: 8161–8163.
Tang, W., Zhang, S., Tan, H. and Zhao, Z. 2010. *Mol Biotechnol* 45: 121–128.
Tang, X., Feng, H. and Chen, W. N. 2013. *Metabol Engin* 16: 95–102.
Tao, X., Zheng, D., Liu, T., Wang, P., Zhao, W., Zhu, M., Jiang, X., Zhao, Y. and Wu, X. 2012. *PLoS One* 7: e31235.
Tapia, V. E., Anschau, A., Coradini, A. L., T, T. F. and Deckmann, A. C. 2012. *AMB Express* 2: 64.
Tehlivets, O., Scheuringer, K. and Kohlwein, S. D. 2007. *Biochim Biophys Acta* 1771: 255–270.
Theuretzbacher, F., Blomqvist, J., Lizasoain, J., Klietz, L., Potthast, A., Horn, S. J., Nilsen, P. J., Gronauer, A., Passoth, V. and Bauer, A. 2015. *Bioresour Technol* 194: 7–13.
Thompson, P. 2012. *Agriculture* 2: 339–358.
Tiukova, I., Eberhard, T. and Passoth, V. 2014a. *Biotechnol Appl Biochem* 61: 40–44.
Tiukova, I. A., de Barros Pita, W., Sundell, D., Haddad Momeni, M., Horn, S. J., Stahlberg, J., de Morais, M. A., Jr. and Passoth, V. 2014b. *Biotechnol Appl Biochem* 61: 51–57.
Tiukova, I. A., Petterson, M. E., Tellgren-Roth, C., Bunikis, I., Eberhard, T., Pettersson, O. V. and Passoth, V. 2013. *PLoS One* 8: e58455.
Unrean, P., and Khajeeram, S. 2015. *Bioresources Bioprocess* 2: 41
US Energy Information Administration. 2016. Monthly Energy Review, January 2016: http://www.eia.gov/totalenergy/data/monthly/#renewable
Valadi, H., Larsson, C. and Gustafsson, L. 1998. *Appl Microbiol Biotechnol* 50: 434–439.
Valle-Rodríguez, J. O., Shi, S., Siewers, V. and Nielsen, J. 2014. *Applied Energy* 115: 226–232.
Vallee, B. L. 1998. *Sci Am* 278: 80–85.
van Dijken, J. P. and Scheffers, W. A. 1986. *FEMS Microbiol Rev* 32: 199–224.
Vanholme, B., Desmet, T., Ronsse, F., Rabaey, K., Breusegem, F. V., Mey, M. D., Soetaert, W. and Boerjan, W. 2013. *Front Plant Sci* 4: 174.
van Urk, H., Postma, E., Scheffers, W. A. and van Dijken, J. P. 1989. *J Gen Microbiol* 135: 2399–2406.
Van Vleet, J. H., Jeffries, T. W. and Olsson, L. 2008. *Metabol Engin* 10: 360–369.
Voronovsky, A. Y., Rohulya, O. V., Abbas, C. A. and Sibirny, A. A. 2009. *Metabol Engin* 11: 234–242.
Wang, J., Li, R., Lu, D., Ma, S., Yan, Y. and Li, W. 2009. *World J Microbiol Biotechnol* 25: 921–925.
Wang, P. M., Zheng, D. Q., Liu, T. Z., Tao, X. L., Feng, M. G., Min, H., Jiang, X. H. and Wu, X. C. 2012. *Bioresour Technol* 108: 203–210.
Westhuizen, T. J. and Pretorius, I. S. 1992. *Antonie van Leeuwenhoek* 61: 249–257.
Wilkie, A. C., Riedesel, K. J. and Owens, J. M. 2000. *Biomass Bioenerg* 19: 63–102.
Woolfit, M. and Wolfe, K. 2005. *Nat Genet* 37: 566–567.

Wright, J., Bellissimi, E., de Hulster, E., Wagner, A., Pronk, J. T. and van Maris, A. J. 2011. *FEMS Yeast Res* 11: 299–306.

Yanase, S., Hasunuma, T., Yamada, R., Tanaka, T., Ogino, C., Fukuda, H. and Kondo, A. 2010. *Appl Microbiol Biotechnol* 88: 381–388.

Yu, K. O., Jung, J., Kim, S. W., Park, C. H. and Han, S. O. 2012. *Biotechnol Bioeng* 109: 110–115.

Yu, X., Zheng, Y., Dorgan, K. M. and Chen, S. 2011. *Bioresour Technol* 102: 6134–6140.

Zhang, F., Rodriguez, S. and Keasling, J. D. 2011. *Curr Opin Biotechnol* 22: 775–83.

Zheng, D.-Q., Wang, P.-M., Chen, J., Zhang, K., Liu, T.-Z., Wu, X.-C., Li, Y.-D. and Zhao, Y.-H. 2012. *BMC Genomics* 13: 479.

Zheng, Y., Yu, C., Cheng, Y. S., Zhang, R., Jenkins, B. and VanderGheynst, J. S. 2011. *Bioresour Technol* 102: 1489–1495.

Zhou, Y. J., Buijs, N. A., Siewers, V. and Nielsen, J. 2014. *Front Bioeng Biotechnol* 2: 32.

Zhu, J. Y. and Pan, X. J. 2010. *Bioresour Technol* 101: 4992–5002.

Zhu, Z., Zhang, S., Liu, H., Shen, H., Lin, X., Yang, F., Zhou, Y. J., Jin, G., Ye, M., Zou, H. and Zhao, Z. K. 2012. *Nat Commun* 3: 1112.

High Performance SBR-Technology for Unsterile Fermentation of Ethanol and Other Chemicals by Yeasts

Reinhard Pätz and Jau-Henryk Richter-Listewnik

Abstract General systems of fermentation in first generation bio-ethanol plants, e.g. with sugar or starch containing raw materials, require fermentation times of 35–70 h. To reduce this long production time, a new process called High-Performance-Sequencing-Batch-Reactor-Technology (HSBR-Technology) has been developed. The new concept is based on the integration of several strategies for increasing productivity, especially to achieve a higher efficiency and a shorter fermentation time, as well as to lower operation costs for energy demand. The integrative approach considers all stages of production: fermentation, bioreactor design, alternative distillation as well as further solutions for processing. The high productivity of the process is ensured by a combination of process parameters, such as the high concentration of active production yeast, a high feed concentration (e.g. glucose between 140 and 200 g l^{-1}), a short fermentation time (7–12 h), and as a result of our work a high ethanol concentration between 70 and 100 g l^{-1}. The process is controlled by a computer-aided measurement of the CO_2 concentration in the exhaust gas. Therefore, non-productive fermentation times and dysfunctions of the process are avoided. The productivity depends on the used substrate. For instance glucose led to an increased productivity for ethanol of at least 6.3–8.3 g l^{-1} h^{-1}.

Keywords Ethanol · Bio-ethanol · Fermentation · SBR · Fed-batch · Gas stripping

1 Introduction

An instinctive fermentation of alcohol is one of the fundamentals of developments of mankind. First alcoholic drunk was meth, fermented honey water. Later the production of beer starts in Sumer and in Egypt they prepared beer. Beer was a

R. Pätz (✉) · J.-H. Richter-Listewnik
Bioprocess Technology, University of Applied Sciences, Bernburger Str. 55, 06366 Köthen, Germany
e-mail: reinhard.paetz@hs-anhalt.de

T. Satyanarayana and G. Kunze (eds.), *Yeast Diversity in Human Welfare*, DOI 10.1007/978-981-10-2621-8_16

daily food with a low concentration of ethanol and with residual sugars. So it was a sweet drink rich of energy and always better than polluted and infected water. Higher concentrated beverage was prepared for holy ceremonies. In Caucasus area men started with wine making. About 2000 years before ancient the Chinese invented distillation which came as process over Arabian area to Europe. At first it was a solvent for drugs, but later people used for drinking.

First cars drove with alcohol as a fuel. Henry Ford meant that alcohol would be the "Fuel of the Future". But he was wrong. The development of the petrol chemistry worldwide and lignite industry in Germany made gasoline and diesel cheaper and cheaper. Alcohol was also produced on a chemical way. The chemical industry used it as a solvent and as a reagent for chemical reactions. The first shock came with ideas of the Club of Rome at the beginning of the 70s in last century. One of the most important ideas was that resources should be renewable. Some countries started with a gasoline compensation program especially USA, Brazil and Sweden.

Nowadays world ethanol production is about 100 billion liters (in 2010) (Vohra et al. 2014). More than 2/3 of all ethanol is used for fuels in different mixtures. Thus alcohol replaces gasoline in fuel mixtures between 15% (E15) and 85% (E85). For a chemically use about 20% of the ethanol production were necessary. For this purposes alcohol should be free of water. Finally about 10% of the produced alcohol is for drinking in different dilutions with water. Chemically used ethanol could be produced chemically or biotechnologically. For application as fuel the modern way is by biotechnologically fermentations. Thus it is termed bio-ethanol. Bio-ethanol can be produced from several substrates and conversion technologies. The main substrate is sucrose from sugarcane. As plant material sugarcane grow in tropical areas such as Brazil, India and other countries. In United States, Europe and China bio-ethanol are produced by starch containing crops like corn, wheat or barley (about 60%) or sugar beet (Gupta and Verma 2015). In all cases the production of bio-ethanol competes with the food industry for substrates. This kind of substrate is used to define first generation bio-ethanol. First generation bio-ethanol is made by sugar or starch.

Bio-ethanol plants of second generation considers about the use of other carbohydrates. Besides pure cellulose we find hemicelluloses and lignocellulosic materials which have to be hydrolyzed to get low molecular sugars, and lignin in the case of lignocellulose. Possible hydrolyses are expensive so that the produced ethanol cannot realize the same price related to ethanol produced from sugar. The production of lignocelluloses as raw material for ethanol syntheses has an expensive price, also by the use of straw or wood. Only lignocellulosic residues are low price substrates, but till now there is no preferred hydrolysis procedure. Historically only the Bergius process seems to be applicable for all different lignocelluloses containing wastes. This process is ruthless but successful especially as a low temperature process (Patent Green Sugar 2012). Sugar as product of starch hydrolysis is defined and may be used by defined microbial cultures. In contrast hydrolysis products of lignocelluloses are mixtures of different substrates. Besides hexoses like glucose a lot of different disaccharides exist depending on different hydrolyses methods. Thus we may find cellobiose and maltose which may not be used by all fermenting microorganisms. Hydrolysis products of pentoses are arabinose, xylose, ribose and other. Their fermentation is either only possible with slow growing microorganisms because of the diauxic substrate utilization or in more than one step processes. Some hydrolysis methods lead to chemically changed sugar molecules with inhibiting properties like furfural as product of acidic dehydration of pentoses.

All ideas from genetically modified microorganisms should consider two important facts:

1. Biomass should be saleable as feed or for other applications. This byproduct increases the economic potential of the whole process for hundreds of years in Europe. Thus we find traditionally combined meat and sausage production nearby drinking ethanol producers.
2. And the process should be able as bulk process stable over some years. Process stability is one of the most economical facts. Ethanol production is a high volume process. It is known that even traditional ethanol production may lead to changes in properties of production monoculture (Blomquist et al. 2010; Passoth et al. 2007).

First and second generation ethanol processes utilize liquid carbon substrates in one or two step fermentation processes. But microorganisms need more than carbon sources for their life, especially nitrogen, sulfur, phosphor as main elements and a lot of micronutrients like magnesium, manganese, potassium, iron, calcium and other. These elements should be added or must be contained in substrate influent stream. In the case of starch as substrate the source of starch like crop or other contain the elements but only insufficient. Thus a mixture of natural raw material und recycled yeast lysate is used as substrate.

The third generation was defined as application of algal biomass and residues from algal biomass for ethanol production. Prerequisite is the effective production of low price algal biomass and low price ethanol fermentation. Till now algal processes were applied for high price products like carotenoids.

When we want to define the fourth, the "next" generation of bioethanol plants than we should consider the so called syngas biotechnology. The disadvantage of second generation lignocellulosic substrate hydrolysis is the unwanted residue of lignin. A biotechnological application of lignin fails because of a slow process technology (till now, see Koncsag et al. 2012). The thermochemical process of synthetic gas production was developed for lignite as basis of the chemical industry. With special catalysts it should be possible to produce ethanol chemically. But syngas is also a possible substrate for biotechnological processes with interesting products and product mixtures (Fischer et al. 2008). There are gaseous substrates which can be used by the LJUNGDAL-pathway. Syngas may be produced from lignocellulosic raw material, lignocellulosic wastes, lignite, plastic wastes and is part of converter gas in steel industries. No other raw material comes from so much different sources and as important fact in some cases it is a question of sustainability.

All generation classifications consider only the biotechnological step. Finally the whole process has to be observed. Schulz and Hebecker (2005) analyzed the energetic side of production. They assessed all steps from raw material pretreatment till waste use and wastewater treatment under thermodynamic consideration. The source material is sugar beet, a typical crop plant of middle Europe. After sugar extraction the syrup directly served as substrate for fermentation, the biotechnological process step. In a three step process the ethanol was enriched to absolute dry ethanol. All wastes besides distiller`s wash were anaerobically treated and the biogas was used to produce electricity. About 30% of energy of all steps was reused inside the process. As a result of this exergetic analysis the defined three main points:

(a) The fermentation step has to be as effective as possible. All substrate should be converted only to ethanol. No byproducts.
(b) The effluent concentration of ethanol should be as high as possible but with high productivity.
(c) The first distillation step should be as effective as possible especially in energy consumption (derived from b).

Our conclusions are concentrated of all these aspects but at first for the fermentation step. Membranes for last step are a real alternative for molecular sieves. Cell free and cell debris free fermentation broth as feed to distillation has the advantage of incrustation free areas in heat exchanger for distillation.

In this chapter we will represent a process which considers points (a) and (b). Further on important points in our opinion are:

(a) Fermentation productivity as function of viability of active biomass and
(b) partially separation of ethanol from fermentation broth for reduction product inhibition during fermentation (by cold distillation/gas stripping).

2 SBR-Technology and the Special Kind of Repeated Fed-Batch

For the highest productivity of fermentation process a technology should be used which has a higher productivity than discontinuous or batch fermentation strategies and with a higher final effluent concentration of the end product. The first mentioned requirement leads to all of the quasi or direct continuous processes without any break for emptying, cleaning and new filling. For large reactors the non-productive steps need much time. Furthermore in continuous processes the resulting product concentration is small because of the outlet which is corresponding to the dilution rate. Feeding technologies are characterized by fewer and rarely breaks. The resulting final product concentration is determined by the chosen feeding strategy. This is also the case for the final biomass concentration. A higher biomass concentration may be reached by biomass retention or biomass recycling after an enrichment step in the downstream processing. The best technology should be a biomass retention coupled with a quasi-continuous process management.

The repeated fed-batch process fulfills the important requirements high product concentration and short fermentation time and, furthermore, implements a biomass retention step from another technology. So the retention step allows a very high biomass concentration without technical devices. In wastewater treatment the sequencing-batch-reactor-technology (SBR) is a well-established process used for technical wastewater and for sewage treatment. For the first time these technology was used by Sir Thomas Wardl in 1898. In the 50s of the 20th century Pasveer and Irvine created the term SBR. Recently this technology is applied for treatment of commercial wastewater and also of sewage. It is a sequential series of batch processes. These batch processes, called cycle, consist of different procedures. All the processes take place in one reactor. A total cycle is demonstrated in Fig. 1.

The whole process cycle consist of four steps:

1. Fermentation: The classical product formation by the consumption of substrate through living microorganisms.
2. Separation of biomass: Normally by sedimentation resulting in a biomass poor water phase and a biomass enriched sediment.
3. Product separation: Upper water phase is pumped out; lowest filling volume.
4. Filling: New substrate is pumped into the reactor; highest filling volume as working volume.

As it can be seen during the cycle the filling volume in the steps is different. So influence able factors are volume which can be chanced and the influent substrate

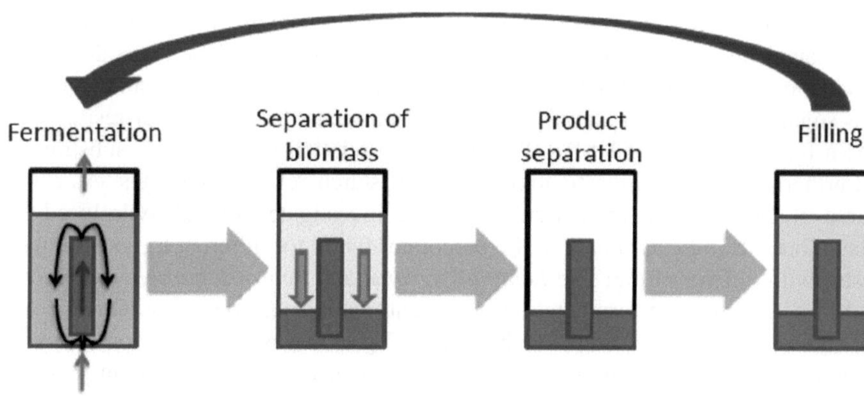

Fig. 1 Illustration of the four step cycle of a SBR

concentration. From biotechnological point of few this technology is a repeated fed-batch operation management with an included sedimentation/separation step (see later).

3 The HSBR Process

For a most effective technological process of ethanol production there are four main criteria:

- Kind and preparation of substrates,
- fermentation technology in form of high performance SBR-Technology,
- final ethanol concentration,
- marketable coproducts.

3.1 Kind and Preparation of Substrates

For our investigations of HSBR technology we use two different pure cultures, *Saccharomyces cerevisiae* which is used from the most producers of ethanol and *Arxula adeninivorans* a producer of alcohols with a wide spectrum of substrates.

Substrate for *Saccharomyces cerevisiae*
The variety of usable substrates for *S. cerevisiae* is small. Besides glucose only maltose and sucrose may utilized effectively. Glucose is the most described substrate. In our work all developments of the reactor system were done with glucose. But under technical conditions industrial substrates are more interesting. For instance this is the case with molasses the most applied substrate for ethanol fermentation over decades. But formerly this substrate contains more sucrose, today the sugar plants work more efficient so the sucrose content of molasses decreased. Another interesting substrate is hydrolysed starch residuals, e.g. C-starch. It contains sugar and a lot of proteins/peptides and minerals. In Table 1 our results were shown for application of HSBR-Technology in relation to glucose as substrate.

Furthermore in our investigations we used starch related substrates. As a new source of carbohydrate related substrates we investigated fungal hydrolysed lignocelluloses. But the application of a percolator failed for different lignocellulolytic wastes with different fungi like *Lentinus tigrinus* and *Sporotrichum thermophile* in kind of solid substrate fermentation with continuous sugar recovery. On this way only a small amount of sugars (about 10–15 g l^{-1}) was reached. For a process technology with a dilution step like HSBR the concentration of influent substrate must be high because the substrate will be diluted immediately after mixing in the reactor. For substrate concentration lower than the K_S value in Monod, or other types of equations for describing biomass growth, no growth takes place.

Table 1 Comparison of different types of ethanol fermentations

Literature	HSBR-technology (8L HSBR)	Ethanol production by simultaneous saccharification and fermentation (Vásquez et al. 2007)	High-level Production of ethanol during fedbatch (Seo et al. 2009)	Repeated-batch fermentation using flocculent hybrid (Choi et al. 2009)	Continuous ethanol production (Choi et al. 2010)
Typ of fermentation	Semi contin. fermentation	Batch	Fed-batch	Repeated batch	Contin. cascade fermentation
Substrate	Glucose	Saccharified cellulose	Glucose	Saccharified casava starch	Saccharified casava starch
Concentration of sugar, average (g/l)	128	60	100–200	150–177	175.195
Biomass, average (g/l)	23.63	n.V.	25	41.2	72
Concentration of ethanol reactor, average (g/l)	105.8	30	160	84.5	78
Exhaust ethanol, average (g/cycle)	12.58	0	0	0	0
Cycle time/fermentation time (h)	12	10	85	24	24
Productivity, average (g/(l*h))	6.30	3	1.88	3.34	3.28

Resulting from kinetical parameters the active substrate concentration at the beginning of the fermentation should not be higher than the K_{IS}-value of the culture.

Substrates for *Arxula adeninivorans*
This yeast may utilize some mono- and disaccharides like glucose, sucrose, lactose, maltose and also small oligosaccharides. The variety of hexoses is higher than of *S. cerevisiae*. Besides these substrates *Arxula* can use pentoses like xyloses. For all these substrates exists the necessary enzymes immanent. In contrast to *S. cerevisiae* there is no catabolic repression. That means that *A. adeninivorans* simultaneously utilize different substrates. No substrate or product inhibition is known. That is why this culture was applied to produce fodder yeast from black liquor which contains a lot of different hexoses and pentoses. The growth rate and the ethanol formation rate are smaller compared with *S. cerevisiae*. But the conditions for product formation during fermentation are more robust. And for the intended technology the fact is important that *A. adeninivorans* has a high sedimentation velocity.

3.2 Fermentation

For the technological part of the development of fermentation some fact may be of high interest.

(a) Choice of production culture and properties of production culture (aggregation behavior),
(b) product influence,
(c) reactor design.

(a) Choice of production culture

As clearly described in literature (e.g. Rudolf et al. 2009) *S. cerevisiae* is most used microbial culture for bio-ethanol production. *S. cerevisiae* can growth under aerobic and anaerobic condition. Under anaerobic conditions yeast produces ethanol with low growth of biomass. Effective growth takes place under aerobic conditions. But even here under aerobic conditions *S. cerevisiae* produces ethanol as byproduct as long as carbon source in form of glucose exist. This is a result of a carbon catabolite repression of the glucose uptake which is important for technological designing (Crabtree effect). *S. cerevisiae* tolerates a wide range of pH, but has its optimum under acidic conditions. Thus is a smaller risk of infection under fermentation conditions. Growth and product formation are coupled as growth associated product formation. *S. cerevisiae* tolerates temperatures up to more than 40 °C with optimum between 30 and 35 °C in relation to strain properties. Higher temperatures do not inactivate cells immediate but influence cell properties, e.g. aggregating behavior (Teich et al. 1990).

The theoretical yield from sugar is 0.51 g g^{-1} with about 2 g ethanol g^{-1}cell mass and hour. In industrial scale these values can be achieved by more than 90% (Skoog and Hahn-Hägerdal 1988). This is also a result of selection because people use *S. cerevisiae* since ancient time for food and beverage production also in industrial scale. *S. cerevisiae* has GRAS-state (Generally Regarded as Safe) and is accepted and wanted in industry. The yeast can utilize a lot of mono- and disaccharides like sucrose, glucose, fructose, galactose and mannose for the production of ethanol as main product (besides biomass) (Erlecke 1986). All other bacteria, yeasts and other fungi are of scientific interest, but not of industrial one. There is a large number of literatures for bio-ethanol, but producer do not accept other cultures because of the high risk of low constancy and stability in production process other than traditional technology.

Another tested alcohol forming yeast is *A. adeninivorans*. The culture was isolated from wastewater of wood-processing industry, can utilize hexoses and pentoses and forms ethanol under semi-anaerobic conditions. The yeast is defined as fodder yeast, the so-called "wood yeast", in Eastern Europe. Thus it is possible to produce ethanol and feed utilizing wastes or wastewaters from wood processing industry. The yeast aggregates fast and has a fast sedimentation velocity, faster than the above mentioned self-aggregation of *S. cerevisiae*.

Kinetic Properties

For industrial application *S. cerevisiae* should be used. The original strain is *S. cerevisiae* strain Kolin EH 04 which has a good productivity for ethanol, is well described and available in our reference stocks. The strain can grow on mono- and disaccharides but shows a substrate inhibition for sucrose and a product inhibition on specific growth rate under aerobic conditions. The values are determined for substrate inhibition $K_{IS} = 154$ g l^{-1} and ethanol (product) inhibition $K_{IEtOH} = 84$ g l^{-1}. That means that over 154 g l^{-1} of sucrose the grow decreases (to half of the possible specific growth rate). The specific growth rate is 0.19 h^{-1}, the yield coefficient for biomass growth Y_{XS} is 0.46 g g^{-1}. Related to other industrial ethanol producers there are some remarkable differences. The substrate removal rate with 0.45 g g^{-1} h^{-1} is lower than for other strain of genus *Saccharomyces* but the Yield coefficient Y_{PS} for ethanol formation from substrate sucrose is higher. The main advantage of the strain is the high ethanol resistance. Under anaerobic conditions the ethanol formation rate depends on ethanol concentration (Fig. 2—right diagram). It can be seen that fast ethanol formation takes place also with 80 g l^{-1}. The difference between maximum of specific ethanol formation rate and start of product inhibition is small. In relation to other authors there is a big difference. Normally there is a typical behavior with high increase in specific ethanol formation rate in the case of small ethanol concentrations and a slow decrease of specific formation rate. The optimum is higher but at a very low ethanol concentration (Fig. 2—left diagram). With our yeast strain there is an increase in specific ethanol formation rate also with higher ethanol concentration. That means a higher productivity connected with higher ethanol concentration.

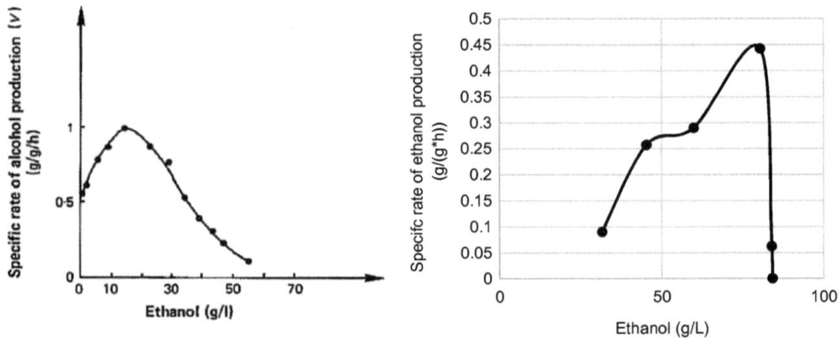

Fig. 2 *Left* typical Specific rate of ethanol production (Baeyens et al. 2015); *right* specific rate of ethanol production of *S. cerevisiae* under oxygen limited conditions (semi-aerobic) in a HSB-Reactor

Fig. 3 Immobilized bodies with *S. cerevisiae*

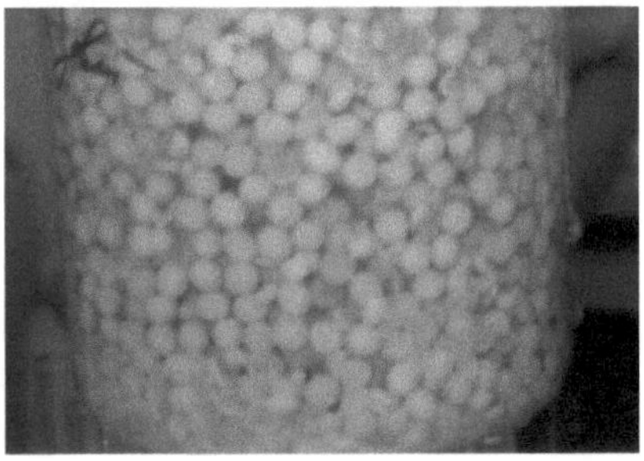

Aggregation

Some scientists try to increase the active biomass concentration by immobilization. There are some different kinds of immobilization. For lab scale investigations particle formation with alginate is described. But cells are growing during ethanol formation (or vice versa). Ethanol formation is directly coupled with growth. This is strong during aerobic growth for *S. cerevisiae* because of the catabolite repression. Under anaerobic and semi-anaerobic conditions the specific growth rate is lower but takes place. So the immobilized bodies will be destroyed by the yeast (Fig. 3).

Surface immobilization is a weak adsorption on a solid surface of any inert material. The use of inert *strong* material increases the viscosity and density and hindered diffusion processes of substrate and products. Besides this it increases the necessary energy input for mixing.

In fact the best method for increasing the active biomass concentration is the self-aggregation of microbial cells. Under anaerobic conditions granula formation is well known in UASB reactors (Lettinga et al. 1980). UASB reactors are used in anaerobic wastewater treatment. Granula are formed like shells with all different microorganisms which are necessary for all wastewater constituents. The density of the granula is high so that granula sediment quickly, when driving force is low. This advantage should be used in ethanol fermentation.

In the case of monoculture fermentation there are two kinds for aggregation during fermentation, the flocculent growth and the aggregation under special growth conditions. Self-aggregation of *Saccharomyces* cells is investigated intensively for brewing processes (Zhao and Bai 2009). Influencing parameters are the yeast itself, concentration of substrate and especially of alcohol, ionic strength and at least the age of the cells. During flocculent growth all cells are aggregated from lag phase to lysis as primary aggregates. The formation of bigger, secondary aggregates takes place when energy input decreases. Special aggregation may be reached under certain outer conditions e.g. change in osmotic pressure. In the case of ethanol formation osmotic pressure related to sugar decrease but related to ethanol increase, resulting in formation of glycerol and other osmo-products. This is connected with changes in cell wall constituents. The ability and velocity of changes of the outer cell surface is genetically determined but is influenced from outer conditions. This property may be used for selection of self-aggregating yeast cells, when all settled yeast cells were restored and single or non-aggregated cells were washed out. Thus a technological selection takes place.

Figure 4 shows the start of self-aggregation at the beginning of technological investigation. There are a lot of small and single cells. After some days of fermentation only big aggregates remain in the reactor (Fig. 4—right side). This phenomenon only can be detected in a longer technological test phase under certain identical conditions. As the main result of this procedure there exists the possibility to get a cell-free supernatant in a sedimentation process. This is important for effective biomass retention to increase active growing biomass and for better durability for distillation column.

(b) Product influence

It is well known that the product formation rate is related to the ethanol concentration. From *S. cerevisiae* it is known that increasing ethanol concentration inhibits

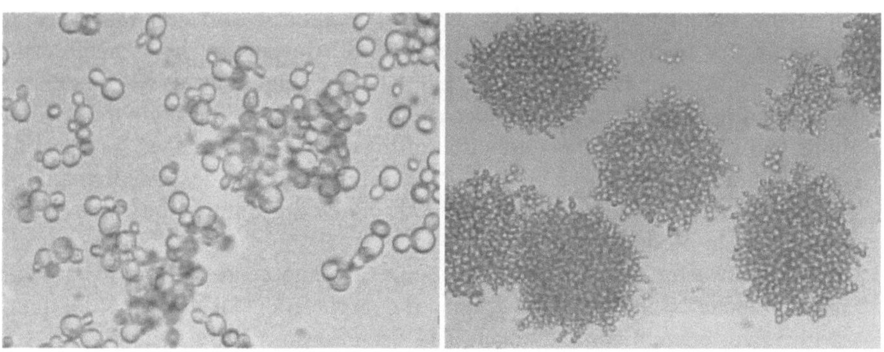

Fig. 4 Self-aggregation of *S. cerevisiae* (*left* at the beginning of the technological investigation (1000-magnification); *right* stable self-aggregated cell aggregates (400-magnification))

the further ethanol formation and decreases the ethanol formation rate shown in Fig. 2. There exists an optimum output. In the case of our applied culture it is about 80 g l^{-1}. With increasing ethanol concentration yeast produce glycerol and ergosterols as osmoprotectants, which decrease the yield of ethanol. With a further increasing ethanol concentration the viability of the culture is decreasing and more and more cells are dying and start with cell lysis. During wine making and beer fermentation this process is wanted because of the formation of typical flavors. But for application in the chemical industry and for the production of biofuels it is not necessary. Thus the optimum of ethanol formation should be earlier than the start of by-product formation of secondary metabolites or lyses products and other secondary products. This only can be reached when biomass has its highest activity in relation to growth under the technical conditions. Activity again may be determined by microbiological methods. This is conventional realized for example in beer brewing (Müller-Auffermann et al. 2015). In technical processes the carbon dioxide formation rate alone or as respiration quotient RQ were used to estimate microbial activity. We used the change in carbon dioxide formation rate in our semi-aerobic fermentation for the regulation of the cycle time. As a result the activity and the productivity increase.

On the other hand a special part of lysis products is necessary as internal nitrogen and vitamin source. These internal components add the medium during fermentation. Bai et al. (2008) showed that "assimilable nitrogen is one of the most important components in the fermentation medium and has been reported to be a limiting nutrient in the VHG (very high gravity) ethanol fermentation using wheat mash."

(c) Reactor technology

The reactor has to fulfill some aspects of technology. Thus the technical design of the reactor is an important fact. The following ideas have to be considered:

- During fermentation mixing should be intensively to get a homogeneous suspension. With increasing biomass content the viscosity will increase and the viscosity is only reduced by carbon dioxide bubbles. Also under these conditions the homogeneity should be ensured to reduce diffusion hindrance for substrate, for microorganisms and for ethanol. But energy input may destroy aggregates by stochastic movement. On the other hand in airlift reactors there is a directed movement in downcomer and in upstreamer with high speed conditions. Bu'Lock et al. (1984) preferred airlift reactors with baffles for yeast cell retention in a reactor cascade. Also a lot of other authors described continuous operations (see Xu et al. 2005). But with increasing dilution rate yeasts show fluctuation in productivity.
- There are two limiting concentrations, the substrate concentration for a beginning substrate inhibition and the ethanol concentration for a product inhibition (with lower ethanol formation rate and decrease in substrate yield). The substrate concentration should be high enough for a noticeable growth. During a cycle the fermentation step should start with the highest possible substrate

concentration for non-inhibited growth. This prediction determines the exchange value. Higher concentrations in feed stream leads to a smaller volume of exchange during the cycle. This results in a constant final concentration of ethanol at the end of every cycle, an important prerequisite for a high growth and formation rate.

- Ethanol is a preserving agent in food production and food storage. For an unsterile fermentation it is necessary to ensure a high concentration. This is the case in our technology. After sedimentation the upper part contains no yeast flocs but ethanol, other byproducts like glycerol, non-flocculating yeast cells and if happens, bacterial contaminants. All these substances were pumped out into the distillation step. The lower part contains yeast agglomerates and ethanol in a concentration near inhibition; a high and thus conserving concentration. The ethanol concentration will be diluted during filling phase.
- Also some small part of oxygen is necessary for growth. Ethanol is a typical primary metabolite. This production is tightly coupled with yeast growth. Besides this only under these conditions sterols and fatty acids are formed necessary for stability against higher ethanol concentration (Macy and Miller 1983). In an airlift-loop reactor the oxygen input is connected with energy. A mixture of air and inert gas is used for movement. With an optimization process for ethanol productivity the fresh air part was determined.
- The carbon dioxide content of exhaust air is a remarkable sign of activity of yeast and the set-point for regulation the substrate input. When substrate is consumed out, the carbon dioxide content decreases. This is the start of the sedimentation process and after that of the recovery of the clear product phase from the top of the reactor.

4 Applications of HSBR-Technology

The advantages possibilities of the application of the HSBR-Technology are explained extensively with the example of a high efficient ethanol production. Other investigated possibilities are stated in Chap. 6.

Application for high efficient ethanol formation
The HSBR-Technology was primary developed for an application in the bio-ethanol industry because the technology is able to reduce the fermentation time and the operation costs for demand of energy and investment. The authors were able to demonstrate the advantages of the technology by the use of different substrates for the production of ethanol in different lab-scale reactors. The used substrates include glucose solutions, molasses, several hydrolyzed starch solutions but also hydrolyzed sugar beet chips and residuals from starch industry. Other substrates, e.g. from agriculture residuals, will be possible in combination with a suitable method for hydrolysis or by the use of a suitable production culture. From the HSBR-Technology exist only one requirement to the substrate, the feed for the

fermentation process has to be free of solids. Otherwise the solids can be accumulated in the reactor.

The main advantages of the HSBR-Technology for a high efficient ethanol production compared to conventional bio-ethanol plants are:

(a) Unsterile fermentation process,
(b) process integrated biomass separation,
(c) product removal during the fermentation process,
(d) high productivity by the application of oxygen limited fermentation conditions,
(e) carbon dioxide controlled HSBR-Technology.

(a) Unsterile fermentation process

In all stable fermentations it was demonstrated that no bacterial contamination was able to establish oneself about a period of month. Because of the unsterile reactor there was always a bacterial contamination at the start phase of new HSB-Reactors. After few cycles the concentration of the contamination was significant reduced and after one week there was no contamination recognizable.

The combination of different circumstances leads to an unsterile fermentation of a mono culture. After a short period of aerobe fermentation for fast yeast growing the aeration with fresh air was stopped when a minimal yeast concentration for the HSBR-Process (>20 g l^{-1}) was reached. After that the produced carbon dioxide with only traces of fresh oxygen (air) was pumped through a closed-circuit. Because of the deficiency of oxygen all obligatory microorganisms stopped to grow. Another point is the low pH-Value. For the process adapted *S. cerevisiae* was it able to leave the pH-Value decreasing down to 3.0–3.5. Under these strong acid conditions the most other microorganisms had a very slow specific grow rate. More stress for bacteria was induced by high concentrations of sugar and ethanol. At the start of a fermentation step the sugar concentration was very high (120–170 g l^{-1}), the high concentrated yeast consumed the sugar within some hours (7–12 h), but even when the sugar concentration was very low the ethanol concentration increased dramatically (80–100 g l^{-1}) and after the full consumption of sugar the aeration stopped. So there are very fluctuating concentrations of sugar and ethanol and both of them lead to an inhibition of contaminants. All these strong facts reduce the specific grow rate of not adapted wild yeasts and bacteria very much while the production yeast has a slow but constant grow rate. When the sugar is completely used the production of ethanol and carbon dioxide stopped. The fast degreasing carbon dioxide level in the exhaust gas is the signal to stop the fermentation step and start the sedimentation. The aggregated yeast agglomerate sink down to the lower sector of the reactor and displace all light cells into the upper part of the reactor. This process is visible in glass reactors. After the sedimentation step the upper part of the reactor (max. 70% of the complete volume) is pumped out and replaced with fresh media. The next cycle starts immediately. About 10% of the production yeast was also pumped out of the reactor but these cells will be replaced during the next fermentation step.

(b) Process integrated biomass separation

The aggregation behavior of flocculent yeast growing is described above in 5.2. Investigations of the yeast *S. cerevisiae* show that the fermentation conditions stimulate the yeast aggregation. The main parameter for a high ability to form large agglomerates is the cell surface hydrophobicity. The higher the hydrophobicity is the better the ability for agglomeration is (Lui et al. 2004). This parameter is depending from very much fermentation and yeast parameters: age of cells, growing of cells, pH-value, temperature, shear stress, medium contents of salts, sugar and ethanol.

It is known that glucose, sucrose, galactose and fructose have only a few or no inhibiting effect to the ability of agglomeration (Jin and Speers 1999). The presence of glucose has indeed an easily negative effect to the speed of agglomeration. But our own investigation shows that the fact of the concentration of the sugars in combination with the age of the cells is much more imported for the HSBR-Technology. The fermentation step can be stopped only when all sugar is consumed. Otherwise the cells will produce carbon dioxide at a high level until the sugar is completely consumed. The production of carbon dioxide with this large volume will prevent the sedimentation step. The age of cells is important because only older cells have a high cell surface hydrophobicity. This is explained with larger volume of cells and irregularities at the cell surface which leads to better adhesion properties (Jibiki et al. 1997). Consequently is this an important fact for the good agglomeration in HSB-Reactors because there are a lot of older cells (only about 10% of cells is replaced at each cycle).

The presence of alcohol stimulates the agglomeration, so is the surface hydrophobicity increased by alcohol and the negative electric charge of the cell surface reduced (Jin and Speers 1999). The positive effect of reduced negative electric charge of cell surfaces for better agglomeration was also described from Wang et al. (1994). Both lead to a better ability of agglomeration which is also demonstrated with our investigations. This is a further reason for the good agglomeration of the yeast cells at the end of the fermentation step because in this moment the ethanol concentration is at the highest level.

The influence of temperature is very low between 15–32 °C only at very high temperatures above 60 °C the agglomerates will destroyed (floc-melting) (Jin and Speers 1999). Never the less the influence of temperature is depending from the strain but all examined strains (*S. cerevisiae, A. adeninivorans, R. toruloides*) has a good or very good agglomeration behaviour.

Our own investigations concerning the pH-Value show no strong dependence of ability of the agglomeration of *S. cerevisiae* between pH 3–6. Dengis et al. (1995) examined top and bottom fermenting yeasts and described this large optimal pH range only for bottom yeasts. Top fermenting yeasts have only an optimal range of pH 3–4.5. For the HSBR-Technology a good agglomeration at a low pH-Level (3.0–5.5) is important so both results confirm the good behaviour in the sedimentation step.

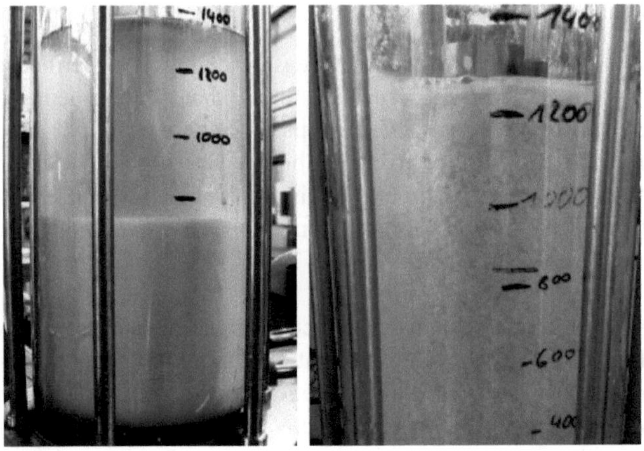

Fig. 5 *Left* HSB-Reactor during sedimentation step, *right* agglomerate structures during refill step

All the described parameters explain the very good ability for agglomeration of the yeast which leads to a fast sedimentation step during the HSBR-Cycle (Fig. 5). But also the large agglomerate is helpful if the yeast is separated by centrifuge, decanter or separator.

(c) Product removal during the fermentation process

The energy input for of the HSB-Reactors is the gas which is blown into the reactor at the bottom (airlift-reactor). The gas consists of the produced carbon dioxide with only traces of fresh oxygen (air) and is pumped through a closed-circuit. During the way through the liquid the gas will be enriched with volatiles organic components and water. The driving force for the enrichment is the difference of vapor pressure from the volatiles in water (higher vapor pressure) and in gas (lower vapor pressure). The saturated gas pass an exhaust cooler and the volatiles will be liquefied again. This process is called gas stripping (Ezeji et al. 2005). By using gas stripping the HSBR-Technology separates a liquid product phase free of biomass with a high product concentration and only small amounts of byproducts. An important additional effect is the reduction of product inhibition of the cells which results in a higher productivity of the whole HSBR-Process. Taylor et al. (2000) calculates furthermore with cheaper costs for the ethanol production (about 0.01 € per litre) and Liu and Hsien-Wen (1990) demonstrated higher biomass concentration, better substrate yield and improved ethanol productivity. Our own investigations show the dependence of the amount of volatiles from the gas stripping flow rates and ethanol concentration in the reactor as the main parameters.

The proportion of the stripped ethanol is much lower than the proportion of ethanol in the reactor but the quality is very good and for the economy calculation of the production process is the collection of the product from the exhaust gas important. Up to 10% of the produced ethanol can be collected by gas stripping, the concentration of ethanol is up to 30 vol.%.

(d) High productivity by the application of oxygen limited fermentation conditions

As described before (see Sect. 5.2) the yeast *S. cerevisiae* is able to produce ethanol with aerobe and anaerobe conditions. Under aerobe conditions the yeast has a very high specific growth rate (about 0.4 1 h^{-1}) and a very high amount of substrate is used for growth. By the use of strict anaerobic conditions the specific growth rate is lower (0.1 1 h^{-1}) and the yeast will use more of the substrate for ethanol production. But the optimal productivity for the ethanol production is between these points. Preliminary batch tests show a significant higher yield and productivity for the ethanol production by a limitation of oxygen but not at anaerobe fermentations. So a small amount of oxygen is for cultivation and regeneration necessary. Especially the fermentation time decreased by 25% compared to aerobe or anaerobe fermentations and the highest possible ethanol concentration and yield was already reached.

These results were used for the optimization of the gas mix for the HSBR-Technology. A stable HSB-Reactor with glucose solution as feed was used to investigate different gas mixtures of air and carbon dioxide enriched circulation gas. The results are shown in Fig. 6. With these reactor (8 L HSB-Reactor) and constant conditions (pH 3.5, temperature 30 °C, glucose solution as feed) and a stable mono culture of *S. cerevisiae* the reactor has an output of 571 g ethanol per cycle. After the fermentation step the reactor has an ethanol content of 605 g but not all ethanol can be removed because a part of the product remains in the reactor together with the yeast.

The formation rate as well as the specific formation rate for ethanol production in Fig. 7 shows also the best conditions with a gas mixture of 1:1 (air:circulation gas).

Very interesting is also the productivity of the HSB-Reactors. As it is shown in Fig. 7 the best productivity is 6.3 g ethanol per litre and hour, which means that one litre of the HSB-Reactor is able to produce more than 55 kg pure ethanol per year.

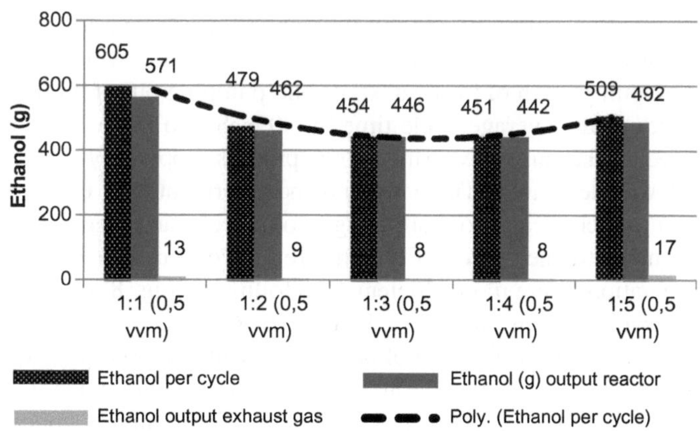

Fig. 6 Production of ethanol per cycle with different gas mixtures, with a gas flow rate of 0.5 vvm (e.g. with 1:2 means 1 part air and 2 parts circulation gas)

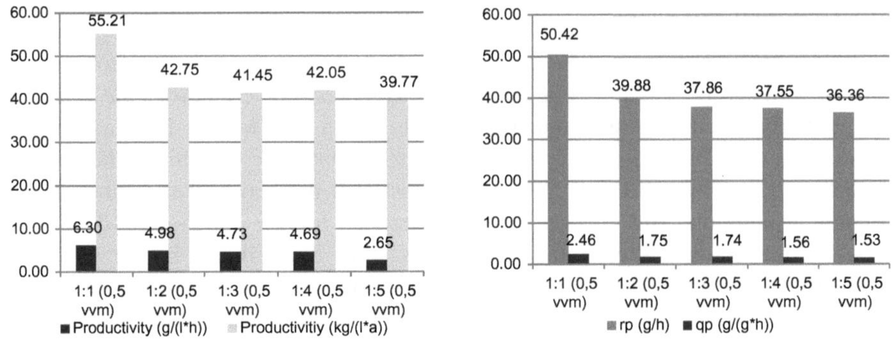

Fig. 7 *Left* productivity per litre per hour and per year, *right* ethanol formation rate (rp) and specific ethanol formation rate (qp)

In the Table 1 the HSBR-Technology is compared with other types of fermentation including fed-batch and repeated batch fermentations. The high productivity and final ethanol concentration are the main advantages of the HSBR-Technology. The high productivity causes of the short fermentation time of about 8 h but the calculation of productivity was done with the complete cycle time of 12 h as it is written in the table. The concentration of ethanol in the HSB-Reactor is really high, this is also a result of the remaining ethanol in the reactor during the removal of product and refill of the reactor.

(e) Carbon dioxide controlled HSBR-Technology

A really great improvement was able with the installation of a carbon dioxide controller for the HSBR-Technology. Until now the processes was controlled by timers and the carbon dioxide content in the exhaust air was only measured and visualised. So it was known that the carbon dioxide increased very much after new substrate was pumped into the reactors. When the sugars were completely consumed the content of carbon dioxide in the exhaust air decreased very fast.

After the installation of the carbon dioxide controller, which controlled the pumps for gas, pumping in of fresh substrate and pumping out of product phase, we was able to reduce the average cycle time by 30%. So no more unproductive time buffer was needed because the fermentation process stops only when the carbon dioxide level was decreased. There was no more fermentation time without limitation of sugar, which means no more hunger to the cells and therefor no lag phases at the beginning of the next cycle. Furthermore the yeast cells had a better training and the fermentations were more efficient. The following Fig. 8 illustrate the typical HSB-Reactor cycles in this case for a 10 L HSB-Reactor.

The calculated yield for the ethanol production with the optimized HSB-Reactor from glucose is 0.43 g g^{-1}. In addition to the reactor output of ethanol there was an ethanol output through gas stripping of about 95 ml and a concentration of 235 g l^{-1}.

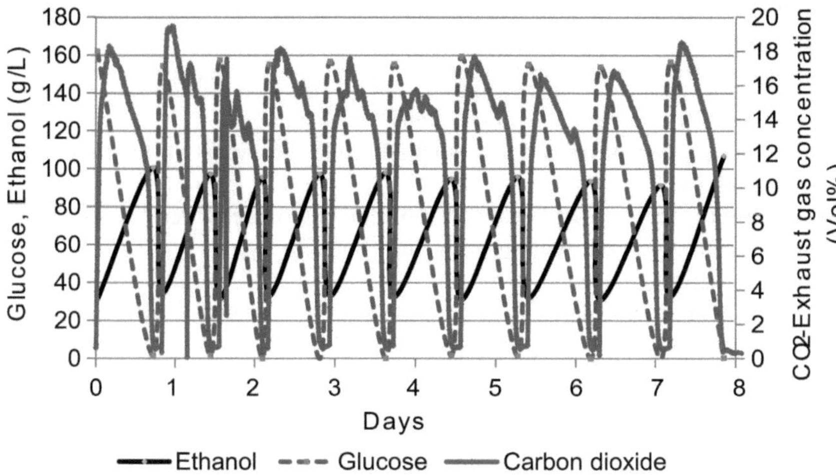

Fig. 8 Carbon dioxide controlled fermentation with typical cycle profile: when fresh substrate is available the carbon dioxide increase very much and degreases very much after the complete consumption of sugars. The ethanol concentration varied between 30 g l^{-1} at the beginning and 95 g l^{-1} at the end of the cycle (*Source* Künstler, Christian)

5 Further Applications of the HSBR-Technology

The HSBR-Technology is useable for a lot of applications to produce products by yeasts. Our investigations have shown that different yeasts can used in the fermentation process. Especially the application of three yeast was investigated. The yeast *S. cerevisiae* for the production of ethanol, the yeast *A. adeninivorans* for the production of C4-alcohols and *Rhodosporidium toruloides* for the production of carotenes.

Further applications are possible.

6 Conclusion

With the realisation of a technology transfer from waste water treatment to an application of production of alcohols we were able to develop a very productive fermentation process with the possibility of a process integrated biomass recovery through self-immobilization. The feasibility of the HSBR-Technology was demonstrated at the university and different public institutes and companies. Therefore different substrates were used, including industrial substrates or residuals of starch and sugar beets. The main advantage of the technology is the high productivity (for ethanol: 6.3 g l^{-1} h^{-1}), in individual cases up to 8.3 g l^{-1} h^{-1} reasoned of adapted self-immobilized yeast with a high constant concentration and thereby the application of high concentrated substrates which results in high product concentrations.

For an industrial application further advantages are interesting:

- Continues working reactor without steps for cleaning,
- high concentration of trained and adapted production yeast all the time without growing phase,
- high activity of yeast and realised replacement of dead yeast through small amounts of oxygen but without reducing productivity,
- short production cycles,
- continuous product separation is possible with a couple of HSB-Reactors,
- energy input and intermixing with gas via fan (pneumatic drive) without moving parts (stirrer),
- very low demand for sterilisation (no regularly sterilisation with steam necessary),
- easy control system, based on concentration of carbon dioxide,
- second product stream free of biomass with a high product concentration through gas stripping with reduction of product inhibition in the reactor.

Besides the technical advantages the HSBR-Technology convinced also with economic aspects. A calculation of an industrial scale HSBR-Facility with 100.000 m^3 per year output of ethanol and a payback period of five years show financial advantage in investment of 19% compared to typical ethanol plants. The calculation for the installation and operation of a pilot scale HSBR-Facility with only one 10 m^3 reactor and an output of 602 m^3 ethanol shows also good operation income. The calculation was done with different substrates (triticale, molasses) and a market price for ethanol of 0.72 Euro per litre. Depending upon the costs of substrates the profit was calculated about 40.000 Euro per year. That means a single pilot plant of 10 m^3 volume is able to recover the cost for installation and operation.

References

Baeyens, J., Kang, Q., Appels, L., Dewil, R., Lv, Y., Tan, T. 2015. *Progress Energy & Combustion Sci.* 47: 60–88.

Bai, F.W., Anderson, W.A., Moo-Young, M. 2008. *Biotechnol. Adv.* 26: 89–105.

Blomquist, J., Eberhard, J.T., Schürer, J., Passoth, V. 2010. *Appl. Microbiol. Biotechnol.* 87: 1487–149.

Bu`Lock, J.D., Comberbach, D.M., Ghommidh, C. 1984. *Chem. Eng. J.* 29: 2399–2411.

Choi, G.-W., Kang, H.-W.; Moon, S.-K. 2009. *Appl. Microbiol. Biotechnol.* 84: 261–269.

Choi, G.-W., Kang, H.-W.; Moon, S.-K., Chung, B.-W. 2010. *Appl. Biochem. Biotechnol.* 60: 1517–1527.

Dengis, P.B., Nélissen, L.R., Rouxhet, P.G. 1995. *Appl. Environ. Microbiol.* 61: 718–728.

Erlecke, K. 1986. Diplomarbeit IHS Köthen.

Ezeji, T.C., Karcher, P.M., Qureshi, N., Blaschek, H.P. 2005. *Bioprocess. Biosyst. Eng.* 27: 207–214.

Fischer, C.R., Klein-Marcusschamer, D., Stephanopoulos, G. 2008. *Metabol. Eng.* 10: 295–304.

Green Sugar 2012. WO/2012/013177.

Gupta, A., Verma, J.P. 2015. *Renew. Sustain. Energy Reviews* 41: 550–567.

Jibiki, M.A., Ishibiki, T., Yamashita, H., Eto, M. 1997. *Techn. Quarterty* 34: 174–178.

Jin, Y.-L., Speers, R.A. 1999. *Food Research International* 31: 421–440.

Koncsag, C.I., Eastwood, D., Collis, A.E.C., Coles, S.R., Clark, A.J., Kirwan, K., Burton, K. 2012. *Resources, Conservation and Recycling* 59: 14–22.

Lettinga, G., van Velsen, A.F.M., Hobma, S.W., de Zeeuw, W., Klapwijk, A. 1980. *Biotechnol. Bioeng.* 22: 699–734

Liu, H.-S., Hsien-Wen, H. 1990. *Chem. Eng. Sci.* 45: 1289–1299.

Lui, Y., Yang, S.-F., Qin, L., Tay, J.-H. 2004. *Appl. Microbiol. Biotechnol.* 64: 410–415.

Macy, J.M., Miller, M.W. 1983. *Archives Microbiol.* 134: 64–67.

Müller-Auffermann, K., Caldera, A., Jacob, F., Hutzler, M. 2015. *Brewingsci.* 68: 46–57.

Passoth, V., Blomquist, J., Schnürer, J. 2007. *Appl. Environ. Microbiol.* 73: 4354–4356.

Rudolf, A., Karhumaa, K., Hahn-Hägerdal, B. 2009. *Yeast Biotechnol.: Diversity and Applications.* DOI:10.1007/978-1-4020-8292-4_23. Springer Sci. Busin. Media B.V.: 490–515.

Schulz, M., Hebecker, D. 2005. *Chem.-Ing.-Tech.* 77: 792–798.

Seo, H.-B., Kim, S.S., Lee, H.-Y., Jung, K.-H. 2009. *Biotechnol. Bioproc. Eng.* 14: 591–598.

Skoog, K., Hahn-Hägerdal, B. 1988. *Enz. Microb. Technol.* 10: 66–80.

Taylor, F., Kurantz, M.J., Goldberg, N., Mcaloon, A.J., Craig Jr, J.C. 2000. *Biotechnol. Progress* 16: 541–547.

Teich, W., Otto, G., Bergmann, H., Pätz, R. 1990. *Acta Biotechnologica* 3: 311–316.

Vásquez, M.P., Da Silva, J.N.C., De Souza Jr., M.B., Pereira Jr., N. 2007. *Appl. Biochem. Biotechnol.* 7: 136–140.

Vohra, M., Manwar, J., Manmode, R., Padgilwar, S., Patil, S. 2014. *J. Environmental Chemical Eng.* 2: 573–584.

Wang, S., Kretzmer. G., Schügerl, K. 1994. *Appl. Microbiol. Biotechnol.* 41: 537–543.

Xu, T.J., Zhao X.Q., Bai, F.W. 2005. *Enz. Microb. Technol.* 37: 634–640.

Zhao, X.Q., Bai, F.W. 2009. *Biotechnol. Adv.* 27: 849–856.

Applications of *Kluyveromyces marxianus* in Biotechnology

Javier A. Varela, Loughlin Gethins, Catherine Stanton, Paul Ross and John P. Morrissey

Abstract *Kluyveromyces marxianus* is a member of the *Saccharomycetales* yeast order and is used for a variety of commercial applications, most notably production of ethanol from food waste streams. Traits such as rapid growth rates, lactose utilization, good tolerance to inhibitory molecules and thermotolerance facilitate these applications. *K. marxianus* is frequently isolated from food and beverage environments and is especially associated with fermented dairy products such as kefir. This history of food association means that *K. marxianus* has GRAS/QPS status, thereby facilitating applications in the food sector. *K. marxianus* strains have the capacity to produce a range of volatile fragrance and flavor (F&F) molecules such as higher alcohols and acetate esters and there is consequent interest in exploiting high-producing strains as F&F cell factories. The availability of genome sequences and the development of molecular tools are facilitating further applications in *K. marxianus* as a novel yeast cell factory for biomolecule production. This chapter will provide an update on the genetics and biology of this yeast, and an overview of commercial applications. It will later focus on three specific areas: *K. marxianus* for bioethanol production; for the production of fragrance and flavors; and the future development of *K. marxianus* as a yeast cell factory.

Keywords *Kluyveromyces marxianus* · Bioethanol · Cell factory · Biotechnology · Fermentation · Flavour · Fragrance

1 Introduction

Kluyveromyces marxianus is a yeast traditionally associated with fermented milk products and decaying fruit (Fonseca et al. 2008; Lane and Morrissey 2010). These two habitats reflect important enzymatic traits that the species has, namely the

J.A. Varela · L. Gethins · P. Ross · J.P. Morrissey (✉)
School of Microbiology, University College Cork, Cork, Ireland
e-mail: J.Morrissey@ucc.ie

L. Gethins · C. Stanton
Teagasc Research Centre, Moorepark, Fermoy, Co. Cork, Ireland

C. Stanton · P. Ross
Alimentary Pharmabiotic Centre, Microbiome Institute,
University College Cork, Cork, Ireland

T. Satyanarayana and G. Kunze (eds.), *Yeast Diversity in Human Welfare*, DOI 10.1007/978-981-10-2621-8_17

capacity to transport and utilize lactose, and to degrade plant fructans to simple sugars. Interestingly, the much better known yeast, *Saccharomyces cerevisiae* lacks both of these traits, which are encoded by the *LAC* genes, *LAC12* and *LAC4*, and the inulinase gene, *INU1*, respectively. *K. marxianus* has other distinctive characteristics, most importantly thermotolerance to temperatures between 44 and 52 °C, and a rapid growth rate that is twice that of *S. cerevisiae* on rich medium. These traits make this yeast quite attractive from a biotechnological perspective. The speed at which *K. marxianus* has emerged as an important yeast for biotechnology is apparent in a

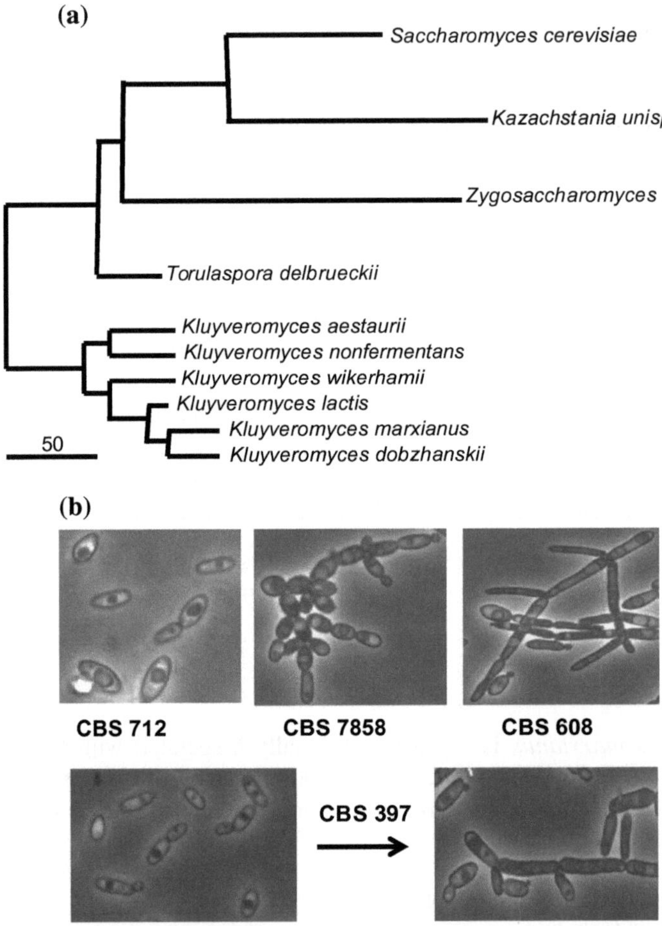

Fig. 1 Phylogeny and morphology of *K. marxianus*. **a** The phylogeny of *K. marxianus* within the genus *Kluyveromyces* and the order *Saccharomycetales* based on 6 different loci is shown. The six species of *Kluyveromyces* and selected other species of industrial interest are included (modified from Lachance 2007). **b** Three different *K. marxianus* strains growing on identical medium show that this fungus can grow in yeast (CBS 712), pseudophyphae (CBS 7858) and hyphal forms (CBS 608). The capacity of some strains to switch morphology is shown by the growth of CBS 397 under two different environmental conditions (images courtesy of Niall Burke, UCC).

bibliometric survey (http://www.ncbi.nlm.nih.gov/pubmed) that shows that >50% of the total number of publications on this yeast have come in the past 7 years (for *S. cerevisiae* in the same period, it is 25%). There are now many publications detailing applications for enzyme production, for fermented foods and beverages, for biofuels, and for cell factory applications. One of the more esoteric applications that has been reported is tequila production which occurs when inulinases produced by wild *K. marxianus* strains break down plant fructans in agave, liberating sugars that can then be fermented to alcohol by *K. marxianus* and *S. cerevisiae*. Some studies have compared starter cultures for these species and identified additional benefits to the organoleptic profile of *K. marxianus*-fermented agave, most likely because of the production of fruity acetate esters (Lopez-Alvarez et al. 2012; Amaya-Delgado et al. 2013; Lopez et al. 2014). The taxonomic position of *K. marxianus* is clear (Fig. 1a): it is a member of the order *Saccharomycetales*, and one of a genus of six species (Lachance 2007). Several but not all of the species in the genus are lac$^+$, but only *K. marxianus* is inulinase$^+$ and thermotolerant (44 °C) and these are often considered diagnostic traits (Lachance 2011). Apart from these characteristics, the species is quite diverse and extensive physiological variation has been reported (Lane et al. 2011). It is also a polymorphic yeast and depending on the strain and the environmental conditions, may grow in yeast, pseudohyphal or hyphal forms (Fig. 1b). For several years, the only genomic information in the public domain was a partial sequence from the Génolevures project (Llorente et al. 2000), but starting with the publication of the strain CBS 6556 in 2012 (Jeong et al. 2012), several have been published, and several more are available in draft format (Gao et al. 2015; Lertwattanasakul et al. 2015; Morrissey unpublished).

2 Advances Made in the Area

Genome sequence availability underpins much of the recent progress and advances with *K. marxianus*. Analysis of the best-annotated genome from the strain DMKU 3-1042 indicates that the genome is approximately 11 Mb, comprising 8 chromosomes ranging from 0.9 to 1.7 Mb in size (Lertwattanasakul et al. 2015). Interestingly, while there is strong synteny between genomes of sequenced strains, there is extensive rearrangement between *K. marxianus* and its closest relative *K. lactis* (Varela and Morrissey, unpublished). It is also worth noting that there are suggestions that some strains may have a different karyotype (Belloch et al. 1998; Fasoli et al. 2015). Comparative analyses showed that *K. marxianus* genome shares 1,552 genes with other 10 yeast species, while encoding 193 unique genes that might be involved in species-specific traits. Furthermore, a high number of sugar transporters, which might relate with the capacity of this yeast to utilize a wide variety of substrates, were identified. Along with these genome sequences, two transcriptome analyses have been published in which the global transcriptional response of *K. marxianus* to industrial-relevant conditions was evaluated. Comparing growth at different temperatures, it was found that at high temperatures

K. marxianus switches its metabolism to cope with the production of reactive oxygen species (Lertwattanasakul et al. 2015). In a different study, the transcriptome of cells growing on inulin was determined (Gao et al. 2015). The authors found that despite being a Crabtree-negative yeast, the transcriptomic profile of *K. marxianus* strain Y179 showed up-regulation of genes associated with central carbon metabolism and ethanol production. While these studies represent the first attempts to obtain a global overview of *K. marxianus* genetic landscape, they also serve as platforms to assist in designing genetic manipulations in this yeast.

K. marxianus is considered a haploid species, though both haploid and diploid strains can be found in culture collections (Lane et al. 2011). The mating-type system resembles that of *K. lactis* and lacks the HO endonuclease found in *S. cerevisiae*. This is important as it means that stable haploid and diploid strains can be maintained as mating-type switching occurs at very low frequency. The potential for genetics in this yeast was demonstrated by Hoshida and colleagues, who generated auxotrophic mutants of a range of strains, successfully crossed these to create diploids, and subsequently sporulated these diploids to produce haploid spores (Yarimizu et al. 2013). It is also necessary to develop molecular tools for understanding the bases of biological processes in yeast. In *K. marxianus*, molecular tools have heretofore been rather limited but the past 5 years have seen this change. The recent developments in molecular tools provide exciting opportunities to explore *K. marxianus* physiology for fundamental and applied research. While relevant and useful, the described tools need to be further improved to cope with the new requirements in the field. Ideally, tools must be universal and allow genetic modification of *K. marxianus* in a simple and versatile manner. In other words, it should be applicable to a great number of strains, regardless of their ploidy, providing fast and reliable results. Finally, it should allow the performing more than one modification at the time, which could be particularly useful when introducing several genes from a metabolic pathway. Current tools are summarised in Table 1.

One essential tool is transformation and targeted genetic integration. The capacity to incorporate and remove DNA sequences from the genome has a significant importance for both fundamental and applied research. While *K. marxianus* can be transformed using routine protocols adapted from other yeasts, targeted genetic integration has proven challenging. Generally speaking, the fate of ectopic DNA transformed into a yeast cell will depend on the presence of nucleases and the efficiency of the DNA repair mechanisms available (Shrivastav et al. 2008). If a fragment with short flanking regions targeting a site in the genome is transformed it could be integrated at the desired spot by homologous recombination (HR) or could be incorporated at a random site through non-homologous end-joining (NHEJ). In *K. marxianus,* NHEJ is more efficient than HR, which leads to random DNA integrations. Some researchers have addressed this problem by increasing the length in the flanking regions to promote HR over NHEJ. In most cases the length used exceeds 1 kb, which is significantly more than what is used in *S. cerevisiae*, where targeted integrations can be achieved using 50 bp flanking regions. This length requirement is a hurdle that must be overcome. One approach to increase gene-targeting strategy is to eliminate the NHEJ system by deleting one of its

Table 1 Development of molecular tools in *K. marxianus*

Markers and genetic integration	References
Non-homologous end joining-mediated functional marker selection for DNA cloning in the yeast *Kluyveromyces marxianus*	Hoshida et al. (2014)
Deletion of a *KU80* homolog enhances homologous recombination in the thermotolerant yeast *Kluyveromyces marxianus*	Choo et al. (2014)
Simultaneous integration of multiple genes into the *Kluyveromyces marxianus* chromosome	Heo et al. (2013)
Identification of auxotrophic mutants of the yeast *Kluyveromyces marxianus* by non-homologous end joining-mediated integrative transformation with genes from *Saccharomyces cerevisiae*	Yarimizu et al. (2013)
Promoter-based gene assembly and simultaneous overexpression	Chang et al. (2013)
Random and targeted gene integrations through the control of non-homologous end joining in the yeast *Kluyveromyces marxianus*	Abdel-Banat et al. (2010)
Sequential gene integration for the engineering of *Kluyveromyces marxianus*	Pecota et al. (2007)
Application of the Cre-loxP system for multiple gene disruption in the yeast *Kluyveromyces marxianus*	Ribeiro et al. (2007)
Promoters and vectors	
Characterization of *Saccharomyces cerevisiae* promoters for heterologous gene expression in *Kluyveromyces marxianus*	Lee et al. (2013)
Evaluation of the tetracycline promoter system for regulated gene expression in *Kluyveromyces marxianus*	Pecota and Da Silva (2005)
Characterizing yeast promoters used in *Kluyveromyces marxianus*	Yang et al. (2015)
Gene expression analysis using strains constructed by NHEJ-mediated one-step promoter cloning in the yeast *Kluyveromyces marxianus*	Suzuki et al. (2015)
Non-homologous end joining-mediated functional marker selection for DNA cloning in the yeast *Kluyveromyces marxianus*	Hoshida et al. (2014)

components. Using long flanking regions, disruption of the *KU80* gene was achieved in the strain CBS 6556 (Choo et al. 2014). This gene encodes a subunit of the Ku heterodimer that binds to the DNA promoting random integration. In the constructed strain, HR was significantly improved from 13 to 70% while no growth defects were found. Nonetheless since the Ku80 protein is part of the normal DNA repair apparatus, it would be preferable to work with wild-type strains. Recently, there has been a revisiting of earlier work on the links between HR/NHEJ and the cell cycle and it has been demonstrated that synchronizing cells and carrying out gene targeting during the S1 phase, greatly increases the likelihood of HR (Tsakraklides et al. 2015). This may have potential for *K. marxianus*. Although not yet published in this yeast, it is only a matter of time before CRISPR-Cas9 systems are developed and this will enhance the capacity to target genes to specific loci in the genome (Mans et al. 2015).

Heterologous expression requires gene cloning methods, promoters and plasmids. Although NHEJ is not desirable when performing gene targeting, it could be used as a tool for gene cloning (Hoshida et al. 2014). It has been reported that a

C-terminal-truncated non-functional *URA3* maker can be repaired by fusing the missing fragment through NHEJ. Using this strategy, if a desired fragment is amplified using primers that complement the truncated *URA3*, then this sequence could be cloned by joining the maker through NHEJ. This technique allows cloning into shuttle plasmids and chromosomal integration into *K. marxianus* genome. A different system called "promoter-based gene assembly and simultaneous over-expression" (PGASO) was also developed for cloning multiple gene fragments (Chang et al. 2013). This method is particularly useful when simultaneous over-expression of multiple genes is required. Since ensuring controlled and s*table* gene expression is critical for many applications, promoters and plasmids are important elements to review. Yang et al. (2015) characterized the strength of 6 constitutive promoters from *S. cerevisiae* and *K. marxianus* at different temperatures. They showed that the *PGK* promoter is the strongest, regardless of the carbon source and temperatures used. Besides, they determined that the rest of the promoters had different levels of activity, important information to consider when tuning expression of a gene of interest. Another approach to evaluate *K. marxianus* promoters was established by Suzuki et al. (2015). In this method a promoter sequence is fused to a reporter gene by NHEJ and then the joined fragments are randomly integrated into the genome. This methodology allowed the researchers to analyze 36 promoters in several culture conditions. In the case of plasmids there are a few cases where plasmids containing autonomously replicating sequences (ARS), have been described. For example, by constructing a genomic DNA library Hoshida et al. (2014), identified a 2.8 kb ARS in *K. marxianus.* When dissecting this DNA fragment to determine the minimal sequence required to provide stable replication, the authors found that a 60 bp region was sufficient for ARS activity. A plasmid containing a centromere sequence was also built in this study.

3 Critical Analysis

Advances in genomics and molecular tools pave the way for developing *K. marxianus* for cell factory applications. This moves the yeast beyond the realm of a component of fermented foods, or a starter culture, to a genuine industrial microbe capable of producing bulk or high value chemicals in large-scale industrial processes. The classic bulk chemical produced in yeast is ethanol and there is considerable interest in developing *K. marxianus* as a platform for production of biofuels, especially bioethanol. As previously mentioned, this yeast has traits of industrial relevance which could boost its use as an alternative platform to *S. cerevisiae*. However, there are also particular limitations, most importantly low ethanol yields relative to *S. cerevisiae* because of the bias towards respiratory metabolism. There is, however, a large variation in this trait between strains the potential to achieve acceptable yields that are close to the theoretical maximum (Lane et al. 2011; Rocha et al. 2011). Because the rate of production (ethanol productivity) is significantly lower in *K. marxianus* than *S. cerevisiae*, it is not likely to be a more efficient platform for conventional first generation ethanol production from starch or sucrose. There is potential, however, for production of second or third generation bioethanol from other substrates such as cellulosic or lignocellulosic biomass, energy crops and waste streams. Some of these applications are possible because *K. marxianus* is capable of using sugars such as lactose and cellobiose, and through the production of extracellular enzymes (inulinase) capable of hydrolyzing plant fructans. Thermotolerance and production of these enzymes also contribute the potential for simultaneous saccharification and fermentation (SSF) where precursor hydrolysis and fermentation take place at the same production stage (Radecka et al. 2015). Multiple screening studies as well as exploitation of culture collections has resulted in the identification of lead strains for bioethanol production. In many cases, these strains have been developed further for applications using classical and GM methods.

The first success with *K. marxianus* for biethanol production came with a process that first introduced on a commercial scale by the Carbery Group in Ireland in 1978 and subsequently adopted by dairies in the US and New Zealand (http://www.agmrc.org/media/cms/rr214_fb445fb28f0d6.pdf). This involved fermenting whey permeate to ethanol. Cheese manufacture generates large amounts of whey (approximately 9 L per kg cheese) that contains lactose (4–6%), proteins, and other nutrients. Traditionally, protein was recovered and the resulting whey permeate treated as a waste stream but the need to comply with environmental standards and to achieve maximum commercial return has led to efforts to further valorise the whey permeate by making use of the lactose (Guimaraes et al. 2010). This sugar can be recovered as a dried powder or fermented to ethanol by microbes. The Carbery process involves a batch fermentation of whey permeate by a proprietary strain of *K. marxianus* that delivers a high yield of ethanol. Assimilation of lactose relies on the products of the *LAC12* gene, encoding a lactose permease, and *LAC4*, encoding a ß-galactosidase. Lac12p, which is a member of the MFS family of sugar transporters, transports lactose (and galactose), which is cleaved by Lac4p to the hexoses glucose and galactose. These are then metabolised via the standard glycolytic and Leloir pathways to pyruvate. Whether this pyruvate is then fermented to ethanol or directed into the TCA cycle for aerobic respiration is strain-dependent also depend on environmental parameters, for example oxygen availability (Lane et al. 2011; Rocha et al. 2011). There have been some efforts to optimise and refine this process and yield is influenced by pH, temperature and lactose concentration (Diniz et al. 2014). The second area where *K. marxianus* has been quite extensively studied for bioethanol is for production of ethanol from Jerusalem artichokes. This is a North American plant that is considered to have strong potential as an energy crop. The plant produces tubers that are rich in inulin, which may have applications as a functional food or as a substrate for bioethanol (Yang et al. 2015). The *K. marxianus* Inu1p protein is very effective in hydrolysing inulin to D-fructose monomers that are then transported into the cell by the Frt1p transporter and processed through the glycolytic pathway. This trait confers some *K. marxianus* strains with the capacity to ferment hydrolysed Jerusalem artichokes to ethanol in a consolidated bioprocess (CBP) where hydrolysis, saccharification and fermentation are carried out simultaneously (reviewed by Yang et al. 2015). Although under certain circumstances, high ethanol yields are obtained, these tend to be at lower inulin concentrations and under conditions not so amenable to large scale industrial processes. A recent transcriptome study addressed these issues and found that varying inulin concentrations and aeration levels dramatically shifted the transcription profile of the entire network of genes involved in ethanol fermentation (Gao et al. 2015). This illustrates the complexity of the process and the need to develop an improved understanding of the metabolic and regulatory networks if it is to become feasible to transfer this process from proof of concept to commercial scale. Further improvements in both these processes are likely to involve selection of more robust strains, improving ethanol tolerance and anaerobic growth, and enhancing the rate of ethanol generation (productivity).

The future for a sustainable bioethanol industry is believed to lie in the production of ethanol from cellulosic material as it enables utilization of non-food crops or crop wastes. In brief, it first involves a saccharification step whereby the cellulosic material is hydrolysed with a combination of endo and exo enzymes that liberate glucose monomers from the cellulosic substrate. These enzymes are often fungal in origin and have optimum temperatures close to 50 °C, whereas the subsequence fermentation of glucose to ethanol is typically carried out by *S. cerevisiae*, which requires temperatures close to 30 °C for optimal growth and fermentation. There would be a considerable advantage if these steps could be combined, thereby allowing simultaneous saccharification and fermentation (SSF). The higher growth range of *K. marxianus* offers this potential and there have been a number of studies that have demonstrated this in SSF processes (for references, see Hong et al. 2007). There has been considerable progress made with selection of strains with high activity and process modifications to establish the parameters that influence yield (Castro and Roberto 2014; Kang et al. 2014). More recently, the focus has shifted to CBP, with the aim that the fermentative yeast could also produce the required enzymes. This could take advantage of the fact that at least some strains of *K. marxianus* can utilise cellobiose but it also requires the heterologous expression of other cellulolytic enzymes. The first successful demonstration of fermentation of cellulosic material to ethanol came with a study in 2010, in which an endoglucanase from *Trichoderma reesei* and a ß-glucosidase from *Aspergillus aculeatus* were expressed on the surface of *K. marxianus* NBRC1777 (Yanase et al. 2010). This strain was able to convert cellulosic ß-glucan to ethanol with a yield of 92% of the theoretical maximum. Interestingly, although wild-type NBRC1777 produces ethanol optimally at 40 °C, the recombinant strain had an optimum production temperature of 48 °C, reflecting the higher activity of the final cellulolytic enzymes at this temperature. Cellulosic biomass derived from pre-treated straw is more complex than ß-glucans and requires a cocktail of enzymes to digest the material to monomers. For example, commercial enzyme preparations often use 5 or more fungal enzymes with a range of activities. To address this challenge, Chang and colleagues expressed a cocktail of five cellulases as well as a transporter in strain KY3 to build a strain, designated KR7, that could successfully hydrolyse crystalline cellulose to produce ethanol (Chang et al. 2013).

Complete fermentation of lignocellulose biomass requires the capacity to ferment both the hexose (C6) monomers derived from cellulose/hemicellulose and the pentose (C5) sugars that comprise up to 50% of hemicellulose. Unlike *S. cerevisiae*, which is poorly able to assimilate pentose sugars, *K. marxianus* can effectively transport and metabolise the pentoses xylose and arabinose via the aldose reductase pathway. The enzymes encoded by the *XYL1*, *XYL2* and *XKS1* genes first reduce xylose to xylitol, then oxidise xylitol to xylulose, and finally phosphorylate xylulose to xylulose-5-P, which enters the pentose phosphate pathway (PPP). Although the conversion of xylose to xylulose is redox balanced overall, the two enzymes involved, xylose reducase (XR) and xylitol hydrolase (XH), use the cofactors NADPH and NAD, respectively. Under aerobic conditions, NADH is recycled to NAD^+, but in the absence of oxygen co-factor imbalance leads to the accumulation of xylitol and low ethanol yields. The consequence is that wild-type strains of *K. marxianus* generally do not efficiently ferment xylose and metabolic reprogramming is required if the yeast is to produce ethanol from pentoses. Initial studies selected the highest producing strains and assessed production parameters under various temperature and sugar regimes. This work demonstrated the potential of strains such as IMB4 (Suryawati et al. 2008), IIPE453 (Kumar et al. 2009) and DMKU3-1042 (Rodrussamee et al. 2011) but also highlighted the need for strain engineering if sufficient yields were to be achieved. The strategies to engineer *K. marxianus* to develop pentose fermenting strains have largely mirrored those adopted in *S. cerevisiae*, namely expression of heterologous xylose metabolizing enzymes and laboratory evolution to select higher yielding variants. One of the first studies replaced the *KmXYL1* gene in a derivative of *K. marxianus* NBRC 1777 with the *Schefferomyces stipitis* xylose reductase gene, which has a dual co-factor specificity (Zhang et al. 2013). This strain (YZB014) was capable of generating ethanol from xylose with a yield of 37% but the rate of production was low and it also accumulated xylitol. A similar study that expressed *SsXYL1*, *SsXYL2* and *S. cerevisiae XKS1* in strain DMB1 generated a strain, DMB3-7, that produced higher amounts of ethanol, though different methodologies preclude direct comparisons of all parameters (Goshima et al. 2013a, b). More recently, this approach was extended to construct a strain that expressed the genes for *Neurospora crassa* xylose reducase (*NcXR*) and the *S. stipitis* xylitol hydrogenase (*SsXDH*) (Zhang et al. 2015a, b). The reason for using these genes is the higher activity of NcXR over ScXR and the specificity of SsXDH for $NADP^+$ (over NAD^+). That study also overexpressed a number of downstream *K. marxianus* genes from the PPP and ethanol fermentation pathways and built a strain (YZX088) that had ethanol yields (g/g) as high as 85%, with productivity that was also comparable to the best engineered *S. cerevisiae* strains. The alternative approach of expressing a xylose isomerase (XI) that directly converts xylose to xylulose with the use of co-factors has also been tried (Wang et al. 2013). The xylose isomerase gene, *XYLA*, from a fungus *Orpinomyces* was expressed in the background of strain NBRC 1777, replacing the endogenous *kmXYL1* and *kmXYL2* genes. The initial engineered strain did not effectively ferment xylose but following laboratory evolution a strain YRL005 that yielded 76% (g/g) of the theoretical ethanol was obtained. As was found in similar studies in *S. cerevisiae*, the evolved strain overexpressed multiple genes in the PPP. Productivity was still low relative to *S. cerevisiae* and the best pentose—fermenting *K. marxianus* strain constructed to date is strain YZX088. Unlike the engineered *S. cerevisiae* strains, all the *K. marxianus* strains described here are capable of fermentation at 42–45 °C.

Higher alcohols, also known as fusel alcohols, are alcohols produced from turnover of amino acids via the Ehrlich pathway (Hazelwood et al. 2008). While these higher alcohols are found in many fermented foods and beverages like wine, beer, bread or some yogurts, these alcohols are also widely produced synthetically as primary products as additives for food or indeed as fragrances for the perfume industry. However, increasing consumer preference for naturally derived products has created an ever growing market niche for these higher alcohols to be of natural origin. Currently there are two methods for natural production, the first one is plant extraction and the second being microbial production. Plant extraction is main source of natural F&Fs, however it is an expensive process that can suffer from limited yield or unstable supply. Different bacteria and yeast can be used in microbial synthesis, but there are clear advantages to using a yeast like *K. marxianus,* which is non-pathogenic, food grade, and carries GRAS (US—Generally Regarded As Safe) and QPS (European Food Safety Authority—Qualified Presumption of Safety) labelling. The Ehrlich pathway consists of three basic enzymatically catalysed reactions: first, transamination of an amino acid to a 2-oxo acid; second, decarboxylation to an aldehyde; and third, either reduction or oxidation to an alcohol or an acid respectively (Ehrlich 1907). Most knowledge of this pathway comes from studies on *S. cerevisiae* but the core genes are conserved in *K. marxianus*. The transaminases encoded by *BAT1/BAT2* and *ARO8/ARO9* act on branched chain amino acids (leucine, isoleucine and valine) and aromatic amino acids (phenyalanine, tyrosine and tryptophan), respectively (Kispal et al. 1996; Iraqui et al. 1998). The most important decarboxylase is encoded by *ARO10* although, at least in *S. cerevisiae*, several other enzymes also have this activity. Expression of *ARO10* is strongly correlated with nitrogen source and is upregulated in the presence of branched-chain or aromatic amino acids (Vuralhan et al. 2003) but transcription of the transaminases does not seem to be linked with the presence of their preferred substrates (Boer et al. 2007). The fate of the aldehyde depends on the redox state of the cells. In *S. cerevisiae*, when glucose is limited in aerobic conditions, the aldehyde is oxidised to a fusel acid, whereas in anaerobic conditions, NADH is used to reduce the aldehyde to the corresponding alcohol such as 2-phenylethanol (phenylalanine); isoamyl alcohol (leucine) or isobutanol (valine) (Dickinson et al. 2003; Boer et al. 2007). There are mutiple aldehyde dehydrogenases and alcohol dehydrogenases that may catalyse these reactions, and as found in a *K. marxianus* transcriptome analysis, genes encoding these enzymes are subject to complex regulation (Gao et al. 2015).

Presently, most work has focused on production of 2-phenylethanol (2-PE) in *K. marxianus* (reviewed in Morrissey et al. 2015). This flavor molecule has a rose-like smell and has a range of applications in the cosmetics and perfume industry to the food and beverage industry. The world annual production of 2-PE was estimated to be over 10,000 tons in 2010, the vast majority of which is produced from harmful carcinogens such as benzene and styrene (Clark 1990; Chen et al. 2011; Hua and Xu 2011). Although the amount of naturally produced 2-PE still pales in comparison to chemical synthesis the vast majority of it is produced by fermentation as plant extraction is not economically viable with prices upwards of $1,000/kg compared to the $5/kg price tag of synthetic production (Etschmann et al. 2002). This has created an interest in producing a naturally derived 2-PE at a more economically affordable price point. Highest yields are obtained in what is essentially a bioconversion process: growth medium is supplemented with phenylalanine, the yeast coverts this to 2-phenylethanol, and the 2-PE is recovered using an in situ product recovery (ISPR) system (to avoid toxicity issues). At least two companies (Puris, South Africa; Lasaffre, France) commercially produce 2-PE from *K. marxianus* but the precise details of the process remain proprietary knowledge. Higher alcohols can be further modified to esters by the addition of acyl group either acetyl coA (to yield an acetate ester) or acyl coA (yielding an ethyl ester). These metabolites are major flavor contributors in all yeast based fermented foods and beverages and controlled production in cell factories would be very desirable. The enzyme responsible for catalysing this reaction is an alcohol acetyltransferase or AATase. In *S. cerevisiae* this activity is mostly shared between the two paralogs *ATF1* and *ATF2,* although as deletion of both of these genes does not completely abolish activity it appears that, as yet, unidentified genes also have a role to play (Verstrepen et al. 2003). There is very limited data available for *K. marxianus* but one recent study found that production of different acetate esters was differentially regulated, implying that there may be more than one enzyme with aromatic acetyl transferase activity in *K. marxianus* (Gethins et al. 2015). This theory would be supported by some of the older *K. marxianus* literature (Kallel-Mhiri and Miclo 1993; Plata et al. 2003).

Alcohol acetyltransferases are also responsible for the synthesis of ethylacetate, which is formed by acetylation of ethanol. Ethyl acetate does have flavour characteristics but the main commercial interest is as an industrial solvent. Because to its moderate polarity, ethyl acetate is an attractive industrial solvent for many applications from cleaning surfaces to extraction and chromatographic recovery of pharmaceuticals. In addition, ethyl acetate is easily degraded by bacteria, making disposal relatively straightforward. As ethyl acetate is derived from primary metabolism its synthesis is not easily increased by manipulation of carbon and nitrogen sources unlike the higher alcohol derived acetate esters. Interestingly, however it has been found that impairing the activity of the TCA cycle by limiting the availability of Iron (Fe) or Copper (Cu) dramatically increases ethyl acetate production (Loser et al. 2012; Urit et al. 2012). Also, since ethanol is the substrate for ethyl acetate production conditions that increase ethanol accumulation, most notably oxygenation, also favour the production of ethyl acetate (Urit et al. 2013). The production of ethyl acetate in *K. marxianus* has been extensively reviewed recently (Loser et al. 2014, 2015).

4 Future Perspectives

Oil-based chemistry processes are the most common means for synthesizing bulk and fine chemicals in our society. Although generally considered as feasible, these procedures are questionable from an environmental and economical point of view. Recently, researchers have focused on developing biotechnological processes that could serve as sustainable alternatives to chemical synthesis. Due to their robustness and versatility yeasts have become interesting candidates for cell factory applications and many examples are available in the literature (for review see Borodina and Nielsen 2014). Although most of the work has focused on *Saccharomyces cerevisiae* there are many cases where other yeasts such as *K. marxianus* have been used. In developing a process, the starting point can either be a biodiversity screen to identify the best natural strain, or an engineering approach based on well-studied platform strains (Porro and Branduardi 2009). Increasingly, the trend is to pursue a hybrid strategy: identify natural strains with traits of interest and understand the genetic basis of the trait; reverse engineer the trait into a platform strain (inverse engineering); apply laboratory evolution to optimise the strain (evolutionary engineering) (Crook and Alper 2012; Bachmann et al. 2015) In order to improve relevant aspects such as yield, substrate/product spectrum and stress tolerance, a combination of computational and genetic tools is required. First, a global view of cellular metabolism must be obtained. To accomplish this, genome-scale metabolic models, which are useful in predicting metabolic fluxes, can be constructed. Then, to obtain a more realistic overview of the system, different sets of data (e.g. transcriptomics, proteomics and metabolomics) must be integrated. Using this model, metabolic fluxes can be simulated using flux balance analysis and therefore targets for reprogramming cellular metabolism can be identified. These tools are particularly useful when trying to redirect metabolism towards a product of interest, identifying potential competing metabolic branches/pathways or optimizing cofactor usage and redox balance (Kavscek et al. 2015). Another key element required to utilize *K. marxianus* for cell factory applications is developing molecular tools that enable genome editing. These tools must be powerful enough to allow efficient targeted integration of DNA fragments in the genome. In *K. marxianus* these tools have not been fully developed but, as described earlier, considerable progress has been made. Future developments are also likely to involve synthetic biology. Two examples where this type of approach has already been implemented in *K. marxianus* were the assembly of the cellulose enzyme cocktail discussed earlier (Chang et al. 2013) and the introduction of a pathway to synthesise hexanoic acid (Cheon et al. 2014). Significant tool development will be required to further progress synthetic biology in *K. marxianus*.

5 Conclusions

Microbial cell factories represent a feasible alternative to chemical synthesis but more improvements are required to meet industrial standards. While yeasts are interesting candidates for many applications, the majority of the research efforts have focused in *S. cerevisiae, which* might not be suitable for all purposes. Lately, *K. marxianus* has arisen as a new potential host for cell factory applications. The development of new genetic and mathematical tools will be critical for engineering this yeast to obtain platform strains that can be exploited for different industrial applications. Although in the present scenario many technological aspects need to be addressed we can speculate about the future for *K. marxianus* cell factories. Using a systems biology approach, it will be possible to understand the complex relationships between metabolic components. By applying efficient molecular tools, different genes/reactions could be removed from the system and other desired components could be integrated. As a result, a chassis strain, in which different metabolic pathways can be installed as independent circuits, could be obtained. Since modifying this strain might be simple, its construction would lead to new industrial applications for *K. marxianus*. *K. marxianus* is likely to develop as a platform, not to replace *S. cerevisiae,* but to provide opportunities to expand the range of biomolecules produced in yeast systems as a whole.

References

Abdel-Banat B.M., Nonklang S., Hoshida H. and Akada R. 2010. *Yeast* **27:** 29–39.

Amaya-Delgado L., Herrera-Lopez E.J., Arrizon J., Arellano-Plaza M. and Gschaedler A. 2013. *World J Microbiol Biotechnol* **29:** 875–881.

Bachmann H., Pronk J.T., Kleerebezem M. and Teusink B. 2015. *Curr Opin Biotechnol* **32:** 1–7.

Belloch C., Barrio E., Garcia M.D. and Querol A. 1998. *Yeast* **14:** 1341–1354.

Boer V.M., Tai S.L., Vuralhan Z., Arifin Y., Walsh M.C., Piper M.D.W., De Winde J.H., Pronk J. T. and Daran J.-M. 2007. *FEMS Yeast Res.* **7:** 604–620.

Borodina I. and Nielsen J. 2014. *Biotechnol J* **9:** 609–620.

Castro R.C. and Roberto I.C. 2014. *Appl Biochem Biotechnol* **172:** 1553–1564.

Chang J.J., Ho F.J., Ho C.Y., Wu Y.C., Hou Y.H., Huang C.C., Shih M.C. and Li W.H. 2013. *Biotechnol Biofuels* **6:** 19.

Chen X.-M., Kobayashi H., Sakai M., Hirata H., Asai T., Ohnishi T., Baldermann S. and Watanabe N. 2011. *J Plant Physiol* **168:** 88–95.

Cheon Y., Kim J.S., Park J.B., et al. 2014. *J Biotechnol* **182–183:** 30–36.

Choo J.H., Han C., Kim J.Y. and Kang H.A. 2014. *Biotechnol Lett* **36:** 2059–2067.

Clark G.S. 1990. *Perfum Flavor* **15:** 37–44.

Crook N. and Alper H.S. 2012. Classical Strain Improvement. Engineering Complex Phenotypes in Industrial Strains, pp. 1–33. John Wiley & Sons, Inc.

Dickinson J.R., Salgado L.E. and Hewlins M.J. 2003. *J Biol Chem* **278:** 8028–8034.

Diniz R.H., Rodrigues M.Q., Fietto L.G., Passos F.M. and Silveira W.B. 2014. *Biocatal Agric Biotechnol* **3:** 111–117.

Ehrlich F. 1907. *Berichte der deutschen chemischen Gesellschaft* **40:** 1027–1047.

Etschmann M., Bluemke W., Sell D. and Schrader J. 2002. *Appl Microbiol Biotechnol* **59:** 1–8.

Fasoli G., Tofalo R., Lanciotti R., Schirone M., Patrignani F., Perpetuini G., Grazia L., Corsetti A. and Suzzi G. 2015. *Int J Food Microbiol* **214:** 151–158.
Fonseca G.G., Heinzle E., Wittmann C. and Gombert A.K. 2008. *Appl Microbiol Biotechnol* **79:** 339–354.
Gao J., Yuan W., Li Y., Xiang R., Hou S., Zhong S. and Bai F. 2015. *Biotechnol Biofuels* **8:** 115.
Gethins L., Guneser O., Demirkol A., Rea M.C., Stanton C., Ross R.P., Yuceer Y. and Morrissey J.P. 2015. *Yeast* **32:** 67–76.
Goshima T., Tsuji M., Inoue H., Yano S., Hoshino T. and Matsushika A. 2013a. *Biosci Biotechnol Biochem* **77:** 1505–1510.
Goshima T., Negi K., Tsuji M., Inoue H., Yano S., Hoshino T. and Matsushika A. 2013b. *J Biosci Bioeng* **116:** 551–554.
Guimaraes P.M., Teixeira J.A. and Domingues L. 2010. *Biotechnol Adv* **28:** 375–384.
Hazelwood L.A., Daran J.M., van Maris A.J., Pronk J.T. and Dickinson J.R. 2008. *Appl Environ Microbiol* **74:** 2259–2266.
Heo P., Yang T.J., Chung S.C., et al. 2013. *J Biotechnol* **167:** 323–325.
Hong J., Wang Y., Kumagai H. and Tamaki H. 2007. *J Biotechnol* **130:** 114–123.
Hoshida H., Murakami N., Suzuki A., Tamura R., Asakawa J., Abdel-Banat B.M., Nonklang S., Nakamura M. and Akada R. 2014. *Yeast* **31:** 29–46.
Hua D. and Xu P. 2011. *Biotechnol Adv* **29:** 654–660.
Iraqui I., Vissers S., Cartiaux M. and Urrestarazu A. 1998. *Mol Gen Genet* **257:** 238–248.
Jeong H., Lee D.-H., Kim S.H., Kim H.-J., Lee K., Song J.Y., Kim B.K., Sung B.H., Park J.C. and Sohn J.H. 2012. *Eukaryotic cell* **11:** 1584–1585.
Kallel-Mhiri H. and Miclo A. 1993. *FEMS Microbiology Letters* **111:** 207–212.
Kang Q., Appels L., Tan T. and Dewil R. 2014. *The Scientific World Journal* **2014**.
Kavscek M., Strazar M., Curk T., Natter K. and Petrovic U. 2015. *Microb Cell Fact* **14:** 94.
Kispal G., Steiner H., Court D.A., Rolinski B. and Lill R. 1996. *J Biol Chem* **271:** 24458–24464.
Kumar S., Singh S.P., Mishra I.M. and Adhikari D.K. 2009. *J Ind Microbiol & Biotechnol* **36:** 1483–1489.
Lachance M.-A. 2011. Kluyveromyces van der Walt. The Yeasts, Vol. 2 (Kurtzman CP, Fell JW & Boekhoet T, eds.), pp. 471–482. Elsevier, Amsterdam.
Lachance M.A. 2007. *FEMS Yeast Res* **7:** 642–645.
Lane M.M. and Morrissey J.P. 2010. *Fungal Biology Rev.* **24:** 17–26.
Lane M.M., Burke N., Karreman R., Wolfe K.H., O'Byrne C.P. and Morrissey J.P. 2011. *Antonie Van Leeuwenhoek* **100:** 507–519.
Lee K.S., Kim J.S., Heo P., et al. 2013. *Appl Microbiol Biotechnol* **97:** 2029–2041.
Lertwattanasakul N., Kosaka T., Hosoyama A., et al. 2015. *Biotechnol Biofuels* **8:** 47.
Llorente B., Malpertuy A., Blandin G., Artiguenave F., Wincker P. and Dujon B. 2000. *FEBS Lett* **487:** 71–75.
Lopez C.L., Beaufort S., Brandam C. and Taillandier P. 2014. *World J Microbiol Biotechnol* **30:** 2223–2229.
Lopez-Alvarez A., Diaz-Perez A.L., Sosa-Aguirre C., Macias-Rodriguez L. and Campos-Garcia J. 2012. *J Biosci Bioeng* **113:** 614–618.
Loser C., Urit T. and Bley T. 2014. *Appl Microbiol Biotechnol* **98:** 5397–5415.
Loser C., Urit T., Keil P. and Bley T. 2015. *Appl Microbiol Biotechnol* **99:** 1131–1144.
Loser C., Urit T., Forster S., Stukert A. and Bley T. 2012. *Appl Microbiol Biotechnol* **96:** 685–696.
Mans R., van Rossum H.M., Wijsman M., Backx A., Kuijpers N.G., van den Broek M., Daran-Lapujade P., Pronk J.T., van Maris A.J. and Daran J.M. 2015. *FEMS Yeast Res* **15:** fov004.
Morrissey J.P., Etschmann M.M.W., Schrader J. and de Billerbeck G.M. 2015. *Yeast* **32:** 3–16.
Pecota D.C. and Da Silva N.A. 2005. *Biotechnol Bioeng* **92:** 117–123.
Pecota D.C., Rajgarhia V. and Da Silva N.A. 2007. *J Biotechnol* **127:** 408–416.
Plata C., Millán C., Mauricio J.C. and Ortega J.M. 2003. *Food Microbiol.* **20:** 217–224.
Porro D. and Branduardi P. 2009. *Microb Cell Fact* **8:** 51.

Radecka D., Mukherjee V., Mateo R.Q., Stojiljkovic M., Foulquie-Moreno M.R. and Thevelein J. M. 2015. *FEMS Yeast Res* **15:** fov053.

Ribeiro O., Gombert A.K., Teixeira J.A. and Domingues L. 2007. *J Biotechnol* **131:** 20–26.

Rocha S.N., Abrahao-Neto J. and Gombert A.K. 2011. *Antonie Van Leeuwenhoek* **100:** 619–630.

Rodrussamee N., Lertwattanasakul N., Hirata K., Suprayogi, Limtong S., Kosaka T. and Yamada M. 2011. *Appl Microbiol Biotechnol* **90:** 1573–1586.

Shrivastav M., De Haro L.P. and Nickoloff J.A. 2008. *Cell research* **18:** 134–147.

Suryawati L., Wilkins M.R., Bellmer D.D., Huhnke R.L., Maness N.O. and Banat I.M. 2008. *Biotechnol Bioengin* **101:** 894–902.

Suzuki A., Fujii H., Hoshida H. and Akada R. 2015. *FEMS Yeast Res* **15**.

Tsakraklides V., Brevnova E., Stephanopoulos G. and Shaw A.J. 2015. *PLoS One* **10:** e0133434.

Urit T., Stukert A., Bley T. and Loser C. 2012. *Appl Microbiol Biotechnol* **96:** 1313–1323.

Urit T., Manthey R., Bley T. and Löser C. 2013. *Engineering in Life Sciences* **13:** 247–260.

Verstrepen K.J., Derdelinckx G., Dufour J.-P., Winderickx J., Pretorius I.S., Thevelein J.M. and Delvaux F.R. 2003. *FEMS Yeast Res.* **4:** 285–296.

Vuralhan Z., Morais M.A., Tai S.L., Piper M.D. and Pronk J.T. 2003. *Appl Environ Microbiol* **69:** 4534–4541.

Wang R., Li L., Zhang B., Gao X., Wang D. and Hong J. 2013. *J Ind Microbiol Biotechnol* **40:** 841–854.

Yanase S., Hasunuma T., Yamada R., Tanaka T., Ogino C., Fukuda H. and Kondo A. 2010. *Appl Microbiol Biotechnol* **88:** 381–388.

Yang C., Hu S., Zhu S., Wang D., Gao X. and Hong J. 2015. *World J Microbiol Biotechnol* **31:** 1641–1646.

Yarimizu T., Nonklang S., Nakamura J., et al. 2013. *Yeast* **30:** 485–500.

Zhang B., Li L., Zhang J., Gao X., Wang D. and Hong J. 2013. *J Ind Microbiol Biotechnol* **40:** 305–316.

Zhang B., Zhang J., Wang D., Gao X., Sun L. and Hong J. 2015. *Data Brief* **5:** 179–186.

Zhang J., Zhang B., Wang D., Gao X., Sun L. and Hong J. 2015. *Metab Eng* **31:** 140–152.

Applications of *Blastobotrys* (*Arxula*) *adeninivorans* in Biotechnology

Felix Bischoff, Alexandre Chamas, Katarzyna Litwińska, Falko Matthes, Erik Böer and Gotthard Kunze

Abstract The yeast *Blastobotrys adeninivorans* (syn. *Arxula adeninivorans*) is a dimorphic, asexual hemiascomycete which is phylogenetically very distant from *Saccharomyces cerevisiae*. It has been shown to be most useful in a wide range of biotechnological applications: as gene donor in enzyme production, as host for heterologous gene expression, and as powerful biological component in biosensors. *B.* (*A.*) *adeninivorans* major advantage is its metabolic flexibility, which enables the utilization of a wide range of different carbon and nitrogen sources. For example, recent analyses of the genome and its transcriptome revealed a new pathway for the assimilation of n-butanol via butyraldehyde and butyric acid as well as new insights into the previously reported purine degradation pathway. Additionally, the synthesis of several secretory enzymes with great biotechnological potential, such as two tannases (Atan1p and Atan2p) and three new cutinases (Acut1p, Acut2p and Acut3p) were identified. Due to these characteristics, *B.* (*A.*) *adeninivorans* can be exploited as a gene donor for the production of enzymes with attractive biotechnological applications. Furthermore, its unusual thermo- and halotolerance as well as differential morphology-dependent glycosylation and the secretion characteristics render *B.* (*A.*) *adeninivorans* attractive as host for heterologous gene expression. Successful expression of bacterial alcohol dehydrogenase (*ADH*) genes from *Rhodococcus ruber* and *Rhodococcus erythropolis* enables *B.* (*A.*) *adeninivorans* to be used as biocatalyst for the synthesis of chiral alcohols as building blocks for the chemical industry. The combination of robustness with its great ability for heterologous gene expression makes *B.* (*A.*) *adeninivorans* a superior choice for the biological component in biosensor applications. Different *B.* (*A.*) *adeninivorans*-based biosensors detecting hormones like estrogens, androgens and glucocorticoides as well as dioxin have been developed and consequently improved in the last decade. This chapter will provide a comprehensive overview on the biology and the biotechnological applications of this yeast.

F. Bischoff · A. Chamas · K. Litwińska · F. Matthes · E. Böer · G. Kunze (✉)
Leibniz Institute of Plant Genetics and Crop Plant Research (IPK), Corrensstr. 3, 06466 Gatersleben, Germany
e-mail: kunzeg@ipk-gatersleben.de

T. Satyanarayana and G. Kunze (eds.), *Yeast Diversity in Human Welfare*, DOI 10.1007/978-981-10-2621-8_18

Keywords *Blastobotrys adeninivorans* · *Arxula adeninivorans* · Heterologous gene expression · Thermotolerance · Halotolerance · n-Butanol · Tannase · Cutinase · Biosensor

1 Introduction

Yeasts are simply organized, unicellular eukaryotes with the ability of rapid adaptation to alternating environmental conditions. Besides the well characterized baker's yeast *Saccharomyces cerevisiae* a wide range of non-conventional yeast species exists, that are exhibiting attractive growth properties and biochemical characteristics. For this reason they can be exploited for industrial applications in the field of biotechnology as well as suitable model organisms for plant and animal research. Expression and transformation platforms have been developed to fortify the practical use of yeast species like the traditional baker's yeast *Saccharomyces cerevisiae*, *Kluyveromyces lactis*, *Pichia pastoris*, *Yarrowia lypolytica*, *Hansenula polymorpha* and *Arxula adeninivorans* for research and biotechnological applications. Using these platforms yeast can be easily deployed as hosts for the production of recombinant proteins or valuable metabolites as building blocks for the chemical industry. Additionally, they are used as gene donors due to their partly wide substrate spectrum (Gellissen 2005; Gellissen et al. 2005; Wolf 1996; Wolf et al. 2003).

In 1984, Middelhoven et al. described the isolation of a yeast species, designated as *Trichosporon adeninovorans*, from soil by enrichment culturing. In particular, the strain CBS 8244T was found to exhibit unusual biochemical activities being able to assimilate a range of amines, adenine and several other purine compounds as sole energy and carbon source. In parallel, a second strain, LS3 (PAR-4), with similar characteristics (Gienow et al. 1990) was isolated from wood hydrolysates in Siberia (Kapultsevich, Institute of Genetics and Selection of Industrial Microorganisms, Moscow, Russia). Seven additional strains were isolated, three of them from chopped maize herbage ensiled at 25 or 30 °C in the Netherlands and four from humus-rich soil in South Africa (Van der Walt et al. 1990). The new genus *Arxula* was proposed for these strains, because all representatives are ascomycetous, anamorphic and arthroconidial. Furthermore, they share properties like nitrate assimilation and xerotolerance (Van der Walt et al. 1990). The phylogenetic analysis of the ascosporic yeast genera *Sporopachydermia*, *Stephanoascus*, *Trichomonascus*, *Wickerhamiella* and *Zygoascus* and the associated anamorphic genera *Arxula*, *Blastobotrys*, *Sympodiomyces* and *Trigonopsis* was accomplished by Kurtzmann and Robnett (2007). They compared sequences derived from the large-subunit rDNA genes, the mitochondrial small-subunit rDNA genes, and the genes for cytochrome oxidase II and deduced that *Arxula*, *Blastobotrys* and *Sympodiomyces* are members of the *Trichomonascus* clade, with the genus *Blastobotrys* having taxonomic priority for anamorphic states (Table 1).

The genus *Blastobotrys* includes now both type species of the genus *B. terrestris* (Van der Walt and Johanssen) Kurtzman and Robnett comb. nov. (Basionym:

Table 1 Taxonomy of *B. adeninivorans* (synonym: *A. adeninivorans*) (Kurtzmann and Robnett 2007)

Superkingdom	Eukaryota
Kingdom	*Fungi*
Phylum	*Ascomycota*
Subphylum	*Saccharomycotina*
Class	*Saccharomycetes*
Order	*Saccharomycetales*
Family	*Saccharomycetaceae*
Genus	*Blastobotrys*
Species	*Blastobotrys adeninivorans*

Arxula terrestris) and *B. adeninivorans* (Middelhoven, Hoogkamer Te-Niet and Kreger van Rij) Kurtzman and Robnett comb. nov. (Basionym: *Arxula adeninivorans*).

So far, approx. 170 articles related to *A. adeninivorans* have been published. Although this yeast is well characterised and its genome sequence has been known since 2014, it is still hardly known in the public. The purpose of this chapter is to provide novel information on this interesting and useful organism.

2 Physiology and Temperature Dependent Dimorphism

Gienow et al. (1990), Middelhoven (1993) and Middelhoven et al. (1984, 1991, 1992) provided a detailed physiological description of *A. adeninivorans*. Likewise *H. polymorpha* this yeast is assimilating nitrate using nitrate reductase and nitrite reductase. *A. adeninivorans* exhibits a wide spectrum of substrates which can be utilized as sole carbon and energy source including most sugars, polyalcohols, organic acids, adenine, uric acid, butylamine, pentylamine, putrescine, soluble starch, melibiose, melicitose, propylamine or hexylamine. L-rhamnose, inulin, lactose, lactate and methanol on the other hand are compounds which are not assimilated by this yeast. Except of creatine and creatinine all nitrogen compounds are suitable nitrogen sources. Moreover, several nitrogen compounds, like amino acids, purine derivatives and many primary n-alkylamines and terminal diamines, are metabolized as sole source of energy, carbon and nitrogen. Furthermore, metabolic intermediates of alcohols, dialcohols, carboxylic acids, dicarboxylic acids and other nitrogen-free analogous compounds are assimilated by *A. adeninivorans*. On top of that, this yeast degrades some phenols, hydroxybenzoates, tannic acid and is able to assimilate urotropine as sole nitrogen source (Middelhoven and van Doesburg 2007).

A. adeninivorans synthesizes a vast number of secretory enzymes including RNases, proteases, glucoamylase, lipase, tannase, some acid phosphatases, trehalase, some cellobiases, invertase, β-glucosidase, xylosidase, some cutinases and phytase, which are summarized with some of their biochemical properties in Table 2.

Table 2 Properties of secretory enzymes of *A. adeninivorans* (Böer et al. 2004b, 2005b; Büttner et al. 1987, 1988, 1989, 1990a, c, 1991a, b; Büttner and Bode 1992; Büttner et al. 1992a, b; Kunze and Kunze 1994b; Sano et al. 1999; Wartmann et al. 1995b; Bischoff et al. 2015)

Enzyme	Optimum		K_M value	Molecular mass (kDa)
	Temperature (°C)	pH		
Glucoamylase	60–70	4.0–5.0	1.2 g/L for starch	225
(EC 3.2.1.3)			11.1 mM for maltose	
Acid phosphatase				
(EC 3.1.3.2)				
I	50–55	5.2–5.5	3.5 mM for *p*-nitrophenyl-phosphate	320
II	50–55	5.2–5.5	5 mM for *p*-nitrophenyl-phosphate	250
Trehalase	45–55	4.5–4.9	0.8–1.0 mM for trehalose	250
(EC 3.2.1.28)				
Cellobiase				
(EC 3.2.1.21)				
I	60–63	4.5	4.1 mM for cellobiose	570
II	60–63	4.5	3.0 mM for cellobiose	525
Invertase	50–60	4.5	40–60 mM for sucrose	600
(EC 3.2.1.26)			36 mM for raffinose	
β-D-Xylosidase	60	5.0	0.23–0.33 mM for *p*-nitrophenyl-β-xylopyranoside	60
(EC 3.2.1.37)				
3-Phytase	75	4.5	0.23 mM for phytate	*ND*
(EC 3.1.3.8)				
Lipase	30	7.5	0.4 mM for *p*-nitrophenyl-caprate	100
(EC 3.1.1.3)				
Tannase	35–45	5.0–6.5	0.14 mM for gallotannin	320
(EC 3.1.1.20)				
Cutinase				
(EC 3.1.1.74)				
ACut1p	20–35	4.5–6.0	1.6 mM for *p*-nitrophenyl-butyrat	21.3[a]
ACut2p	20–45	4.0–6.0	1.46 mM for *p*-nitrophenyl-butyrat	21.3[a]
ACut3p	20–45	4.5–6.5	1.93 mM for *p*-nitrophenyl-butyrat	59.7[a]

[a]Native molecular mass determined with size exclusion chromatography; *ND* not-determined

A. adeninivorans and the Siberian wild-type strain LS3 in particular, shows some special features like thermotolerance and temperature-dependent dimorphism, which have an important impact on the biotechnological application of this yeast.

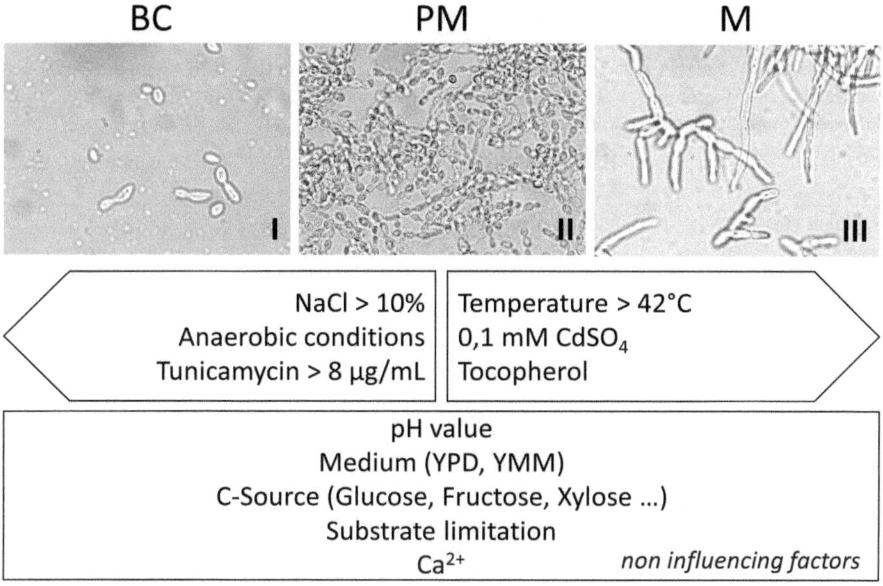

Fig. 1 Cell morphology of *A. adeninivorans* LS3 grown at 30 °C (I), 42 °C (II) and 45 °C (III). The cells were cultured on YEPD medium for 18 h. Influencing and non-influencing factors are indicated in the boxes below. Budding cells (BC) appear at 30 °C, whereas pseudo mycelium (PM) can be observed at 42 °C. At 45 °C mycelium (M) is the predominant growth form

Without previous adaptation *A. adeninivorans* LS3 can grow at temperatures up to 48 °C and survives at 55 °C for some hours (Böttcher et al. 1988; Wartmann et al. 1995a). At temperatures above 42 °C LS3 exhibits a transition from budding cells to mycelia forms, which is reversed to budding cells when the temperature is decreased below 42 °C (Fig. 1). Mutant strain *A. adeninivorans* 135 with altered dimorphism were selected by Wartmann et al. (2000), which is growing as mycelia already at 30 °C. This mutant was used to examine the morphology-, and not temperature-related effects on gene expression and protein accumulation. It was found that budding cells and mycelia differ in cell dry weight and in their contents of RNA and soluble protein, which are lower in mycelia during the middle and the final phases of the exponential growth. In contrast, the concentration of secreted proteins including glucoamylase and invertase is two-fold higher in mycelia compared to budding cells. The summary in Table 3 indicates that morphology, rather than temperature, is the relevant factor. Moreover, a strong correlation between the morphological status and the iron uptake was found to exist. Two transport systems with different iron affinity are responsible for iron uptake by *A. adeninivorans*. Budding cells accumulate iron up to seven-fold higher than mycelia at high Fe(II) concentrations (>2 µM), whereas at concentrations <2 µM both cell types accumulate similar amounts of iron. The expression of the *AFET3* gene, which encodes a copper-dependent Fe(II) oxidase (Afet3p), was found to strongly depend on iron concentration as well as on the morphological state, but just in a

Table 3 DNA and RNA content, dry weight and amount of soluble protein of *A. adeninivorans* LS3 cultured at 30 °C (budding cell) and at 45 °C (mycelium) and of *A. adeninivorans* 135 cultured at 30 °C (mycelium) in yeast minimal medium (Tanaka et al. 1967) with 1% maltose as carbon source (Wartmann et al. 2000). The values are means ±SD from three separate cultures each with three batches in parallel

	Budding Cell LS3–30 °C	Mycelium LS3–45 °C	Mycelium 135–30 °C
Content (fg)			
DNA	25.3 ± 0.5	23.6 ± 2.4	24.8 ± 0.9
RNA (45 h)	118.0 ± 15.2	56.4 ± 9.1	44.7 ± 8.8
Max. RNA	142.5 ± 18.0 (45 h)	73.0 ± 21.0 (36 h)	57.5 ± 9.6 (36 h)
Dry weight (pg, 45 h)	18.2 ± 0.2	22.4 ± 0.8	23.3 ± 0.3
Soluble protein (fg, 45 h)	169.2 ± 16.3	107.5 ± 20.2	76.9 ± 0.3
Max. soluble protein (fg)	234.0 ± 5.5 (60 h)	150.1 ± 17.9 (60 h)	125.8 ± 11.6 (60 h)

minor way. However, the greater influence of morphology on posttranslational modifications of Afet3p was found. O-glycosylation occurred only in budding cells, but both cell types showed N-glycosylation to some extent. Differential glycosylation of heterologous proteins could be used as a tool to study the influence of O-glycosylation on biological activity or immunological tolerance (Wartmann et al. 2002).

There are also other factors influencing the dimorphism of *A. adeninivorans*. Compounds like Cd^{2+} and tocopherol lead to mycelia, whereas NaCl, anaerobic conditions and tunicamycin are enhancing the formation of budding cells. Other factors like Ca^{2+}, pH-value, carbon source and its concentration show no influence on cell morphology (Fig. 1).

Osmotolerance is another interesting property of *A. adeninivorans*. In presence of ionic (NaCl), osmotic (PEG400) and water stress (ethylene glycol) LS3 is able to grow to osmolarities up to 3.32 osmol kg^{-1} H_2O. At concentrations lower than 3.4 M NaCl showed only limited influence on the growth behaviour. Supplementation with higher concentration (>3.4 M) of NaCl led to a decrease of the specific growth rate, a longer adaptation time as well as a lower cell count during stationary growth phase (Yang et al. 2000). Likewise in other yeast species, the osmotolerance is mediated by compounds of the high osmolarity glycerol (HOG) pathway, which is activated by an elevated osmolarity in the cells′ environment, leading to an increased synthesis of the compatible solutes glycerol, erythritol and mannitol (Böer et al. 2004a).

3 Genetics and Molecular Biology

A. adeninivorans has a DNA content comparable to that of ascomycetous yeasts such as *S. cerevisiae* (Gienow et al. 1990; Samsonova et al. 1996; Wartmann et al. 2000). Furthermore the fact, that after nitrosoguanidine mutagenesis a high number of auxotrophic mutants is obtained (Samsonova et al. 1989, 1996), along with the quantitative analysis of chromosomal DNA and the determination of genome size (Gienow et al. 1990) show *A. adeninivorans* to be a haploid organism.

DNA reassociation studies as well as karyotyping were performed in order to assess the complexity of the *A. adeninivorans* genome (Gienow et al. 1990; Kunze and Kunze 1994a). Genome size measurements, as performed in reassociation kinetics experiments of *A. adeninivorans* chromosomal DNA, resulted in 16.1 and 16.9 GDa for *A. adeninivorans* strains LS3 and CBS 8244T, respectively, rendering them the largest genomes reported so far amongst all yeast species, including *S. cerevisiae* (9.2 GDa). Also the amount of repetitive sequences (33.1% in LS3 and 35.9% in CBS 8244T) exceeds that of other yeasts. Finally, karyotyping demonstrated the existence of four chromosomes with sizes ranging between 1.6 and 4.6 Mb.

As previously mentioned, mutagenesis using UV light or *N*-methyl-*N*′-nitro-*N*-nitrosoguanidine treatment led to a relatively high frequency of auxotrophic mutants and mutants with different catabolite repression (resistant to 2-deoxy-D-glucose) which have subsequently been selected and characterized by Böttcher and Samsonova (1983); Büttner et al. (1990b) and Samsonova et al. (1989, 1996). Heterozygous diploids were generated from auxotrophic mutants of strains LS3 and CBS 8244T using PEG-induced spheroplast fusion as a first step to establish genetic maps (Büttner et al. 1990b; Samsonova et al. 1996). The resulting diploids were then segregated using benomyl treatment, allowing for linkage analysis of a set of markers that were assigned to four linkage groups. Additionally, specific probes for all 32 auxotrophic mutation markers were labelled and hybridized to genomic yeast DNA separated by pulsed field gel electrophoresis, confirming the predicted number of chromosomes (Samsonova et al. 1996).

The whole genome sequence of *A. adeninivorans* has been analysed and published (Kunze et al. 2014). Both mitochondrial and nuclear genomes were sequenced using the Sanger and 454 pyrosequencing approaches with different shotgun, plasmid and BAC libraries. The mitochondrial genome has a size of 31,662 bp and encodes 24 tRNA and 15 protein coding genes including the seven NADH: ubiquinone dehydrogenase subunits of complex I, the genes encoding the RNA component of RNase P and the two subunits of the mitochondrial ribosomal RNA. As already described and shown by pulsed field gel electrophoresis, the sequencing approach revealed four chromosomes Arad1A, Arad1B, Arad1C and Arad1D with a size of 1,659,397, 2,016,785, 3,827,910, and 4,300,524 Bp, respectively. Regional centromeres could be identified for all of them (Fig. 2). Table 4 shows the comparison of genome data of different representative hemiascomycetes. With 914 introns within 6116 genes, *A. adeninivorans* is one of the most intron-rich hemiascomycetes sequenced to date. Examples of genes containing at least one intron are *ARFC1, AHOG1* and *AHSB4*. The comparison of 5′-splice site (DS/GUARGU), branch site (HRCUAAC) and 3′-splice site (HAG/R) sequences demonstrates that the consensus sequences are similar to that of *S. cerevisiae* and filamentous fungi (Böer et al, 2005a).

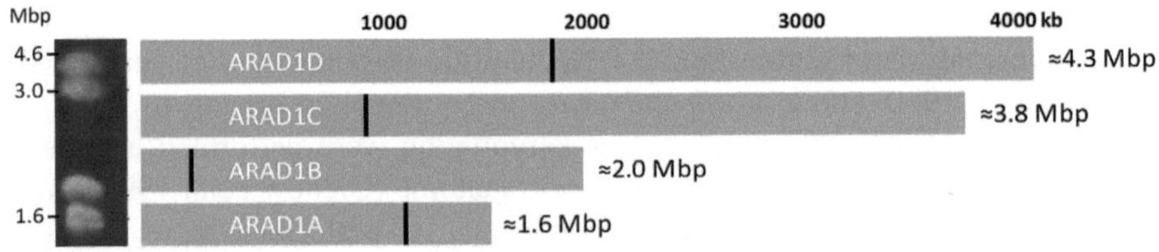

	Chromosome			CDS			Pseudo-genes	i-genes	Intrones	tRNA	5S rRNA	ncRNA
	Size	G+C (%)	Coding (%)	#	G+C (%)	Mean size (nt)						
Arad1A	1 659 397	48.2	73.2	871	49.6	1395	3	84	106	13	4	9
Arad1B	2 016 785	48.4	72.6	1051	49.8	1394	5	109	135	31	8	5
Arad1C	3 827 910	48.0	75.8	1991	49.2	1457	11	260	343	54	6	16
Arad1D	4 300 524	48.1	73.6	2203	49.3	1437	14	250	330	49	12	15
Total	11 804 616	48.1	74.1	6116	49.4	1430	33	703	914	147	30	45

G+C – guanine and cytosine; CDS – coding DNA sequence; i-genes – intron-containing genes; ncRNA – non coding RNA

Fig. 2 Schematic representation of the four chromosomes of *A. adeninivorans* LS3 as they can be seen on the PFGE. The positions of the putative centromere regions are indicated by *black lines*. The summary of the statistical properties of the four chromosomes given in the table can be found in Kunze et al. (2014)

Table 4 Annotated features of *A. adeninivorans* when compared to other representative hemiascomycetes (Kunze et al. 2014)

Species strain	*S. cerevisiae* S288c	*L. thermotolerans* CBS 6340	*D. hansenii* CBS 767	*Y. lipolytica* E150	*A. adeninivorans* LS3
Chromosome number	16	8	7	6	4
Genome					
Ploidy	n	2n	n	n	n
Size (Mbp)	12.1	10.4	12.2	20.5	11.8
Average G + C content (%)	38.3	47.3	36.3	49.0	48.1
Genome coding coverage (%)	70.0	72.3	74.2	46	74.1
CDS					
Total CDS (pseudo)	5769	5094 (46)	6272 (129)	6449 (137)	6116 (33)
Average G + C content (%)	39.6	49.2	38.0	52.9	49.4
Average size (aa)	485	492	479	476	477
i-genes	287	278	420	984	703
Introns	296	285	467	1119	914
Total tRNA genes	274	229	205	510	147
Total snRNA	6	5	5	6	5
Total snoRNA	77	43	*ND*	*ND*	37
rRNA clusters	1 (internal)	1 (internal)	3 (internal)	6 (subtelomeric)	1 (internal)
Total dispersed 5S rRNA genes	0	0	0	116	30

G + C guanine and cytosine; *CDS* coding DNA sequence; *aa* amino acids; *i-genes* intron-containing genes; *snRNA* small nuclear ribonucleic acid; *snoRNA* small nucleolar ribonucleic acid; *ND* not-determined

4 Biochemical Properties and *A. adeninivorans* as Gene Donor

A. adeninivorans is described as having a wide substrate spectrum that includes the assimilation of many nitrogenous and aromatic compounds such as nitrate and nitrite, purines, tannins and benzoic acid derivatives. The ability to degrade purine compounds is reported in all kingdoms and can occur either aerobically or anaerobically in separate pathways. In the aerobic pathway, the critical step in the degradation of purine bases is the oxidation of hypoxanthine and xanthine to uric acid, catalysed by xanthine oxidase and/or dehydrogenase. The various purine-degradative pathways are unique and differ from other metabolic pathways because they may serve quite different purposes, depending on the organism or tissue. While some organisms degrade the naturally occurring purines to CO_2 and ammonia, others perform only some of the steps of the purine degradation pathways, resulting in partial degradation of purines or certain intermediary catabolites. Purine catabolism which is shown in Fig. 3 is a characteristic feature of *A. adeninivorans*. The purine nucleosides (adenosine, inosine, xanthosine and guanosine) are transported across the membrane and into the cytoplasm by a purine permease. They are then converted to adenine, hypoxanthine, xanthine and guanine, further degraded to uric acid and, after transport into the peroxisomes, to urea. All corresponding genes of this pathway are localized on different chromosomes and are induced by adenine and other pathway intermediates (Jankowska et al. 2013a, b). It has been shown using a strain lacking xanthine oxidoreductase activity that the true inducer of the urate oxidase gene is in fact uric acid (Jankowska et al. 2013b). Interestingly, an adenosine deaminase, needed to transform adenosine to inosine in animals and human, is absent (Fig. 3). This pathway allows *A. adeninivorans* to use all of these purine derivatives as nitrogen and carbon sources (Jankowska et al. 2013a, b).

Tannin, a plant polyphenol molecule, is widely distributed in the plant kingdom where it protects plants against attack by parasites and herbivores. It inhibits the activity of enzymes by binding and precipitation and is to a greater or lesser extent recalcitrant to biodegradation (Field et al. 1991). While tannins are growth inhibitors for most microorganisms, a few bacteria, fungi and yeasts such as *D. hansenii*, *Mycotorula japonica* or *Candida* sp. are capable of exploiting tannins as a carbon and/or energy source for growth (Aguilar et al. 2007; Bhat Singh and Sharma 1998; Lekha and Lonsane 1997). *A. adeninivorans* is one of these yeasts that use tannic acid and gallic acid as carbon sources (Sietmann et al. 2010). Genes encoding tannases (*ATAN1*—ARAD1A06094g, *ATAN2*—ARAD1A19822g), gallate decarboxylase (*AGDC*—ARAD1C45804g) and catechol 1,2-dioxygenase (*ACDO*—ARAD1D18458g) have been identified and His-tagged recombinant enzymes and corresponding gene mutants were used to confirm the activity of these enzymes (data not shown). This demonstrated that the tannic acid catabolism pathway enables this yeast to assimilate tannic acid and other hydroxylated derivatives of benzoic acid by non-oxidative decarboxylation. Interestingly,

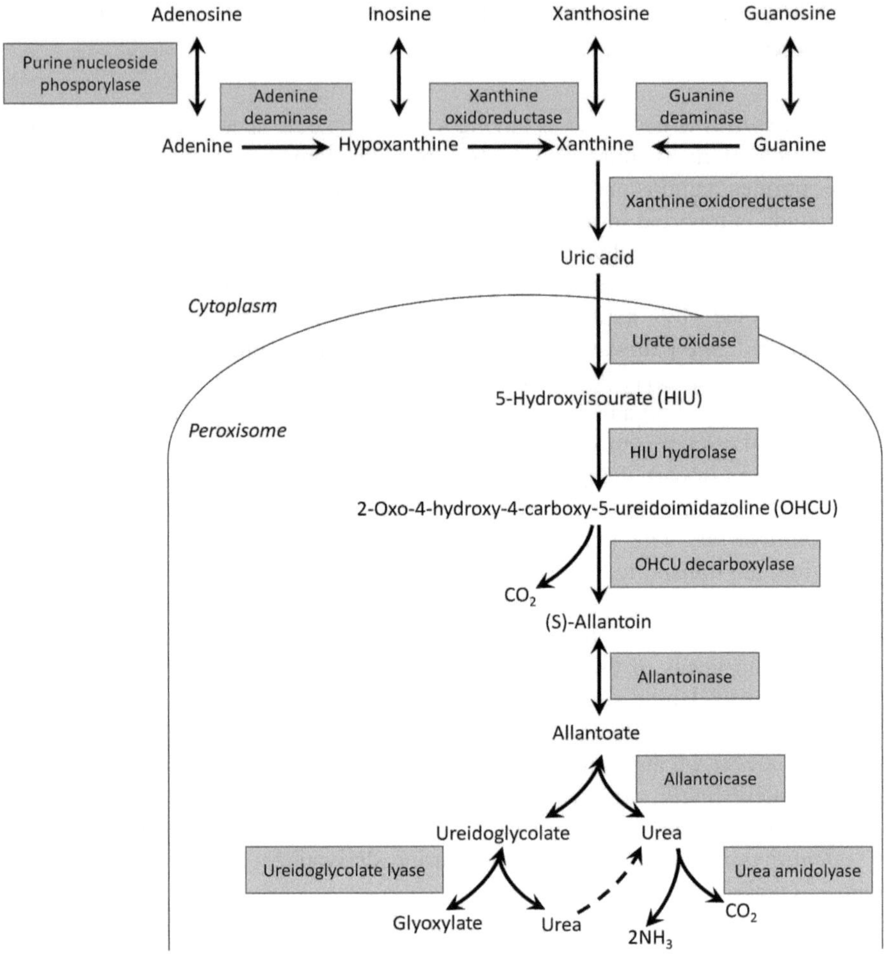

Fig. 3 Proposed purine degradation pathway in *A. adeninivorans*. The presence of OHCU decarboxylase has not been confirmed yet. Purine nucleoside phosphorylase uses adenosine, inosine, xanthosine as well as guanosine as substrate

A. adeninivorans is thus the first eukaryote known to synthesize two tannases, one extracellular (Atan1p) (Böer et al. 2009) and one cell-wall localized (Atan2p—data not shown) which permits effective degradation of extracellular tannic acid as well as the release of gallic acid from both condensed and hydrolysable tannins. Its biochemical parameters (pH optimum at approx. 6.0, temperature optimum 35–40 °C) and the almost complete extracellular localization ($\geq$97%) make Atan1p the superior enzyme for industrial applications. Constructed tannase producing strains are able to accumulate up to 51,900 U L^{-1} in 42 h with a dry cell weight of 162 g L^{-1} (Böer et al. 2011).

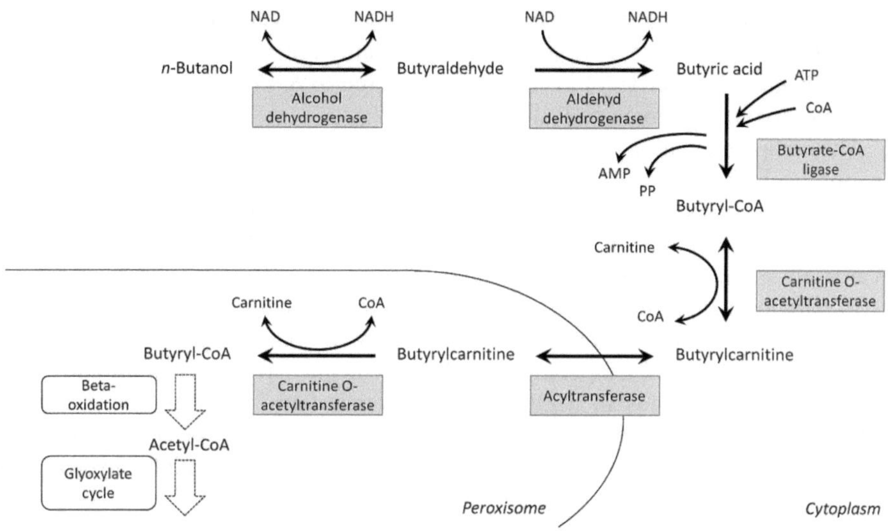

Fig. 4 n-Butanol degradation pathway of *A. adeninivorans*. n-Butanol is oxidized to butyric acid via butyraldehyde by an alcohol dehydrogenase (Aadh1p, *AADH1*—ARAD1B16786g) and one of two aldehyde dehydrogenases (Aald2p, *AALD2*—ARAD1B17094g; Aald5p, *AALD5*—ARAD1C17776g). Butyric acid is transported into peroxisome and further degraded in the β-oxidation

A genome mining approach (Kunze et al. 2014) was performed to find more interesting features of *A. adeninivorans* leading to the discovery of the n-butanol degradation pathway which had not been reported to exist in eukaryotes (Fig. 4). The collected data suggest that n-butanol is oxidized to butyraldehyde by an alcohol dehydrogenase (Aadh1p, *AADH1*—ARAD1B16786g) with a high substrate specificity, and then to butyric acid in a one-way reaction by two aldehyde dehydrogenases (Aald2p, *AALD2*—ARAD1B17094g; Aald5p, *AALD5*—ARAD1C17776g). The last steps involve an acyl-CoA ligase, a cytoplasmic acyl-CoA carnitine acyltransferase and a peroxisomal acyl-CoA carnitine acyltransferase for butyryl-carnitine synthesis via a butyryl-CoA intermediate that is transported from the cytoplasm to peroxisomes or mitochondria for ß-oxidation. The one-way reactions of aldehyde dehydrogenase and acyl-CoA ligase are special features of this pathway (Kunze et al. 2014).

Several genes like *AEFG1, AFET3, AHOG, AHSB4, AINV, ALIP1, ALYS2, APHO1, ARFC3, ATAL, AXDH* and *TEF1* as well as the complete rDNA repeat were isolated from gene libraries containing either cDNA or chromosomal DNA from *A. adeninivorans* strain LS3 via PCR using primers for conserved sequences (Böer et al. 2004a, b, 2005b, c; El Fiki et al. 2007; Kaur et al. 2007; Kunze and Kunze 1996; Rösel and Kunze 1995, 1996; Steinborn et al. 2005; Stoltenburg et al. 1999; Wartmann et al. 2001, 2002, 2003a). Additional genes like *GAA* encoding glucoamylase were identified from a heterologously expressed cDNA library with

S. cerevisiae and *K. lactis* as hosts using an anti-glucoamylase antibody as probe for product detection. More than 90% of the glucoamylase was found to be secreted, with a 20 times higher level in *K. lactis* compared to *S. cerevisiae* using a similar construct for transformation (Bui et al. 1996a, b).

The complementation of respective *E. coli* and *S. cerevisiae* mutants was used as an approach for the isolation of additional genes, namely the *ALYS2, AILV1, ALEU2* and *ATRP1* genes which are suitable selection markers for the *A. adeninivorans*-based platform (Kunze and Kunze 1996; Steinborn et al. 2007b; Wartmann et al. 1998, 2003b).

In parallel, genes encoding for biotechnologically important secretory enzymes like lipases found in *A. adeninivorans* were homologously expressed and characterized. This group of temperature-sensitive proteins with a pH optimum at 7.5 hydrolyses ester bounds in triglycerides and other fatty acid esters with highest efficiency for middle-sized chains between C8 and C10 (Böer et al. 2005b).

The *AINV* gene encoding for an invertase, which is preferentially hydrolyzing β-D-fructofuranosides, has potential to be applied in hydrolysis of sugar cane molasses or sugar beet molasses on an industrial scale. High concentrations of this enzyme have been obtained in recombinant *A. adeninivorans* strains carbon source independent using the strong constitutive *TEF1* promoter (Böer et al. 2004b).

An example for an interesting intracellular protein is the temperature-sensitive xylitol dehydrogenase. This enzyme oxidizes polyols like xylitol and D-sorbitol. Furthermore it can catalyse the reduction of D-xylulose, D-ribulose and L-sorbose. Its biochemical parameters—optimum at low temperatures and weak basic pH values—increase the potential for applications of this enzyme in food manufacturing processes. The respective *AXDH* gene has already been isolated and successfully overexpressed in *A. adeninivorans* (Böer et al. 2005c).

An enzyme with acidic pH optimum, transaldolase encoded by the *ATAL* gene, is another example with potential industrial applications. This temperature-sensitive enzyme, which is using D-erythrose-4-phosphate and D-fructose-6-phosphate as preferred substrates, could be useful in C–C bonding and enantio-selective synthesis of novel sugars (El Fiki et al. 2007).

Recently three genes *ACUT1, ACUT2* and *ACUT3* encoding for cutinases were isolated from the *A. adeninivorans* genome (Bischoff et al. 2015). Since cutinases are typically found in plant pathogenic fungi, only a few yeast species like *Cryptococcus* spp. (Masaki et al. 2005; Suzuki et al. 2013) and *Pseudozyma antarctica* (Shinozaki et al. 2013; Watanabe et al. 2014) have been found to produce cutinase like enzymes so far. The catalytic triade was identified as S-D-H with a conserved G-Y-S-Q-G domain containing the nucleophilic serine which is interacting with histidine and aspartic acid. Thus the three cutinases belong to the serine hydrolase family with α/β-structure. Recombinant variants outfitted with a 6xhistidine tag were expressed in *A. adeninivorans* and subsequently purified and characterized. All three cutinases show a pH optimum in the slightly acidic range from 4.0 to 6.5. The temperature optimum is in the range between 20 and 30 °C due to the low temperature stability of the enzymes. The substrate spectrum revealed the highest activity for short chain fatty acid esters of p-nitrophenol and glycerol

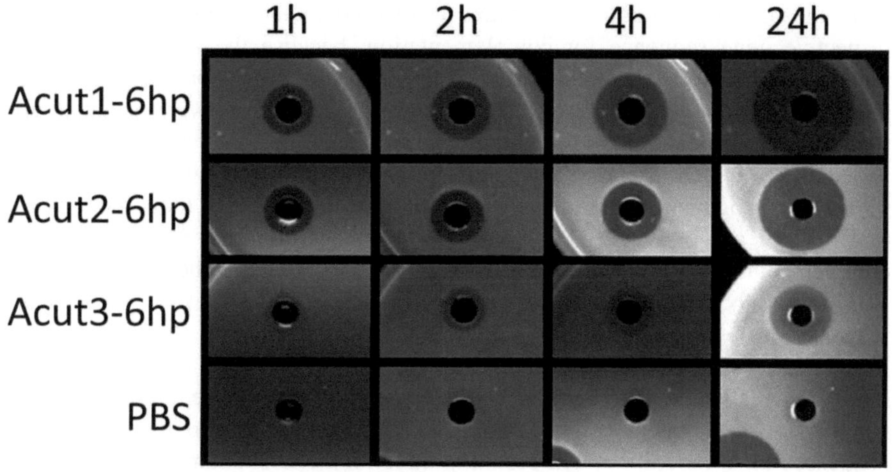

Fig. 5 Degradation of polycaprolactone (PCL) is shown as clear-zone formation on turbid agar plates containing PCL after incubation with Acut1-6hp, Acut2-6hp and Acut3-6hp at pH 5.5 and 25°C. Tests with PBS buffer without enzyme served as a negative control

(C4 and C6). However, the activity for p-nitrophenyl acetate (C2) was low (approx. 10% compared to p-nitrophenyl butyrate), which clearly distinguishes the three cutinases from ordinary carboxyl esterases and lipases. Additionally, the degradation of the natural polymer cutin from apple peels and of the model substrate polycaprolactone was observed (Fig. 5), which is proof of cutinolytic activity. However, the natural function of these cutinases for the yeast and their gene regulation remains unclear.

5 *A. adeninivorans* as Suitable Host for Heterologous and Homologous Gene Expression

5.1 *Transformation and Expression System*

An attractive gene expression platform for *A. adeninivorans* has been established and developed over the last decades starting from a first transformation system based on *S. cerevisiae* and *A. adeninivorans*-derived *LYS2* genes for selection (Kunze et al. 1990; Kunze and Kunze 1996). Improvements of the transformation system were introduced by Rösel and Kunze (1998) using a vector type (pAL-HPH1) that employed a 25S rDNA fragment from *A. adeninivorans* targeting sequence for stable integration and the *E. coli*-derived *hph* gene (conferring hygromycin B resistance) under the control of the *A. adeninivorans*-derived *TEF1* promoter for dominant selection. Two to ten stable integrations of the hygromycin resistance cassette could be found within the ribosomal DNA.

In the next step auxotrophic strains were used in the combination with their respective gene sequence for complementation, because the usage of dominant marker genes leads to the undesired need for toxic compounds or antibiotics during strain development. Firstly the *AILV1* and *ALEU2* genes were isolated after selecting the respective auxotrophic strains incapable of synthesizing leucine and isoleucine, which were obtained after treatment with N-methyl-N′-nitro-N-nitrosoguanidine. Strains generated by transformation of these mutants with pAL-AILV1 and pAL-ALEU2m containing the *AILV1* and the *ALEU2m* gene for complementation were found to harbour 1–3 copies of the heterologous DNA within the rDNA sequences (Steinborn et al. 2005; Wartmann et al. 1998, 2003b). However, at a frequency of 10^{-6}, the *ailv1* and *aleu2* mutant strains reverted to leucine/isoleucine prototrophy (Samsonova et al. 1989, 1996), which was a great disadvantage. This was eliminated by generating a *Δatrp1* gene disruption mutant using a DNA fragment containing the *ALEU2m* gene flanked by *ATRP1* gene sequences of some 750 bp. The resulting auxotrophic host strain *A. adeninivorans* G1212 [*aleu2 atrp1::ALEU2*] excels in mitotic stability during cultivation in both rich and minimal medium. The first strain complemented with the *ATRP1* gene as selection marker and the 25S rDNA contained just a single copy of the pAL-ATRP1 DNA (Steinborn et al. 2007b). For this reason a novel vector element containing the *ATRP1* coding sequence under control of a 53 bp truncated version of the *ALEU2* promoter was constructed that provides multicopy integrations (8 or more copies) in *A. adeninivorans* G1212 [*aleu2 atrp1::ALEU2*]. In addition the vector design enables the integration of a small vector fragment that consists of yeast DNA only (yeast integration-expression cassette—YIEC) providing high transformation frequencies and a high mitotic stability (Steinborn et al. 2007a).

The construction of expression plasmids followed a two-step cloning strategy. First the heterologous genes are inserted between the respective *A. adeninivorans*-derived promoter and fungal terminator elements like *PHO5* from *S. cerevisiae* and *trpC* from *Aspergillus nidulans*. Subsequently the resulting expression modules (*A. adeninivorans* promoter—heterologous gene—fungal terminator) are integrated into the respective *A. adeninivorans* expression plasmid. For this purpose the modules are flanked by unique restriction sites (*Apa*I–*Sal*I, *Apa*I–*Xho*I, *Spe*I–*Sac*II, *Spe*I–*Not*I) allowing a directional integration.

A new set of plasmids was introduced into the latest version of the *A. adeninivorans* expression platform to circumvent the two-step cloning strategy. Those plasmids already contain the selection marker as well as several combinations of promoter and terminator elements. Figure 6 shows one example plasmid which provides two integration sites for genes with either *TEF1* promoter and *PHO5* terminator or *TIF5a* promoter and *60SRPL3* terminator. The genes are integrated after linearization with *Pac*I for *TEF1/PHO5* or *Spe*I for *TIF5a/60SRPL3* using isothermal assembly, which avoids complications with restriction sites present in the gene sequence.

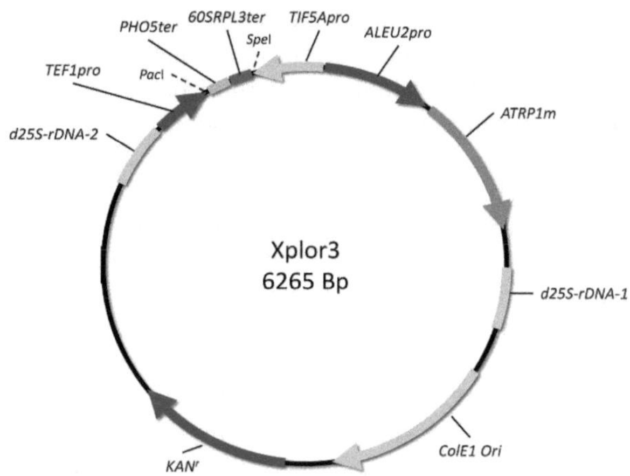

Fig. 6 Xplor3 plasmid for expression of two genes under the control of constitutive *TEF1* and *TIF5* promoters

5.2 *Heterologous Gene Expression*

A. adeninivorans is not only a donor of genes encoding for industrially relevant proteins but also a host for heterologous gene expression. An increasing number of heterologous genes have been expressed in *A. adeninivorans*. First examples were the *XylE* gene encoding catechol 2,3-dioxygenase from *Pseudomonas putida* under the control of *AILV1* promoter and *GFP* as well as the *HSA* gene under the control of the strong constitutive *TEF1* promoter (Kunze et al. 1990; Kunze and Kunze 1996). One to two copies of heterologous DNA were found within the 25S rDNA region after transformation of wild-type and mutant strains with pAL-HPH1 and pAL-ALEU2m. Recombinant GFP was localized in the cytoplasm rendering the cells fluorescent, whereas expression of *HSA* led to secretion of 95% recombinant HSA into the culture medium at levels of 50 mg L^{-1} after 96 h. In this instance, no difference in secretion levels could be observed comparing budding cells and mycelia, which demonstrates a morphology-independent productivity (Wartmann and Kunze 2003; Wartmann et al. 2003b).

Besides the *TEF1* promoter, the strong constitutive *AHSB4* promoter was successfully assessed for suitability and was found to exhibit similar expression levels (Wartmann et al. 2003a).

A next example is the expression of a *MF1-IL6* fusion under control of the strong *TEF1* promoter in *A. adeninivorans* budding cells and mycelia. Unlike other yeast species (*S. cerevisiae*, *H. polymorpha*), *A. adeninivorans* was correctly processing the MF1-IL6 precursor, which led to the accumulation of recombinant interleukin-6 (IL-6) to more than 95% in the culture medium. A productivity of 210 and 145 mg L^{-1} was observed in cultivations on shaking flask scale with budding cells and mycelia, respectively (Böer et al. 2007).

A metabolic engineering approach had been done by introducing several genes of the synthesis of polyhydroxyalkanoate (PHA) biosynthetic pathway of *Ralstonia eutropha*. The genes *phbA, phbB* and *phbC* encoding β-ketothiolase, NADPH-linked acetoacetyl-CoA reductase and PHA synthase were established simultaneously enabling *A. adeninivorans* to synthesize poly-3-hydroxybutyrate (PHB) as well as poly-3-hydroxyvalerate (PHV). The first recombinant yeast strains were able to accumulate up to 2.2% PHV and 0.019% PHB with ethanol as carbon source (Terentiev et al. 2004).

Also promoter assessment studies were performed using *lacZ* from *E. coli*, the *GFP* from *Aequorea victoria*, the *phyK* from *Klebsiella spec.* and the *XylE* from *Ps. putida* as reporter genes. In this instance, the *GAA, AHOG1, AINV, AXDH* as well as the *ATAL* promoter were analysed on aspects like carbon source, osmolarity of the medium or morphological stage (Böer et al. 2004a, b, 2005c; El Fiki et al. 2007; Hahn et al. 2006; Wartmann and Kunze 2000).

Another successful example is the expression of alcohol dehydrogenase (*ADH*) from *Rhodococcus ruber* and *Rhodococcus erythropolis* as biocatalyst for the synthesis of enantiomerically pure secondary alcohols (Kasprzak et al. 2016). *R. erythropolis* alcohol dehydrogenase was used for synthesis of 1-(S)-phenylethanol and ethyl (R)-4-chloro-3-hydroxybutanoate either with purified enzyme or with permeabilized cells of genetically modified *A. adeninivorans*, co-expressing glucose dehydrogenase from *Bacillus megaterium* for enzyme-coupled co-factor regeneration. One of the major advantages of yeast cells over *E. coli* is that they are more robust in organic solvents because of their cell wall construction. Ethyl (R)-4-chloro-3-hydroxybutanoate is an important chiral building block in the synthesis of pharmaceuticals such as (−)-macrolactin A (Marino et al. 2002), l-carnitine, (R)-γ-amino-β-hydroxybutyric acid (GABOB) (Song et al. 1995), (+)-negamycin or chiral 2,5-cyclohexadienone synthon, therefore *A. adeninivorans* expressing *R. erythropolis* alcohol dehydrogenase as well as *B. megaterium* glucose dehydrogenase is of great interest for industrial application. 1-(R)-phenylethanol, another chiral building block for the chemical and pharmaceutical industry, was synthetized using *A. adeninivorans* co-expressing *Lactobacillus brevis ADH* and *GDH* from *B. megaterium* or *Bacillus pumilus G6PDH* as co-factor regeneration system (Rauter et al. 2014). The combination of *ADH* and *G6PDH* showed highest activity. However, together with *GDH* a higher stability was obtained which in the end leads to higher product formation due to a higher rate of reusability. In this context *A. adeninivorans* was used for optimization studies in enzymatic synthesis of 1-(S)-phenylethanol with permeabilized cells co-expressing the *ADH* of *R. ruber* and the *B. megaterium GDH* (Rauter et al. 2014). A comparison of the reaction with permeabilized cells, permeabilized immobilized cells and immobilized crude extract of those cells showed that immobilization of permeabilized cells is the best option for high reusability in synthesis reactions which is a key factor for the cost-effective use of enzymes in industrial applications. A summary of the different industrially relevant alcohol dehydrogenases expressed in *A. adeninivorans* is given in Table 5.

Despite this great potential for industrial applications little research was done on definition of culture and media conditions. Stöckmann et al. (2014) investigated the

Table 5 *A. adeninivorans* as host for heterologous expression of bacterial alcohol dehydrogenases (*ADH*) for the synthesis of chiral secondary alcohols

	ReADH	RrADH	LbADH
Donor	*Rhodococcus erythropolis*	*Rhodococcus ruber*	*Lactobacillus brevis*
pH optimum			
Oxidation	9.0	9.0	9.0
Reduction	7.0	6.5	6.5
Temperature optimum (°C)	45	40	30
Native molecular mass (kDa)	144	*ND*	*ND*
Co-factor	NADH/NAD	NADH/NAD	NADH/NAD NADPH/NADP
Regeneration system	GDH *(B. megaterium)* Isopropanol	GDH *(B. megaterium)* Isopropanol	GDH *(B. megaterium)* G6PDH (*B. pumilus)*
Substrates			
Oxidation	1-(S)-Phenylethanol 1,6-Hexanediol	2.5-Hexanediol 2-Nonanol 1-Phenylethanol	*ND*
Reduction	Acetophenone 2,5-Hexatiedione Ethyl 4-chloroacetoacetate	2.5-Hexanedione Acetophenon Hexanal Phenylacetaldehyd Pentanal	*ND*
Process			
Product	1-(S)-Phenylethanol Ethyl (R)-4-chloro-3-hydroxybutanoate	1-(S)-Phenylethanol	1-(R)-Phenylethanol
Condition	Isolated enzymes Permeabilized cells	Isolated enzymes Immobilized cells Permeabilized cells	Isolated enzymes Immobilized cells Permeabilized cells
Enantio-selectivity	99.9%	99.9%	99.9%

ND not determined

growth characteristics and culture conditions in a more rational approach. Cultures of *A. adeninivorans* were inhibited at pH-values below 2.8 and the phosphorus demand has been determined as 1.55 g phosphorus per 100 g dry cell weight. An optimized SYN6 medium was developed which is buffered at pH 6.4 with 140 mmol MES L^{-1}. It provides non-limited cultivation conditions without by-product formation in shake flask cultivations. A maximal specific growth rate of 0.32 h^{-1} and short fermentations of 15 h were achieved. The rational definition of conditions for a non-limiting oxygen and phosphorus supply as well as the pH stabilization of the medium to non-inhibiting values provide basic conditions for *A. adeninivorans* cultures characterized by short fermentation times, a complete aerobic metabolism without anaerobic by-products or overflow-metabolites, maximized growth rates, and maximized biomass yields.

6 *A. adeninivorans* as Bio Compound in Biosensors for the Detection of Hormonal Activities

6.1 A. adeninivorans*-Based Biosensors*

The utilization of *A. adeninivorans* for the development of biosensors targeted for the detection of several hormones and pharmaceuticals has been consequently extended in the last years. Since the first evocation of an estrogen-detecting biosensor via a cell-based in vitro assay in 2006 (Hahn et al. 2006), 7 new studies describing the design of biosensors dedicated to estrogens, androgens, progestogens, glucocorticoids and pharmaceuticals using recombinantly produced human hormone receptors have been published (Kaiser et al. 2010; Pham et al. 2012, 2013, 2015, 2016; Gerlach et al. 2014; Chamas et al. 2015). Although most of these published works discuss the development of cell-based in vitro assays followed by enzymatically mediated detection, amperometric detection and fluorometric detection are also described. Additionally, the electrochemical detection of estrogens using an estrogen binding protein from yeast origin recombinantly produced in *A. adeninivorans* was also reported (Vijayan et al. 2015). Detection of hormones or pharmaceuticals in aqueous samples became of major interest as accumulating studies have shown the negative impact endocrine disruptors can have for the aquatic life (Rempel and Schlenk 2008). All natural and synthetic molecules which can bind to the vertebrate hormone receptors present a potential risk also for human health if released in an uncontrolled way into the environment (Kabir et al. 2015). Among the recently developed whole-cell biosensors, the ones using *A. adeninivorans* as host present some of the best performance and four of them are already commercially available for routine testing.

6.2 Principle of Action

In the majority of the recently developed *A. adeninivorans* based biosensors, a similar strategy was used for the detection of hormones. The first step was the construction of a modified *A. adeninivorans* strain by genomic integration of two expression modules responsible for the constitutive production of the desired human hormone receptor and for the ligand-induced expression of a *reporter* gene. The human hormone receptor will bind to the ligand and this binding event will be transformed in a measurable signal with the help of a *reporter* gene. To obtain the inducible mechanism of the *reporter* gene, a modified version of the *A. adeninivorans*-based *GAA* promoter was constructed by inserting two 15 bp sequences called hormone response element (*HRE*) at the position-107 of the promoter. *HRE* is a conserved DNA-binding site for the dimerized hormone receptor and only when this dimerized receptor is bound to the *HRE* the expression of the downstream gene can occur. In the case of the estrogen-related assays, *HRE* was replaced by the slightly different estrogen response element (*ERE*) sequence as it shows higher affinity to the estrogen receptor than *HRE* (Hahn et al. 2006). Hence, the two expression modules used for yeast transformation were *TEF1* promoter—*hormone receptor* gene—*PHO5* terminator and *GAA*[*2xHRE* or *2xERE*]$^{-107}$ promoter—*reporter* gene—*PHO5* terminator. The mechanism of the biosensor is described in Fig. 7. If present in the cultivation medium, the hormone is transported into the cytoplasm of the modified *A. adeninivorans* cell by passive diffusion through the membrane. When bound to the constitutively expressed hormone receptor (HR) it induces a conformational change that allows the homodimerization of the receptor in the cytoplasm. The nuclear localization signal present in the receptor will then permit receptor homodimer translocation into the nucleus. Once in the nucleus, the ligand bound receptor dimer can bind to the *HRE* and activates the transcription of the *reporter* gene. Three different *reporter* genes have so far been utilized to produce a measurable signal: the *Klebsiella*-derived *phytase K* gene (*phyK*), the *A. adeninivorans*-derived *tannase* gene (*ATAN1*) and the *Discosoma*-derived *dsRED* gene. Both *phyK* and *ATAN1* genes produce an extracellularly located enzyme which can be utilized in an enzymatic reaction. Successful transformation of the substrate into the desired reaction product can then be detected either by spectrometric or amperometric methods. While *ATAN1* was successfully utilized for estrogen detection (Kaiser et al. 2010), *phyK* was solely used in all other biosensors utilizing enzymatically-based detection. Finally, the

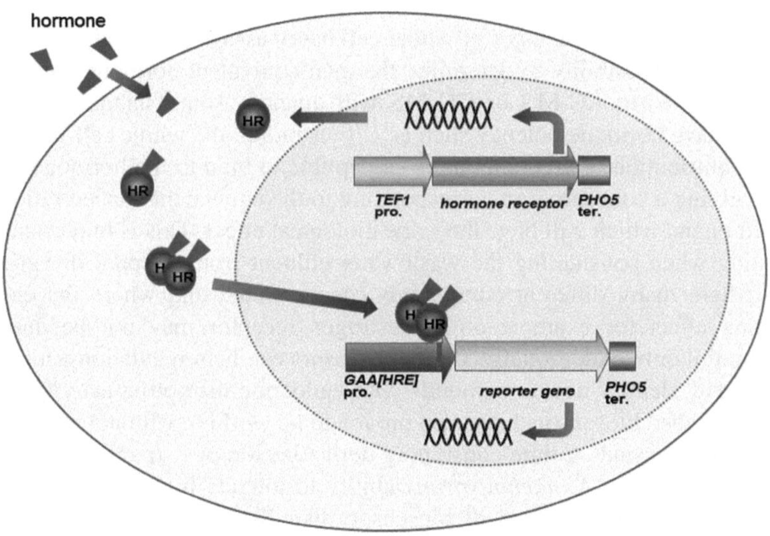

Fig. 7 Principle of *A. adeninivorans*-based hormone detection assays. The hormone receptor is constitutively produced in the cytoplasm of the yeast cell. If present, hormones that enter the cytoplasm can bind to the hormone receptor. After ligand-induced homodimerization, the receptor dimer translocates into the nucleus and binds to the *HRE* present in the *GAA* promoter, thus inducing the production of the reporter protein

dsRED gene is responsible for the production of the cytoplasmic fluorescent protein dsRED which can be spectrofluorometrically detected at an emission wavelength of 582 nm if exposed to an excitation wavelength of 542 nm.

Several improvements have been made to the general mechanism of the assay in order to reliably detect pharmaceuticals. Two genes were integrated in the genome of *A. adeninivorans* cells to be expressed constitutively: the *human arylhydrocarbon receptor* gene (*hAhR*) and the *human arylhydrocarbon receptor nuclear translocator* gene (*hARNT*). Only the hAhR protein can bind to pharmaceuticals but it requires heterodimer formation together with hARNT in order to translocate into the nucleus. Additionally, the *HRE* sequence in the *GAA* promoter was replaced by the *cyp1A1*-derived core sequence that moderates the interaction between the promoter and the hAhR-hARNT heterodimer. *Reporter* gene in this case was *phyK* both for amperometric and enzymatically-mediated detection.

6.3 Performance of the Assays and Applications

Some of the major characteristics and performances of the developed *A. adeninivorans*-based biosensors are presented in Table 6. For each of them, the concentration of respective standard ligand giving a half-maximal response (EC_{50}) as well as the limit of detection (LoD) towards this same ligand is indicated. It is also of note that for each class of target compounds, real samples from diverse origins were included in the studies.

One of the major advantages of whole-cell based assays over analytical detection methods is the possibility to determine the total equivalent hormone concentration of a sample. When LC-MS or GC-MS will quantify some standard compounds known to have hormone potency such as 17β-estradiol, the whole cell-based assays add the contribution of all compounds susceptible to bind to the hormone receptor, thereby giving a concentration corresponding to the equivalent concentration of the standard ligand which will have the same biological effect. This is in particular very interesting when considering the wastewater effluent from hospital or agricultural origin where many different compounds can be found and where the endocrine disrupting effect for example on the estrogen receptor may not be due to the 17β-estradiol only. Additionally, these biosensors can help regulation authorities to discover and identify new compounds with endocrine disrupting activity.

The described biosensors have also proven to be working with untreated samples and do not necessitate a time-consuming derivatization or extraction process. This is mainly due to the *A. adeninivorans* ability to tolerate high salt conditions and allows on-site implementation of biosensors like *Estramonitor* or *Pharma* in the wastewater treatment plant for direct monitoring of hormones or pharmaceuticals in effluent water samples.

Table 6 Description of the recently developed *A. adeninivorans* based biosensors

target compound	biosensor	reporter gene	detection method	tested ligand	EC_{50}	LoD	nature of real samples	measured total equivalent concentrations in real samples	source
estrogens	A-YES	*PhyK*	biochemical	17β-estradiol, 17α-estradiol, estrone, estriol, testosterone, 5α-dihydrotestosterone, progesterone, diethylstilbestrol, hexestrol, menstranol, 17α-ethynylestradiol, 4-tert-octylphenol, 4-n-nonylphenol, bisphenol A, kepone, 8-prenylnaringenin, coumestrol, genestin, zeranelone	17β-estradiol: 33.2 ng/l	17β-estradiol: 2.8 ng/l	wastewater swine urine	up to 70 ng/l	(Kaiser et al., 2010)
	A-YES	*ATAN1*	biochemical	17β-estradiol, 17α-estradiol, estrone, estriol, testosterone, 5α-dihydrotestosterone, progesterone, diethylstilbestrol, hexestrol, menstranol, 17α-ethynylestradiol, 4-tert-octylphenol, 4-n-nonylphenol, bisphenol A, kepone, 8-prenylnaringenin, coumestrol, genestin, zeranelone	17β-estradiol: 39.4 ng/l	17β-estradiol: 3.1 ng/l	wastewater swine urine	up to 70 ng/l	(Kaiser et al.,2010)
	Estramonitor	*PhyK*	amperometric	17β-estradiol, 17α-estradiol, estrone, estriol, 17α-ethynylestradiol, bisphenol A	17β-estradiol: 54.6 ng/l	17β-estradiol: 9.92 ng/l	wastewater	up to 107 ng/l	(Pham et al., 2013)
androgens	A-YAS	*PhyK*	biochemical	5α-dihydrotestosterone, 17β-testosterone, androstenedione, dehydroepiandrosterone (DHEA), DHEA-3-sulfate, 17b-estradiol, estrone, flutamide, cyproterone acetate, 17α-methyltestosterone, 17β-nandrolone, 17β-boldenone, 17β-trenbolone, fenitrothion, methoxychlor, 4,4'-DDE, tributylin chloride, bis(tributylin) oxide, triphenyl chloride	5α-DHT: 277 ng/l	5α-DHT: 56.5 ng/l	cattle urine	up to 266 ng/l	(Gerlach et al., 2014)
progestogens	A-YPS	*PhyK*	biochemical	progesterone, medroxyprogesterone acetate, dihydrotestosterone, nandrolone, trenbolone, cyproterone acetate, 17β-estradiol, mifepristone	progesterone: 147 ng/l	progesterone: 40.9 ng/l	wastewater	up to 500 ng/l	(Chamas et al., 2015)
	A-YPFS	*dsRED*	fluorescence	progesterone, medroxyprogesterone acetate, dihydrotestosterone, nandrolone, trenbolone, cyproterone acetate, 17β-estradiol, mifepristone	progesterone: 231 ng/l	progesterone: 65.3 ng/l	NI	NI	(Chamas et al., 2015)
glucocorticoids	A-YGS	*PhyK*	biochemical	cortisol, corticosterone, cyproterone acetate, dexamethasone, mifepristone, prednisolone, prednisone, methylprednisolone, triamcinolone	dexamethasone: 0.81 μM	dexamethasone: 0.29 μM	wastewater	under the LoD	(Pham et al., 2016)
	A-YGFS	*dsRED*	fluorescence	betamethasone, cortisol, corticosterone, cyproterone acetate, dexamethasone, mifepristone, prednisolone, progesterone, methylprednisolone, triamcinolone, 17β-estradiol	dexamethasone: 9.42 μM	dexamethasone: 0.47 μM	NI	NI	(Pham et al., 2016)
pharmaceuticals	Pharma	*PhyK*	amperometric	β-naphtoflavone	β-naphtoflavone : 123.1 ng/l	β-naphtoflavone: 44.27 ng/l	NI	NI	(Pham et al., 2015)
	Pharma	*PhyK*	biochemical	ampiciline, carbaryl, carbeniciline, diclofenac, lansoprazole, indirubin, methylcholantrene, omeprazole, spectinomycine, β-naphtoflavone	β-naphtoflavone: 35.5 ng/l	ND	wastewater	up to 32 ng/l	(Pham et al., 2015)

NI not included, *ND* not determined

The family of *A. adeninivorans*-based biosensors will expand in the next years as compounds like dioxins or bisphenols will be targets of new assays. The construction of such biosensors with modified recombinant human receptors is currently performed. Another perspective will be the combination of already developed biosensors with chromatographic assays for prior separation of compounds, especially when it is of interest to know, which compounds in the sample contribute to the total equivalent hormone concentration. That's why future studies will aim to first separate a complex sample with the help of thin-layer chromatography before applying one of the developed *A. adeninivorans* biosensors in order to determine how many compounds possess hormonal activity and for which proportion of the total signal these compounds are responsible. Comparing the obtained retention factors with the retention factors of known chemicals will allow a first identification. Subsequently, transfer of the unknown compounds for analytical identification by mass spectrometry will then lead to complete description of a complex sample.

7 Conclusions

The dimorphic, asexual hemiascomycete *A. adeninivorans* is an attractive organism for both, basic and applied research and has been shown to have a high potential in interesting biotechnological applications due to the very broad range of substrates, which can be used as carbon and/or nitrogen sources, the growth and secretion characteristics as well as the thermo- and osmotolerance. *A. adeninivorans* has been intensively exploited as host for heterologous gene expression leading to transformants which, for example, can be used in the chemical industry as biocatalyst for the synthesis of chiral alcohols. Furthermore, the genome of this yeast contains a large pool of genes encoding specialized enzymes with useful biochemical properties making *A. adeninivorans* an interesting gene donor. In addition, *A. adeninivorans* emerged to be a powerful tool in bio-based assays for the determination of hormonal activities in environmental samples due to its high robustness combined with high ability in heterologous gene expression.

References

Aguilar, C.N., Rodriguez, R., Gutierrez-Sanchez, G., Augur, C., Favela-Torres, E., Prado-Barragan, L.A., Ramirez-Coronel, A., and Contreras-Esquivel, J.C. 2007. *Appl. Microbiol. Biotechnol.* **76**: 47–59.

Bhat, T. K., Singh, B., and Sharma, O. P. 1998. *Biodegradation.* **9**: 343–357.

Bischoff, F., Litwinska, K., Cordes, A., Baronian, K., Bode, R., Schauer, F., and Kunze, G. 2015. *Appl. Environ. Microbiol.* **81**: 5497–5510.

Böer, E., Wartmann, T., Dlubatz, K., Gellissen, G., and Kunze, G. 2004a. *Curr. Genet.* **46**: 269–276.

Böer, E., Wartmann, T., Luther, B., Manteuffel, R., Bode, R., Gellissen, G., and Kunze, G. 2004b. *Antonie van Leeuwenhoek.* **86**: 121–134.

Böer, E., Gellissen, G., and Kunze, G. 2005a. *Arxula adeninivorans* In: Gellissen, G. (Ed) *Production of Recombinant Proteins,* Wiley-VCH Verlag GmbH & Co. KGaA. pp. 89–110.

Böer, E., Mock, H.P., Bode, R., Gellissen, G., and Kunze, G. 2005b. *Yeast* **22**: 523–535.

Böer, E., Wartmann, T., Schmidt, S., Bode, R., Gellissen, G., and Kunze, G. 2005c. *Antonie van Leeuwenhoek* **87**: 233–243.

Böer, E., Steinborn, G., Matros, A., Mock, H.P., Gellissen, G., and Kunze, G. 2007. *FEMS Yeast Res.* **7**: 1181–1187.

Böer, E., Bode, R., Mock, H.P., Piontek, M., and Kunze, G. 2009. *Yeast.* **26**: 323–337.

Böer, E., Breuer, F. S., Weniger, M., Denter, S., Piontek, M., and Kunze, G. 2011. *Appl. Microbiol. Biotechnol.* **92**: 105–114.

Böttcher, F. and Samsanova, I.A. 1983. Systemische Fungizide und antifungale Verbindungen In: Lyr, H. and Polter, C. (Eds.) *Systemische Fungizide und antifungale Verbindungen*, Akademie Verlag. pp. 255–258.

Böttcher, F., Klinner, U., Köhler, M., Samsonova, I.A., Kapultsevich, J. and Bliznik, X. 1988. Verfahren zur Futterhefeproduktion in zuckerhaltigen Medien. DD 278 354 A1.

Bui, D. M., Kunze, I., Förster, S., Wartmann, T., Horstmann, C., Manteuffel, R., and Kunze, G. *Appl. Microbiol. Biotechnol.* **44**: 610–619.

Bui, D.M., Kunze, I., Horstmann, C., Schmidt, T., Breunig, K. D., and Kunze, G. 1996. *Appl. Microbiol. Biotechnol.* **45**: 102–106.

Büttner, R., and Bode, R. 1992c. *J. Basic Microbiol.* **32**: 159–166.

Büttner, R., Bode, R., and Birnbaum, D. 1987. *J. Basic Microbiol.* **27**: 299–308.

Büttner, R., Bode, R., Scheidt, A., and Birnbaum, D. 1988. *Acta Biotechnol.* **8**: 517–525.

Büttner, R., Scheit, A., Bode, R., and Birnbaum, D. 1989. *J. Basic Microbiol.* **29**: 67–72.

Büttner, R., Bode, R. and Birnbaum, D.. 1990a. *Wiss. Z. Ernst-Moritz-Arndt-Univ. Greifswald, Math.-nat.wiss. Reihe* **39**: 1–21.

Büttner, R., Bode, R., Samsonova, I. A., and Birnbaum, D. 1990b. *J. Basic Microbiol.* **30**: 227–231.

Büttner, R., Bode, R., and Birnbaum, D. 1991a. *J. Basic Microbiol.* **31**: 423–428.

Büttner, R., Bode, R., and Birnbaum, D. 1991b. *Zbl. Mikrobiol.* **146**: 399–406.

Büttner, R., Schubert, U., Bode, R., and Birnbaum, D. 1990c. *Acta Biotechnol.* **10**: 361–370.

Büttner, R., Bode, R., and Birnbaum, D. 1992a. *Zbl. Mikrobiol.* **147**: 225–230.

Büttner, R., Bode, R., and Birnbaum, D. 1992b. *Zbl. Mikrobiol.* **147**: 291–296.

Chamas, A., Nieter, A., Pham, H. T., Giersberg, M., Hettwer, K., Uhlig, S., Simon, K., Baronian, K., and Kunze, G. 2015. *Anal. Bioanal. Chem.* **407**: 8109–8120.

Field, J., Leyendeckers, M.J., Sierra-Alvarez, R., Lettinga, G., and Habets, L. H. 1991. *Biotechnol. Bioeng.* **37**: 247–255.

Fiki, A.E., Metabteb, G. E., Bellebna, C., Wartmann, T., Bode, R., Gellissen, G., and Kunze, G. 2007. *Appl. Microbiol. Biotechnol.* **74**: 1292–1299.

Gellissen, G. 2005. *Production of Recombinant Proteins - novel microbial and eukaryotic expression systems,* Wiley-VCH Verlag GmbH & Co. KGaA.

Gellissen, G., Kunze, G., Gaillardin, C., Cregg, J. M., Berardi, E., Veenhuis, M., and van der Klei, I. 2005. *FEMS Yeast Res.* **5**: 1079–1096.

Gerlach, T., Knaust, J., Kaiser, C., Körner, M., Hettwer, K., Uhlig, S., Simon, K., Baronian, K., and Kunze, G. 2014. *Sci. Total Environ.* **490**: 1073–1081.

Gienow, U., Kunze, G., Schauer, F., Bode, R., and Hofemeister, J. 1990. *Zbl. Mikrobiol.* **145**: 3–12.

Hahn, T., Tag, K., Riedel, K., Uhlig, S., Baronian, K., Gellissen, G., and Kunze, G. 2006. *Biosens. Bioelectron.* **21**: 2078–2085.

Jankowska, D.A., Faulwasser, K., Trautwein-Schult, A., Cordes, A., Hoferichter, P., Klein, C., Bode, R., Baronian, K., and Kunze, G. 2013a. *J. Appl. Microbiol.* **115**: 1134–1146.

Jankowska, D.A., Trautwein-Schult, A., Cordes, A., Hoferichter, P., Klein, C., Bode, R., Baronian, K., and Kunze, G. 2013b. *J. Appl. Microbiol.* **115**: 796–807.

Kabir, E.R., Rahman, M.S., and Rahman, I. 2015. *Environ. Toxicol. Pharmacol.* **40**: 241–258.

Kaiser, C., Uhlig, S., Gerlach, T., Körner, M., Simon, K., Kunath, K., Florschütz, K., Baronian, K., and Kunze, G. 2010. *Sci. Total Environ.* **408**: 6017–6026.

Kasprzak, J., Bischoff, F., Rauter, M., Becker, K., Baronian, K., Bode, R., Schauer, F., Vorbrodt, H.M., and Kunze, G. 2016. *Biochem. Eng. J.* **106**: 107–117.

Kaur, P., Lingner, A., Singh, B., Böer, E., Polajeva, J., Steinborn, G., Bode, R., Gellissen, G., Satyanarayana, T., and Kunze, G. 2007. *Antonie van Leeuwenhoek* **91**: 45–55.

Kunze, G., and Kunze, I. 1994a. *Antonie van Leeuwenhoek* **65**: 29–34.

Kunze, I., and Kunze, G. 1994b. *J. Eur. Microbiol.* **212**: 24 – 28.

Kunze, I., and Kunze. G. 1996. *Arxula adeninivorans* In: Wolf, K. (Ed) *Nonconventional yeasts biotechnology*. Springer Verlag Berlin Heidelberg, pp. 389–409.

Kunze, G., Pich, U., Lietz, K., Barner, A., Büttner, R., Bode, R., Conrad, U., Samsonova, I.A., and Schmidt, H. 1990. Wirts-Vektor-System und Verfahren zu seiner Herstellung. DD 298 821 A5.

Kunze, G., Gaillardin, C., Czernicka, M., Durrens, P., Martin, T., Böer, E., Gabaldon, T., Cruz, J. A., Talla, E., Marck, C., Goffeau, A., Barbe, V., Baret, P., Baronian, K., Beier, S., Bleykasten, C., Bode, R., Casaregola, S., Despons, L., Fairhead, C., Giersberg, M., Gierski, P.P., Hahnel, U., Hartmann, A., Jankowska, D., Jubin, C., Jung, P., Lafontaine, I., Leh-Louis, V., Lemaire, M., Marcet-Houben, M., Mascher, M., Morel, G., Richard, G.F., Riechen, J., Sacerdot, C., Sarkar, A., Savel, G., Schacherer, J., Sherman, D.J., Stein, N., Straub, M.L., Thierry, A., Trautwein-Schult, A., Vacherie, B., Westhof, E., Worch, S., Dujon, B., Souciet, J. L., Wincker, P., Scholz, U., and Neuveglise, C. 2014. *Biotechnol. Biofuels.* **7**: 66.

Kurtzman, C. P., and Robnett, C.J. 2007. *FEMS Yeast Res.* **7**: 141–151.

Lekha, P.K., and Lonsane, B.K. 1997. *Adv. Appl. Microbiol.* **44**: 215–260.

Marino, J.P., McClure, M.S., Holub, D. P., Comasseto, J.V., and Tucci, F.C. 2002. *J. Am. Chem. Soc.* **124**: 1664–1668.

Masaki, K., Kamini, N.R., Ikeda, H. and Iefuji, H. 2005. *Appl. Environ. Microbiol.* **71**: 7548–7550.

Middelhoven, W.J. 1993. *Antonie van Leeuwenhoek* **63**: 125–144.

Middelhoven, W.J., and van Doesburg, W. 2007. *Antonie van Leeuwenhoek.* **91**: 191–196.

Middelhoven, W.J., Hoogkamer-Te Niet, M.C., and Kreger-Van Rij, N.J. 1984. *Antonie van Leeuwenhoek.* **50**: 369–378.

Middelhoven, W.J., de Jong, I.M., and de Winter, M. 1991. *Antonie van Leeuwenhoek* **59**: 129–137.

Middelhoven, W.J., Coenen, A., Kraakman, B., and Sollewijn Gelpke, M.D. 1992. *Antonie van Leeuwenhoek* **62**: 181–187.

Pham, H.T.M., Giersberg, M., Uhlig, S., Hanke, G., Simon, K., Kunath, K., Baronian, K., and Kunze, G. 2012. *Sens. Actuators B-Chem.* **161**: 137–145.

Pham, H.T.M., Kunath, K., Gehrmann, L., Giersberg, M., Tuerk, J., Uhlig, S., Hanke, G., Simon, K., Baronian, K., and Kunze, G. 2013. *Sens. Actuat B-Chem.* **185**: 628–637.

Pham, H.T.M., Giersberg, M., Gehrmann, L., Hettwer, K., Tuerk, J., Uhlig, S., Hanke, G., Weisswange, P., Simon, K., Baronian, K., and Kunze, G. 2015. *Sens. Actuat B-Chem.* **211**: 439–448.

Pham, H.T.M., Chamas, A., Nieter, A., Giersberg, M., Rutten, T., Gehrmann, L., Hettwer, K., Tuerk, J., Uhlig, S., Simon, K., Baronian, K., and Kunze, G. 2016. *Sens. Actuat B-Chem.* **223**: 540–549.

Rauter, M., Prokoph, A., Kasprzak, J., Becker, K., Baronian, K., Bode, R., Kunze, G., and Vorbrodt, H.M. 2014. *Appl. Microbiol. Biotechnol.* **99**: 4723–4733.

Rempel, M.A., and Schlenk, D. 2008. *Int. Rev. Cell Mol. Biol.* **267**: 207–252.

Rösel, H., and Kunze, G. 1995. *Curr. Genet.* **28**: 360–366.

Rösel, H., and Kunze, G. 1996. *Yeast.* **12**: 1201–1208.

Rösel, H., and Kunze, G. 1998. *Curr. Genet.* **33**: 157–163.

Samsonova, I.A., Böttcher, F., Werner, C., and Bode, R. 1989. *J. Basic Microbiol.* **29**: 675–683.

Samsonova, I.A., Kunze, G., Bode, R., and Böttcher, F. 1996. *Yeast* **12**: 1209–1217.

Sano, K., Fukuhara, H., and Nakamura, Y. *Biotechnol. Lett.* **21**: 33–38.

Shinozaki, Y., Morita, T., Cao, X., Yoshida, S., Koitabashi, M., Watanabe, T., Suzuki, K., Sameshima-Yamashita, Y., Nakajima-Kambe, T., Fujii, T. and Kita-moto, H.K. 2013. *Appl. Microbiol. Biotechnol.* **97:** 2951–2959.

Sietmann, R., Uebe, R., Böer, E., Bode, R., Kunze, G., and Schauer, F. 2010. *J. Appl. Microbiol.* **108**: 789–799.

Song, C. E., Lee, J.K., Lee, S.H., and Lee, S.-g. 1995. *Tetrahedron: Asymmetry* **6**: 1063–1066.

Steinborn, G., Gellissen, G., and Kunze, G. 2005. *FEMS Yeast Res.* **5**: 1047–1054.

Steinborn, G., Gellissen, G., and Kunze, G. 2007a. *FEMS Yeast Res.* **7**: 1197–1205.

Steinborn, G., Wartmann, T., Gellissen, G., and Kunze, G. 2007b. *J. Biotechnol.* **127**: 392–401.

Stöckmann, C., Palmen, T.G., Schroer, K., Kunze, G., Gellissen, G., and Büchs, J. 2014. *J. Ind. Microbiol. Biotechnol.* **41**: 965–976.

Stoltenburg, R., Lösche, O., Klappach, G., and Kunze, G. 1999. *Curr. Genet.* **35**: 8–13.

Suzuki, K., Sakamoto, H., Shinozaki, Y., Tabata, J., Watanabe, T., Mochizuki, A., Koitabashi, M., Fujii, T., Tsushima, S. and Kitamoto, H.K. 2013. *Appl. Microbiol. Biotechnol.* **97:** 7679–7688.

Tanaka, A. O., Nobuko; Fukui, Saburo. 1967. *J. Ferment. Technol.* 617–623.

Terentiev, Y., Breuer, U., Babel, W., and Kunze, G. 2004. *Appl. Microbiol. Biotechnol.* **64**: 376–381.

Van der Walt, J.P., Smith, M.T., and Yamada, Y. 1990. *Antonie van Leeuwenhoek* **57**: 59–61.

Vijayan, V., Giersberg, M., Chamas, A., Mehrotra, M., Chelikani, V., Kunze, G., and Baronian, K. 2015. *Biosens. Bioelectron.* **66**: 379–384.

Wartmann, T., and Kunze, G. 2000b. *Appl. Microbiol. Biotechnol.* **54**: 619–624.

Wartmann, T., and Kunze, G. 2003. Expression of the *HSA* and *GFP* gene in budding cells and mycelia of *A. adeninivorans* In: Wolf, K., Breunig, K. and Barth, G (Eds) *Non-Conventional Yeasts in Genetics, Biochemistry and Biotechnology*, Springer-Verlag.

Wartmann, T., Kruger, A., Adler, K., Duc, B. M., Kunze, I., and Kunze, G. 1995a. *Antonie van Leeuwenhoek* **68**: 215–223.

Wartmann, T., Kunze, I., Duc, B.M., Manteuffel, R., and Kunze, G. 1995b. *Microbiol. Res.* **150**: 113–120.

Wartmann, T., Rosel, H., Kunze, I., Bode, R., and Kunze, G. 1998. *Yeast* **14**: 1017–1025.

Wartmann, T., Erdmann, J., Kunze, I., and Kunze, G. 2000a. *Arch. Microbiol.* **173**: 253–261.

Wartmann, T., Gellissen, G., and Kunze, G. 2001. *Curr. Genet.* **40**: 172–178.

Wartmann, T., Stephan, U.W., Bube, I., Böer, E., Melzer, M., Manteuffel, R., Stoltenburg, R., Guengerich, L., Gellissen, G., and Kunze, G. 2002. *Yeast* **19**: 849–862.

Wartmann, T., Bellebna, C., Böer, E., Bartelsen, O., Gellissen, G., and Kunze, G. 2003a. *Appl. Microbiol. Biotechnol.* **62**: 528–535.

Wartmann, T., Stoltenburg, R., Böer, E., Sieber, H., Bartelsen, O., Gellissen, G., and Kunze, G. 2003b. *FEMS Yeast Res.* **3**: 223–232.

Watanabe, T., Shinozaki, Y., Yoshida, S., Koitabashi, M., Sameshima-Yamashita, Y., Fujii, T., Fukuoka, T. and Kitamoto, H.K. 2014. *J. Biosci. Bioeng.* **117:** 325–329.

Wolf, K. 1996. *Nonconventional yeasts in biotechnology*, Springer-Verlag.

Wolf, K., Breunig, K. and Barth, G. 2003. *Non-conventional yeasts in genetics, biochemistry and biotechnology,* Springer-Verlag.

Yang, X.X., Wartmann, T., Stoltenburg, R., and Kunze, G. 2000. *Antonie van Leeuwenhoek* **77**: 303–311.

Index

T. Satyanarayana and G. Kunze (eds.), *Yeast Diversity in Human Welfare*, DOI 10.1007/978-981-10-2621-8

MIX
Papier aus verantwortungsvollen Quellen
Paper from responsible sources
FSC® C105338

If you have any concerns about our products,
you can contact us on
ProductSafety@springernature.com

In case Publisher is established outside the EU,
the EU authorized representative is:
Springer Nature Customer Service Center GmbH
Europaplatz 3, 69115 Heidelberg, Germany

Printed by Libri Plureos GmbH
in Hamburg, Germany